Forensic Entomology

INTERNATIONAL DIMENSIONS AND FRONTIERS

CONTEMPORARY TOPICS in ENTOMOLOGY SERIES

THOMAS A. MILLER EDITOR

Forensic Entomology

INTERNATIONAL DIMENSIONS AND FRONTIERS

Jeffery Keith Tomberlin
DEPARTMENT OF ENTOMOLOGY
TEXAS A&M UNIVERSITY

M. Eric Benbow
DEPARTMENT OF ENTOMOLOGY AND DEPARTMENT OF OSTEOPATHIC MEDICAL SPECIALTIES
MICHIGAN STATE UNIVERSITY

CRC Press
Taylor & Francis Group
Boca Raton London New York

CRC Press is an imprint of the
Taylor & Francis Group, an **informa** business

CRC Press
Taylor & Francis Group
6000 Broken Sound Parkway NW, Suite 300
Boca Raton, FL 33487-2742

First issued in paperback 2020

© 2015 by Taylor & Francis Group, LLC
CRC Press is an imprint of Taylor & Francis Group, an Informa business

No claim to original U.S. Government works

ISBN 13: 978-0-367-57588-5 (pbk)
ISBN 13: 978-1-4665-7240-9 (hbk)

Dedication

We would like to dedicate this book to Sung T'zu and Pierre Mégnin as they planted the seeds from which a field emerged.

Contents

Part I
History, Accomplishments, and Challenges of Forensic Entomology in Australasia

James F. Wallman

Wang Jiangfeng

**Nazni Wasi Ahmad, Lee Han Lim, Chew Wai Kian, Roziah Ali, John Jeffery,
and Heo Chong Chin**

Kabkaew L. Sukontason and Kom Sukontason

Meenakshi Bharti

James F. Wallman and Melanie S. Archer

Part II
History, Accomplishments, and Challenges of Forensic Entomology in Europe

Martin J. R. Hall

Preface

Forensic entomology is the use of insects and other arthropods in legal investigations including, but not limited to, cases of medicolegal, stored product, or urban relevance. Although entomological evidence is often considered for investigations involving stored products (e.g., a bug in a can of corn or beetles in cereal) or urban settings (e.g., bedbugs or termite infestations), here we focus on the aspects of entomology used in medicolegal investigations and the research that has led to such applications. Insect evidence in these types of cases is often used to assist in entomology-based estimates of the time of colonization that can be related to myiasis and/or a minimum postmortem interval. Such investigations are broadly defined as violent crimes that include murder but can also involve cases of neglect and abuse. In most instances, forensically relevant inferences are made by the collection, identification, and study of arthropods associated with a decomposing body. Often the body is human, but there are also important cases in wildlife and veterinary forensics that involve criminal activities such as poaching.

Forensic entomology has evolved considerably since the famous case in thirteenth-century China, where the activity of flies associated with a weapon was used to identify a potential suspect. And, like many disciplines, forensic entomology has metamorphosed from a science initially reliant on anecdotes and simple observation to one built on solid research and established principles in biology. Recent trends suggest that researchers are beginning to test these principles to better understand the variation that occurs in nature and utilize this information to assist in solving crimes, the cases particularly associated with death and abuse investigations. Today, increasing emphasis is placed on bridging research in applied areas such as forensic entomology with other sciences in the more basic realm of research (Tomberlin et al. 2011a,b). Indeed, theory and the conceptual underpinnings of resource pulse ecology (i.e., carrion), island biogeography, and disease ecology, to name a few, are becoming increasingly recognized as critical to the new dimensions and frontiers of forensic entomology research.

The establishment of forensic entomology as a field and its continued diversification, expansion, and applications all over the world can arguably be attributed to a combination of globalization and continued technological advancements that has allowed researchers and practitioners to establish lines of communication and advance novel avenues of research never before possible. Some of these lines of research are employing sophisticated new genome-based technology for use in both understanding necrophagous insect population biology and employing microbial communities in developing time lines of forensic importance, among other applications. In addition, because of such advancements, sciences that were at one time considered peripheral to forensic entomology research are now being recognized as integral to research: microbiology (Stokes et al. 2009; Barnes et al. 2010), engineering (Rains et al. 2008), and chemistry (Frederickx et al. 2012) are just a few examples of the scientific disciplines now being regularly integrated into forensic entomology research. Because of this diversification, greater opportunity exists for research and associated collaborations among disciplines, regions, and nations globally.

Forensic entomology research has been conducted for many years from countries worldwide (Tomberlin et al. 2012a). On the basis of a search of the literature from 1999 to 2003, researchers from 22 nations had published over 100 articles on forensic entomology as related to medicolegal investigations (Tomberlin et al. 2004). At that time, 31% of these articles were published in *Forensic Science International*, 14% in the *Journal of Forensic Sciences*, and 9% in the *Journal of Medical Entomology* (Tomberlin et al. 2004). The trends in publishing over the last 10 years are both similar and different compared to this earlier study, with a pattern of increased globalization in forensic entomology research productivity reflected as journal publications. We conducted a more recent survey of the literature that is presented in detail as part of Chapter 31, but the general trend of publications is depicted in Figure 1, which shows the global distribution of forensic entomology publications from 1974 to 2012.

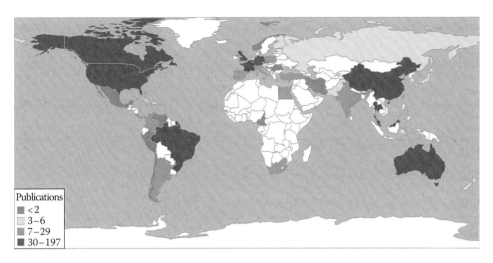

Publications
■ <2
□ 3–6
■ 7–29
■ 30–197

Figure 1   (**See color insert.**) Presentation of publications on forensic entomology throughout the world from 1974 to 2012. (Figure courtesy of Jennifer Pechal.)

The results of this recent literature survey also suggest that the publishing atmosphere is changing for forensic entomology, much like that for all sciences. The development of online and open access journals has revolutionized scientific productivity and communication. Most of the top journals in entomology and the forensic sciences have online access that enables researchers to locate publications related to their work with a quick online literature search from anywhere in the world. Such advancements can be considered a double-edged sword. Although researchers now have the opportunity to publish quickly and access literature from around the world, it also represents a grand challenge for all researchers to keep up with studies that are being published at a rapid pace. There is another challenge of recognizing the drawbacks of the rapidly expanding field of online, open access journals that have been discussed in a *Nature* special issue (2013), but such journals are important to maintaining rigorous and peer-reviewed published literature.

This development has not only resulted in an increase of forensic entomology publications overall but also facilitated an increase in the diversity of nations publishing new and exciting data from even the most remote regions of the Earth. Many of the initial research articles specific to forensic entomology originated from Europe (Bergeret 1855; Mégnin 1894). During the early to middle twentieth century, an influx of forensic entomology publications came from the United States with consistent productivity from Europe, Asia, and South America, and countries such as Poland, Malaysia, Thailand, Argentina, and Brazil, to name a few. Today, researchers from these nations and many others from these regions are now publishing on a regular basis and are having a profoundly positive impact on the science of forensic entomology. Researchers from these countries are publishing their articles in the top forensic journals, such as *Journal of Forensic Sciences* (Bharti and Singh 2003), *Forensic Science International* (Turchetto et al. 2001), and *International Journal of Legal Medicine* (Gagliano-Candela and Aventaggiato 2001), which are available to the forensic sciences community as a whole. In addition, papers related to forensic entomology are also now being published in journals that have not had a traditional forensic focus, such as *Tropical Biomedicine* (Heo et al. 2008), *Trends in Ecology and Evolution* (Tomberlin et al. 2011), *Parasitology Research* (Sukontason et al. 2010), *Naturwissenschaften* (Amendt et al. 2004), and *Animal Behaviour* (Tomberlin et al. 2012b). These expansions of forensic entomology publications from around the world and into journals with more general scientific breadth suggest that the state of forensic entomology research may be changing and expanding. This remarkable growth, expansion, and diversification resulted in the concept for this book.

Remaining current and up-to-date with such rapid and expanding developments in forensic entomology will continue to challenge most researchers and practitioners. To facilitate individuals along this path, we have developed this book into two portions: the first portion reviews the history of forensic entomology, accomplishments, and future challenges in nations around the world, and the second portion provides perspective of other scientific disciplines now shaping the questions being addressed in the growing field of forensic entomology.

There were four major goals of this book. One goal was to bring together an internationally recognized group of forensic entomologists and researchers to provide countrywide and regionally relevant syntheses on the current state and future of forensic entomology worldwide. We understand that "history" is held in the eye of the beholder and that each author (or authors) has given their perspective of forensic entomology for their nation. Every effort was made to include the most relevant and historical references for each country as a means to provide one of the largest lists of global forensic entomology references available; however, because of space limitations not all references could be described in tremendous detail and some were undoubtedly overlooked in this ambitious effort. We as editors take full responsibility for any omissions. And, although different views might exist among researchers, practitioners, and other experts of the forensic sciences, our intention was for the readers to gain a greater appreciation for the information provided and the major advances that have been made in the field of forensic entomology on a global scale. To that end, to increase consensus, each chapter was reviewed anonymously by an author of another chapter or by a practicing forensic entomology expert from the nation being discussed. A second goal was to give individuals who might represent the future of forensic entomology the opportunity to wave the banner and contribute a new perspective for their respective country while paying tribute to those who have paved the way to the current state of the discipline. By doing so, these "new" individuals gain exposure and the opportunity to interact with others from their nation and around the world and encourage others to participate and help continue the evolution of our field. Our third goal was to highlight both established and newly emerging areas of forensic entomology research that provide the foundation and future of this exciting discipline. The fourth goal was to celebrate the success of forensic entomology as a discipline, identify key challenges to current work and practice, and provide an internationally cohesive perspective to emerging multidisciplinary dimensions and frontiers of forensic entomology.

An additional accomplishment emerged from this exercise as well. The world of forensic entomology consists of researchers from many nations, and this is something represented by the list of contributors. We learned through this experience that while some individuals use one term to describe a biological phenomenon, others often use a different term. We felt it was important to allow the authors to use the terms and definitions of their choice. Doing so will provide the reader with a broader appreciation of the growth of the field and diversification of terminology and applications. This diversity in semantic usage also indicates a possible need to better standardize the language and approaches used by forensic entomologists worldwide.

In the end, we hope that this book will provide the reader a greater appreciation of the history of forensic entomology and the scientists who built the foundation of forensic entomology in each country. We also encourage others from around the world to join us in celebrating the current activities and new dimensions provided by contemporary researchers and practitioners. In doing so, we feel that the team of authors that contributed to this book will inspire new and exciting collaborations globally with people who are well established in the field and those just now crossing the threshold of research and application within forensic entomology.

Jeffery K. Tomberlin
Department of Entomology, Texas A&M University
M. Eric Benbow
Department of Entomology and Department of Osteopathic Medical Specialties,
Michigan State University

REFERENCES

Amendt, J., R. Krettek, and R. Zehner. 2004. Forensic entomology. *Naturwissenschaften* 91: 51–65.

Barnes, K.M., D.E. Gennard, and R.A. Dixon. 2010. An assessment of the antibacterial activity in larval excretion/secretion of four species of insects recorded in association with corpses, using *Lucilia sericata* Meigen as the marker species. *Bulletin of Entomological Research* 100: 635–640.

Bergeret, M. 1855. Infanticide, momification du cadaver. Decouverte du cadaver d'un enfant nouveau-ne dans une dheminee ou il setait momifie. Determination de l'epoque de la naissance par la presence de numphes et de larves d'insectes dans le cadaver et par l'etude de leurs metamorphoses. [Infanticide, mummification of the cadaver. Discovery of the cadaver of a newborn child in a mummified state. Determination of the age of the birth by the presence of nymphs and insect larvae in the cadaver and the study of their metamorphoses.] *Annales d'Hygiène Publique et de Médecine Légale* 4: 442–452.

Bharti, M. and D. Singh. 2003. Insect faunal succession on decaying rabbit carcasses in Punjab, India. *Journal of Forensic Sciences* 48: 1–11.

Editorial. 2013. The future of publishing. Special Issue. Nature 495. Issue 7442.

Frederickx, C., J. Dekeirsschieter, Y. Brostaux, J.-P. Wathelet, F.J. Verheggen, and E. Haubruge. 2012. Volatile organic compounds released by blowfly larvae and pupae: New perspectives in forensic entomology. *Forensic Science International* 219: 215–220.

Gagliano-Candela, R. and L. Aventaggiato. 2001. The detection of toxic substances in entomological specimens. *International Journal of Legal Medicine* 114: 197–203.

Heo, C.C., A.M. Mohamad, J. John, and O. Baharudin. 2008. Insect succession on a decomposing piglet carcass placed in a man-made freshwater pond in Malaysia. *Tropical Biomedicine* 25: 23–29.

Mégnin P. 1894. La Faune des Cadavres: Application de L'entomologie à la Médecine Légale, Masson, Paris.

Rains, G.C., J.K. Tomberlin, and D. Kulasiri. 2008. Using insect sniffing devices for detection. *Trends in Biotechnology* 26: 288–294.

Stokes, K.L., S.L. Forbes, and M. Tibbett. 2009. Freezing skeletal muscle tissue does not affect its decomposition in soil: Evidence from temporal changes in tissue mass, microbial activity and soil chemistry based on excised samples. *Forensic Science International* 183: 6–13.

Sukontason, K., N. Bunchu, T. Chaiwong, K. Moophayak, and K.L. Sukontason. 2010. Forensically important flesh fly species in Thailand: Morphology and developmental rate. *Parasitology Research* 106: 1055–1064.

Tomberlin, J.K., M.E. Benbow, A.M. Tarone, and R.M. Mohr. 2011a. Basic research in evolution and ecology enhances forensics. *Trends Ecology and Evolution* 26: 53–55.

Tomberlin, J.K., J.H. Byrd, J.R. Wallace, and M.E. Benbow. 2012a. Assessment of decomposition studies indicates need for standardized and repeatable methods in forensic entomology. *Journal of Forensic Research* 3: 147.

Tomberlin, J.K., T.L. Crippen, A.M. Tarone, B. Singh, K. Adams, Y.H. Rezenom, M.E. Benbow et al. 2012b. Interkingdom response of flies to bacteria mediated by fly physiology and bacterial quorum sensing. *Animal Behaviour* 84: 1449–1456.

Tomberlin, J.K., R. Mohr, M.E. Benbow, A.M. Tarone, and S.L. Vanlaerhoven. 2011b. A roadmap for bridging basic and applied research in forensic entomology. *Annual Review of Entomology* 56: 401–421.

Tomberlin, J.K., J. Wallace, and J.H. Byrd. 2004. The state of forensic entomology. In, *American Academy of Forensic Sciences*, pp. 257–258. Dallas, TX: American Academy of Forensic Sciences.

Turchetto, M., S. Lafisca, and G. Costantini. 2001. Postmortem interval (PMI) determined by study sarcophagous biocenoses: Three cases from the province of Venice (Italy). *Forensic Science International* 120: 28–31.

MATLAB® and Simulink® are registered trademarks of The MathWorks, Inc. For product information, please contact:

The MathWorks, Inc.
3 Apple Hill Drive
Natick, MA 01760-2098 USA
Tel: 508 647 7000
Fax: 508-647-7001
E-mail: info@mathworks.com
Web: www.mathworks.com

Acknowledgments

We are indebted to the outstanding authors of this volume and the anonymous reviewers of the chapters who have made this one of the most comprehensive and cohesive collection of essays representing a global effort in forensic entomology.

We are grateful to the National Institute of Justice for providing funding (2010-DN-BX-K243) that allowed the editors to collaborate and develop this text. It should be noted that the points of view in this document are those of the authors and do not necessarily represent the official position or policies of the U.S. Department of Justice.

A special acknowledgment is extended to all individuals that donated their remains for use in research discussed in this text. Without such giving individuals, much of this work could not have been accomplished.

We thank our families (Laura, Celeste, and Jonah Tomberlin and Melissa, Arielle, and Alia Benbow) for their support and encouragement during the many hours that we dedicated to this endeavor and for their tolerance and understanding when we have been away pursuing science and "bugs."

We would also like to give special thanks to Jen Pechal and Jonathan Cammack for their extra efforts in editing and proofing various aspects of this book that have greatly improved the final version.

We are also greatly appreciative of the efforts of Martin Hall and James Wallman for their willingness to review chapters for the Europe and Asia sections, respectively, and preparing introductions to these sections.

Editors

Dr. Jeffery K. Tomberlin is an associate professor and codirector of the Forensic & Investigative Sciences Program and principal investigator of the Forensic Laboratory for Investigative Entomological Sciences (FLIES) facility (forensicentomology.tamu.edu) in the Department of Entomology at Texas A&M University. Research in the FLIES facility examines species interactions on ephemeral resources such as vertebrate carrion, decomposing plant material, and animal wastes to better understand the mechanisms regulating arthropod behavior related to arrival, colonization, and succession patterns. The goals of his program are to refine current methods used by entomologists in forensic investigations. His research is also focused on waste management in confined animal facilities and the production of alternate protein sources for use as livestock, poultry, and aquaculture feed. Since arriving on campus at Texas A&M University in 2007, six PhD and eight MS students have completed their degrees under his supervision. Dr. Tomberlin welcomes those who are interested in collaborating or gaining experience in forensic entomology or other areas of his research to visit the FLIES facility.

Dr. Tomberlin has been very active within the forensic science community. He, along with a colleague, initiated the first forensic entomology conference in North America as well as the formation of the North America Forensic Entomology Association, of which he served as the first president. He is also a Fellow in the American Academy of Forensic Sciences and has served as the chair of the Pathology/Biology Section. Dr. Tomberlin is also one of 17 entomologists board certified by the American Board of Forensic Entomology (ABFE).

Dr. M. Eric Benbow is currently an assistant professor in the Departments of Entomology and Osteopathic Medical Specialties at the Michigan State University. The research in his laboratory focuses on microbial–invertebrate community interactions in aquatic ecosystems, disease systems, and carrion ecology and evolution. All of these research foci use basic science to inform applications in forensics. Dr. Benbow was part of the inaugural executive committee for the North American Forensic Entomology Association (NAFEA) where he served as the editor-in-chief of the annual NAFEA newsletter and NAFEA Webmaster (www.nafea.net) for 8 years. He was the president of the NAFEA from 2012 to 2013 and has served as an expert witness and worked on several cases that involved insects as evidence during investigations or litigation. Dr. Benbow has authored or coauthored over 95 peer-reviewed articles, book chapters, and proceedings, many of which relate to forensic entomology. He is regularly invited as a speaker at international and national academic meetings related to forensic entomology and has led workshops at the international level discussing experimental design, statistical analyses, and the importance of novel basic ecological concepts in advancing the field of forensic entomology. He continues a sustained research program in forensics that supports undergraduate and graduate students and postdoctoral associates. Dr. Benbow continues to mentor and comentor students and postdoctoral associates through forensic entomology research, facilitating professional opportunities for all students interested in forensic entomology. He sees the future of forensic entomology to fundamentally be in the hands of students and early career scientists worldwide.

Contributors

Nazni Wasi Ahmad
Medical Entomology Unit
Institute for Medical Research
Kuala Lumpur, Malaysia

Jacqueline A. Aitkenhead-Peterson
Department of Soil & Crop Sciences
Texas A&M University
College Station, Texas

Michael B. Alexander
Department of Soil & Crop Sciences
Texas A&M University
College Station, Texas

Roziah Ali
Medical Entomology Unit
Institute for Medical Research
Kuala Lumpur, Malaysia

Jens Amendt
Institute of Forensic Medicine
Goethe University
Frankfurt, Germany

Gail Anderson
School of Criminology
Simon Fraser University
British Columbia, Canada

Gunnar Andersson
Department for Chemistry, Environment and
  Feed Hygiene
National Veterinary Institute
Uppsala, Sweden

Melanie S. Archer
Department of Forensic Medicine
Monash University
Southbank, VIC, Australia

Daria Bajerlein
Department of Animal Taxonomy and
  Ecology
Adam Mickiewicz University
Poznań, Poland

M. Eric Benbow
Department of Entomology and Department of
  Osteopathic Medical Specialties
Michigan State University
East Lansing, Michigan

Meenakshi Bharti
Department of Zoology &
  Environmental Sciences
Punjabi University
Punjab, India

Luc Bourguignon
Laboratory Microtraces & Entomology
National Institute for Criminalistics and
  Criminology
Brussels, Belgium

Yves Braet
Laboratory Microtraces & Entomology
National Institute for Criminalistics and
  Criminology
Brussels, Belgium

Jason H. Byrd
William R. Maples Center for Forensic
  Medicine
College of Medicine
University of Florida
Gainesville, Florida

Joan A. Bytheway
Forensic Science Department
College of Criminal Justice
Sam Houston State University
Huntsville, Texas

Carlo P. Campobasso
Department of Medicine & Health Sciences
  (DiMeS)
University of Molise
Campobasso, Italy

David O. Carter
Division of Natural Sciences and Mathematics
  Forensic Sciences Unit
Chaminade University of Honolulu
Honolulu, Hawaii

Valerie J. Cervenka
Department of Natural Resources
Saint Paul, Minnesota

Damien Charabidze
Forensic Taphonomy Unit
Lille University
Lille, France

Daniel Cherix
Department of Ecology and Evolution
UNIL Sorge
Lausanne, Switzerland

Heo Chong Chin
Faculty of Medicine
Universiti Teknologi MARA
Selangor, Malaysia

Tawni L. Crippen
Southern Plains Agricultural Research Center
Agricultural Research Service, U.S.
  Department of Agriculture
College Station, Texas

Ian Dadour
Centre for Forensic Science (M420)
University of Western Australia
Nedlands, Western Australia

Falko P. Drijfhout
School of Physical and Geographical
  Sciences
Keele University
Staffordshire, United Kingdom

Robin Fencott
Freelance Software Consultant
London, United Kingdom

Ana M. García-Rojo
Laboratory of Forensic Entomology
General Commissariat of Scientific Police
Madrid, Spain

Emmanuel Gaudry
Unité nationale d'investigations criminelles
Institut de Recherche Criminelle de la
  Gendarmerie Nationale
Rosny-Sous-Bois, France

M. Denise Gemmellaro
Department of Entomology
Rutgers University
New Brunswick, New Jersey

Wesley A. C. Godoy
Departamento de Entomologia e Acarologia,
  Escola Superior de Agricultura Luiz de
  Queiroz
Universidade de São Paulo (USP)
Piracicaba, São Paulo, Brazil

Martin Grassberger
Forensic Medicine Institute of Pathology and
  Microbiology
Rudolfstiftung Hospital and Semmelweis
  Clinic
Vienna, Austria

Martin J. R. Hall
Department of Life Sciences
Natural History Museum
London, United Kingdom

Andrew J. Hart
Specialist Forensic Service
Metropolitan Police Service
London, United Kingdom

Michelle Harvey
School of Life and Environmental Sciences
Deakin University
Victoria, Australia

Françoise Hubrecht
Laboratory Microtraces & Entomology
National Institute for Criminalistics and
  Criminology
Brussels, Belgium

John Jeffery
Medical Entomology Unit
Institute for Medical Research
Kuala Lumpur, Malaysia

Wang Jiangfeng
Department of Forensic Medicine
Soochow University
Suzhou, Jiangsu, China

Chew Wai Kian
Medical Entomology Unit
Institute for Medical Research
Kuala Lumpur, Malaysia

Szymon Konwerski
Natural History Collections
Adam Mickiewicz University
Poznań, Poland

Simonetta Lambiase
Department of Public Health Experimental and
  Forensic Medicine
University of Pavia
Pavia, Italy

Hélène N. LeBlanc
University of Ontario Institute of
  Technology
Ontario, Canada

Lee Han Lim
Medical Entomology Unit
Institute for Medical Research
Kuala Lumpur, Malaysia

Anders Lindström
Department for Chemistry, Environment and
  Feed Hygiene
National Veterinary Institute
Uppsala, Sweden

Concepción Magaña-Loarte
Laboratorio de Antropología y Odontología
  Forense
Instituto Anatómico Forense
Madrid, Spain

Szymon Matuszewski
Laboratory of Criminalistics
Department of Criminalistics
Adam Mickiewicz University
Poznań, Poland

Jean-Philippe Michaud
Royal Canadian Mounted Police (RCMP)
High River, Alberta, Canada
Départment de Biologie
Université de Moncton
Moncton, New Brunswick, Canada

Hannah E. Moore
School of Physical and Geographical
  Sciences
Keele University
Staffordshire, United Kingdom

Gaétan Moreau
Départment de Biologie
Université de Moncton
Moncton, New Brunswick, Canada

Thiago C. Moretti
Departamento de Biologia Animal
Universidade Estadual de Campinas
  (UNICAMP)
Campinas, São Paulo, Brazil

Beryl Morris
School of Biological Sciences
University of Queensland
St Lucia, Australia

Jennifer L. Pechal
Department of Entomology
Michigan State University
East Lansing, Michigan

Christine J. Picard
Department of Biology
Indiana University, Purdue University in
  Indianapolis (IUPUI)
Indianapolis, Indiana

Glen C. Rains
Department of Entomology
University of Georgia
Tifton, Georgia

Marta I. Saloña-Bordas
Department of Zoology and Animal Cell
  Biology
Universidad del País Vasco UPV/EHU
Bilbao, Spain

Michelle R. Sanford
Harris County Institute of Forensic Sciences
Houston, Texas

Kenneth G. Schoenly
Department of Biological Sciences
California State University, Stanislaus
Turlock, California

Baneshwar Singh
Department of Forensic Sciences
Virginia Commonwealth University
Richmond, Virginia

Kabkaew L. Sukontason
Department of Parasitology
Chiang Mai University
Chiang Mai, Thailand

Kom Sukontason
Department of Parasitology
Chiang Mai University
Chiang Mai, Thailand

Krzysztof Szpila
Chair of Ecology and Biogeography
Department of Animal Ecology
Nicolaus Copernicus University
Toruń, Poland

Aaron M. Tarone
Department of Entomology
Texas A&M University
College Station, Texas

Jeffery K. Tomberlin
Department of Entomology
Texas A&M University
College Station, Texas

Sherah L. VanLaerhoven
Department of Biology
University of Windsor
Ontario, Canada

Sofie Vanpoucke
Laboratory Microtraces & Entomology
National Institute for Criminalistics and
   Criminology
Brussels, Belgium

Martin H. Villet
Department of Zoology and Entomology
Rhodes University
Grahamstown, South Africa

John R. Wallace
Department of Biology
Millersville University
Millersville, Pennsylvania

James F. Wallman
Institute for Conservation Biology &
   Environmental Management
School of Biological Sciences
University of Wollongong
NSW, Australia

Daniel J. Wescott
Department of Anthropology
Texas State University
San Marcos, Texas

Amoret P. Whitaker
Department of Life Sciences
Natural History Museum
London, United Kingdom

PART I

History, Accomplishments, and Challenges of Forensic Entomology in Australasia

CHAPTER 1

Introduction to Asian and Australasian Chapters

James F. Wallman

This first chapter of the book acts as an introduction to Chapters 2 through 6, which concern the history and activities of workers in forensic entomology in Asia, Australia, and New Zealand. While these regions cover a part of the world that is exceptionally varied both ecologically and culturally, there are also certain underlying evolutionary affinities that unite them. I highlight here the main themes that have been dealt with in these chapters. China, the topic of Chapter 2, is notable in that it can be considered the "birthplace" of forensic entomology, given that the earliest known documentation of flies (Diptera) being used to solve crimes comes from the Far East (Giles 1924). It is therefore entirely appropriate that China be the first country in this book to have its achievements in forensic entomology addressed.

In strict zoogeographic terms (relating to the geographic distribution of animal species), the chapters that follow represent the Oriental (Indomalaya), southeastern Palaearctic, and Australasian regions. Such zoogeographic regions help in understanding the taxonomic relationships of forensically important insects, and thus their evolutionary origins. China (Chapter 2) falls within both the southeastern Palaearctic and Oriental regions (there are noted ecological differences between lowland southern China and the rest of the country); Malaysia (Chapter 3), Thailand (Chapter 4), and India (Chapter 5) occupy the Oriental region; while Australia and New Zealand are major parts of the Australasian region. The Oriental region is denoted by a subtropical or tropical climate, which extends south into northern Australia. Southern Australia, New Zealand, and northern China are, by contrast, in temperate climatic zones. The insects, and especially the flies, that form the focus of the work addressed in Chapters 2 through 6 therefore represent a diverse fauna, but also one with some common elements. These characteristics pose challenges for the workers in these areas in ensuring not only that all of the entomological evidence collected from scenes of crime is adequately and correctly documented taxonomically, but also that any relevant data on the same species available from neighboring regions are considered.

Although much forensic entomology research has been based on exploring the biological characteristics of the insects and their relatives attracted to human remains, the inescapable fact remains that, without confidence in the identity of the species involved, any subsequent forensic conclusion is questionable. Successful legal proceedings against an accused would prove difficult, if not impossible, were the identity of the victim to be uncertain, but uncertainty appears commonplace in the identification of the species collected from a victim's body. Fortunately, there has been considerable progress in the identification of the arthropod species of forensic importance in the countries detailed in this first part of the book. Fundamentally, there have been useful advances in the documentation of carrion fly morphology, especially in Malaysia (Omar 2002), Thailand (Sukontason et al. 2006), Australia (Wallman 2001), and New Zealand (Dear 1985). However, much of the diversity in forensic arthropod taxa remains undocumented, especially in tropical regions. Furthermore, the traditional taxonomic study of insects is unpopular among young entomologists because it is detailed, time consuming, and seemingly unhelpful for their employment prospects. Larval morphology has been especially neglected. However, forensic entomology casework should not be undertaken without a thorough familiarity with the taxonomy of the carrion insects of the geographic region concerned. This includes understanding subtleties, such as hybridization potential, and geographic variation in morphology and genetic composition.

Forensic arthropod identification has of course in recent years also been extended significantly into the molecular realm. Such approaches have been more readily embraced than morphology, but molecular research on arthropods is problematic without a rigorous taxonomic underpinning. Workers in China, Malaysia, and Australia in particular have examined a wide range of mitochondrial and nuclear DNA markers of insects of forensic importance, although only a relatively small range of taxa has been examined in comparison with the true scope of the regional faunas, and mostly only blow flies (Diptera: Calliphoridae) and flesh flies (Diptera: Sarcophagidae) have been analyzed (Wallman et al. 2005; Song et al. 2008; Tan et al. 2009). In practice, some very closely related species, such as within the blow fly genera *Chrysomya* and *Calliphora*, may require the application of multiple gene regions for their reliable diagnosis based on DNA alone. There remains considerable opportunity for expanded taxon sampling in the molecular study of forensically important flies in all parts of Asia, Australia, and New Zealand, as well as the need to take better account of genetic intra- and interspecific variability.

Even given certainty about identification in a forensic entomological investigation, data concerning the rate of development of species are fundamental to the application of insects in this way. However, this continues to be an area of research also sorely in need of attention throughout Asia, Australia, and New Zealand, as well as in other parts of the world. It is commonly assumed that the development of species will be similar if those species are closely related and exist sympatrically with one another. However, this has been shown to be unsubstantiated, at least in common *Chrysomya* species (Nelson et al. 2009) and probably others. Where developmental data do exist, it is also generally unlikely that they have been acquired using sufficient replication and randomized sampling. The importance of such a rigorous approach to the acquisition of reference data for use in casework has been highlighted by several workers (Richards and Villet 2008; Johnson and Wallman 2014). However, *ad hoc* approaches continue with little justification, and often no interaction between research groups, even in the same country. For example, independent developmental studies, using quite diverse techniques, have been carried out on the cosmopolitan blow fly species *Chrysomya megacephala* (Fabricius) (Diptera: Calliphoridae) in China (Wand et al. 2002), Malaysia (Rashid et al. 2008), Thailand (Sukontason et al. 2008), India (Bharti et al. 2007), and Australia (Nelson et al. 2009). The divergence of the methodologies used here hinders comparison of the results, thus limiting the exchange of data that could otherwise be helpful in forensic cases. The need for increased cooperation between workers is another theme that features in the chapters that follow, and all workers would benefit, as in other areas of science, from greater collaborative efforts and consistency in their experimental approaches. As described in the Preface, one goal of this book is to facilitate this kind of collaborative exchange and communication among scientists from around the world.

Finally, there appears to have been a strong focus in recent forensic entomology research, in Asian countries in particular, on the accumulation of data from actual human case studies. This has been a valuable supplement to more extensive work done on documenting the insects attracted to a range of nonhuman animal models (including, interestingly, monkeys in Malaysia [Lee and Marzuki 1993]). Cases have been documented across the range of habitats found in Asian countries, with a typical emphasis on either forested or urban places. The fact that Asian countries are mostly tropical or subtropical means that insects are active throughout the year and human remains become infested quickly. This therefore increases the likelihood that the exposed bodies of victims of foul play in these countries will become the focus of entomological analysis. Of special interest is that documentation of the insects on dead bodies has also been done systematically as part of processing corpses received by mortuaries (Lee et al. 2004; Sukontason et al. 2007; Goyal 2012). The less rigid mortuary protocols in many Asian countries no doubt help facilitate access to cadavers for research, but the ongoing lack of taxonomic discrimination of infesting insect species likely limits the degree of interpretation of such work. Some similar documentation of corpse fauna has been done in Australia and New Zealand in the past (Smeeton et al. 1984), but human ethics requirements have presumably been an impediment more recently.

Overall, there is an unquestionably solid level of interest and scientific talent available to further develop forensic entomology in Asia, Australia, and New Zealand. Many exciting and novel research ideas have been pioneered in this part of the world. It is especially encouraging to see such progress in countries that have only been developing the science of forensic entomology relatively recently. The climatic and zoogeographic affinities that these countries share make it imperative that scientists work harder, not only to expand the entomological knowledge of each specific region, but to look for areas of overlap that can be developed for the benefit of all forensic practitioners and any legal proceedings featuring insects.

REFERENCES

Bharti, M., D. Singh, and Y.P. Sharma. 2007. Effect of temperature on the development of forensically important blowfly, *Chrysomya megacephala* (Fabricius) (Diptera:Calliphoridae). *Entomon* 32: 149–151.

Dear, J.P. 1985. Calliphoridae (Insecta: Diptera). *Fauna of New Zealand* 8: 1–86.

Giles, H.A. 1924. The "Hsi Yüan Lu" or "instructions to coroners." *Proceedings of the Royal Society of Medicine* 17: 59–107.

Goyal, P.K. 2012. An entomological study to determine the time since death in cases of decomposed bodies. *Journal of Indian Academy of Forensic Medicine* 34: 10–12.

Johnson, A.P. and J.F. Wallman. 2014. Effect of massing on larval growth rate. *Forensic Science International* 241: 141–149.

Lee, H.L., M. Krishnasamy, A.G. Abdullah, and J. Jeffery. 2004. Review of forensically important entomological specimens in the period of 1972–2002. *Tropical Biomedicine* 21: 69–75.

Lee, H.L. and T. Marzuki. 1993. Preliminary observation of arthropods on carrion and its application to forensic entomology in Malaysia. *Tropical Biomedicine* 10: 5–8.

Nelson, L.A., M. Dowton, and J.F. Wallman. 2009. Thermal attributes of *Chrysomya* species. *Entomologia Experimentalis et Applicata* 133: 260–275.

Omar, B. 2002. Key to third instar larvae of flies of forensic importance in Malaysia. In: *Entomology and the Law: Flies as Forensic Indicators*, edited by Greenberg, B. and J.C. Kunich. Cambridge, United Kingdom: Cambridge University Press.

Rashid, R.A., K. Osman, M.I. Ismail, R.M. Zuha, and R.A. Hassan. 2008. Determination of malathion levels and the effect of malathion on the growth of *Chrysomya megacephala* (Fabricius) in malathion-exposed rat carcass. *Tropical Biomedicine* 25: 184–190.

Richards, C.S. and M.H. Villet. 2008. Factors affecting accuracy and precision of thermal summation models of insect development used to estimate post-mortem intervals. *International Journal of Legal Medicine* 122: 401–408.

Smeeton, W.M.I., T.D. Koelmeyer, B.A. Holloway, and P. Singh. 1984. Insects associated with exposed human corpses in Auckland, New Zealand. *Medicine, Science and the Law* 24: 167–174.

Song, Z.K., X.Z. Wang, and G.Q. Liang. 2008. Species identification of some common necrophagous flies in Guangdong Province, southern China based on the rDNA internal transcribed spacer 2 (ITS2). *Forensic Science International* 175: 17–22.

Sukontason, K., P. Narongchai, C. Kanchai, K. Vichairat, P. Sribanditmongkol, T. Bhoopat, H. Kurahashi et al. 2007. Forensic entomology cases in Thailand: A review of cases from 2000 to 2006. *Parasitology Research* 101: 1417–1423.

Sukontason, K., S. Piangjai, S. Siriwattanarungsee, and K.L. Sukontason. 2008. Morphology and developmental rate of blowflies *Chrysomya megacephala* and *Chrysomya rufifacies* in Thailand: Application in forensic entomology. *Parasitology Research* 102: 1207–1216.

Sukontason, K.L., C. Kanchai, S. Piangjai, W. Boonsriwong, N. Bunchu, D. Sripakdee, T. Chaiwong, B. Kuntalue, S. Siriwattanarungsee, and K. Sukontason. 2006. Morphological observation of puparia of *Chrysomya nigripes* (Diptera: Calliphoridae) from human corpse. *Forensic Science International* 161: 15–19.

Tan, S.H., M.A. Edah, S. Johari, B. Omar, H. Kurahashi, and M. Zulqarnain. 2009. Sequence variation in the cytochrome oxidase subunit I and II genes of two commonly found blow fly species, *Chrysomya megacephala* (Fabricius) and *Chrysomya rufifacies* (Macquart) (Diptera: Calliphoridae) in Malaysia. *Tropical Biomedicine* 26: 173–181.

Wallman, J.F. 2001. Third-instar larvae of common carrion-breeding blowflies of the genus *Calliphora* (Diptera: Calliphoridae) in South Australia. *Invertebrate Taxonomy* 15: 37–51.

Wallman, J.F., R. Leys, and K. Hogendoorn. 2005. Molecular systematics of Australian carrion-breeding blowflies (Diptera: Calliphoridae) based on mitochondrial DNA. *Invertebrate Systematics* 19: 1–15.

Wang, J. F., Y. C. Chen, C. Hu, J. X. Min, and J. T. Li. 2002. Chronometrical morphology of *Chrysomya megacephala* and its application on the determination of postmortem interval. *Acta Parasitology et Medica Entomologica Sinica* 9: 33–38.

CHAPTER **2**

China

Wang Jiangfeng

CONTENTS

2.1 HISTORY OF FORENSIC ENTOMOLOGY IN CHINA

2.1.1 Ancient History

China has a long history of using insects as evidence in criminal investigations. The book *Washing Away of Wrongs* (洗冤集录, Tz'u 1247), published during the Song dynasty, presents more than 15 cases where insect evidence was used in relation to criminal investigations, ranging from estimating the postmortem interval (PMI) to descriptions of damage to human remains by arthropods. The most famous of which was the identification of a murderer due to flies (Diptera) gathering on the blade of the perpetrator's sickle, which was usually used to cut rice, but in this case, used to commit murder.

However, what is not well known is that prior to Tz'u (1247) there were other uses of forensic entomology in China. For instance, in the Houhan dynasty (He ad 960), Yan, a prefectural governor named in Hangzhou, came across a woman crying on the roadside. He inquired as to why she was upset, and she indicated her husband had burned to death; due to her unusual behavior, the governor suspected foul play. Yan asked his entourage to examine the corpse, and they observed flies on the head. In ancient times, the hair of Chinese males was as long as females; the investigators separated the hair and observed that an iron nail had been driven into the man's head. Without the flies aggregating around the wound, they would not have found the nail. Consequently, the woman was investigated for the murder of her husband.

Other examples of the association of insect larvae with remains were recorded in the book, *Character and Sentence of Chinese* (说文解字, Xu AD 100), that noted "maggots are worms produced by flies in the decayed meat." In another book, *Records of Natural Science* (续博物志, Li 1368), the authors recorded that "there will be maggots when objects decompose, maggots turn into flies which then produce maggots, and the cycle continues." However, it was not until the publishing of the book *Compendium of Materia Medica* (本草纲目) (Li 1596) that they realized this relationship was achieved by fly oviposition on the resource. In this book, Li clarified this relationship by stating, "Flies are everywhere, they come out in summer, hide in winter. They like warm and hate cold, maggots burrow into the subsoil and become flies as silkworm."

In somewhat related cases, many records regarding poisonous insects killing people were documented in ancient China. Inscriptions on oracle bones from the Yin dynasty ruins (seventeenth to eleventh centuries BC), recorded the Chinese character "蛊", which means killing humans through virulent insects. Other ancient references on using insects to poison people include *Master Zuo's Spring and Autumn Annals* (左传, compiled ~fifth century BC) and *Chou's Codex* (周礼, prepared by Chou in eleventh century BC).

2.1.2 Modern History

Modern forensic entomology in China was influenced by the West. In the early 1990s, at Zhejiang University in south China (Hu 2000) and Zhou in north China began to carry out forensic entomology research (Zhou et al. 1997). As of 2014, approximately 200 research papers and 60 cases studies from China have been published. In addition, 10 monographs and textbooks containing information about forensic entomology also have been published. Two in particular focused specifically on the field of forensic entomology. These monographs were *Forensic Entomology* (Hu et al. 2000) and *Modern Forensic Entomology* (Cai 2010). In addition, two science fiction books, *Fly Evidence* (Zhang 2010) and *Accusation of Insects: The Forensic Investigation Cases and Stories* (Yang 2002) have been published along with the production of five films about using insects to kill people or solve forensic cases. Currently, a majority of entomology and forensic medicine seminars include topics related to forensic entomology. Furthermore, some universities have developed courses on the subject, such as an elective course at the Guangdong Police College. For the past five years, there has been an annual forensic training conference hosted by the Ministry of Public Security, which includes training in forensic entomology. In Guangdong Province, there are more than 8000 police officers trained each year, with a majority of them receiving training in forensic entomology.

However, unlike Europe or North America, China is still lacking a specific forensic entomology academic society. And, employment as a forensic entomologist within police agencies is still limited, with approximately 10 individuals currently holding such positions. Nevertheless, Chinese police and other departments are making a conscious effort to use forensic entomology as part of their investigations, but such practice has been limited to difficult cases. In China, forensic entomology is predominately used to estimate the PMI of human remains and occasionally to determine the cause of death (e.g., abuse or neglect). At present, there are approximately 60 researchers in forensic entomology, mainly associated with medical schools, police schools, and entomology departments in agriculture universities.

2.2 ACCOMPLISHMENTS OF FORENSIC ENTOMOLOGY IN CHINA

China has made great strides in the field of forensic entomology. Most advances have been with regard to developmental biology, succession, DNA analysis, entomotoxicology, and case applications. These topics are expanded below.

2.2.1 Developmental Biology

Fan (1997) and Xue (2002) developed the accepted taxonomy of flies in China, which laid the foundation for forensic entomology in the country. Work on the biology of pest fly species associated with sanitation issues also provided a preliminary base for Chinese forensic entomology. In the book *Fauna Sinica, Insecta, Diptera, Calliphoridae* by Fan (1997), the duration and development of the following blow flies (Diptera: Calliphoridae) were determined: *Lucilia sericata* (Meigen), *Lucilia cuprina* (Wiedemann), *Lucilia illustris* (Meigen), *Aldrichina grahami* (Aldrich), *Calliphora vicina* Robineau-Desvoidy, *Calliphora nigribarbis* Vollenhoven, *Chrysomya megacephala* (Fabricius), *Chrysomya phaonis* (Séguy), *Phormia regina* (Meigen), and *Protophormia terraenovae* Robineau-Desvoidy; however, these data are not considered robust enough to be used to estimate the PMI of human remains. As for Coleoptera, taxonomists have described many of those associated with carrion, but there has been little research on development of these arthropods.

Ma (1996) and Wang (1999) studied the development of many forensically important species for their respective dissertation research projects. The species studied were *A. grahami* (Wang et al. 2002b), *C. megacephala* (Wang et al. 2002b), *L. sericata* (Wang et al. 2000d), *Boettcherisca peregrina* Robineau-Desvoidy (Diptera: Sarcophagidae) (Wang et al. 2001c), *Parasarcophaga crassipalpis* (Macq.) (Diptera: Sarcophagidae) (Ma and Hu 1997), *Musca domestica* L. (Diptera: Muscidae), *Muscina stabulans* (Fallén) (Diptera: Muscidae) (Wang et al. 2001a), and *Dermestes maculatus* Degeer (Coleoptera: Dermestidae) (Ma and Hu 1997). Data generated from these studies are readily applied in forensic investigations in China. Wang 2010 has also carried out research on the growth and development of *Chrysomya rufifacies* (Macquart) (Diptera: Calliphoridae), *Ophyra spinigera* Stein (Diptera: Muscidae), *Chrysomya chani* Kurahashi (Diptera: Calliphoridae), *Hermetia illucens* L. (Diptera: Stratiomyidae), *Necrobia ruficollis* (Fabricius) (Coleoptera: Cleridae), *Saprinus splendens* (Paykull) (Coleoptera: Histeridae), and *Creophilus maxillosus* L. (Coleoptera: Staphylinidae), which occur on human remains in late stages of decomposition. Feng and Liu (2012b) published a development study on *Megaselia scalaris* (Loew) (Diptera: Phoridae). Nevertheless, there are still many arthropod species associated with decomposing human remains that have yet to be studied in China.

Wang et al. (2002b) studied the relationship between dipteran larval development and the changes in morphological features such as the cephalopharyngeal skeleton and spiracle gap. Wang also suggested that modifications in the posterior spiracles and alimentary canal can be partitioned by larval stadium. He demonstrated that the number of spiracular slits is just one feature that can be used to distinguish instars and species, and suggests that color patterns and distance between spiracles can also be useful in describing developmental changes. Furthermore, the color and quantity of food in the alimentary canal can be used to indicate the developmental duration of the post-feeding stage (Wang 1999; Wang et al. 2000b, 2001a, 2002b,c). Spiracular and cephalopharyngeal morphology have also been studied in detail for some of the key forensic species (Zhao 2009a; Zhao et al. 2009b). Li (2007) studied the variation of the anterior and posterior spiracles and cephalopharyngeal skeleton of *L. sericata* larvae in different development time and its value to the determination of PMI based on flies. The same has also been done for *P. crassipalpis* (Wang et al. 2007). The morphological changes of dermestid beetles over time, including the exuviae of different larval instars, can also be used to assist in estimates of the PMI (Figure 2.1).

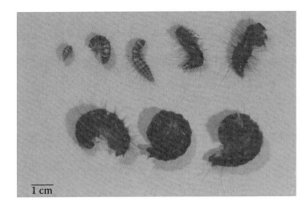

Figure 2.1   Exuvia of different instars of *Dermestes* sp. (Coleoptera: Dermestidae) associated with human remains could be used to determine the postmortem interval.

The pupal stage accounts for approximately half of the immature development of blow flies, and pupae are commonly collected from crime scenes. Wang and his graduate students examined the morphological changes of pupal tissue at different temperatures in *L. sericata* (Wang et al. 2000a), *A. grahami* (Wang et al. 2002a), *C. megacephala* (Wang et al. 2001d), *B. peregrina* (Wang 1999), *C. rufifacies* (Ma 2013), *Hedruris spinigera* (Ma 2013) and *Hermetia illucens* (Ma 2013). They documented the segmentation patterns, appearance of appendages, pneustocera, compound eye, sulci, and development of antennae, bristles, compound eyes, and aristae. Similar work was also done with *L. cuprina* (Wang 2007).

Ma et al. (1996) measured the thickness and number of chitin layers in the larval cuticle of *A. grahami* by electron microscopy; this method was a useful way to estimate insect age. Zhu et al. (2006, 2013a) found that the hydrocarbon profile changed significantly with age in the larvae of *C. rufifacies*. They also determined the changes in the hydrocarbon composition of the puparia of *C. megacephala* under natural environmental conditions (Zhu et al. 2007, 2013a). The relative abundance of nearly all branched alkanes and alkenes decreased significantly with the weathering time, which was due to different weathering rates of various hydrocarbons (Zhu et al. 2013a). Zhu et al. (2003, 2013b) found significant age-dependent increases in pteridine fluorescence in adults of *C. megacephala* and *B. peregrine*. Li (2006) studied the relationship of the protein, lipid, and total soluble sugar content and weight of sarcosaprophagous fly larvae with development times in different temperatures. The nonspecific esterase changes over time have also been studied in larvae of *P. crassipalpis* (Wang et al. 2007).

2.2.2 Body Decomposition and Insect Succession on Human Remains

Interpreting arthropod succession is a primary method used to estimate the PMI of human remains and this has been done in China (Table 2.1). Currently, such research has been conducted in 11 districts. Among them, Zhou et al. (1997) and Yang et al. (1998) have applied such data in casework. Chen et al. (2009) published one of the more comprehensive studies of succession where such patterns were determined for six human corpses.

A majority of arthropod succession research is being conducted at the Forensic Entomology Institute of Guangdong Police College. Wang and colleagues have examined arthropod succession on nearly 50 swine carcasses (Chen 2006; Wan 2007; Wang et al. 2008), simulating a number of scenarios, including, but not limited to carcasses placed on, or in the ground during different seasons. From their work, they have determined the following: (1) the variation of arthropod taxa present, (2) dominant arthropod taxa, and (3) variation of the arthropod community structure over time. At present, new research projects subsidized by Natural Science Foundation of China are under way.

Table 2.1   Body Decomposition and Insect Succession in China

Citation	City (Province)	Location	Experimental Model	Season or Month	Insect Species and Other Information
Zhou et al. 1997	Beijing (capital)	39°09'N, 116°03'E	Human corpse	March to August	Coleoptera: 38 species
Yang et al. 1998	Beijing (capital)	39°09'N, 116°03'E	Human viscera	March to November	Diptera: 26 species
Ma et al. 1997	Hangzhou (Zhejiang)	30°03'N, 120°02'E	Pork	Annual	Total: 33 species
					Diptera: 17 species
					Coleoptera: 14 species
Li et al. 2002	Harbin (Heilongjiang)	45°45'N, 126°38'E	Pork		Insect: 29 species
					Diptera: 8 species
					Coleoptera: 19 species
Chen et al. 2004	Guizhou (Guizhou)	26°35'N, 06°42'E	Pork lung	Annual	Diptera: 27 species
Wang et al. 2003	Chengtu (Sichuan)	30°67'E, 104°06'N	Rabbit		Diptera: 5 species
Chen 2006	Zhongshan (Guangdong)	22°31'N, 113°22'E	Swine	Autumn and winter	Total: 38 species
Wan et al. 2007	Guangzhou (Guangdong)	22°57'N, 113°22'E	Swine	Summer	The death time in a day's influence over insect succession corpse decomposition
Chang et al. 2006	Hohhot (Inner Mongolia)	40°29'N, 111°47'E	Rabbit, Dog	July to October	Diptera: 10 species
Jiang et al. 2011	Yongzhou (Hunan)	26°41'N 111°28'E	Rabbit	July to September	Total: 26 species
					Diptera: 14 species
					Coleoptera: 8 species
Wang et al. 2008	Pearl River Delta (Guangdong)	22°31'N, 113°22'E	Adult swine	Annual	Insect: 42 species
					Diptera: 17 species
					Coleoptera: 16 species
					Other: 9 species
Wu et al. 2008	Guangzhou (Guangdong)	22°57'N, 113°22'E	Rabbit	Spring and summer	Insect: 20 species
					Diptera: 10 species
					Coleoptera: 7 species

(Continued)

Table 2.1 Body Decomposition and Insect Succession in China (*Continued*)

Citation	City (Province)	Location	Experimental Model	Season or Month	Insect Species and Other Information
Dong et al. 2009	Sanmenxia (Henan)	34°47'N, 111°12'E	Rabbit	July to October	3 families, 13 species
Chen et al. 2009	Guiyang (Guizhou)	26°35'N, 106°42'E	Human corpse	Annual	Diptera: 11 species
Nie et al. 2010	Xi'an (Shanxi)	34°14'N, 108°54'E	Rabbit	Spring	Total: 16 species
					Diptera: 10 species
					Coleoptera: 4 species
Shi et al. 2010	Guangzhou (Guangdong)	22°57'N, 113°22'E	Rabbit	Summer	Effects of malathion on the insect succession

Figure 2.2 Researchers examining arthropods associated with human remains in China.

This research will use adult swine carcasses and human bodies (Figure 2.2) to determine the associated microbial communities and volatile succession.

2.2.3 Insect Identification (Classification) and DNA Technology

Efforts have been made to develop a color atlas of forensically important arthropods. These images have been utilized in other textbooks (Wang et al. 2010). Other methods for visual documentation of immature arthropods have also been utilized, including two-dimensional (Li et al. 2005), scanning electron microscopy (Wang et al. 2000c; Feng and Liu 2012a,b) and analytical chemistry (Li, 2006;

Ye et al. 2007). Molecular research on forensically important arthropods has also been accomplished in China. A total of 55 published papers and 4 PhD dissertations have been produced since 2000; these studies have used restriction fragment length polymorphism (RFLP), amplified fragment length polymorphism (AFLP), randomly amplified polymorphic DNA (RAPD), inter-simple sequence repeat (ISSR), sequence characterized amplified region (SCAR), and DNA sequence analysis of cytochrome oxidase subunit I (COI), cytochrome oxidase subunit II (COII), 16SrDNA, 16SrRNA, NADH dehydrogenase subunit 1 (ND1), NADH dehydrogenase subunit 5 (ND5), and internal transcribed spacer (ITS) (Lin et al. 2007; Ying et al. 2007; Song et al. 2008; Zheng et al. 2010; Guo et al. 2011; Quan et al. 2011; Tang et al. 2012; Xiong et al. 2012; Yang et al. 2012).

2.2.4 Entomotoxicology

Entomotoxicology research in China has mainly focused on the influence of toxins on the growth and development of insects. Tian (2004) determined that morphine could accelerate the growth of larvae of *C. megacephala*. Morphine also resulted in an increase in maximum body length and weight. Such impact on development could result in a maximum PMI deviation of 18 hours. Zhao et al. (2005) determined that morphine increased the development rate of *L. sericata* while feeding, but not for larvae in the postfeeding period, and accelerated pupal development. The effect of morphine on development could result in a maximum deviation of 80 hours when using *L. sericata* larval or pupal development data to estimate the PMI of a decedent. Dai (2005) determined that diazepam accelerated the growth rate of larvae of *L. sericata*. Larval duration was reduced by 55 hours, but pupal duration and eclosion rate were not significantly affected. Wang (2006) determined that diazepam had similar effects on *C. megacephala* with the larval duration shortened up to 60 hours. Liu et al. (2009) determined that malathion suppressed larval and pupal development by approximately 36 hours and decreased maximum body length of *C. megacephala* larvae by approximately 1.1 mm. Lu (2012) reported that ketamine suppressed the growth and development rate of larvae of *C. megacephala* in a dose-and-time-dependent manner. Shi et al. (2010) showed that malathion and arsenic delayed the decomposition of rabbit carcasses by suppressing arthropod colonization. Zou, et al. (2013) studied the effects of ketamine on the growth and development of *Lucilia sericata*.

2.2.5 Food Contamination

China's emphasis on food security is increasing. Wang (unpublished data) has handled a dozen cases involving food infested with arthropods. For example, in one case, a person complained that a fly was present in their dumplings. The dumplings and fly were sent to Wang's laboratory where the fly was identified as an adult *L. sericata*. Based on where the dumplings were produced, in combination with the insect species and associated pollen evidence, Wang concluded that it was a case of fraud (the pollen was from the south and the dumpling was from the north of China).

2.3 CHALLENGES

2.3.1 Existing Hurdles

There are still many hurdles for forensic entomology in China. Although most regions have forensic entomologists, more are required to enhance the use of arthropods in criminal investigations. Postgraduates focusing their efforts on forensic entomology still have a difficult time locating jobs, and in many instances change their profession. An additional concern is the magnitude of research still required to provide a solid basic science foundation for forensic entomology in many countries around the world, and this is equally important for China.

China is located in the Palaearctic and Oriental Regions of the world, thus doubling the area where an entomological inventory is necessary for a comprehensive scientific evaluation of the related arthropods of forensic importance. Furthermore, a predominant amount of research is focused on the development of forensically important arthropods; however, the number and species present are still not completely known for many regions of China.

Although much has been accomplished, forensic entomology in China is still in its infancy. Future research examining the arrival patterns of flies to remains as related to the true PMI is needed. Researchers in China and abroad could come together and determine specific research goals. Doing so would alleviate some of the burden with research, while opening the door for future collaborations and streamlining progress toward a goal of increased accuracy and application of forensic entomology throughout China.

2.3.2 Prospects

Entomology is irreplaceable and unique in the forensic sciences, and greater efforts are needed to accomplish the following:

1. Develop a forensic entomology society in China, strengthen police training, establish professional standards, and enhance the qualifications for practitioners.
2. Improve and standardize methods for fundamental research in the basic science that supports forensic entomology.
3. Develop bridges with scientific experts in new technologies while strengthening bonds with experts in existing fields such as microbiology and computer science.

ACKNOWLEDGMENTS

I would like to express my gratitude to Dr. Zhu Guanghui, Dr. Guo Yadong, and master students Ma Ting and Wang Yu, who helped me collect data during the writing of this chapter. Thanks are extended to my family for their continuous support and encouragement.

REFERENCES

Cai, J.F. 2010. *Modern Forensic Entomology*. People's Medical Publishing House, Beijing, China.

Chang, Y.F., J.F. Cai, Z.H. Deng, L.M. Lan, X.Z. Wang, W.B. Liang, R.L. Deng et al. 2006. The constitution and succession of sarcosaprophagous flies community in hohhot. *Forensic Science and Technology* 30: 14–17.

Chen, L.S. 2004. The species and distribution of the necrophagous flies in Guizhou Province. *Acta Entomologica Sinica* 54: 849–853.

Chen, L.S., L.C. Qiu, J.C. Guo, and Y. Lin. 2009. Initial research on process of necrophagous flies participating in decomposing corpse parenchyma in four seasons. *Chinese Journal of Forensic Medicine* 24: 188–190.

Chen, Q.S. 2006. Study on the characteristics of cadaver decomposition and the disciplinarian of arthropod successional patterns in cadaver at suburban districts of Zhongshan. MSc dissertation, Sun Yat-sen University, Guangdong, China.

Chou, G. ~11th century BC. *Chou's Codex* Ancient Chinese Book published in Chou Dynasty. Reprint, Zhongzhou Ancient Books Publishing House, Zhengzhou, Henan, China, 2010.

Dai, J. 2005. Effect of diazepam on the growth and development of *Lucilia sericata* (Diptera Calliphoridae) and its forensic importance. MSc dissertation, Hebei Medical University, Hebei, China.

Dong, Y.C., Y. Zhu, L. Wang, C.Z. Dong, and Z.G. Liao. 2009. Primary study on common species sarcosaphagous flies. *Forensic Science and Technology* 33: 14–16.

Fan, Z.D. 1997. *Fauna Sinica, Vol. 6 Insecta, Diptera, Calliphoridae*. Science Press, Beijing, China.

Feng, D.X. and G.C. Liu. 2012a. A study on morphology of immature stages of *Diplonevra peregrina* (Wiedemann) (Diptera: Phoridae). *Microscopy Research and Technique* 75: 1425–1431.

Feng, D.X. and G.C. Liu. 2012b. Morphology of immature stages of *Megaselia spiracularis* Schmitz (Diptera: Phoridae). *Microscopy Research and Technique* 75(9): 1297–1303.

Guo, Y.D., J.F. Cai, Y.F. Chang, X. Li, Q.L. Liu, X.H. Wang, M. Zhong, J.F. Wen, and J.F. Wang. 2011. Identification of forensically important sarcophagid flies (Diptera: Sarcophagidae) in China, based on COI and 16SrDNA gene sequences. *Journal of Forensic Sciences* 56: 1534–1540.

He, N. 960. Selection of doubtful cases. An ancient Chinese book published in Houhan Dynasty and reprinted in 1986. Fudan University Press, Shanghai, China.

Hu, C. 2000. *Forensic Entomology*. Chongqing Press, Chongqing, China.

Jiang, Y., J.F. Cai, L. Yang, L. Yang, W.P. Yi, L.M. Lan, X. Li, and J.B. Li. 2011. A study of sarcosaphagous insects from arthropod in Yongzhou district of Hunan Province. *Chinese Journal of Applied Entomology* 48: 191–196.

Li, K. 2006. Fundamental research on the application of molecular identification and biochemical characters of necrophagous flies in the determination of postmortem interval. PhD dissertation, Zhejiang University, Zhejiang, China.

Li, K., G.Y. Ye, and C. Hu. 2005. Identification of early larvae of common carrion-breeding flies by two-dimensional gel electrophoresis finger map. *Acta Entomologica Sinica* 48: 576–581.

Li, L., Y.F. Zhang, and Y.K. Ma. 2002. The major necrophagous insects and their regular Activity on Carcass in Harbin. *Journal of Northeast Forestry University* 30: 93–96.

Li, S. 1368. *Records of Natural Science*. Ancient Chinese Book. Published in Song Dynasty and reprinted in 1991. Bashu Press, Chengdu, Sichuan, China.

Li, S.Z. 1596. *Compendium of Materia Medica*. The Northern Literature and Art Publishing House, Harbin, Heilongjiang, China.

Li, Y.Y. 2007. Effect of temperature on the growth and development of the larvae of *Lucilia sericata* (Diptera: Calliphoridae) and its significance in forensic medicine. MSc dissertation, Hebei Medical University, Hebei, China.

Lin, H., S.B. Wang, X.X. Miao, H. Wu, and Y.P. Huang. 2007. Identification of necrophagous fly species using ISSR and SCAR markers. *Forensic Science International* 168: 148–153.

Liu, X.S., Y.W. Shi, H.Y. Wang, and R.J. Zhang. 2009. Determination of malathion levels and its effect on the development of *Chrysomya megacephala* (Fabricius) in South China. *Forensic Science International* 192: 14–18.

Lu, Z. 2012. Effect of ketamine on the development of *Chrysomya megacephala* (Diptera Calliphoridae). *Chinese Journal of Parasitology and Parasitic Diseases* 30(5): 361–366.

Ma, T. 2013. Study of several sarcosaphagous insects to estimate postmortem interval of the high decomposition bodies. MSc dissertation, South China Agricultural University, Guangdong, China.

Ma, Y.K. 1996. Research on several sarcosaprophagous insects in Hangzhou and their application in forensic entomology. PhD dissertation, Zhejiang University, Zhejiang, China.

Ma, Y.K., and C. Hu. 1997. A preliminary study on the species and biological characters of necrophagous insects in Hangzhou. *Journal of Zhejiang Agricultural University* 23: 19–24.

Nie, T.F., Z.M. Wei, F.Y. Tian, F. Tian, and Z.M. Lian 2010. Primary study on sarcosaphagous insects on rabbit carcass in Xi'an. *Chinese Bulletin of Entomology* 47: 587–591.

Quan, Z., J.F. Cai, M.Q. Zhang, H. Feng, Y.D. Guo, L.M. Lan, and Y.Q. Chen. 2011. Molecular identification of forensically significant beetles (Coleoptera) in China based on COI gene. *Revista Colombiana de Entomología* 37: 95–102.

Shi, Y.W., X.S. Liu, H.Y. Wang, and R.J. Zhang. 2010. Effects of malathion on the insect succession and the development of *Chrysomya megacephala* (Diptera: Calliphoridae) in the field and implications for estimating postmortem interval. *American Journal of Forensic Medicine & Pathology* 31: 46–51.

Song, Tz'u. 1247. *Washing Away of Wrongs*. Reprint, Shanghai: Shanghai Classics Publishing House, 1981.

Song, Z.K., X.Z. Wang, and G.Q. Liang. 2008. Species identification of some common necrophagous flies in Guangdong Province, southern China based on the rDNA internal transcribed spacer 2 (ITS2). *Forensic Science International* 175: 17–22.

Tang, Z.C., Y.D. Guo, X.W. Zhang, J. Shi, K.T. Yang, X.L. Li, Y.Q. Chen, and J.F. Cai. 2012. Identification of the forensically important beetles *Nicrophorus japonicus, Ptomascopus plagiatus* and *Silpha carinata* (Coleoptera: Silphidae) based on 16srRNA gene in China. *Tropical Biomedicine* 29: 493–498.

Tian, J. 2004. The application of insect to determine the postmortem interval Effect of Morphine in tissues on development of *Chrysomya megacephala* (Diptera) and implication of this effect on estimation of postmortem intervals using arthropod development patterns. MSc dissertation, Hebei Medical University, Hubei, China.

Wan, X.B. 2007. The study of the insects invasion and succession pattern on swine cadavers executed at the different time point in one day. MSc dissertation, Sun Yat-sen University, Guangdong, China.

Wang, H. 2007. Pupal morphogenesis of *Lucilia cuprina* (Diptera:Calliphoridae) in different constant temperature and its significance in forensic medicine. MSc dissertation, Hebei Medical University, Hubei, China.

Wang, H.Y., Y.W. Shi, X. S. Liu, and R.J. Zhang. 2010. Growth and development of *Boettcherisca peregrine* under different temperature conditions and its significance in forensic entomology. *Journal of Environmental Entomology* 32: 166–172.

Wang, J.F. 1999. Research on several sarcosaprophagous flies morphology and development and their application in the determination of PMI. PhD dissertation, ZheJiang University, Zhejiang, China.

Wang, J.F. 2010. Forensic entomology. In *Modern Forensic Medicine*. Zhengzhou University Press, edited by X.Y. Wu, Henan, China.

Wang, J.F., Y.C. Chen, C. Hu, J.X. Min, and J.T. Li. 2000a. Chronology of development within puparium of *Lucilia sericata* in different constant temperature and its application in the determination of post-mortem interval. *Acta Scientiarum Naturalium Universitatis Sunyatseni* 39(6(A)): 250–254.

Wang, J.F., Y.C. Chen, C. Hu, J.X. Min, and J.T. Li. 2000b. Chronometrical Morphology of *Lucilia sericata* and its application in the determination of postmortem interval. *Acta Scientiarum Naturalium Universitatis Sunyatseni* (S2): 208–213.

Wang, J.F., Y.C. Chen, C. Hu, J.X. Min, and J.T. Li. 2000d. Effect of temperature on the body-length change of *Lucilia sericata*. *Acta Scientiarum Naturalium Universitatis Sunyatseni* (S3): 141–145.

Wang, J.F., Y.C. Chen, C. Hu, J.X. Min, and J.T. Li. 2001d. Chronology of development within puparium of *Chrysomya megacephala* in different constant temperature and its application in the determination of postmortem interval. *Acta Parasitologica et Medica Entomologica Sinica* 8: 232–236.

Wang, J.F., Y.C. Chen, C. Hu, J.X. Min, and J.T. Li. 2002c. Chronometrical morphology of *Chrysomya megacephala* and its application on the determination of postmortem interval. *Acta Parasitology et Medica Entomologica Sinica* 9: 33–38.

Wang, J.F., Y.C. Chen, C. Hu, J.X. Min, and J.T. Li. 2002d. Effect of temperature over the body-length change of *Chrysomya megacephala*. *Acta Parasitology et Medica Entomologica Sinica* 9: 100–105.

Wang, J.F., C. Hu, and Y.C. Chen. 2001a. Morphological changes of *Boettcherisca peregrina* larvae and its application in the determination of postmortem interval. Advances in urban entomology—proceedings of the sixth National Symposium on Urban Entomology, 106–111.

Wang, J.F., C. Hu, Y.C. Chen, J.X. Min, B.J. Hu, and J.T. Li. 2000c. Egg chorionic ultra-structures of four necrophagous flies species. *Academic Journal of Sun Yat-sen University of Medical Sciences* (S1): 10–12.

Wang, J.F., C. Hu, Y.C. Chen, J.X. Min, B.J. Hu, and J.T. Li. 2001b. The activity and oviposition habits in different environment of four important sarcosaprophagous flies. *Journal of Law & Medicine* 8: 218–223.

Wang, J.F., C. Hu, Y.C. Chen, J.X. Min, and J.T. Li. 2002a. Application of the pupal morphogenesis of *Aldrichina grahami* (Aldrich) to the deduction of postmortem interval. *Acta Entomologica Sinica* 45: 696–699.

Wang, J.F., C. Hu, Y.C. Chen, J.X. Min, and J.T. Li. 2002b. Chronometrical morphology of *Aldrichina graham* and its application in the determination of postmortem interval. *Acta Entomologica Sinica* 45: 265–270.

Wang, J.F., Z.G. Li, Y.C. Chen, Q.S. Chen, and X.H. Yin. 2008. The succession and development of insects on pig carcasses and their significances in estimating PMI in south China. *Forensic Science International* 179: 11–18.

Wang, J.F., J.X. Min, and C. Hu. 2001c. Preliminary studies on the morphological changes of *Boettcherisca peregrine* pupae and its application to the determination of daily age. Advances in Urban Entomology—proceedings of the sixth National Symposium on Urban Entomology, 112–115.

Wang, J.F., X.H. Yin, and Y.C. Chen. 2007a. Molecular identification of five common species of necrophagous flies in China. *Acta Entomologica Sinica* 50: 423–428.

Wang, L., Z.M. Li, and M.Y. Zhang. 2007. The biological characteristics of *Parasarcophaga crassipalpis* and its significance in forensic entomology. *Journal of Hebei Medical University* 28: 457–460.

Wang, L.J. 2006. Effect of diazepam in tissues on development of *Chrysomya megacephala* (Diptera) and implication of this effect on estimation of postmortem intervals using arthropod development patterns. MSc dissertation, Hebei Medical University, Hebei, China.

Wang, Y., M. Liu, and D.H. Sun. 2003 A Study on sarcosaphagous insects species variety with seasons in Chengdu. *Journal of Forensic Medicine* 19: 86–88.

Wu, D.P., R.Q. Mao, M.F. Guo, J. Zhou, F.L. Jia, G.S. Ou, and R.J. Zhang. 2008. Species and succession of necrophagous insect community in spring and summer seasons in Guangzhou. *Acta Sci Natur Univ Sunyatseni* 47: 56–60.

Xiong, F., Y.D. Guo, B.H. Luo, J.L. Zhang, J.F. Cai, X.L. Li, and Z. Yang. 2012. Identification of the forensically important flies (Diptera: Muscidae) based on cytochrome oxidase subunit I (COI) gene in China. *AJB* 11: 10912–10918.

Xu, S. AD 100. *Character and Sentence of Chinese.* An ancient Chinese book published in Han Dynasty and reprinted in 1963. Zhonghua Press, Beijing, China.

Xue, W.Q. 2002. *Fauna Sinica Insecta* (Diptera: Muscidae). Science Press, Beijing, China.

Yang, J.B., J.J. Jiang, and J.F. Wang. 2012. Application of mtDNA COI and ND5 genes in the identification of common necrophagous flies. *Chinese Journal of Forensic Medicine* 27: 201–204.

Yang, Y.P. 2002. *Accusation with Insects.* Jiangsu People's Publishing House, Jiangsu, China.

Yang, Y.P., J.C. Ren, L. Liu, Y.W. Li, R. Yan, X.J. Wang, and X.Z. Zhang. 1998. Study on necrophagous flies living on human corpses in Beijing and its application on forensic medicine practice. *Chinese Journal of Forensic Medicine* 13: 159–162.

Ye, G.Y., K. Li, J.Y. Zhu, G.H. Zhu, and C. Hu. 2007. Cuticular hydrocarbon composition in pupal exuviae for taxonomic differentiation of six necrophagous flies. *Journal of Medical Entomology* 44: 450–456.

Ying, B.W., T.T. Liu, and H. Fan. 2007. The application of mitochondrial DNA cytochrome Oxidase II gene for the identification of forensically important blowflies in western China. *American Journal of Forensic Medicine & Pathology* 28: 308–313.

Zhang, Y. 2010. *Flies as Evidences.* International Cultural Publishing Company, Beijing, China.

Zhao, B. 2009a. Chronometrical morphology changes of larval of three species necrophagous flies and their valuation in forensic medicine. MSc dissertation, Hebei Medical University, Hubei, China.

Zhao, B., L. Wang, H. Wang, W.J. Wang, L.L. Qi, and Z.M. Li. 2009b. Morphological changes of *Lucilia sericata* larvae and its significance in forensic medicine. *Chinese Journal of Vector Biology & Control* 20: 534–537.

Zhao, W.A., S.A. Hu, M.Y. Zhang, X.Y. Feng, and B.X. Wang. 2005. Effects of morphine on *Lucilia sericata* growth accumulated degree hour and deduction of decedent post-mortem interval. *Chinese Journal of Ecology* 24: 111–114.

Zheng, X.L., J.L. Hu, S.P. Kunnon, and X.G. Chen. 2010. Identification of necrophagous fly species from 12 different cities in China using ISSR and SCAR markers. *Asian Pacific Journal of Tropical Medicine* 3: 510–514.

Zhou, H.Z., Y.P. Yang, J.C. Ren, L. Liu, S.Y. Wang, R. Yan, and Y.W. Li. 1997. Studies on forensic entomology in Beijing District. *Chinese Journal of Forensic Medicine* 12: 79–83.

Zhu G.H., G.Y. Ye, and C. Hu. 2003. Determining the adult age of the oriental latrine fly, *Chrysomya megacephala* (Fabricius) (Diptera: Calliphoridae) by pteridine fluorescence analysis. *Insect Science* 10: 245–255.

Zhu G.H., G.Y. Ye, C. Hu, X.H. Xu, and K. Li. 2006. Development changes of cuticular hydrocarbons in *Chrysomya rufifacies* larvae: Potential for determining larval age. *Medical and Veterinary Entomology* 20: 438–444.

Zhu G.H., X.H. Xu, X. Yu, Y. Zhang, and J.F. Wang. 2007. Puparial case hydrocarbons of *Chrysomya megacephala* as an indicator of the postmortem interval. *Forensic Sciences International* 169: 1–5.

Zhu G.H., G.Y. Ye, K. Li, C. Hu, and X.H. Xu. 2013b. Determining the age of adult *Boettcherisca peregrina* by pteridine fluorescence. *Medical Veterinary Entomology* 27: 59–63.

Zhu G.H., X.J. Yu, L.X. Xie, H. Luo, D. Wang, J.Y. Lv, and X.H. Xu. 2013a. Time of death revealed by hydrocarbons of empty puparia of *Chrysomya megacephala* (Fabricius) (Diptera: Calliphoridae): A field experiment. PLoS ONE 8(9): e73043.

Zou, Y., M. Huang, and R.T. Huang. 2013. Effect of ketamine on the development of *Lucilia sericata* (Meigen) (Diptera: Calliphoridae) and preliminary pathological observation of larvae. *Forensic Science International* 226: 273–281.

Zuo, Q. M. ~5th century BC. *Master Zuo's spring and Autumn Annals.* Published in Chou Dynasty and reprinted in 2007, Zhong Hua Press, Beijing, China.

CHAPTER 3

Malaysia

**Nazni Wasi Ahmad, Lee Han Lim, Chew Wai Kian, Roziah Ali,
John Jeffery, and Heo Chong Chin**

CONTENTS

3.1 HISTORY AND THE FORENSIC ENTOMOLOGY CASE STUDIES

In Malaysia, forensic entomology dates back to 1953 when the pathologist Dr. H. M. Nevin shipped young fly larvae collected from the body of a woman shot by bandits in Penang to Dr. J. A. Reid who was at the Institute for Medical Research (IMR), Kuala Lumpur. These specimens were identified as early second-stage larvae of the blow fly *Chrysomya megacephala* (Fabricius) (Diptera: Calliphoridae), which were between 16 and 24 hours old and agreed fairly well with the police evidence that the woman had been killed in less than 24 hours prior to discovery (Reid 1953).

Following Reid (1953), there was a lapse of 20 years without research or casework involving forensic entomology, until 1982 when Lee and Cheong (1982) reported the recovery of *Hermetia* sp. (Diptera: Stratiomyidae) larvae from a highly decomposed female body. Subsequently, Lee et al. (1984) wrote the first comprehensive review on forensic entomological cases (1973–1983) received by the Medical Entomology Unit, IMR. They reported that *C. megacephala* was the dominant fly found associated with cadavers. Following this, several case reports had been published, including Lee (1994) who reported the presence of the drone fly larvae, *Eristalis* sp. (Diptera: Syrphidae) recovered from a decomposed corpse of a newborn baby floating in an irrigation canal. His report confirmed the association of an aquatic environment with *Eristalis* sp. larvae in Malaysia. In the same year, Omar et al. (1994a) reported the presence of *Synthesiomyia nudiseta* (van der Wulp) (Diptera: Muscidae) in decomposing corpses found indoors in Peninsular Malaysia. Their findings were useful in determining body movement in forensic investigations as this species usually infest corpses found indoors. Hamid et al. (2003) and Salleh et al. (2007) subsequently reviewed 12 and 8 forensic entomology cases from Kuala Lumpur Hospital and Hospitals of the National University of Malaysia (Hospital Universiti Kebangsaan Malaysia) from 2001 and 2002, respectively. Their conclusion indicated that *C. megacephala* was the dominant fly colonizing human cadavers in Malaysia.

Nazni et al. (2008) reported the recovery of *Piophila casei* (Linnaeus) (Diptera: Piophilidae) from two human cadavers found indoors in Malaysia and concluded *P. casei* were important forensic indicators for minimum postmortem interval (mPMI) determination as they are found during active decay and skeletonized stages. Besides using dipteran species collected from human remains to make mPMI estimations, other researchers such as Kumara et al. (2009a) reported the occurrence of adult and larvae of the beetle *Dermestes ater* De Geer (Coleoptera: Dermestidae) infesting a human corpse in an advanced stage of decomposition. The corpse was found in a house in the residential area in Penang, and the mPMI was estimated to be 14 days prior to the discovery of the body based on the police investigation. Morphological descriptions of the second and third instars of *Hypopygiopsis violacea* Macquart (Diptera: Calliphoridae) as a diagnostic feature for the first time isolated from forensic cases have been described by Firdaus et al. (2010). Recently, Kumara et al. (2010a) recorded the larvae of two species of Oriental scuttle flies larvae namely *Megaselia curtineura* (Brues) and *Megaselia spiracularis* Schmitz (Diptera: Piophilidae) on human cadavers. Kumara et al. (2010b) also reported the pupae of *Desmometopa* sp. (Diptera: Milichiidae) collected from a human corpse found indoors on Penang Island, which was a new record for the country. In a study of seven forensic cases, *Piophila foveolata* Megnin (Diptera: Piophilidae) specimens were collected. The presence of this species has been mainly associated with burnt or buried cadavers in the skeletonized stage where the estimated mPMI ranged from 6 weeks to 9 months (Smith 1986; Martin-Vega 2011). It is interesting to note that the scenario was similar in forensic cases in Malaysia. More recently, nine cases involved the larvae of *Hemipyrellia tagaliana* (Bigot) (Diptera: Calliphoridae), which represented a new finding. It is noteworthy that seven cases reported that *Lucilia cuprina* (Wiedemann) (Diptera: Calliphoridae) infested corpses, as this synantrophic species is commonly known as a household nuisance pest in Malaysia (Tahir et al. 2007).

Because of the wealth of information obtained from analyzing and reporting forensic specimens, a comprehensive review of forensic entomological specimens received by the Medical Entomology, IMR, in the last three decades (1972–2002) was compiled by Lee et al. (2004). According to them, a total of 448 forensic entomological cases were received from 1972 to 2002. In their review, the blow fly species *C. megacephala* and *Chrysomya rufifacies* (Macquart) were the most common species associated with cadavers from different ecological habitats. Furthermore, to validate mPMI estimates using entomological findings with pathological findings, Kavitha et al. (2008) reported that the methods used by the entomologist and pathologist for estimating PMI were significantly correlated.

Between 1972 and 2013 (Figure 3.1), 1000 forensic cases have been received by the Medical Entomology Unit, IMR from forensic pathology and police departments, with the majority of the cases coming from the forensic pathology departments. Among these cases, only 6% have been used for judiciary trial. In comparison over the past 20 years, the staffs at the Natural History Museum, United Kingdom, have undertaken forensic entomology work in more than 100 police investigations and about 10% of those cases required them to attend court to deliver expert evidence (Amoret and Martin 2012).

In Malaysia, the most important component of biological decay of a cadaver as related to the entomofauna is led by dipteran species belonging to Calliphoridae, Muscidae, Sarcophagidae, Phoridae, and Piophilidae. Coleopteran species belonging to the following families of Silphidae, Dermestidae, Histeridae, and Staphylinidae represent the second major group of arthropods identified from human remains (Kumara et al. 2010b; Nazni et al. unpublished data; Table 3.1).

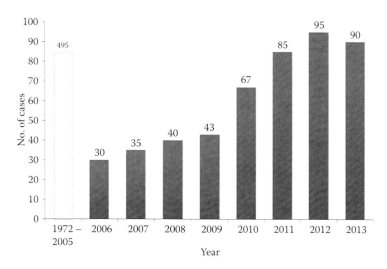

Figure 3.1 Entomological cases received from the Institute for Medical Research since 1972–2013. Grey color in the bar denotes unpublished cases since 2005.

Table 3.1 Diversity of Forensically Important Arthropods Recovered from Forensic Cases Received by the Institute for Medical Research from 1972 to 2013.

Family	Species
Calliphoridae	*Chrysomya megacephala* (Fabricius, 1794)
	Chrysomya (*Achoetandrus*) *rufifacies* (Macquart, 1843)
	Chrysomya nigripes Aubertin, 1932
	Chrysomya villeneuvi Patton, 1922
	Chrysomya pinguis (Walker, 1858)
	Chrysomya bezziana (Villeneuve, 1914)
	Chrysomya chani Kurahashi, 1979
	Chrysomya sp.
	Hemipyrellia ligurriens (Wiedemann, 1830)
	Hemipyrellia tagaliana (Bigot, 1877)
	Hemipyrellia sp.
	Lucilia cuprina (Wiedemann, 1830)
	Lucilia sp.
	Calliphora fulviceps (Wulp, 1881)
Sarcophagidae	*Sarcophaga* (*Liopygia*) *ruficornis* Fabricius, 1794
	Sarcophaga (*Liosarcophaga*) *dux* Thomson, 1869
	Sarcophaga (*Boettcherisca*) *peregrina* Robineau-Desvoidy, 1830
	Sarcophaga (*Boettcherisca*) sp.
	Parasarcophaga sp.
Muscidae	*Hydrotaea* (*Ophyra*) *spinigera* Stein, 1910
	Hydrotaea (*Ophyra*) *chalcogaster* (Wiedemann, 1830)
	Hydrotaea sp. (Diptera: Muscidae)
	Synthesiomyia nudiseta (van der Wulp 1883)
	Musca domestica Linnaeus, 1758

(Continued)

Table 3.1   Diversity of Forensically Important Arthropods Recovered from Forensic Cases Received by the Institute for Medical Research from 1972 to 2013. (*Continued*)

Family	Species
Stratiomyidae	*Hermetia illucens* (Linnaeus, 1758)
	Hermetia sp.
Piophlidae	*Piophila casei* (Linnaeus, 1758)
	Piophila faveolata (Meigen, 1830)
Phoridae	*Megaselia scalaris* (Loew, 1866)
	Megaselia rufipes (Meigen, 1804)
	Megaselia sp.
Syrphidae	*Eristalis* sp.
Cleridae	*Necrobia ruficolis* Fabricius, 1775
	Necrobia rufipes (DeGeer, 1775)
	Necrobia violacea (Linnaeus, 1758)
Silphidae	*Diamesus osculans* (Vigors, 1825)
	Necrophila (*Chrysosilpha*) *formosa* (Laporte, 1832)
Scarabidae	Unknown sp.
Blattellidae	*Blatella germanica* (Linnaeus, 1767)
	Unknown sp.
Syrphidae	*Leucopodella* sp.
Formicidae	*Dorylus* sp.
Ephydridae	Unknown spp.
Psocoptera	Unknown spp.
Chalcididae	Unknown spp.
Culicidae	*Aedes albopictus* (Skuse, 1895)
Pthiridae	*Phthirus pubis* (Linnaeus, 1758)
Cypridinidae	myodocopid Ostracoda
Unidentified family (eggs)	Unknown sp.
Unidentified family (larvae)	Unknown sp.

3.2 INSECT SUCCESSION STUDIES IN MALAYSIA USING ANIMAL MODELS

Although there have been many decomposition studies conducted in different parts of the world and under different environmental conditions, few have been conducted in tropical and subtropical regions (Goff 2003). Little is known about arthropod succession on vertebrate carrion in Malaysia. There is a knowledge gap especially in the biology, bionomics, and ecology of insects recovered from the cadavers, as no complete set of studies has been conducted on the wave of succession of arthropod fauna on carrion under different ecoregions and conditions in Malaysia.

Several early studies on succession were conducted in outdoor situations in Malaysia to establish baseline data for estimating the mPMI of human remains. Lee and Marzuki (1993) conducted the pioneer succession study on the necrophagous community on silvered-leaf monkey (*Presbytis cristata* [Raffles]) carcasses in a lowland forest fringe in Malaysia. They found that calliphorines were the dominant flies. This was followed by Omar et al. (1994b) who placed a long-tailed macaque (*Macaca fascicularis*, Raffles) carcass on the soil of a rubber plantation. They also reported the presence of *Hermetia illucens* (Linnaeus) (Diptera: Stratiomyidae) and *Hydrotaea* (*Ophyra*) *spinigera* Stein (Diptera: Muscidae) larvae and pupae. Regarding insect successional sequence on carrions, Heo et al. (2008b) reported that insect colonization on pig carrion placed in an oil palm plantation began with calliphorids (*C. megacephala*, *C. rufifacies*, *Hemipyrellia ligurriens*), then followed by sarcophagids, muscids (*Hydrotaea spinigera*), stratiomyiids (*H. illucens*), phorids (*Megaselia scalaris*), and finally sepsids, *Allosepsis indica* (Wiedemann) (Diptera: Sepsidae). The species of the first arriving calliphorid on carrion largely depends on the type of ecoregion surrounding the carrion. For example, carrion placed near a jungle fringe might be first colonized by *Hypopygiopsis* spp., a relatively large-size blow fly that inhabits in the tropical rain forest (Firdaus et al. 2010; Heo et al. 2012a), while carcasses placed in the highlands (e.g., >1500 m a.s.l.) might be colonized first by *Lucilia porphyrina* (Walker) (Diptera: Calliphoridae) (Aisyah et al. 2013).

Because of the paucity of information on the ecological aspects of the Malaysian sarco-saprophagous flies, Omar et al. (1994c) conducted a study to elucidate the patterns of arthropod succession on monkey and cat carcasses at five different sites in Selangor and Kuala Lumpur and reported that *H. ligurriens*, *Hypopygiopsis* sp. (Diptera: Calliphoridae) and *Lucilia sinensis* Aubertin preferred to lay their eggs in the nose and mouth of the carcasses, while *Chrysomya* sp. preferred to oviposit their eggs all over the body, and amongst the fur of the animals. They also reported the dispersal behavior of postfeeding larvae of Malaysian native flies, and stated that late third instar larvae of *C. rufifacies*, *Chrysomya villeneuvi* Patton, *Chrysomya nigripes* Aubertin, and *H. spinigera* remained at or nearby the carcasses, while the late third instars of *C. megacephala*, *Chrysomya bezziana* (Villeneuve), *L. sinensis*, and *H. ligurriens* would disperse from the remains and pupate in the soil.

Swine has been considered the closest mammal to human with regard to decomposition studies and thus has been the model of choice in forensic entomology studies (Payne 1965). Hence, Heo et al. (2007, 2008a,b,c, 2009a) were the first in Malaysia to use swine as their study model. Smith (1986) divided arthropods associated with corpses into four categories: necrophagous, predator, omnivore, and adventives, and Heo et al. (2011a) showed that the dipterans in the family Micropezidae, Neriidae, Sepsidae, and Ulidiidae fall within the adventives group as they primarily arrive at carcasses to feed on associated fluid. However, no eggs or immatures of these groups were recovered from their study. Similarly, Chen et al. (2008a) reported the occurrence of signal flies, *Scholastes* sp. (Diptera: Platystomatidae) on a swine carcass in an oil palm plantation in Tanjung Sepat and monkey carcass in forested area in Gombak within five minutes and an hour of placement of carcasses in the field, respectively. Both animal carcasses were visited by *Scholastes* sp. during the fresh decomposition period. Zuha et al. (2009) reported a new record on the occurrence of *Myospila pudica pudica* (Stein) (Diptera: Muscidae) visiting a monkey carcass in a forested area in Bangi, Selangor.

Heo et al. (2008a) conducted an insect succession study on a decomposing piglet carcass placed in a human-made freshwater pond in Malaysia. Their study highlighted that there were five stages of decomposition (submerged-fresh, early-floating, floating decay, bloated-deterioration, and sunken-remains) of the piglet carcass in the aquatic environment, and flies observed on the floating carcass were *C. megacephala*, followed by *C. rufifacies*, *Hydrotaea spinigera*, and *Musca domestica* Linnaeus (Diptera: Muscidae). They concluded that insect activities on the floating carcass were not intense compared to a carcass placed on the land.

Owing to lack of information on the association between burned animal or humans and insect succession, Heo et al. (2008b) conducted a study to elucidate insect succession and rate of decomposition between a partially burned swine carcass and an unburned swine carcass in an oil palm plantation in Tanjung Sepat, Selangor, Malaysia. The results showed that there was no significant difference between the rate of decomposition and sequence of faunal succession on both swine carcasses. The only difference noted was in the number of adult flies with more attracted to the unburnt carcasses.

Another important finding by Heo et al. (2008f) was the collection of two adult flies of *Bengalia emarginata* Malloch (Diptera: Calliphoridae) around the vicinity of a swine carcass. *Bengalia emarginata* has never been recorded from Malaysia and the authors suggested vigilance for this fly, as it may play an important role in forensic investigation. In another study, Heo et al. (2008d) reported the first record of *M. domestica* oviposition activity on a freshly dead pig in an oil palm plantation in Tanjung Sepat, Selangor. However, no housefly larvae were recovered from the carcass. In contrast, Chen et al. (2010) in their study using a monkey carcass reported the presence of third instar larvae of *M. domestica* 33 days after placement of the remains in the field. Their observation indicated that *M. domestica* larvae did colonize the carcass in the presence of other predatory muscid fly larvae, of *H. spinigera*, which were found on the remains of the carcass. Studies conducted by Azwandi and Abu Hassan (2009) in the wet and dry seasons in Kedah, a state in the northern region of Malaysia, also indicated that *C. megacephala* was recognized as the earliest fly to arrive, present on day one, at a monkey carcass. They also reported that adult *C. nigripes* were abundant for approximately 2 weeks after placement of the monkey carcasses in the field.

Nazni et al. (2011) conducted the first comprehensive study aimed at comparing arthropod succession associated with monkey remains in indoor versus outdoor conditions. Nazni et al. (2011) showed that flies arrived at a monkey carcass indoors within 30 minutes, and the first blow fly, *H. violacea*, arrived and oviposited eggs after 1.5 hour in the oral cavity. Within hours, many other calliphorids arrived and began ovipositing in the ear and the neck region. Nazni et al. (2011) noted that fly eggs, which had been oviposited in the oral cavity (first oviposition site), hatched within 6 hours after placement of the remains in the field. Such information is vital when incorporating hatching time of dipteran eggs into estimates of mPMI.

3.3 ECOLOGY AND BEHAVIOR OF FORENSIC FLIES IN MALAYSIA

Hanski (1981) conducted a study in 1978 on the distribution of carrion flies in tropical rain forests from eight different forest types in Gunung Mulu National Park in Sarawak, east Malaysia. He discovered 575 specimens of flies of which 22 species were carrion flies. He stated that common species preferred different altitudes and forest types. All *Chrysomya* sp. (*C. megacephala*, *C. defixa* [Walker], *C. nigripes*, *Chrysomya pinguis* [Walker], and *C. villeneuvi*) were found in the lowland forest below 800 m, but some preferred the alluvial forest, whereas others occurred more numerously in the mixed dipterocarp forest. He reported that *L. porphyrina* was the only abundant calliphorid between 800 and 1600 m, and two *Calliphora* sp. (*Calliphora atripalpis* Malloch and *Calliphora fulviceps* Wulp) were confined to the upper montane forests.

In west Malaysia, Omar et al. (2003a) studied the distribution and bionomics of different species of Muscidae and Calliphoridae in three different locations around Kuala Lumpur and Gombak areas to elucidate the synanthropic behavior of Malaysian flies. They reported that the asynanthropic flies were *Bengalia labiata* Robineau-Desvoidy, *H. violacea*, *Hypopygiopsis fumipennis* Walker, *C. defixa*, and *C. nigripes*, while hemisynanthropic flies in ascending order of synanthropic were *C. villeneuvi*, *Chrysomya chani* Kurahashi, *M. sorbens* Wiedemann (Diptera: Muscidae), *H. ligurriens*, *M. domestica*, *C. rufifacies*, *L. cuprina* (Wiedemann) (Diptera: Calliphoridae), and *C. megacephala*.

Omar et al. (2003b) conducted another study using decomposing prawns and ox liver to better understand the distribution of flies, their food preference, and sex ratio at various altitudes in seven locations along the Titiwangsa transect (Kuala Lumpur, Malaysia). They successfully collected 32 fly species belonging to seven families (i.e., Calliphoridae, Muscidae, Anthomyiidae, Sarcophagidae, Lauxaniidae, Otitidae, and Tephritidae). In a fly bait preference study conducted in Kuantan, Heo et al. (2008e) showed that *C. megacephala* were highly attracted to defrosted pork and salted fish compared to prawns and mango fruits.

Omar et al. (2003c) also showed the distribution of blow flies in a five-story building in Kuala Lumpur using cattle liver as bait. They demonstrated that blow fly density decreased with increasing level of the building, and no nocturnal activity at the ground floor was observed, although the fly density was highest during the daytime. Heo et al. (2011b) conducted a study on the oviposition capability of forensic flies on the rooftop of a building that consisted of 21 floors (or ca. 101.6 m from ground) and determined that four species successfully located the chicken liver bait, (*C. megacephala*, *Megaselia scalaris*, *Parasarcophaga ruficornis* [Fabricius] [Diptera: Sarcophagidae], and *Parasarcophaga dux* [Thomson] [Diptera: Sarcophagidae]). Similarly, Syamsa et al. (2012) reported the occurrence of *S. nudiseta* on a human corpse found on the top floor of a 13-story building in Kuala Lumpur and the mPMI was estimated at approximately 5–9 days.

Nazni et al. (2007) investigated the distribution and abundance of diurnal and nocturnal dipteran species using multiple baits in the Federal Territory, Putrajaya, Malaysia and succeeded in collecting 23 species of flies belonging to six families (Calliphoridae, Chrysomyinae, Muscidae, Sarcophagidae, Tachinidae, and Ullilidae). Their findings showed that *C. megacephala* was the most dominant species, followed by Sarcophagidae and *M. domestica*. As expected, the diurnal period had greater numbers of flies compared to the nocturnal period: *Parasarcophaga misera* Walker, *Musca ventrosa* Wiedemann, *Anthomyia iliocata*, *Dichaetomyia* sp., *Phumosia testacea* Senior-White, *Cosmina* sp., *Stomorhina discolor* (Fabricius), Ullitidae, and Tachinidae were strictly diurnal, whereas only one species (*Lispe* cf. *leucospila*) was strictly nocturnal.

Zuha et al. (2008) examined fly artifacts (e.g., fecal and regurgitant specks) produced by the forensically important blow fly, *C. megacephala*. Documentation of the characteristics of fly artifacts is essential to interpret bloodstains because the presumptive test for blood using Saugur and Luminol could not differentiate stains from flies or victims (Bevel and Gardner 2002). Thus, Zuha et al. (2008) highlighted the important feature of fecal spots, regurgitant specks, and swiping stains produced by *C. megacephala*. This information could be useful for forensic scientists in differentiating blood spatter produced from the victims compared to those formed by flies, especially in the process of crime event reconstruction.

Besides studying the ecological behavior, Chen et al. (2008b) and Kumara et al. (2009b) provided life table developmental data of several blow flies. These data may assist in improving the accuracy of mPMI estimation. Recently, Zuha et al. (2012) investigated the interaction effects of temperature and food on the development of scuttle fly, *Megaselia scalaris*, and highlighted the importance of considering these interactions when deciding the most suitable medium for rearing fly larvae.

3.4 OTHER CARRION INSECTS IN MALAYSIA

The biodiversity and distributional ecology of carrion beetles (Scarabaeidae and Staphylinidae) in tropical rain forest in Sarawak, east Malaysia, have been reported by Hanski (1983), and Hanski and Hammond (1986). Their study collected and identified 66 species of dung- and carrion-feeding beetles in Sarawak, Borneo (Coleoptera: Scarabaeidae) (Hanski 1983) and 110 species of Staphylinidae with exclusive to the Aleocharinae, one species of silphid, and six species of Histeridae (Hanski and Hammond 1986). Both studies reported that the number of species began to decrease with increasing altitude.

Chen et al. (unpublished data) recorded a total of 24 Coleoptera species belonging to 12 families from monkey carcasses placed outdoors and indoors in a lowland forested area and a montane forested area of Peninsular Malaysia. The species of beetles collected from the carcasses increased over the course of decomposition of the remains in both outdoor and indoor locations. It is important to note that the majority of beetle specimens from these studies conducted in the lowland forest and montane, forested areas were adults, only larvae of *Necrophila (Chrysosilpha) formosa* (Laporte) (Coleoptera: Silphidae) and *Philonthus* sp. (Coleoptera: Staphylinidae) were collected. Their study indicated that beetles preferred to colonize heavily decomposed monkey carcasses. Beetles in montane, forested areas were more diverse in species compared to lowland, forested area. Staphylinids dominated carcasses placed in a montane forested area, while hybosorids dominated carcasses placed in a lowland forested area. Overall, only four species visited carcasses located in both ecological habitats, whereas the other 20 species were perhaps geographically isolated.

With the advancement of knowledge in forensic entomology, this discipline has now gained its rightful position in forensic case investigations. The authors have received many samples from forensic cases throughout the country, and among these beetle adults and larvae were the most often documented. Therefore, it is imperative that beetle taxonomy be further improved to ensure the correct identification and hence mPMI estimations. It is noteworthy that most of the beetles obtained are from skeletonized cadavers, indicating that beetles normally invade during the late stages of decomposition. Two species of beetles commonly obtained from forensic cases in Malaysia belong to the Family Silphidae: *N. formosa*, and *Diamesus osculans* (Vigors). Beetle larvae belonging to the Silphidae were also found in forensic cases and were not identifiable to species level. Adult beetles were also collected that belonged to the Staphylinidae (unable to identify to species level) and Cleridae with two species identified, namely *Necrobia ruficolis* Fabricius and *Necrobia rufipes* (De Geer).

Heo et al. (2009a) reported six species of ants (Hymenoptera: Formicidae) recovered from swine carcasses placed in an oil palm plantation in Tanjung Sepat, Selangor. The ants collected belonged to three subfamilies: Formicinae (*Oecophylla smaragdina* Fabricius and *Anoplolepis gracilipes* Smith), Myrmicicnae (*Tetramorium* sp. and *Pheidologeton* sp.), and Ponerinae (*Odontoponera* sp. and *Diacamma* sp.). According to Heo et al. (2009a) and Nazni et al. (2011), ants were typically observed shortly after death and were present throughout the decay process. In addition, these ants predated on fly eggs, larvae, pupae, adults, and some fed on tissues, while Chen et al. (2014a) stated that different ant species were obtained from monkey carcasses placed in different ecological habitats. *Cardiocondyla* sp. was only found on carcasses placed in the coastal area, while *Pheidolelongipes*, *Hypoponera* sp., and *Pachycondyla* sp. were solely found on carcasses placed in the highland area. On the other hand, *Pheidologeton diversus* (Jerdon) and *Paratrechina longicornis* (Latreille) were found in all ecological habitats that were examined. These data suggest that ants may serve as geographic indicators for different ecological habitats in Malaysia.

3.5 DEVELOPMENT OF TAXONOMY KEYS FOR ADULTS AND LARVAE OF FORENSIC FLIES IN MALAYSIA

Precise and understandable taxonomic keys are of importance in the identification of forensically important fly adults and larvae. The first preliminary taxonomic key of forensically important flies in Malaysia was published by Singh et al. (1979). Later, Kurahashi et al. (1997) reviewed 118 species of adult flies belonging to Calliphoridae from Malaysia and Singapore. The localities, distributions, and ecological data for each species were also recorded. Omar (2002) compiled taxonomic features of nine species of fly larvae commonly found with human corpses in Malaysia. The taxonomic features of adult flies of Calliphoridae and Sarcophagidae obtained from forensic entomological studies in Thailand and Malaysia were published by Kurahashi (2002a,b). Recently, Nazni et al. (2011) published a simplified pictorial key of forensic flies for the novice to practice forensic entomology and perform simple identification. Heo et al. (2012a) compiled and revised an identification key for third instars of forensic interest, which included several species of Sarcophagidae and Calliphoridae such as *H. fumipennis* and *L. porphyrina*. An account of the morphological description of third instar larvae of *Boettcherisca highlandica* Kurahashi and Tan and their role in forensics have been given by Heo et al. (2012b) and Aisyah et al. (2013). New locality records of blow flies in Peninsular Malaysia and Borneo have been continuously documented, especially those necrophilous and necrophagous species that were attracted to putrid baits. Examples of Calliphoridae that were recently reported include *Isomyia paurogonita* Fang and Fan (Heo et al. 2013) and *C. thanomthini* Kurahashi and Tumrasvin, which were new records for Peninsular Malaysia (Aisyah et al. 2013).

3.6 ENTOMOTOXICOLOGY

Entomotoxicology is a relatively new branch of forensic entomology in Malaysia. Studies done in Malaysia have indicated that the presence of malathion (Mahat et al. 2009) and gasoline (Rashid et al. 2010) influenced the initial oviposition and development stages of dipterans. Recently, Mahat et al. (2013) indicated that paraquat did not influence initial oviposition and larval mortality and did not kill or repel the adults of *C. megacephala*. Conversely, Rashid et al. (2012) examined the effect of an alkaloid, mitragynine, on *C. megacephala* growth rate. Mitragynine is an alkaloid in *Mitragyna speciosa* Korth, which is also known as Ketum by local Malaysians. The results of this study showed that the life cycle of specimens of *C. megacephala* in 60-g Ketum extract was delayed by up to 24 hours than the control group and had the lowest survival rate among the test groups. Another study by Rashid et al. (2013) showed that the presence of mitragynine in *C. rufifacies* delayed its developmental rate by 3 days.

In a recent murder case, police found skeletal remains and a huge number of pupae in a luggage bag. After entomotoxicological analysis, a highly poisonous organophosphate (presumably sarin) was isolated and identified from the pupal cases by using the LC-MS/MS QExactive ORBITRAP system (Rozie et al. 2014). If confirmed, this will be the first report of sarin isolation from pupal cases.

3.7 MOLECULAR FORENSIC ENTOMOLOGY

Over the years, species identification by sequencing fragments of the mitochondrial DNA (mtDNA) has become a valuable tool for identifying blow fly species with forensic importance, especially species of Sarcophagidae. Malaysian researchers are strengthening their capacity to use molecular approaches for identifying flies of forensic importance.

Tan et al. (2009) have reported the first molecular work in Malaysia to distinguish two forensically important flies, *C. megacephala* and *C. rufifacies*. They successfully sequenced cytochrome oxidase subunit I (*COI*) and subunit II (*COII*) genes with 2303 bp. Low intraspecific sequence variations were observed from both species: 0.26% and 0.17% for *C. megacephala* and *C. rufifacies*, respectively. Phylogenetic analysis also demonstrated that these two species were clearly separated in different branches supported with a high bootstrap value. More recently, Tan et al. (2010) sequenced the full length of *COI* and *COII* genes from 20 species of flesh flies (Sarcophagidae). An intraspecific sequence variation was low between species (~2%). A phylogenetic tree analysis based on 2294 bp of combined *COI* and *COII* genes also showed strongly supported monophyletic grouping for all species.

Recently, Rozie et al. (2014) reported the use of mitochondrial and nuclear DNA to distinguish the forensically important flies of the subgenus *Boettcherisca* (Sarcophagidae) in Malaysia. Regardless of the low sequence variation between the three different species examined, *Boettcherisca javanica* Lopes, *B. highlandica*, and *Boettcherisca karnyi* (Hardy) could be clearly separated in the phylogenetic analysis.

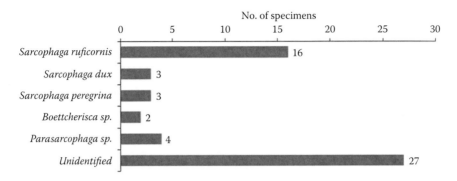

Figure 3.2 Flesh flies (Sarcophagidae) identified from human cadavers from 2006 to 2013.

A Molecular Forensic Laboratory has been established in the IMR in Malaysia. Currently, these facilities can be used to identify sarcophagid flies to species level, allowing for more precise description of their associated distribution patterns (Figure 3.2).

In a more recent study in 2013, *C. megacephala* was shown to have very low intraspecific sequence variation (>1%). Specimens for this species were collected throughout Peninsular Malaysia and later analyzed using the *COI* gene and the internal transcribed spacer 2 (ITS-2) region. Phylogenetic analysis showed that the species were well separated. This study also reported that the sequences obtained from the ITS-2 region could differentiate the specimens geographically using intraspecific variation. Specimens of *C. megacephala* collected from the highlands have a high value of sequence variation compared to lowland specimens (1.18%). The polymerase chain reaction-restriction fragment length polymorphism (PCR-RFLP) technique was also used to distinguish *C. megacephala* from other forensically important flies' species. Partial sequence of the cytochrome oxidase gene was digested using restriction enzymes and different profiles were obtained according to the species. High resolution melting analysis (HRMA) was also conducted to differentiate between *C. megacephala* and *C. rufifacies*. To date, besides Malewski et al. (2010), this is only the second study of fly identification to use real-time PCR based on melting curve analysis. However, this technique was not able to determine the geographically isolated species.

3.8 AQUATIC FORENSIC ENTOMOLOGY

There are two distinct regions of Malaysia, which are Peninsular Malaysia to the west and Malaysia Borneo to the east. Between these two regions is the South China Sea, which is the largest water body in Malaysia. The Strait of Malacca lies between Sumatera (Indonesia) and the west coast of Peninsular Malaysia. There are many rivers and freshwater lakes in Malaysia, with the Rajang River in Sarawak (Malaysia Borneo) being the longest river (760 km) and Pahang River being the longest in the Peninsular (435 km). Because of the abundance of water bodies, cases where bodies are found near or in the water are common. These cases include submerged or floating corpses found in freshwater, brackish, and saltwater ecoregions.

The role of freshwater and marine fauna in forensic investigations has received very little attention and the use of aquatic insects in forensic minimum postmortem submergence interval (mPMSI) estimates has been rare (Haskell et al. 1989; Catts and Goff 1992; Sorg et al. 1997; Tomberlin and Adler 1998; Davis and Goff 2000). Heo et al. (2008a) placed a piglet in a freshwater pond and found that oviposition activities of *C. megacephala* started only after 4 days postmortem and the developing larvae were unable to complete their life cycles into the adult stage. In a recent forensic investigation, some noninsect forensic samples (Figure 3.3) had been received and proved to be challenging

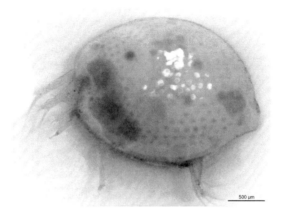

Figure 3.3 Myodocopids that were recovered from the throat of the corpse.

to identify. According to the information provided by the police, the body was found floating in a marine environment and the samples were extracted from the throat of the cadaver. Through networking and collaboration with Martin Hall, Miranda Lowe, and Martin Angel from the Natural History Museum, United Kingdom, the specimens were identified as myodocopid Ostracoda. Myodocopids are marine scavengers and these accounts for their presence on the cadaver. The shape of the carapace suggests that they were cypridinids (Cypridinidae), which could be a novel species. Very few researchers study myodocopids. Although the organism was identified, an mPMI could not be estimated due to paucity of related information on this group of marine invertebrates. The use of these arthropods for estimating the time of death is therefore more difficult and highly dependent on the conditions of the aquatic habitat. Presently, there are no validated invertebrate successional models for corpses found in aquatic habitats (Merrit and Wallace 2001).

3.9 FUTURE TRENDS IN FORENSIC ENTOMOLOGY

Forensic entomology has been practiced in Malaysia for the past six decades. In spite of a wealth of data and research findings, many gaps in knowledge remain. Although many studies were conducted on succession of insect on carcasses, many of these actually focused on the postcolonization interval (post-CI) activity. Tomberlin et al. (2011) recommended that both pre- and post-CI can be important for estimating a mPMI, depending on the nature of the case and condition of the remains. The study of Nazni et al. (2011) was probably the only report on pre-CI activity of forensic flies on monkey carcasses in Malaysia. This is one area of research that should be examined further. Although DNA techniques are fairly widely used in Malaysia, these techniques were mainly used for species identification of forensic flies only. The interaction of genetics with environment is seldom investigated. In many laboratory studies on developmental time, it has often been observed that the same species of fly reared under similar environmental conditions display different developmental period lengths; this can lead to confusion and miscalculations of mPMI estimates. It has been observed that the same fly species from different geographical origins may have different developmental times under identical environmental conditions (Greenberg 1990; Grassberger and Reiter 2001). This can be due to genetic differences among individuals, environment, and the interaction of these two factors (Tarone and Foran 2006). The recent understanding of epigenetic (DNA methylation) and polyphenism (phenotypic plasticity) could better explain the various behaviors demonstrated by necrophagous flies in nature. This aspect should be examined to ensure more accurate mPMI estimations. To date, most of the studies in Malaysia have concentrated on forensic fly identifications using conventional taxonomy or molecular techniques, and estimating the mPMI still depends on morphological characteristics such as body length of the immature. Other age-determination methods, such as the use of pteridine should be investigated. Preliminary results have shown that pteridine concentration in larvae of Calliphoridae, Sarcophagidae, and Muscidae correlated significantly with age (Rozie et al. 2014). This warrants further research not only on pteridine, but also other potentially useful age indicators, such as temporal gene expression and cuticular hydrocarbon signatures, among others.

In addition to studies on the role of fly larvae in decomposition of the carcass/cadaver, the role of microbes, or epinecrotic communities of the necrobiome (Benbow et al. 2013), in relation to arthropod activity is another area worthy of investigation to determine if such information can be utilized to further improve mPMI estimations. Furthermore, the presence of soil arthropods is of great potential to be utilized in mPMI determination. The study of mites associated with carrion, which is also known as forensic acarology, is showing promising outcomes. Preliminary studies have been conducted by Aisyah et al. (2013) in three ecoregions in Peninsular Malaysia and the results showed that Macrochelidae can been detected throughout the decomposition stages of rabbit carcasses, while soil mites, Oribatida, were found at the peak of their population during the fresh stage of decay and declined in abundance as decomposition proceeded (Figure 3.4). A list of the recovered Acari from rabbit carcasses is shown in Figure 3.5.

The use of fly puparia parasitoids (parasitic microhymenopteran) is another direction that can be explored for assisting in estimates of the mPMI. Studies in Malaysia have revealed *Exoristobia philippinensis* Ashmead (Hymenoptera: Encyrtidae) is a natural parasitoid that colonizes *C. rufifacies* and *P. ruficornis* puparia (Heo et al. 2009b, 2011b). Further, *Dirhinus himalayanus* Westwood (Hymenoptera: Chalcididae) was found to colonize the puparia of *P. ruficornis*, which was a new host record for this parasitoid (Heo et al. 2011b).

In conclusion, estimation of the mPMI is one of the most important functions in the practice of forensic entomology, and the precision and accuracy of such entomologically based mPMI interpretation is the utmost imperative factor to determine whether it is qualified to being accepted by the court. To achieve this goal, developing standard procedures is urgently needed, and the practices in forensic entomology should be supported by a valid scientific foundation. A solid scientific base means the experimental data in the laboratory or in the field should be replicated to offer a range of estimated variation and acknowledged weaknesses in the research. Malaysia is gratefully blessed

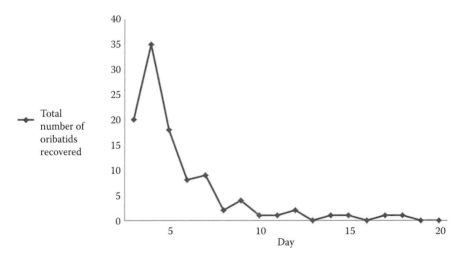

Figure 3.4 Abundance of oribatid specimens recovered from soil underneath the rabbit carcass placed in the Cameron Highlands, Malaysia. Aisyah, S., B. Latif, H. Kurahashi, D. E. Walter, and C. C. Heo. Flies and mites associated with rabbit carcasses in Malaysia. 11th Annual Conference of North American Forensic Entomology Association. OH: University of Dayton, 14–17 July 2013.

Order: Suborder	Superfamily / Family	Genus and Species
Acariformes: Oribatida	Microzetidae	-
	Oppiidae	-
	Haplozetidae	-
	Oripodoidea	-
Sarcoptiformes: Astigmata	Acaridae	*Tyrophagus sp.*(to be confirmed)
Parasitiformes: Mesostigmata	Macrochelidae	*Macrocheles sp.* *Macrocheles muscadomestica* *Macrocheles glaber*
	Parasitidae	*Gamasodes sp.* several other unidentified
	Pachylaelapidae	*Pachyseius sp.* several other unidentified
	Eviphididae	-
	Uropodoidea	-
	Ologamasidae	-
	Laelapidae	*Gaeolaelaps sp.*
	Parholaspididae	*Holaspulus sp.*

Figure 3.5 Acari associated with decomposition of rabbit carcasses in Peninsular Malaysia Aisyah, S., B. Latif, H. Kurahashi, D. E. Walter, and C. C. Heo. Flies and mites associated with rabbit carcasses in Malaysia. 11th Annual Conference of North American Forensic Entomology Association. OH: University of Dayton, 14–17 July 2013.

with megabiodiversity and thus more ecological studies on necrophagous guilds are needed to depict the complex trophic cascades that describe the decomposition process in the tropics. Future research and continued improvements in understanding carrion flies, microbes and factors affecting these communities in Malaysia will contribute to a more comprehensive application of forensic entomology in tropical areas of the world.

ACKNOWLEDGMENTS

We thank Datuk Dr. Noor Hisham Bin Abdullah, Director General of Health, and Dr. Shahnaz Murad, Director of the Institute for Medical Research, for their permission to publish this chapter. Many thanks to Khairul Asuad Muhammed for his assistance in Diptera larvae identification, Azahari Abdul Hadi for the identification of coleopteran and larvae, and Heah Sock Kiang for establishing 15 species of forensically important fly colonies in the insectarium. A special note of thanks to all the staff of Medical Entomology Unit, IMR for their assistance in the field work.

REFERENCES

Aisyah, S., B. Latif, H. Kurahashi, D. E. Walter, and C. C. Heo. 2013. Flies and mites associated with rabbit carcasses in Malaysia. 11th Annual Conference of North American Forensic Entomology Association. OH: University of Dayton, 14–17 July 2013.

Amoret P. W. and J. R. H. Martin. 2012. Where did we go wrong? A review of past cases. 9th Meeting of the European Association for Forensic Entomology, Faculty of Biology and Earth Science, Nicolaus Copernicus University, Toruń, Poland, 18th–21st April 2012.

Azwandi, A. and A. Abu Hassan. 2009. A preliminary study on the decomposition and dipteran associated with exposed carcasses in an oil palm plantation in Bandar Baharu, Kedah, Malaysia. *Tropical Biomedicine* 26(1): 1–10.

Benbow, M. E., A. J. Lewis, J. K. Tomberlin, and J. L. Pechal. 2013. Seasonal necrophagous insect community assembly during vertebrate carrion decomposition. *Journal of Medical Entomology* 50: 440–450.

Bevel, T. and R. M. Gardner. 2002. *Bloodstain Pattern Analysis: With an Introduction to Crime Scene Reconstruction.* 2nd ed. FL: CRC Press LLC. pp. 7, 87, 221.

Catts, E. P. and M. L. Goff. 1992. Forensic entomology in criminal investigations. *Annual Review of Entomology* 37: 253–272.

Chen, C. D., C. C. Heo, D. McAlpine, H. Kurahashi, W. A. Nazni, M. A. Marwi, J. Jeffery, H. L. Lee, B. Omar, and M. Sofian-Azirun. 2008a. First report of the signal fly, *Scholastes* sp. (Diptera: Platystomatidae) visiting animal carcasses in Malaysia. *Tropical Biomedicine* 25(3): 264–266.

Chen, C. D., H. L. Lee, W. A. Nazni, R. Ramli, J. Jeffery, and M. Sofian-Azirun. 2010. First report of the house fly larvae, *Musca domestica* (Linnaeus) (Diptera: Muscidae) associating with the monkey carcass in Malaysia. *Tropical Biomedicine* 27(2): 355–359.

Chen, C. D., W. A. Nazni, H. L. Lee, R. Hashim, N. A. Abdullah, R. Ramli, K. W. Lau, C. C. Heo, T. G. Goh, A. A. Izzul, and M. Sofian-Azirun. 2014a. A preliminary report on ants (Hymenoptera: Formicidae) recovered from forensic entomological studies conducted in different ecological habitats in Malaysia. *Tropical Biomedicine* 31(2): 381–386.

Chen, C. D., W. A. Nazni, H. L. Lee, J. Huijbregts, M. Maruyama, T. G. Goh, W. K. Lau, R. Ramli, A. A. Izzul, and M. Sofian-Azirun. First report on beetle (Coleoptera) succession in forensic entomological studies in Malaysian lowland and montane forested areas. (Submitted for review).

Chen, C. D., W. A. Nazni, H. L. Lee, J. Jeffery, W. M. Wan-Norjuliana, A. G. Abdullah, and M. Sofian-Azirun. 2008b. Larval growth parameters and growth rates of forensically important flies, *Hypopygiopsis violacea* Macquart, 1835 and *Chrysomya rufifacies* Macquart, 1842. *Proceeding of The Asean Congress of Tropical Medicine and Parasitology* 3: 97–100.

Davis, J. B. and M. L. Goff. 2000. Decomposition patterns in terrestrial and intertidal habitats on Oahu Island and Coconut Island, Hawaii. *Journal of Forensic Science* 45: 836–842.

Firdaus, M. S., M. A. Marwi, R. A. Syamsa, R. M. Zuha, Z. Ikhwan, and B. Omar. 2010. Morphological descriptions of second and third instar larvae of *Hypopygiopsis violacea* Macquart (Diptera: Calliphoridae), a forensically important fly in Malaysia. *Tropical Biomedicine* 27(1): 134–137.

Goff, M. L. 2003. The maggot and the law. In: Being Human: science, culture and fear. The Royal Society of New Zealand, Miscellaneous Service 63: 45–51.

Grassberger, M and C. Reiter, C. 2001. Effect of temperature on *Lucilia sericata* (Diptera: Calliphoridae) development with special reference to the isomegalen- and isomorphen-diagram. *Forensic Science International* 120: 32–36.

Greenberg, B. 1990. Behavior of postfeeding larvae of some Calliphoridae and a muscid (Diptera). *Annals of Entomological Society of America* 83: 1210–1214.

Hamid, N. H., B. Omar, M. A. Marwi, F. M. Ahmad, A. Saleh, M. Halim, S. F. Siew, and M. Norhayati. 2003. A review of forensic specimens sent to Forensic Entomology Laboratory Universiti Kebangsaan Malaysia for the year 2001. *Tropical Biomedicine* 20: 27–31.

Hanski, I. 1981. Carrion flies (Calliphoridae) in tropical rain forest in Sarawak, South-east Asia. *Sarawak Museum Journal* 29: 191–200.

Hanski, I. 1983. Distributional ecology and abundance of dung and carrion-feeding beetles (Scarabaeidae) in tropical rain forests in Sarawak, Borneo. *Acta Zoologica Fennica* 167: 1–45.

Hanski, I. and P. Hammond. 1986. Assemblages of carrion and dung Staphylinidae in tropical rain forests in Sarawak, Borneo. *Annales Entomologici Fennici* 52: 1–19.

Haskell, N. H., D. G. McShaffrey, D. A. Hawley, R. E. Williams, and J. E. Pless. 1989. Use of aquatic insects in determining submersion interval. *Journal of Forensic Science* 34: 622–632.

Heo, C. C., S. Aisha, H. Kurahashi, and B. Omar. 2013. New locality record of *Isomyia paurogonita* Fang & Fan, 1986 (Diptera: Calliphoridae) from Peninsular Malaysia and Borneo. *Tropical Biomedicine* 30(1): 159–163.

Heo, C. C., H. Kurahashi, M. A. Marwi, J. Jeffery, C. D. Chen, R. M. Zuha, and B. Omar. 2008f. A new record of *Bengalia emarginata* Malloch, 1927 (Diptera: Calliphoridae) from Malaysia. *Tropical Biomedicine* 25(3): 262–263.

Heo, C. C., H. Kurahashi, M. A. Marwi, J. Jeffery, and B. Omar. 2011a. Opportunistic insects associated with pig carrions in Malaysia. *Sains Malaysiana* 40(6): 601–604.

Heo, C. C., B. Latif, S. Aisyah, H. Kurahashi, and J. K. Tomberlin. 2012b. Descriptions of third instar larvae of *Boettcherisca highlandica* Kurahashi & Tan, 2009 (Diptera: Sarcophagidae): A highlander of forensic importance in Malaysia. 11th Annual Scientific Meeting of the College of Pathologists, Academy of Medicine, Malaysia. Crowne Plaza Mutiara Hotel, Kuala Lumpur, Malaysia. 8–10 June 2012.

Heo, C. C., B. Latif, S. Aisyah, W. A. Nazni, and B. Omar. 2012a. Morphological description on the larvae of *Hypopygiopsis fumipennis* (Walker, 1856) (Diptera: Calliphoridae) and a revised key for third instar larvae of forensic importance in Malaysia. 9th Meeting of the European Association for Forensic Entomology, Nicolaus Copernicus University, Torun, Poland. 18–21 April 2012.

Heo, C. C., B. Latif, B, H. Kurahashi, W. A. Nazni, and B. Omar. 2011b. Oviposition of forensically important dipterans on a high-rise building, with a new host record of parasitoids in Malaysia. 9th Annual Conference of North American Forensic Entomology Association. Texas A&M University, College Station, TX, 20–22 July 2011.

Heo, C. C., M. A. Marwi, R. Hashim, N. A. Abdullah, C. D. Chen, J. Jeffery, H. Kurahashi, and B. Omar. 2009a. Ants (Hymenoptera: Formicidae) associated with pig carcasses in Malaysia. *Tropical Biomedicine* 26(1): 106–109.

Heo, C. C., M. A. Marwi, J. Jeffery, I. Ishak, and B. Omar. 2008e. Flies specimens collected from agricultural park, Teluk Cempedak and Bukit Pelindung in Kuantan, Pahang. *Jurnal Sains Kesihatan Malaysia* 6(2): 93–99.

Heo, C. C., M. A. Marwi, J. Jeffery, H. Kurahashi, and B. Omar. 2008d. On the occurrence of *Musca domestica* L. oviposition activity on pig carcass in Peninsular Malaysia. *Tropical Biomedicine* 25(3): 252–253.

Heo, C. C., M. A. Marwi, J. Jeffery, and B. Omar. 2008a. Insect succession on a decomposing piglet carcass placed in a man-made freshwater pond in Malaysia. *Tropical Biomedicine* 25(1): 23–29.

Heo, C. C., M. A. Marwi, J. Jeffery, and B. Omar. 2008c. On the predation of fly, *Chrysomya rufifacies* (Macquart) by spider, *Oxyopes* sp. Latreille (Oxyopidae). *Tropical Biomedicine* 25(1): 93–95.

Heo, C. C., M. A. Marwi, A. F. M. Salleh, J. Jeffery, H. Kurahashi, and B. Omar. 2008b. Study of insect succession and rate of decomposition on a partially burned pig carcass in an oil palm plantation in Malaysia. *Tropical Biomedicine* 25(3): 202–208.

Heo, C. C., M. A. Marwi, A. F. M. Salleh, J. Jeffery, and B. Omar. 2007. A preliminary study of insect succession on a pig carcass in a palm oil plantation in Malaysia. *Tropical Biomedicine* 24(2): 23–27.

Heo, C. C., W. A. Nazni, H. L. Lee, J. Jeffery, C. D. Chen, K. W. Lau, B. Omar, and M. Sofian-Azirun. 2009b. Predation on pupa of *Chrysomya rufifacies* (Macquart) (Diptera: Calliphoridae) by parasitoid, *Exoristorbia philippinensis* Ashmead (Hymenoptera: Encyrtidae) and *Ophyra spinigera* larva (Diptera: Muscidae). *Tropical Biomedicine* 26(3): 369–372.

Kavitha, R., C. D. Chen, H. L. Lee, W. A. Nazni, I. Sa'diyah, and M. A. Edah. 2008. Estimated post-mortem interval (PMI) of pathologist and entomologist in Malaysia: A comparison. *Proceeding of The ASEAN Congress of Tropical Medicine and Parasitology* 3: 21–27.

Kumara, T. K., H. A. Abu, M. R. Che Salmah, and S. Bhupinder. 2010b. A report on the pupae of *Desmometopa* sp. (Diptera: Milichiidae) recovered from a human corpse in Malaysia. *Tropical Biomedicine* 27 (1): 131–133.

Kumara, T. K., A. Abu Hassan, M. R. Che Salmah, and S. Bhupinder. 2009a. The infestation of *Dermestes ater* (De Geer) on a human corpse in Malaysia. *Tropical Biomedicine* 26(1): 73–79.

Kumara, T. K., A. Abu Hassan, M. R. Che Salmah, and S. Bhupinder. 2009b. Larval growth of the muscid fly, *Synthesiomyia nudiseta* (Wulp), a fly of forensic importance, in the indoor fluctuating temperatures of Malaysia. *Tropical Biomedicine* 26(2): 200–205.

Kumara, T. K., H. R. H. L. Disney, and H. A. Abu. 2010a. First records of two species of Oriental scuttle flies (Diptera: Phoridae) from forensic cases. *Forensic Science International* 195: e5–e7.

Kurahashi, H. 2002a. Key to the calliphorid adults of forensic important in the Oriental Region. In *Entomology and Law: Flies as Forensic Indicators*, edited by Greenberg, B. and Kunich, J. C., 127–138. United Kingdom: Cambridge University Press.

Kurahashi, H. 2002b. Key to the sarcophagid of the Oriental Region. In *Entomology and Law: Flies as Forensic Indicators*, edited by Greenberg, B. and Kunich, J. C, 138–142. United Kingdom: Cambridge University Press.

Kurahashi, H., N. Benjaphong, and B. Omar. 1997. Blow flies (Insecta: Diptera: Calliphoridae) of Malaysia and Singapore. *The Raffles Bulletin of Zoology Supplement* (5): 1–88.

Lee, H. L. 1994. Larvae of *Eristalis* spp (Family: Syrphidae) found in human cadaver in Malaysia. *Journal of Bioscience* 5(1&2): 67–68.

Lee, H. L., A. G. Abdullah, and W. H. Cheong. 1984. The use of fly larvae from human corpses in determining the time of death: A review and some technical considerations. *Journal Med Hlth Lab Tech Malaysia* 9: 15–17.

Lee, H. L. and W. H. Cheong. 1982. Larva of *Hermetia* species from a dead human female. *Southeast Asian Journal of Tropical Medicine and Public Health* 13(2): 289–290.

Lee, H. L., M. Krishnasamy, A. G. Abdullah, and J. Jeffery. 2004. Review of forensically important entomological specimens in the period of 1972–2002 *Tropical Biomedicine*. 21(2): 69–75.

Lee, H. L. and T. Marzuki. 1993. Preliminary observation of arthropods on carrion and its application to forensic entomology in Malaysia. *Tropical Biomedicine* 10: 5–8.

Mahat, N. A., C. L. Yin, and P. T. Jayaprakash. 2013. Influence of paraquat on *Chrysomya megacephala* (Fabricius) (Diptera: Calliphoridae) infesting minced-beef substrates in Kelantan, Malaysia. *Journal of Forensic Science.* 59(2): 529–32. doi: 10.1111/1556-4029.12355

Mahat, N. A., Z. Zafarina, and P. T. Jayaprakash. 2009. Influence of rain and malathion on the oviposition and development of blow flies (Diptera: Calliphoridae) infesting rabbit carcasses in Kelantan, Malaysia. *Forensic Science International* 192: 19–28.

Malewski, T., A. Draber-Mońko, J. Pomorski, M. Łoś, and W. Bogdanowicz. 2010. Identification of forensically important blowfly species (Diptera: Calliphoridae) by high-resolution melting PCR analysis. *International Journal of Legal Medical.* 124(4): 277–285. doi: 10.1007/s00414-009-0396-x.

Martin–Vega, D.2011. Skipping clues: Forensic importance of the family Piophilidae (Diptera). *Forensic Science International* 212(1–3): 1–5. doi: 10.1016/j.forsciint.2011.06.016.

Merrit, R. W. and J. R. Wallace. 2001. The role of aquatic insects in forensic investigations. In *Forensic Entomology: The Utility of Arthropods in Legal Investigations*, edited by Byrd, J. H. and J. L. Castner, 177–222. Boca Raton, FL: CRC Press.

Nazni, W. A., J. Jeffery, I. Sa'diyah, W. M. Noorjuliana, C. D. Chen, S. A. Rohayu, A. H. Hafizam, and H. L. Lee. 2008. First report of maggots of family: Piophilidae recovered from human cadavers in Malaysia. *Tropical Biomedicine* 25(2): 173–175.

Nazni, W. A., H. L. Lee, C. D. Chen, C. C. Heo, A. G. Abdullah, W. M. Wan-Norjuliana, W. K. Chew, J. Jeffery, H. Rosli, and M. Sofian-Azirun. 2011. Comparative insect fauna succession on indoor and outdoor monkey carrions in a semi-forested area in Malaysia. *Asian Pacific Journal of Tropical Biomedicine* 1(2): S232–S238.

Nazni, W. A., H. Nooraidah, J. Jeffery, A. H. Azahari, I. Mohd Noor, I. Sa'diyah, and H. L. Lee. 2007. Distribution and abundance of diurnal and nocturnal dipterous flies in the Federal Territory, Putrajaya. *Tropical Biomedicine* 24(2): 61–66.

Omar, B. 2002. Key to third instar larvae of flies of forensic importance in Malaysia. In *Entomology and Law: Flies as Forensic Indicators*, edited by Greenberg, B. and J. C. Kunich. United Kingdom: Cambridge University Press.

Omar, B., M. A. Marwi, A. Ahmad, R. M. Zuha, and J. Jeffery. 2003a. Synanthropic index of flies (Diptera: Muscidae and Calliphoridae) collected at several locations in Kuala Lumpur and Gombak, Malaysia. *Tropical Biomedicine* 20(1): 77–82.

Omar, B., M. A. Marwi, M. S. Chu, J. Jeffery, and H. Kurahashi. 2003b. Distribution of medically important flies at various altitudes of Titiwangsa Range near Kuala Lumpur, Malaysia. *Tropical Biomedicine* 20(2): 137–144.

Omar, B., M. A. Marwi, A. H. Mansar, M. S. Rahman, and P. Oothuman. 1994a. Maggots of *Synthesiomyia nudiseta* (Wulp) (Diptera: Muscidae) as decomposers of corpses found indoors in Malaysia. *Tropical Biomedicine* 11: 145–148.

Omar, B., M. A. Marwi, P. Oothuman, and H. F. Othman. 1994c. Observation on the behaviour of immatures and adults of some Malaysian sarcosaprophagous flies. *Tropical Biomedicine* 11: 149–153.

Omar, B., M. A. Marwi, S. Sulaiman, and P. Oothuman. 1994b. Dipteran succession in monkey carrion at a rubber tree plantation in Malaysia. *Tropical Biomedicine* 11: 77–82.

Omar, B., M. Yasin, M. A. Marwi, and J. Jeffery. 2003c. Observation on the distribution of *Lucilia cuprina* (Diptera: Calliphoridae) in a building of a housing estate at kelompok Embun Emas, AU3, Kuala Lumpur. *Tropical Biomedicine* 20(2): 181–183.

Payne, J. A. 1965. A summer carrion study of the baby pig *Sus scrofa* Linnaeus. *Ecology* 46: 592–602.

Rashid, A. R., A. S. Siti, F. R. Siti, A. R. Reena, H. S. S. Sharifah, F. Z. Nurul, and W. A. Nazni. 2013. Forensic implications of Blowfly *Chrysomya rufifacies* (*Calliphoridae*: Diptera) development rates affected by Ketum extract. World Academy of Science, Engineering and Technology. *International Journal of Medical Science and Engineering* 7: 7.

Rashid, R. A., O. Khairul, R. M. Zuha, and C. C. Heo. 2010. An observation on the decomposition process of gasolin ingested monkey carcasses in a secondary forest in Malaysia. *Tropical Biomedicine* 27(3): 373–383.

Rashid, R. A., N. F. Zulkifli, R. A. Rashid, S. F. Rosli, S. H. S. Sulaiman, and W. A. Nazni. 2012. Effects of Ketum extract on blowfly Chrysomya megacephala development and detection of mitragynine in larvae sample. Business, Engineering and Industrial Application (ISBEIA). *IEEE Symposium.* pp. 337–341. 23–26 September 2012.

Rumiza, A.R., Osman, K., Mohd Iswadi, I., Raja Muhammad, Z., and Rogaya, A.H. 2008. Determination of malathion levels and the effect of malathion on the growth of *Chrysomya megacephala* (Fibricius) in malathion-exposed rat carcass. *Tropical Biomedicine* 25(3): 184–190.

Reid, J. A. 1953. Notes on houseflies and blowflies in Malaya. *Bulletin of Institute for Medical Research Malaya* 7: 1–26.

Rozie S, Maizatul HO and Nazni WA (2014). Toxicological substances detection from empty puparial cases—sample 060/13. Presented at the 50th Anniversary Golden Jubilee of the Malaysian Society of Parasitology and Tropical Medicine and the 6th ASEAN Congress of Tropical Medicine and Parasitology, Intercontinental Kuala Lumpur 5–7 Mac 2014 Poster 112 page 290.

Salleh, A. F. M., M. A. Marwi, J. Jeffery, N. A. A. Hamid, R. M. Zuha, and B. Omar. 2007. Review of forensic entomology cases from Kuala Lumpur Hospital and Hospital Universiti Kebangsaan Malaysia, 2002. *The Journal of Tropical Medicine and Parasitology* 30: 51–54.

Singh, I., H. Kurahashi, and R. Kano. 1979. A preliminary key to the common calliphorid flies of Peninsular Malaysia (Insecta: Diptera). *Bulletin of Tokyo Med Dental University* 26: 5–24.

Smith, K. G. V. 1986. *A Manual of Forensic Entomology.* London, United Kingdom: British Natural History Museum.

Sorg, M. H., J. H. Dearborn, E. I Monahan, H. F. Ryan, K. G. Sweeney, and E. David. 1997. Forensic taphonomy in marine contexts. In *Forensic Taphonomy: The Postmortem Fate of Human Remains,* edited by Haglund, W. D. and M. H. Sorg, 567–604. Boca Raton, FL: CRC Press.

Syamsa, R. A., F. M. S. Ahmad, R. M. Zuha, A. Z. Khairul, M. A. Marwi, A. W. Shahrom, and B. Omar. 2012. An occurrence of *Synthesiomyia nidiseta* (Wulp) (Diptera: Muscidae) from a human corpse in a high rise building in Malaysia: A case report. *Tropical Biomedicine* 29(1): 107–112.

Tahir, N. A., A. H. Ahmad, N. A. Hashim, N. Basari, and C. M. Md. Rawi. 2007. The seasonal abundance of synantrophic fly populations in two selected food outlets in Pulau Pinang, Malaysia. *Jurnal Biosains* 18(1): 81–91.

Tan, S. H., M. A. Edah, S. Johari, B. Omar, H. Kurahashi, and M. Zulqarnain. 2009. Sequence variation in the cytochrome oxidase subunit I and II genes of two commonly found blow fly species, *Chrysomya megacephala* (Fabricius) and *Chrysomya rufifacies* (Macquart) (Diptera: Calliphoridae) in Malaysia. *Tropical Biomedicine* 26(2), 173–181.

Tan, S. H., M. Rizman-Idid, E. Mohd-Aris, H. Kurahashi, and M. Zulqarnain. 2010. DNA-based characterisation and classification of forensically important flesh flies (Diptera: Sarcophagidae) in Malaysia. *Forensic Science International* 199: 43–49.

Tarone, A. M. and D. R. Foran. 2006. Components of developmental plasticity in a Michigan population of *Lucilia sericata* (Diptera: Calliphoridae). *Journal of Medical Entomology* 43: 1023–1033.

Tomberlin, J. K. and P. H. Adler. 1998. Seasonal colonization and decomposition of rat carrion in water and on land in an open field in South Carolina. *Journal Medical Entomology* 35: 704–709.

Tomberlin, J. K., R. Mohr, M. E. Benbow, A. M. Tarone, and S. VanLaerhoven. 2011. A roadmap for bridging basic and applied research in forensic entomology. *Annual Review of Entomology* 56: 401–2.

Zuha, R. M., H. Kurahashi, C. C. Heo, K. Osman, R. A. Rashid, R. A. Hassan, S. R. Abdullah, A. F. M. Salleh, M. A. Marwi, and B. Omar. 2009. *Myospila pudica pudica* (Stein, 1915) (Diptera: Muscidae) in Peninsular Malaysia and its occurrence on a monkey carrion. *Tropical Biomedicine* 26(2): 216–218.

Zuha, R. M., T. A. Razak, W. A. Nazni, and B. Omar. 2012. Interaction effects of temperature and food on the development of forensically important fly, *Megaselia scalaris* (Loew) (Diptera: Phoridae). *Parasitology Research* 111 (5): 2179–2187.

Zuha, R. M., M. Supriyani, and B. Omar. 2008. Fly artifact documentation of *Chrysomya megacephala* (Fabricius) (Diptera: Calliphoridae) – a forensically important blowfly species in Malaysia. *Tropical Biomedicine* 25(1): 17–22.

CHAPTER 4

Thailand

Kabkaew L. Sukontason and Kom Sukontason

CONTENTS

4.1 INTRODUCTION

Thailand represents over 500,000 km^2 within the Indochina peninsula (15°0′0″N, 100°0′0″E) and is the home to more than 66 million people. The capital city, Bangkok, has over eight million inhabitants, and is situated on the delta of the Chao Phraya River. Weather conditions are best described by two seasons based primarily on rainfall: wet and dry. Because of the tropical conditions in Thailand, arthropods are numerous and active throughout the year.

Although a developing nation, crime still occurs and an investigative force is necessary. However, disciplines such as forensic entomology are a relatively new science being applied in Thailand. As such, the application of entomology in death investigations has been historically limited but has seen a recent increase in interest. For the purpose of this chapter, a review of past case studies will be presented along with an assessment of accomplishments and a discussion of the future for this critical science as part of the forensic sciences and their growth in Thailand. A majority of the forensic entomology research occurring in Thailand takes place within the Department of Parasitology at Chiang Mai University, in Chiang Mai, while some occurs at Chulalongkorn University and Khon Kaen University, in the central and northeastern regions, respectively.

4.2 CASE HISTORIES

Human death investigations in Thailand commonly find dipterous larvae associated with corpses and/or death scenes; however, historically, this information was typically not applied as evidence. Mankosol (1986) and Fongsiripaibul (1987) were the earliest forensic practitioners conducting experiments examining dipteran development for determining the minimum postmortem interval (mPMI) of human remains. The first forensic entomology case in Thailand was published in 2001 (Sukontason et al. 2001a). The investigators examined a mummified corpse discovered in a forest in Chiang Mai in northern Thailand. Subsequent forensic entomology cases from northern Thailand were examined by investigators from the Faculty of Medicine, Chiang Mai University. These cases were published along with a review of 30 cases occurring from 2000 to 2006 (Sukontason et al. 2003, 2007a). A catalog of the insects associated with human remains from the northeastern region was developed concurrently with these contributions of forensic entomology in northern Thailand (Sritavanich et al. 2007, 2009). Investigations pertaining to necrophagous insect succession on swine carcasses in various parts of Thailand have documented the similarity of insect taxa between swine and human remains (Table 4.1) (Champathet 2003; Vitta et al. 2007; Sukjit 2011).

In Thailand, the application of insects as evidence in forensic investigation has been limited, predominately due to a lack of relevant information regarding the development of larval flies. A recent survey of casework in Thailand was conducted by the authors with other colleagues. Cases were categorized according to the condition of the remains or habitat where they were located (Sukontason et al. 2003, 2007a): forest, aquatic habitats, burnt remains, myiasis, mummified remains, and remains located indoors. In Sections 4.2.1 through 4.2.6, we discuss casework associated with each of these circumstances.

4.2.1 Forest

In most forensic entomology cases in northern Thailand, the remains were discovered in forested or suburban areas. In a series of 30 case studies, Sukontason et al. (2007a) determined that 18 were located in forested areas. Blow flies (Diptera: Calliphoridae) in the genus *Chrysomya* were the predominant insect taxa colonizing the remains, with *Chrysomya megacephala* (Fabricius) and *C. rufifacies* (Macquart) (Diptera: Calliphoridae) being the two most common species.

4.2.2 Aquatic Habitats

The authors are aware of two cases where human remains were discovered in aquatic habitats. In the first, bloated remains were discovered floating on the water surface of a reservoir. Wounds were present on the neck and back of the corpse. It was concluded that these wounds originated from sharp force trauma. During autopsy, large populations of fly larvae were collected and later identified as third instars of *C. megacephala* and *C. rufifacies*. *C. rufifacies* larvae measuring 1.2 cm were considered the oldest larvae present and, based on the developmental rate of this species reared under similar temperature conditions, were estimated to be 5-days-old.

In the second case, remains in a similar state of decomposition were discovered in an aquatic habitat in northern Thailand; however, no injuries were documented. Following the autopsy, it was concluded that the decedent had drowned. Two collections of fly larvae were obtained: from the corpse and the death scene. With regard to the remains, larvae were located in the eyes, mouth, and nose. The first group was represented by second instars of *C. megacephala* (~0.8 cm) and *C. rufifacies* (0.6 cm), and third instars of *Hemipyrellia ligurriens* (Wiedemann) (Diptera: Calliphoridae) (~0.6 cm), another blow fly common to northern Thailand. The other group of larvae, which

Table 4.1 Species of Flies Associated with Human Remains and Pig Carcasses in Thailand

Family	Species of Fly	Human Remains[a]	Pig Carcasses[b]
Calliphoridae	*Chrysomya bezziana* Villeneuve	✓	✓
	C. chani Kurahashi	✓	✓
	C. megacephala (Fabricius)	✓	✓
	C. nigripes Aubertin	✓	✓
	C. pinguis (Walker)		✓
	C. rufifacies (Macquart)	✓	✓
	C. villeneuvi Patton	✓	✓
	C. thanomthini Kurahashi et Tumrasvin		✓
	Hemipyrellia ligurriens (Wiedemann)	✓	✓
	Hypopygiopsis infumata (Bigot)		✓
	Lucilia cuprina (Wiedemann)	✓	✓
Muscidae	*Atherigona* spp.		✓
	Hydrotaea spinigera (Stein)	✓	✓
	Musca domestica L.		✓
	Musca sorbens Wiedemann		✓
	Synthesiomyia nudiseta (van der Wulp)	✓ (indoor)	✓
Phoridae	*Megaselia scalaris* (Loew)	✓	✓
Piophilidae	*Piophila casei* (L.)	✓	✓
Sarcophagidae	*Boettcherisca peregrina* (Robineau-Desvoidy)		✓
	Parasarcophaga (*Liosarcophaga*) *dux* (Thomson)		✓
	Liopygia ruficornis (Fabricius)	✓ (indoor)	✓

[a] Data from Sukontason et al. (2007a).
[b] Data from Sukjit (2011) collected as adult flies.

was low in number, was identified as flesh flies (Diptera: Sarcophagidae), as the anterior spiracles had 28 irregularly arranged lobes. The larval length of the *C. rufifacies* indicated that they were 2-days-old, based on the development rate of this species at the relevant laboratory temperature.

Examining arthropod evidence from human remains located in aquatic habitats can be quite challenging because aquatic insects have not evolved to consume decomposing remains, and therefore forensic entomologists are restricted to collecting evidence that is found on floating remains (Merritt and Wallace 2010). While the species of fly collected from these floating cases are the same as what has been collected at terrestrial death scenes, it must be considered if the remains had been previously submerged, only being colonized by terrestrial flies once floating to the surface. Thus, additional time must be added to the mPMI estimate to account for the submersion interval; little research has examined this process in Thailand. Consequently, the postmortem submersion interval estimates in various aquatic habitats are speculative at best, when using terrestrial insects such as blow flies.

4.2.3 Burnt Remains

Two unidentified males were discovered in a wooded area, tied together with. Based on the Crow–Glassman scale (CGS) of burn injuries (Glassman and Crow 1996), these corpses were assigned a CGS level 4 rating; the remains were considered to be in an advanced state of decomposition. Several arthropods were identified from the remains, including second instars of *C. megacephala*, third instars of *Hydrotaea spinigera* (Stein) (Diptera: Muscidae), and an unidentified flesh fly larva. Larval and adults of *Dermestes maculatus* (DeGeer) (Coleoptera: Dermestidae) were also collected from the remains. On the basis of the development data for *C. megacephala*, it was concluded that the two males had been dead a minimum of 7 days.

In a similar case, third instars and puparia of *C. rufifacies*, third instars of *H. spinigera*, larvae of *Sargus* sp. (Diptera: Stratiomyidae), adult *D. maculatus*, and third instars of *Piophila casei* (L.) (Diptera: Piophilidae) were all collected from burnt human remains. However, this case illustrates the limited amount of research that had been completed on forensic entomology in Thailand at that time. Although arthropods were collected from the remains, an mPMI estimate could not be rendered due to limited information associated with the development rates of Thai populations of many of these arthropods.

Burning human remains may lead to delayed oviposition by arthropods (Avila and Goff 1998). Such events require additional investigation because understanding the delaying effect of burning could lead to more refined estimates of the associated mPMI. Such studies could also lay a foundation for quantifying arthropod succession associated with vertebrate carrion that could then be used to estimate the mPMI based on the arthropod community structure.

4.2.4 Myiasis

One case involving myiasis has been documented in Thailand. In this particular case, an unidentified, deceased male was discovered indoors within a suburban area. The corpse showed no sign of decomposition or mummification, indicating that he had died within 1 day prior to discovery. However, during the investigation, a large squamous cell carcinoma lesion on the lower right leg was found to be heavily infested with third instar fly larvae. It was concluded, based on the size of the carcinoma lesion, that it had existed for approximately 1 month prior to death. Two species of blow flies were found in the lesion: third instars of *C. megacephala* and *C. rufifacies*, both about 1.4 cm in length. On the basis of the size of these larvae, it was estimated that they had been present on the individual prior to death. On the basis of the development rate of both fly species under the natural temperatures in Chiang Mai in December 2001 and 2002, both fly specimens were estimated to be approximately 5-days old. If the investigators had not been made aware of the time of death based on the medical examiner's inspection of the remains, due to the age of the larvae, they might have concluded that the individual had been dead much longer.

4.2.5 Mummified Remains

Three cases involving mummified remains have been documented in Thailand. In the first case, mummified and partially skeletonized remains were discovered hanging from a rope in a tree near a road along the edge of a dense forest. When the remains were discovered, they were beginning to disarticulate as both feet were missing; several entomological specimens were collected, including third instars of *H. spinigera* and *P. casei* (Sukontason et al. 2001b). Third instars of *Megaselia scalaris* (Loew) (Diptera: Phoridae) larvae also were collected and were the most common specimens on the remains; a second instar of *Sargus* sp. was collected. In addition, third instars of two species of flesh fly were found; however, they could not be identified. The forensic practitioner suspected the individual to have been dead for approximately 4 months; however, the mPMI estimate based on the entomological evidence was approximately 4.5 months.

The second case involved mummified remains of an unidentified male located in a forested area; the cause of death was unknown. The entomological evidence consisted of third instars of *C. nigripes* Aubertin (Diptera: Calliphoridae) and *H. spinigera*; larvae and adults of *D. maculatus* also were collected. The mPMI was estimated at approximately 3 months.

For the last case, remains of an unidentified male were discovered hanging from a rope in a tree in a cemetery located in a rural region of Thailand. Several species of insects were collected from the remains. The most abundant was *C. nigripes*, either second or third instars. Moreover, numerous puparia were found aggregated, adhering side by side, on the tibia of the skeletonized corpse. Third instars of *H. spinigera* and *P. casei* were found in much smaller numbers; a few larvae of *D. maculatus* also were collected. The mPMI based on the entomological evidence indicated that the man had been dead for many months; however, a specific mPMI could not be given due to limited data available on the development of these species.

Like many areas of forensic entomology in Thailand, information pertaining to arthropods associated with mummified human remains is limited, thus making the application of forensic entomology difficult, especially when estimating an mPMI.

In Thailand, arthropods, such as *C. nigripes*, can be found associated with every stage of decomposition regardless of if fresh (Ahmad and Ahmad 2009), in advance decay (Ahmad and Ahmad 2009), or mummified. In Guam, this species is considered a secondary colonizer of decomposing vertebrate carrion and is known to have a longer duration than most other arthropods (Bohart and Gressitt 1951). On the basis of such information, *C. nigripes* is not an ideal candidate for estimating an mPMI of human remains in late stages of decomposition, as its colonization pattern is highly variable.

The presence of *D. maculatus* in association with human remains in Thailand is in agreement with published accounts from other continents, including the following: Asia (Lee et al. 2004), Africa (Tantawi et al. 1996), Europe (Schroeder et al. 2002), North America (Valdes-Perezgasga et al. 2010), and South America (Horenstein et al. 2012). However, little is known about the biology of beetles frequenting carrion in Thailand, and more research is justified.

4.2.6 Indoor Cases

Two cases involving indoor death scenario were encountered by our investigations in Thailand. A deceased 42-year-old male was found in his bedroom. No wounds were found during the autopsy, and the cause of death was unknown. The mPMI estimate was based on the development of blow fly larvae collected from the face, back, arm, and leg of the decedent. The blow flies were identified as third instars of *C. megacephala* and *C. rufifacies*. Moreover, third instars of *Synthesiomyia nudiseta* (van der Wulp) (Diptera: Muscidae) were collected. Based on the analysis of the arthropod evidence, the mPMI of the decedent was estimated at 3–5 days.

In the other indoor case, a mummified 49-year-old male was discovered in 2006. The remains were located indoors, and the air conditioner was running with the thermostat set at 25°C. No wounds were present on the decedent. Larvae associated with the human remains were identified as third instar *S. nudiseta*. In addition, third instar and puparia identified as flesh flies were collected and reared to adults. These were later identified as *Liopygia ruficornis* (Fabricius). However, an mPMI was not rendered in this case due to lacking development data.

C. megacephala and *C. rufifacies* are the most common fly species encountered at outdoor death scenes in Thailand. However, through casework, it has been determined that these species will oviposit on human remains located indoors, corresponding with a report from Malaysia (Kumara et al. 2012). Colonization of human remains by *S. nudiseta* indicates that this fly is of forensic importance and seems to be most commonly encountered indoors. This behavior has been documented in other countries including Malaysia (Kumara et al. 2012; Syamsa et al. 2012), the southeastern United States (Williams 2008), and Spain (Velasquez et al. 2013). This fly was also documented at a death scene in a high-rise building in Malaysia (Syamsa et al. 2012). In a review of 22 cases in Spain, *S. nudiseta* was determined to prefer remains located indoors, but was documented to colonize remains located outdoors as well (Velasquez et al. 2013). As for *L. ruficornis*, literature records reveal little in terms of its preferred environment; however, the little information available indicates that it has a preference for remains located indoors (Al-Mesbah et al. 2011).

4.3 FORENSIC ENTOMOLOGY ACCOMPLISHMENTS IN THAILAND

Many forensically important fly species exist in Thailand, not only in mountainous regions but also in lowland areas. Forensic entomology has been studied intensively, especially in the northern part—both lowland and mountainous areas, and this research indicates the two most common blow fly species are *C. megacephala* and *C. rufifacies* (Sukontason et al. 2007a).

To date, most research has focused on the biology (e.g., morphology of immature stages, developmental rate, distribution) of these two species; however, other species have been investigated as well, because they are frequently encountered on decomposing human remains (e.g., *C. nigripes, C. villeneuvi, Lucilia cuprina* (Wiedemann) and *H. ligurriens*). The development rate of larvae of *C. megacephala, C. rufifacies*, and the flesh fly *Parasarcophaga (Liosarcophaga) dux* Thomson were documented (Sukontason et al. 2008, 2010).

There also has been a primary research focus on the development of taxonomic keys for identifying these species and determining their development rates, as both are essential for estimating the mPMI of human remains. To date, keys for the identification of forensically important fly species in Thailand have been published and have included the following: eggs (Sukontason et al. 2007b), larvae (Klong-klaew et al. 2012), puparia (Sukontason et al. 2007c), and adults (Kurahashi and Bunchu 2011).

Forensic entomologists can utilize such keys for identification of flies of forensic importance, along with development curves to predict larval age. However, according to Thai law, only the forensic doctor is authorized to do the autopsy. For practical purposes, forensic doctors are permitted to ask for help from forensic entomologists. The data associated with fly larvae discovered with human remains have been discussed and summarized by both forensic doctors and entomologists; however, only forensic doctors report to the court.

4.4 CHALLENGES FOR FORENSIC ENTOMOLOGY IN THAILAND

The high biodiversity of fly species is one of the major problems facing forensic entomologists in tropical areas like Thailand. Most fly species are found in the forests and high mountainous areas, which corresponds with the location of most corpses that have been discarded due to unlawful activity. The more surveys of arthropods associated with decomposing carrion, the more likely new insect species will be described. Biological data for these species are needed to properly assist law enforcement.

The difficulty of morphological identification is a major problem. To date, the use of adult morphology for identification has been accepted. Moreover, identification of some groups of flies, especially flesh flies, is quite difficult because adult male genitalia are needed for the identification of species belonging to this family. The process of capturing an adult and curating it properly is a time-consuming and tedious work. Furthermore, there are only a few taxonomists that specialize in this area. In most cases, immatures are collected from human remains and identification of flesh fly larvae cannot be accomplished without the availability of proper molecular techniques. Thus, additional work examining both morphological and molecular markers for proper identification is needed in Thailand.

Another limitation of forensic entomology in Thailand is that development data for the arthropods of forensic importance are lacking, with only the most common species (*C. megacephala* and *C. rufifacies*) having been studied. Many fly species, especially those in the forested areas, remain understudied and cannot be utilized as evidence in association with death investigations.

The last great challenge in forensic entomology in Thailand is the variety of crime scenes encountered. Corpses with drug overdose or insecticide intoxication may delay or speed up fly development, while burnt corpses experience extremely high temperatures that can delay colonization of the remains. Such situations require more investigation to precisely estimate the mPMI. While each of the challenges listed is equally important, they all tie into a single theme—the need for increased forensic entomology research and law enforcement training in Thailand. Through the continued efforts of researchers within Thailand, it is therefore promising that these hurdles can be overcome, and Thailand will continue to serve as a leading forensic entomology research force in Asia.

REFERENCES

Ahmad, A. and A.H. Ahmad. 2009. A preliminary study on the decomposition and dipteran associated with exposed carcasses in an oil palm plantation in Bandar Baharu, Kehah, Malaysia. *Tropical Biomedicine* 26: 1–10.

Al-Mesbah, H., Z. Al-Osaimi, and O.M.E. El-Azazy. 2011. Forensic entomology in Kuwait: The first case report. *Forensic Science International* 206: e25–e26.

Avila, F.W. and M.L. Goff. 1998. Arthropod succession patterns onto burnt carrion in two contrasting habitats in the Hawaiian islands. *Journal of Forensic Sciences* 43: 581–586.

Bohart, G.E. and J.L. Gressitt. 1951. Filth-inhabiting flies of Guam. *Bulletin—Bernice P. Bishop Museum* 204: 1–151.

Champathet, Y. 2003. Insects found from carcass of piglet. M.Sc. thesis, Kasetsart University, Bangkok, Thailand.

Fongsiripaibul, V. 1987. Estimating time of death by fly's cycle. *Journal of Central Hospital* 24: 1–10.

Glassman, D.M. and R.M. Crow. 1996. Standardization model for describing the extent of burn injury to human remains. *Journal of Forensic Sciences* 41: 152–154.

Horenstein, M.B., B. Rosso, and M.D. Garcia. 2012. Seasonal structure and dynamics of sarcosaprophagous fauna on pig carrion in a rural area of Cordoba (Argentina): Their importance in forensic science. *Forensic Science International* 217: 146–156.

Klong-klaew, T., K. Sukontason, P. Sribanditmongkol, K. Moophayak, S. Sanit, and K.L. Sukontason. 2012. Observations on morphology of immature *Lucilia porphyrina* (Diptera: Calliphoridae), a fly species of forensic importance. *Parasitology Research* 111: 1965–1975.

Kumara, T.K., R.H.L. Disney, A. Abu Hassan, M. Flores, T.S. Hwa, Z. Mohamed, M.R. CheSalmah, and S. Bhupinder. 2012. Occurrence of oriental flies associated with indoor and outdoor human remains in the tropical climate of north Malaysia. *Journal of Vector Ecology* 37: 62–68.

Kurahashi, H. and N. Bunchu. 2011. The blow flies recorded from Thailand, with the description of a new species of *Isomyia* Walker (Diptera: Calliphoridae). *Japanese Journal of Systematic and Entomology* 17: 237–278.

Lee, H.L., M. Krishnasamy, A.G. Abdullah, and J. Jeffery. 2004. Review of forensically important entomological specimens in the period of 1972–2002. *Tropical Biomedicine* 21: 69–75.

Mankosol, R. 1986. Estimation time of death in putrefied body. *Siriraj Hospital Gazette* 38: 855–857.

Merritt, R.W. and J.R. Wallace. 2010. The role of aquatic insects in forensic investigations. In *Forensic Entomology: The Utility of Arthropods in Legal Investigations*, edited by J.H. Byrd and J.L. Castner, 271–320. Boca Raton, FL: CRC Press.

Schroeder, H., H. Klotzbach, L. Oesterhelweg, and K. Puschel. 2002. Larder beetles (Coleoptera, Dermestidae) as an accelerating factor for decomposition of a human corpse. *Forensic Science International* 27: 231–236.

Sritavanich, N., T. Jamjanya, A. Chamsuwan, and Y. Hanboonsong. 2009. Biology of hairy maggot blow fly, *Chrysomya rufifacies* and its application in forensic medicine. *Khon Kaen University Research Journal* 9: 10–15 (in Thai with English abstract).

Sritavanich, N., T. Jamjanya, Y. Hanboonsong, and A. Jamsuwan. 2007. An application of forensic entomology to evaluate post-mortem interval. *Khon Kaen University Research Journal* 7: 1–5.

Sukjit, S. 2011. Diversity and succession of carrion arthropods on pig *Sus scrofa domestica* carcasses under different conditions in Nan Province, Thailand. M.Sc. thesis, Chulalongkorn University, Thailand.

Sukontason, K., N. Bunchu, T. Chaiwong, K. Moophayak, and K.L. Sukontason. 2010. Forensically imporflesh fly species in Thailand: Morphology and developmental rate. *Parasitology Research* 106: 1055–1064.

Sukontason, K., P. Narongchai, C. Kanchai, K. Vichairat, P. Sribanditmongkol, T. Bhoopat, H. Kurahashi et al. 2007a. Forensic entomology cases in Thailand: A review of cases from 2000 to 2006. *Parasitology Research* 101: 1417–1423.

Sukontason, K., S. Piangjai, S. Siriwattanarungsee, and K.L. Sukontason. 2008. Morphology and developrate of blowflies *Chrysomya megacephala* and *Chrysomya rufifacies* in Thailand: Application in forensic entomology. *Parasitology Research* 102: 1207–1216.

Sukontason, K., K.L. Sukontason, S. Piangjai, T. Chaiwong, N. Boonchu, and H. Kurahashi. 2003. Hairy magof *Chrysomya villeneuvi* (Diptera: Calliphoridae), a fly species of forensic importance. *Journal of Medical Entomology* 40: 983–984.

Sukontason, K., K. Sukontason, K. Vichairat, S. Piangjai, S. Lertthamnongtham, R.C. Vogtsberger, and J.K. Olson. 2001a. The first documented forensic entomology case in Thailand. *Journal of Medical Entomology* 38: 746–748.

Sukontason, K.L., N. Bunchu, T. Chaiwong, B. Kuntalue, and K. Sukontason. 2007b. Fine structure of the eggshell of the blow fly, *Lucilia cuprina*. *Journal of Insect Sciences* 7: Article 9.

Sukontason, K.L., C. Kanchai, S. Piangjai, W. Boonsriwong, N. Bunchu, D. Sripakdee, T. Chaiwong, B. Kuntalue, S. Siriwattanarungsee, and K. Sukontason. 2006. Morphological observation of puparia of *Chrysomya nigripes* (Diptera: Calliphoridae) from human corpse. *Forensic Science International* 161: 15–19.

Sukontason, K.L., R. Ngern-klun, D. Sripakdee, and K. Sukontason. 2007c. Identifying fly puparia by clearing technique: Application to forensic entomology. *Parasitology Research* 101: 1407–1416.

Sukontason, K.L., K. Sukontason, S. Piangjai, W. Choochote, R.C. Vogtsberger, and J.K. Olson. 2001b. Scanning electron microscopy of the third-instar *Piophila casei* (Diptera: Piophilidae), a fly species of forensic importance. *Journal of Medical Entomology* 38: 756–759.

Syamsa, R.A., F.M.S. Ahmad, R.M. Zuha, A.Z. Khairul, M.A. Marwi, A.W. Shahrom, and B. Omar. 2012. An occurrence of *Synthesiomyia nudiseta* (Wulp) (Diptera: Muscidae) from a human corpse in a high-rise building in Malaysia: A case report. *Tropical Biomedicine* 29: 107–112.

Tantawi, T.I., E.M. eL-Kady, B. Greenberg, and H.A. el-Ghaffar. 1996. Arthropod succession on exposed rabbit carrion in Alexandria, Egypt. *Journal of Medical Entomology* 33: 566–580.

Valdes-Perezgasga, M.T., F.J. Sanchez-Ramos, O. Garcia-Martinez, and G.S. Anderson. 2010. Arthropods of forensic importance on pig carrion in the Coahuilan semidesert, Mexico. *Journal of Forensic Sciences* 55: 1098–1101.

Velasquez, Y., T. Ivorra, A. Grzywacz, A. Martínez-Sánchez, C. Magaña, A. García-Rojo, and S. Rojo. 2013. Larval morphology, development and forensic importance of *Synthesiomyia nudiseta* (Diptera: Muscidae) in Europe: A rare species or just overlooked? *Bulletin of Entomology Research* 103: 98–110.

Vitta, A., W. Pumidonming, U. Tangchaisuriya, C. Poodendean, and S. Nateeworanart. 2007. A preliminary study on insects associated with pig (*Sus scrofa*) carcasses in Pitsanulok, northern Thailand. *Tropical Biomedicine* 24: 1–5.

Williams, R.E. 2008. Case histories of the use of insects in investigations. In *Entomology and Death: A Procedural Guide*, edited by N.H. Haskell and R.E. Williams, 10–39. Clemson, SC: East Park Printing.

CHAPTER 5

India

Meenakshi Bharti

CONTENTS

5.1 INTRODUCTION

The application of forensic entomology in legal situations is quite prevalent in Europe (Smith 1986), Africa (Williams and Villet 2006), and North America (Byrd and Castner 2010). The scientists of these continents use a multidisciplinary approach in forensic entomology, drawing on the principles of ecology, toxicology, and molecular biology that are highlighted in this chapter. However, the situation in developing countries, such as India, is currently not as advanced. The field is still in its infancy and needs the establishment of a coordinated effort to develop a research system based on scientific studies with direct, practical applications to the field of forensic entomology. Indian scientists have been working toward understanding the diverse aspects of the field for a few decades, but no comprehensive review on the status of research is yet available. Therefore, this chapter reviews the advancements and future perspectives of forensic entomology in India.

5.2 REVIEW OF CARRION INSECT COMMUNITY COMPOSITION STUDIES

Forensically significant conclusions are often drawn through analysis of succession patterns of insects that colonize a carcass as decomposition progresses. Baseline data of carrion insect communities in India were generated for the first time by Bharti and Singh (2003). They carried out an exhaustive analysis of insect colonization of rabbit carcasses in different seasons. They recognized four phases of the decomposition, based on the condition of the carcass: fresh, bloated, decay, and dry. These phases were similar to those reported by Payne (1965) in North America. A total of 38 insect species, belonging to 4 orders and 13 families, were cataloged. The principal members of the carrion insect community belonged to the orders Diptera, Coleoptera, and Hymenoptera. The decay phase of carcass decomposition was dominated by Calliphoridae, Sarcophagidae, Muscidae, Silphidae, and Staphylinidae, whereas Tenebrionidae, Dermestidae, Cleridae, and Histeridae dominated the "dry" phase. The species richness and diversity of the community were observed to be highest during the decay stage. Furthermore, a maximum of 24 species were collected from different stages of decay during the spring as compared to only 12 and 17 species during winter and summer, respectively.

Bharti (2012) carried out studies to correlate the effect of altitude on the diversity of forensically important blow flies (Diptera: Calliphoridae) in the Himalayan region. The study was conducted with bovine carcasses at altitudes of 350, 970, 2057, and 2511 m in different regions of the Himalayas. Blow fly diversity was highest at an altitude of 2057 m and least at 2511 m. The widespread and common species of the lower altitudes in Himalaya, *Chrysomya megacephala* (Fabricius) and *Chrysomya rufifacies* (Macquart), were replaced by *Calliphora vicina* Robineau-Desvoidy and *Chrysomya villeneuvi* Patton at higher altitudes.

Apart from community composition studies, Kulshrestha and Chandra (1987) attempted to calculate the postmortem interval (PMI) of human cadavers based on associated arthropod specimens. They reared immature stages of flies collected from 25 bodies at constant temperatures of 30°C and 40°C, and from the results estimated the minimum postmortem interval (mPMI) of the infested remains. One of the major drawbacks of this work was that the reared insects were not classified to the species level. Kashyap and Pillay (1989) examined and analyzed 16 cadavers to evaluate the reliability of entomological methods in the estimation of the mPMI in relation to other approaches. They found the entomological methods to be more reliable than other methods. Kulshrestha and Satpathy (2001) observed that dermestid (*Dermestes maculatus* [De Geer]) (Coleoptera: Dermestidae) and clerid (*Necrobia rufipes* [De Geer]) (Coleoptera: Cleridae) beetles were the main insect inhabitants of dry remains in the plains of India, and provided significant entomological evidence. Similarly, Aggarwal et al. (2003) examined the insects associated with 20 human corpses and determined the mPMI from the entomological evidence. However, the study failed to provide a checklist of insect taxa found during the investigation. Following suit, Goyal (2012) carried out studies on 47 human cadavers in different stages of decay. He recorded the presence of calliphorids, sarcophagids, muscids, silphids, and dermestids on the cadavers, but failed to provide a detailed list of the carrion insect community. A detailed list of the insects colonizing remains is a prerequisite for calculating the mPMI, as arthropods have predictable habitats and distributions, and their presence or absence from a carrion community provides important information about when, where, and how a particular death has occurred (Tomberlin et al. 2011).

5.3 COMMUNITY DYNAMICS

The insect community present on a resource changes over three timescales: in a circadian periodicity, throughout the duration of decomposition, and annually. Being ectothermic, most carrion animals are less active in colder conditions, such as at night and in winter (Villet 2011). Knowledge about the circadian cycles of forensically important insects is a prerequisite for accurate estimation of the mPMI. Previous work suggested that blow flies do not lay eggs at night, which can greatly impact mPMI estimates. Singh and Bharti (2001b, 2008) conducted experiments to study the nocturnal oviposition behavior of blow flies and larviposition behavior of flesh flies using a meat bait that was placed on a wooden platform 2 m above the ground. To prevent flies from walking to the bait, the pole supporting the platform was coated with a band of sticky material, and the bait was kept on the platform from 10 p.m. to 3 a.m. daily, for a week during March and September 1999. They reported nocturnal oviposition by *C. vicina*, *Ch. megacephala*, and *Ch. rufifacies*, and larviposition by *Sarcophaga albiceps* Meigen and *Sarcophaga hirtipes* Wiedemann. Joseph and Parui

(1980) investigated the attraction of the filth fly families Calliphoridae, Muscidae, Sarcophagidae, Phoridae, Drosophilidae, Syrphidae, Sepsidae, Borboridae, Milichiidae, Stratiomyidae, and Phychodidae, and found that the favorite resource for Calliphoridae, Muscidae, and Sarcophagidae was dead and decomposing meat.

Bharti and Singh (2002) observed that two blow fly species, *Ch. megacephala* and *Ch. rufifacies*, colonized rabbit carcasses during all seasons, whereas *Ch. vicina* were active only during the winter and spring (Figures 5.1 and 5.2). Similar to all arthropods, seasonal occurrence of carrion insects depends directly on temperature, a factor that is important to the upper and lower physiological limits of specific species. Bharti et al. (2007) studied the life cycle of *Ch. megacephala* at four constant temperatures (22°C, 25°C, 28°C, and 30°C). Andhale et al. (2013) observed that immature development of *Ch. megacephala* was 14 days during the rainy season at an average temperature of 31°C. Bharti (2009) also studied the development of *Ch. vicina* at 20°C, 25°C, and 30°C. It was observed that at

1 mm

Figure 5.1 *Chrysomya megacephala,* female head.

1 mm

Figure 5.2 *Chrysomya rufifacies,* female head.

30°C, the larvae of *C. vicina* failed to pupate. As this species in question is found only during cooler months of the year, the temperature of ≥30°C could be marked as the upper threshold of this species. Karunamoorthy and Lalitha (1987) determined the best rearing medium for the Australian sheep blow fly *Lucilia cuprina* (Wiedemann) to be lean beef and also determined the development of the immature stages to be 9–11 days at 32 ± 2°C.

Gola and Lukose (2007) investigated the effect of nine drugs and toxins on the development of *Calliphora* species. Librium, Diazepam, Phenobarbitone, Prednisolone, and alcohol were detected in all the stages of insect development. Verma (2013) studied the effects of codeine and sodium pentothal on the growth rate of *Ch. rufifacies* and observed an increased development rate. These represent the first studies of entomotoxicology in India.

Singh and Bala (2011) studied the survivorship of larval blow flies after submergence in water to estimate the minimum time since submergence (TSS). They found an inverse relationship between submergence period and survival rate in the larvae of *Ch. megacephala* and *Ch. rufifacies*. The survival rate also depended on the age of the larvae; young larvae (10-hour-old) had lowest survival, even when submerged for 2 hours; 30- to 70-hour-old larvae tolerated submergence for a maximum of 4 hours. The lower oxygen requirement by the older larvae was correlated with the increase in survival rate (El-Kady et al. 1999). Singh and Bala (2010) compared the larval dispersal of two blow fly species, *Ch. megacephala* and *Ch. rufifacies*, and found that the larvae of *Ch. megacephala* dispersed farther from the resource than *Ch. rufifacies*, which preferred to remain near the food site.

Dasgupta and Roy (1969) investigated the parasitic behavior of the blow fly *Lucilia illustris* Meigen. Fish, toads, frogs, wall lizards, pigeons, and shrews acted as live baits and were infested with eggs of *L. illustris*. In all cases, the eggs hatched and larvae fed on the tissue of the host, ultimately killing it. This information is important as myiasis cases could involve this species.

Insect flight is periodic, and the term *diurnal* is used to describe periodicity involving a 24-hour cycle. Flight periodicity largely depends on the physiological responses of insects to cyclic factors in the environment (Lewis and Taylor 1965). Das et al. (1978) studied the flying activity of *Ch. megacephala* during summer and winter seasons and observed that the fly exhibited bimodal diurnal periodicity during the summer (the fly was active from 09:00 to 11:00 hours and 13:00 to 15:00 hours) and unimodal flight periodicity during the winter (12:00 to 13:00 hours).

5.4 TAXONOMIC STATUS

No exhaustive catalogue of arthropods associated with vertebrate carrion is available for India, but there are taxonomic contributions in the form of checklists and new species descriptions for forensically important groups. Senior-White et al. (1940) provided the first comprehensive record of the blow fly species of British India, including the Oriental region. They provided descriptions of 154 species collected from the Oriental region, of which 99 were recorded from the Indian subregion. Similarly, Van Emden (1965) described the Muscidae of India. Kurahashi and Okadome (1976) reported a new species of *Polleniopsis* (Diptera: Calliphoridae) from the Himalaya. Kurahashi (1977) also recorded a new species of *Hypopygiopsis* (Diptera: Calliphoridae) from India. Nandi (1980) described in detail the immature stages of five species of the genus *Parasarcophaga* (Diptera: Sarcophagidae), namely *Parasarcophaga ruficornis* (Fabricius), *Parasarcophaga albiceps* (Meigen), *Parasarcophaga orchidea* (Boettcher), *Parasarcophaga dux* (Thomson), and *Parasarcophaga brevicornis* (Ho), with an identification key to the immature stages. Ghezta and Kumar (1991) reported a new species of the genus *Strongyloneura* (Diptera: Calliphoridae) from the plains of Northern India. Rognes (1992) discovered *Gulmargia angustisquama* gen.n., sp.n. (Diptera: Calliphoridae) from the Indian Himalaya and also provided descriptions of the male, female, and first instar. Nandi (1994) described a new calliphorid species in the genus *Melinda* from West Bengal. He also reported *Lucilia bazini* Seguy (Diptera: Calliphoridae) for the first time from India (Figure 5.3).

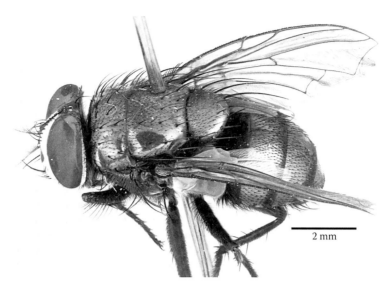

2 mm

Figure 5.3 *Lucilia bazini*, profile.

Singh and Bharti (2000, 2001a) provided a list of forensically important blow flies and ants from the plains of Punjab, and Nandi (2000) catalogued the blow fly biodiversity from northeast India. Although studying the insect fauna of decaying rabbits, Bharti et al. (2001) recorded four species of muscid and a sarcophagid for the first time in India: *Hydrotaea occulta* (Meigen), *Hydrotaea capensis* (Wiedemann), *Atherigona savia* (Pant and Magpayo), and *Sarcophaga princeps* Wiedemann. Nandi (2002a) assessed the calliphorid biodiversity of West Bengal, and in the same year, Nandi (2002b) identified that the Indian sarcophagid fauna was represented by 163 species in 50 genera and 3 subfamilies. Sinha and Nandi (2004) described a new calliphorid, *Chrysomya indica*, from Sunderbans Biosphere Reserve. Sidhu and Singh (2005) recorded the genus *Wilhelmina* from India, with a description of a new species (*W. indica*). Singh and Sidhu (2007) discovered two species of *Melinda* from northwest India. Bharti (2008) provided a checklist of the family Muscidae from India. Bharti and Kurahashi (2009) discovered a feral, derived form of *Ch. megacephala* from India, which is a missing link between the two earlier forms (normal form, now known as *Ch. pacifica* Kurahashi, and a synanthropic derived form *Ch. megacephala*). The finding of this morphological intermediate has raised questions about the ecological and evolutionary history of this fly. Bharti and Kurahashi (2010) recorded *L. calviceps* for the first time from India and also provided key to the Indian species. Bharti (2011) provided an updated checklist of Indian blow flies taxa that were represented by 9 subfamilies, 30 genera, and 120 species. Singh et al. (2012) carried out scanning electron microscope (SEM) studies on the immature stages of the sarcophagid *Parasarcophaga ruficornis*.

5.5 CONCLUSIONS

A review of articles related to forensic entomology reveals that this discipline of forensics is gaining roots in India and its potential use in calculating the mPMI of human remains could be put to utmost use. To establish this field in India, in-depth observations in the area of carrion ecology, molecular biology, and entomotoxicology are required, which will aid in providing services to the forensic community in solving homicidal cases. Thus, a unified, multidisciplinary approach toward the forensic-related research is the need of the hour.

REFERENCES

Aggarwal, A.D., R.K. Gorea, O.P. Aggarwal, and D. Singh. 2003. Forensic entomology: A guide to post-mortem interval. *Journal of Punjab Academy of Forensic Medicine and Toxicology* 3: 7–10.

Andhale, A.V., P. Shinde, and M.S. Arora. 2013. Morphological study of *C. megacephala* in relation to its importance in forensic science in rainy season. *World Journal of Pharmacy Research* 1: 1–5.

Bharti, M. 2008. Current status of family Muscidae (Diptera) from India. *Journal of Entomological Research* 32: 171–176.

Bharti, M. 2009. Studies on life cycles of forensically important flies, *Calliphora vicina* and *Musca domestica nebulo* at different temperatures. *Journal of Entomological Research* 33: 273–275.

Bharti, M. 2011. An updated checklist of blowflies (Diptera: Calliphoridae) from India. *Halteres* 3: 34–37.

Bharti, M. 2012. Altitudinal diversity of forensically important blowflies collected from decaying carcasses in Himalaya. *The Open Forensic Science Journal* 5: 1–3.

Bharti, M. and H. Kurahashi. 2009. Finding of feral derived form (fdf) of *Chrysomya megacephala* (Fabricius) from India with an evolutionary novelty (Diptera: Calliphoridae). *Japanese Journal of Systematic Entomology* 15: 411–413.

Bharti, M. and H. Kurahashi. 2010. *Lucilia calviceps* Bezzi, new record from India (Diptera: Calliphoridae), with a revised key to Indian species. *Halteres* 2: 29–30.

Bharti, M. and D. Singh. 2002. Occurrence of different larval stages of blowflies (Diptera: Calliphoridae) on decaying rabbit carcasses. *Journal of Entomological Research* 26: 343–350.

Bharti, M. and D. Singh. 2003. Insect faunal succession on decaying rabbit carcasses in Punjab, India. *Journal of Forensic Sciences* 48: 1133–1143.

Bharti, M., D. Singh, and Y.P. Sharma. 2007. Effect of temperature on the development of forensically important blowfly, *Chrysomya megacephala* (Fabricius) (Diptera: Calliphoridae). *Entomon* 32: 149–151.

Bharti, M., D. Singh, and I.S. Sidhu. 2001. First record of some carrion flies (Diptera: Cyclorrhapha) from India. *Uttar Pradesh Journal of Zoology* 2: 267–268.

Byrd, J.H. and J.L. Castner. 2010. *Forensic Entomology: The utility of Arthropods in Legal Investigations*, 2nd Ed. Boca Raton, FL: CRC Press.

Das, S.K., P. Roy, and B. Dasgupta. 1978. The flying activity of *Chrysomya megacephala* (Diptera: Calliphoridae) in Calcutta, India. *Oriental Insects* 12: 103–109.

Dasgupta, B. and P. Roy. 1969. Studies on the behavior of *Lucilia illustris* Meigen as a parasite of vertebrates under experimental conditions. *Parasitology* 59: 299–304.

El-Kady, E.M., A.M. Kheirallah, A.N. Kayed, S.I. Dekinesh, and Z.A. Ahmed. 1999. The bioenergetics of the housefly *Musca domestica* and the blowfly *Lucilia sericata*. *Pakistan Journal of Biological Science* 2: 472–477.

Ghezta, R.K. and D. Kumar. 1991. A new species of genus *Strongyloneura* Bigot from India (Insecta, Diptera, Calyptrata: Calliphoridae). *Reichenbachia* 34: 181–184.

Gola, S. and S. Lukose. 2007. A study of presence of ante-mortem administered drugs and to analyze them in the different stages of insect development. *International Journal of Medical Toxicology and Legal Medicine* 10: 12–15.

Goyal, P.K. 2012. An entomological study to determine the time since death in cases of decomposed bodies. *Journal of Indian Academy of Forensic Medicine* 34: 10–12.

Joseph, A.N.T. and P. Parui. 1980. Filth inhabiting flies (Diptera) of Calcutta city. *Bulletin of Zoological Survey of India* 3: 1–12.

Karunamoorthy, G. and C.M. Lalitha. 1987. Rearing of the Australian sheep blowfly *Lucilia cuprina* (Wiedemann). *Cheiron* 16: 132–134.

Kashyap, V.K. and V.V. Pillay. 1987. Efficacy of entomological method in estimation of post mortem interval: A comparative analysis. *Forensic Science International* 40: 245–250.

Kulshrestha, P. and H. Chandra. 1987. Time since death. An entomological study on corpses. *American Journal of Forensic Medicine and Pathology* 8: 233–238.

Kulshrestha, P. and D.K. Satpathy. 2001. Use of beetles in Forensic Entomology. *Forensic Science International* 120: 15–17.

Kurahashi, H. 1977. The tribe Luciliini from Australian and Oriental region. I. Genus *Hypopygiopsis* Townsend (Diptera: Calliphoridae). *Kontyu* 45: 553–562.

Kurahashi, H. and T. Okadome. 1976. A new *Polleniopsis* from Kashmir, India (Diptera: Calliphoridae). *Bulletin of Japan Entomological Academy* 9: 42–44.

Lewis, T. and L.R. Taylor. 1965. Diurnal periodicity of flights by insects. *Transactions of the Royal Entomological Society of London* 116: 393–479.

Nandi, B.C. 1980. Studies on the larvae of flesh flies from India (Diptera: Sarcophagidae). *Oriental Insects* 14: 303–323.

Nandi, B.C. 1994. Studies on calliphorid flies (Diptera: Calliphoridae) from Calcutta and adjoining areas. *Journal of Bengal Natural History Society* 13: 37–47.

Nandi, B.C. 2000. Studies on blowflies (Diptera: Calliphoridae) of Sikkim, India. *Records of the Zoological Survey of India* 98: 1–9.

Nandi, B.C. 2002a. Blowflies (Diptera: Calliphoridae) of West Bengal, India with a note on their biodiversity. *Records of Zoological survey of India* 100: 117–129.

Nandi, B.C. 2002b. *Fauna of India and the Adjacent Countries: Diptera*, Vol. X, *Sarcophagidae*, pp. 608. Calcutta, India: Zoological Survey of India.

Payne, J.A. 1965. A summer carrion study of baby pig Sus scrofa Linnaeus. Ecology 46: 592–602.

Rognes, K. 1992. A new genus of *Helicobosinae* from the Himalayas (Diptera: Calliphoridae), with emended genus and family concepts. *Entomologica Scandinavica* 23: 391–404.

Senior-White, R., D. Aubertin, and J. Smart. 1940. *The Fauna of British India, Including Remainder of the Oriental Region: Diptera*, Vol. VI, *Family Calliphoridae*, pp. 288. London: Taylor & Francis.

Sidhu, I.S. and D. Singh. 2005. First record of genus *Wilhelmina* Schmitz and Villeneuve from India with description of new species (Diptera: Calliphoridae). *Entomon* 30: 255–259.

Singh, D. and B. Bala. 2010. Studies on larval dispersal on two species of blowflies (Diptera: Calliphoridae). *Journal of Forensic Research* 1: 102.

Singh, D. and B. Bala. 2011. Larval survival of forensically important blowflies (Diptera: Calliphoridae) after submergence in water. *Entomological Research* 41: 39–45.

Singh, D. and M. Bharti. 2000. Forensically important blowflies (Diptera: Calliphoridae) of Punjab (India). *Uttar Pradesh Journal of Zoology* 20: 249–251.

Singh, D. and M. Bharti. 2001a. Ants (Hymenoptera: Formicidae) associated with decaying rabbit carcasses. *Uttar Pradesh Journal of Zoology* 21: 93–94.

Singh, D. and M. Bharti. 2001b. Further observations on the nocturnal oviposition behavior of blow flies (Diptera: Calliphoridae). *Forensic Science International* 120: 124–126.

Singh, D. and M. Bharti. 2008. Some notes on the nocturnal larviposition by two species of *Sarcophaga* (Diptera: Calliphoridae). *Forensic Science International* 177: 19–20.

Singh, D., R. Garg, and B. Wadhawan. 2012. Ultramorphological characteristics of immature stages of a forensically important fly *Parasarcophaga ruficornis* (Fabricius) (Diptera: Calliphoridae). *Parasitology Research* 110: 821–831.

Singh, D. and I.S. Sidhu. 2007. Two new species of *Melinda* Robineau-Desvoidy (Diptera: Calliphoridae) from India with a key to Indian species of this genus. *Journal of Bombay Natural History Society* 104: 55–57.

Sinha, S.K. and B.C. Nandi. 2004. Notes on calliphorid flies (Diptera: Calliphoridae) from Sunderbans Biosphere Reserve and their impact on man and animals. *Journal of the Bombay Natural History Society* 101: 415–420.

Smith, K.G.V. 1986. A manual of forensic entomology, pp. 205. London: British Museum (Natural History).

Tomberlin, J.K., R. Mohr, M.E. Benbow, A.M. Tarone, and S.L. VanLaerhoven. 2011. A roadmap for bridging basic and applied research in Forensic Entomology. *Annual Review of Entomology* 56: 401–421.

Van Emden, F. 1965. *Fauna of India: Diptera*, Vol. III, *Muscidae* Part-I. Calcutta, India: Baptist Mission Press.

Verma, K. 2013. Effects of codeine, sodium pentothal and different temperature factors on growth rate development of *Chrysomya rufifacies* for the forensic entomotoxicological purposes. *Journal of Bioanalysis and Biomedicine* 5: 6–12.

Villet, M.H. 2011. African carrion ecosystems and their insect communities in relation to forensic entomology. *Pest Technology* 5: 1–15.

Williams, K.A. and M.H. Villet. 2006. A history of southern African research relevant to forensic entomology. *South African Journal of Science* 102: 59–65.

CHAPTER 6

Australia and New Zealand

James F. Wallman and Melanie S. Archer

CONTENTS

6.1 CASEWORK AND INNOVATION IN ITS METHODOLOGY

The first known Australian or New Zealand case featuring forensic entomology occurred in 1923 (Morgan 2012). Insect evidence recovered from a corpse found in a river suggested that infestation had occurred at a forested site before the body was immersed. However, the analysis appears to have been made by the pathologist and police rather than an entomologist. There are two other early New South Wales records of minimum postmortem interval (mPMI) estimates from 1960 (Clarke 1962) and 1978 (Miller 1991). Regular use of forensic entomology increased from the 1980s onward, and reference to Australian and New Zealand casework can be found in O'Flynn (1980), Smeeton et al. (1984), Crosby et al. (1985), Morris (1993), Levot (2003), Archer and Ranson (2005), Archer et al. (2005, 2006), and Porter (2012). Acceptance of forensic entomology in Australia was also further increased by its inclusion in the Expert Evidence series, which summarizes fields of accepted expertise for the legal community (Morris and Dadour 2011). Today, there are usually between 10 and 15 practitioners, researchers, and students working in Australia and New Zealand at any one time.

Several innovations for use in casework have been developed in Australia, most of which also have application globally. Day and Wallman (2006a) showed that the width of blow fly (Diptera: Calliphoridae) larvae could be used as an accurate alternative to length for inferring their age. They also built on overseas work to show that common preservative solutions affected the size of some Australian blow fly larvae (Day and Wallman 2008). Detection of certain substances in the body on which fly larvae have fed has been made possible by researchers from various laboratories. Roeterdink et al. (2003) showed that gunshot residue is detectable from larvae fed on beef that has been shot. Mingari et al. (2006) and Johns et al. (2010) detected petrol and kerosene from larvae fed on piglet carcasses that had been ignited from these liquids. Gunn et al. (2006) used chemiluminescence to detect morphine from *Calliphora stygia* (Fabricius) (Diptera: Calliphoridae) larvae.

Other novel research has sought to describe contamination of entomological samples by extraneous material, which can occur either at the body discovery site (BDS) or in the mortuary, and can potentially distort mPMI estimates. This occurs at the BDS when entomological material originates from animal carcasses or organic refuse, and, in the mortuary, can result from invertebrate transfer between bodies, or via carrion insect entry into the facility. Case studies from Victoria have been published where contamination was either proven or suspected, and strategies were presented for minimizing the risk. These include making a thorough search at the BDS for contaminants around the body (Archer et al. 2006) and avoiding picking up samples in the mortuary from the floor or benches where contaminants are more likely to be found (Archer et al. 2005).

There has also been regional work on providing tools to solve common operational problems and to test them to ensure court readiness. For example, linear regression equations have been validated for correlating and correcting ambient temperature data to make them representative of BDS ambient temperatures (Archer 2004a; Johnson et al. 2012b). Importantly, the limitations of this technique have also been identified; they have been shown to work only when there is <5°C difference between weather station temperatures from the correlation period and suspected body in situ period, and when the periodicity of correlation data is every 3 hours or less for at least a 2-day correlation period (Johnson et al. 2012b).

Two pioneering models were developed for aging fly larvae in forensic casework in the 1980s, but, although innovative, neither was developed fully enough to incorporate into routine practice. The first model was converted into two computer programs written in FORTRAN IV. It used dry larval weight from unpreserved samples to estimate the time of egg hatch for the blow flies *Calliphora vicina* Robineau-Desvoidy, *Calliphora hilli* Patton, *C. stygia*, and *Lucilia cuprina* (Wiedemann) with a temperature-dependent logistic regression curve (Williams 1984). It could only be used until the prepupal phase, and the time between egg laying and hatch was not included. The second model was presented as a Macintosh program called "Strikedate" written by R. Laughlin in 1992 using growth rate data for *L. cuprina* (Morris 1993). The program used a day-degree approach to work backward to the day of colonization, therefore giving an mPMI estimate. It could also apparently incorporate rainfall, sunshine, and an estimated larval mass heating effect, although it is not clear how this was done. The model could only be used for cases where larvae had already wandered from the corpse, pupariated, or enclosed as adults. As with the previous model, "Strikedate" required subjective choices as to the correct temperature inputs to use. It is encouraging that at least some of the data these early models lacked are now available, but many are still needed, especially basic but reliable growth data for common species.

Another example of casework problem solving has been the recent application of computed tomography (CT) scans to visualize maggot masses and provide information about their locations and sizes. Visualization techniques have been optimized and validated for both casework and research by workers from the University of Wollongong and the Victorian Institute of Forensic Medicine (VIFM) (Johnson et al. 2012a) (Figure 6.1). Analysis of postmortem CT scans for entomological evidence has been routine at the VIFM for forensic entomology cases conducted since late 2010. Findings are initially reported by the entomologist and verified by the VIFM's forensic radiologist. The method has particular application in estimating larval mass volume, and thus associated heat production. It has also been used for locating entomological evidence in inaccessible areas of the body (Johnson et al. 2013a) and in reconstructing larval infestation patterns on bodies at the time they were received in the mortuary (before disturbance by postmortem investigation and/or redistribution in the refrigerator). In another technological innovation, Johnson and Wallman (2013)

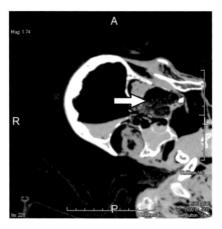

Figure 6.1   CT image showing areas of heterogeneous HU density mainly within the left nasal cavity of a partially skeletonized woman. The numerous clustered ellipsoid objects 1–3 mm in size (indicated by arrow) are pupae and enclosed puparia of sphaerocerid flies.

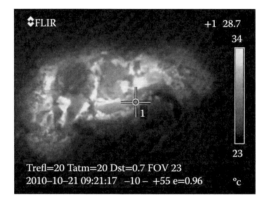

Figure 6.2   **(See color insert.)** Infrared camera image showing heating (temperatures indicated on right) in a 10-kg pig carcass infested with larvae of *Chrysomya rufifacies* (Macquart) and located in a 23°C constant-temperature room.

were the first to document the applicability of thermal imaging to the measurement of larval mass and carcass temperatures (Figure 6.2).

6.2 TAXONOMY AND SYSTEMATICS

The correct identification of the insects featuring in forensic casework is essential to the accuracy of any mPMI estimate. In Australia and New Zealand, studies on the taxonomy and systematics of carrion-breeding insects in recent years have been at least partially inspired by the need to improve the rigor of such identifications. However, there has also been fundamental interest in the evolutionary history and relationships of the species concerned. Historically, much of the early relevant work was conducted because of the involvement of carrion-breeding flies in sheep myiasis (sheep strike). Even today, sheep blow fly strike results in the loss of AU$280 million annually for the Australian agricultural sector alone (Sackett et al. 2006).

This section summarizes the Australian and New Zealand studies devoted to the description, evolutionary relationships, and identification of the insects of forensic importance (although only the literature concerned with carrion flies and beetles has been surveyed). For further information, an online resource using LUCID™ technology is now available that permits the identification of all of the common species and families of flies of forensic importance in Australia (Kavazos et al. 2012). This work expands on the earlier keys by Wallman for the adults (2002a) and third-instar larvae (2002b) of forensic fly species prevalent in southeastern Australia.

6.2.1 Calliphoridae

One of the most forensically important Australian blow flies, *Calliphora augur* (Fabricius), was also among the first insects to be collected by western scientists during the early European exploration of Australia. It was collected in 1770 by Joseph Banks and Daniel Solander on the east coast of Australia during the voyage of Captain Cook's vessel, the *Endeavour* (Marks 1991). *Calliphora augur* was subsequently described by the Danish pupil of Linnaeus, Johan Fabricius, in the first publication on Australian entomology (Fabricius 1775).

Subsequent publications on blow flies by Macquart (1851), Patton (1925), Hardy (1926, 1930, 1932, 1937, 1940), Bezzi (1927), and Malloch (1927, 1932) described adults of the bulk of the endemic Australian carrion-breeding fauna in the four genera: *Calliphora*, *Chrysomya*, *Hemipyrellia*, and *Lucilia*. The diverse genus *Calliphora* was not revisited taxonomically for another 30 years (Kurahashi 1971). Helpful data were provided by Waterhouse and Paramonov (1950) to assist with the morphological differentiation of the introduced *L. cuprina* and *Lucilia sericata* Meigen (*L. cuprina* is the main agent of ovine myiasis in Australia). The prominent Australian calliphorid worker K.R. Norris published several papers of forensic relevance on the taxonomy of this family (Norris 1990, 1994, 1997) in addition to his invaluable ecological reviews (Section 6.3). The adults of blow fly species known or suspected to breed in carrion in southern Australia can be identified using the key of Wallman (2001a).

Third-instar larvae of Australian blow flies were first documented morphologically by Froggatt (1918), although he only described the anterior and posterior spiracles of five species. Fuller (1931) described the biology and all immature stages of the eastern Australian species, *Calliphora ochracea* Schiner, and published descriptions of the third instars of a further eight species of *Calliphora*, *Lucilia*, and *Chrysomya* (Fuller 1932b). She expanded on Froggatt's approach by adding descriptions of the cephaloskeleton to those of the spiracles, but all of these structures are now known to be largely unreliable for species-level identification. More comprehensive descriptions of the third larval instars of all Australian *Chrysomya* species were prepared by Kitching (1976) and Kitching and Voeten (1977), who examined the important cuticular spinulation of Australian species for the first time. O'Flynn and Moorhouse (1980) provided morphological details of the eggs and/or early instars of eight calliphorid species from carrion, including four of the same *Chrysomya* species dealt with by Kitching. Morris (1990) described the life history stages of *Calliphora dubia* Hardy (as *Calliphora nociva* Hardy), a species that features very commonly in forensic cases in southern and western Australia. The third larval instars of southern Australian carrion-breeding *Calliphora* species were revised by Wallman (2001b), including the introduction of more reliable characters.

Despite the abovementioned studies, the morphological identification of the immatures of many blow fly species continues to be difficult or impossible. This has consequently led, in the 1990s and since, to the application of genetic approaches to species identification. In addition to the forensic benefits of these techniques, they have helped refine our understanding of the evolution of these flies. Wallman and colleagues were the first to successfully apply allozyme electrophoresis and mtDNA sequencing to the identification of Australian carrion-breeding blow flies (Wallman and Adams 1997, 2001; Wallman and Donnellan 2001; Wallman et al. 2005; Lessard et al. 2009). The extensive and relatively recent radiation of calliphorids within Australia has required the use of multiple gene loci to differentiate closely related species, such as *Calliphora albifrontalis* Malloch and *C. stygia*. The initial mitochondrial work was consequently extended by applying the ITS2 (rDNA) region to the molecular separation of species (Nelson et al. 2008). Australian workers have also been innovative in testing the mitochondrial "barcoding" (COI) method for molecular identification of forensically important blow flies (Nelson et al. 2007) and in producing whole mitochondrial genomes for nine common Australian *Chrysomya* and *Lucilia* species (Nelson et al. 2012). Some studies on the DNA of Australian blow flies have additionally incorporated populations from outside Australia into their analyses (Harvey et al. 2003a,b, 2008).

In related work, Carvalho et al. (2005) isolated ingested DNA from the larvae of *C. dubia*, a technique that can identify the substrate on which forensically important immature blow flies have fed. Durdle et al. (2009, 2013a) showed that the DNA of a victim or offender can be isolated from the artifacts of human body fluids regurgitated or excreted by adult *L. cuprina* after having consumed them. They also documented the morphology of such artifacts to assist their recognition at the crime scene (Durdle et al. 2013b).

Because the distributions of fly species can be an important supplement to morphological and molecular data in helping to identify them, it is crucial to document range extensions when they come to light. Wallman (1993, 1997) has produced two such new distribution records for Australian blow flies *Calliphora nigrithorax* Malloch and *Chrysomya megacephala* (Fabricius), whereas the occurrence of *Chrysomya incisuralis* (Macquart) in Victoria was noted for the first time by Archer and Wallman (2002).

In New Zealand, several of the species often prominently collected in forensic casework, such as *C. hilli* and *C. stygia*, are in fact Australian natives that arrived in New Zealand in the days of sail. Miller (1939) provided a taxonomic revision of all New Zealand blow flies, followed by a more thorough treatment by Dear (1985). Holloway (1991) provided some taxonomic and distribution data to assist with the differentiation of the closely related *L. cuprina* and *L. sericata* (Holloway 1991), and this topic was further addressed by Bishop (1991, 1993). Gleeson (1995) studied *L. cuprina* in New Zealand from a molecular genetic perspective. The third instars of the common carrion-breeding blow fly species have also been described (Holloway 1985, 1991).

6.2.2 Sarcophagidae

As with the blow flies, early Australian flesh fly (Diptera: Sarcophagidae) taxonomy had an applied focus. One of the earliest species to be described was the sheep myiasis agent and carrion-breeder, *Sarcophaga* (*Sarcorohdendorfia*) *froggatti* Taylor (Taylor 1917). Further descriptions of species known or likely to be carrion breeders followed in the early part of the twentieth century (Johnston and Tiegs 1921a, 1922; Johnston and Hardy 1923; Hardy 1927, 1934, 1943). H. de Souza Lopes (Lopes 1954, 1959; Lopes and Kano 1979) provided a detailed taxonomic treatment, along with keys, for all of the species known from carrion, along with many others with life histories thought to be parasitic. All immature stages of seven species of carrion-breeding *Sarcophaga* were described by Cantrell (1981).

Despite these taxonomic efforts, the identification of flesh flies in Australian forensic investigations continued to prove difficult, especially for immatures and females. This recently prompted a comprehensive revision of the entire Australian fauna of Sarcophaginae by Meiklejohn and colleagues. This work has resulted in new species being described (Meiklejohn et al. 2013c), many new data on species distributions (Meiklejohn et al. 2012a), and a completely new taxonomic key that encompasses the females of many species, along with the more easily identified males (Meiklejohn et al. 2013a). This has also been provided online in LUCID™ format (Meiklejohn et al. 2012b) in addition to the aforementioned LUCID™ key that focuses more broadly on dipteran species of known forensic significance. Concurrent molecular analysis of the Australian flesh flies (Meiklejohn et al. 2011, 2012c, 2013b) has been combined with the earlier data on calliphorids to provide an online DNA database, accessible via BOLD and GenBank, to now represent all Australian sarcophagids and calliphorids of forensic importance. The entire mitochondrial genome of the widespread flesh fly *Sarcophaga* (*Sarcorohdendorfia*) *impatiens* Walker was presented by Nelson et al. (2012).

The New Zealand sarcophagid fauna comprises only a few species, none of them are endemic. All are included in the literature concerned with the species known from Australia.

6.2.3 Other Common Fly Families

Aside from the blow flies and flesh flies, other fly families have just as much potential forensic significance in Australia and New Zealand. Some introduced species, such as *Hermetia illucens* (Linnaeus) (Stratiomyidae), are familiar to workers elsewhere. However, in general, the endemic species of the "secondary" fly families have been much less well studied.

The main endemic species of Muscidae that feature forensically are a handful of species of the genus *Hydrotaea* and the widespread *Australophyra rostrata* (Robineau-Desvoidy), which also occurs in New Zealand. These genera were revised by Pont (1973), after being initially described by diverse authors from the early 1800s onward. Descriptions of some features of the third-instar larvae of *Hydrotaea spinigera* (Stein) were provided by Froggatt (1918) and Fuller (1932b), with the latter paper also including the third instar of *A. rostrata*.

In addition to the muscids, Pont (1977) also revised the Australian species of the related family Fanniidae, among which at least several *Fannia* species commonly occur forensically. Pont emphasized the depauperate and poorly studied nature of this family in both Australia and New Zealand, compared with the northern hemisphere.

The Australian and New Zealand fauna of the scuttle flies (Phoridae) has also been poorly documented from a forensic perspective. However, there has also been a great diversification of phorids in this region, especially in Australia. The number of species likely to breed in carrion is unknown. Disney (2003) provided keys to specimens found in Tasmania, but the mainland taxa also need attention if their forensic utility is to be maximized. The biology and immature stages of the related *Sciadocera rufomaculata* White (Sciadoceridae; the only species of this family known from Australia and New Zealand) was studied by Fuller (1934b).

6.2.4 Coleoptera

Although the Australian beetle fauna in carrion is quite extensive, our ability to apply it meaningfully in forensic investigations is still limited by lack of taxonomic knowledge. Modern revisions of these species are therefore needed, given that most taxonomic treatments are now very old and outdated. The most prominent groups, Staphylinidae and Histeridae, have not received serious attention for decades (Olliff 1886; Blackburn 1903; Lea 1925; Steel 1949). Another common group, the Leiodidae, was examined somewhat more recently by Zwick (1979), whereas the Silphidae were treated by Peck (2001).

6.3 ECOLOGY

Research into carrion insect ecology in Australia and New Zealand has had two main foci: sheep blow fly strike and, increasingly since the 1980s, studies oriented toward forensic entomology. There has been minimal pure ecological research investigating the carrion community in this region, although an early study by Bornemissza (1957) provided cornerstone data describing arthropod succession on guinea pig carrion in Western Australia. He was also one of the first workers to investigate the effect of overlying carcasses on the soil community. Palmer (1980) was another pure ecologist who explored sympatric carrion blow fly resource partitioning in Victoria as well as blow fly succession in Canberra (Palmer 1975). Monzu (1977) pursued similar studies in Western Australia. Unfortunately, neither worker ever published this work. Other pure contributions of forensic interest have elucidated aspects of the ecology of individual carrion species or genera (Fuller 1931; May 1961; Callan 1974; Roberts and Warren 1975; Roberts 1976; Williams 1981; Archer 2000; Hawkeswood and Turner 2008; Barton et al. 2013).

The first major study on carrion ecology in Australia was conducted by M.E. Fuller on behalf of the Council for Scientific and Industrial Research (CSIR) (Fuller 1934c). This work focused on sheep strike flies, and was published near the beginning of several decades of fruitful Australian and New Zealand research, funded mainly by the wool industry and government agricultural departments. Because the main pest fly species also breed in carrion, early research was directed at understanding their control, ecology, behavior, and physiology.

Many of the findings from this agricultural work have been of significant forensic use world-wide, and some of the early key information is particularly well summarized by Norris (1959, 1965) and Kitching (1981). Topics from agricultural research include carrion fly succession and ecology on carcasses (Anderson et al. 1988; Smith et al. 1981; Lang et al. 2006), blow fly population density (Gilmour et al. 1946), competition for food (Nicholson 1950), annual and diel activity times for forensic species (Davidson 1933; Fuller 1934a; Norris 1966; Dallwitz and Wardaugh 1984; Vogt et al. 1985; Dymock and Forgie 1995), habitat preferences (Dymock and Forgie 1993), individual species ecology (Callinan 1980), and important early observations on larval mass thermogenesis (Waterhouse 1947). Investigation of the effect of carcass burial on sheep strike blow fly maturation also has forensic application (Fuller 1932a), as does certain work exploring biological control of sheep blow flies using hymenopteran parasites (Johnston and Tiegs 1921b; Bishop et al. 1996). This includes work conducted in France by CSIR (now CSIRO) on wasp species (Evans 1933).

Ecology studies designed to include answers to questions of specific forensic interest have appeared in increasing numbers since the early 1980s, and the Australian literature has benefited from forensic entomology researchers being resident at various times in the Australian Capital Territory, New South Wales, South Australia, Queensland, Victoria, and Western Australia. There has been no specific forensic entomological research produced to date in the Northern Territory or Tasmania.

The first forensic entomology succession work in Australia was conducted in Queensland with carcasses of various mammalian species obtained opportunistically when they became available (O'Flynn and Moorhouse 1979; O'Flynn 1983a,b). This was followed by further research carried out with a small number of pig, sheep, rabbit, and goat carcasses placed in Guyra, New South Wales, and included data on internal temperatures of carcasses (Morris 1993).

The first systematically conducted forensic succession study to be published was carried out in Victoria with still-born piglet carcasses in a dry sclerophyll forest. This study elucidated the basic pattern of succession and decomposition over 2 years. It was demonstrated that inter-year variation in succession patterns ($n = 5$ per season) was significant, and most likely because of annual temperature and rainfall differences. This hypothesis was supported by the fact that the temperatures and rainfall recorded during the study were shown to significantly affect decomposition rates (Archer 2003, 2004b). It was also confirmed that the main carrion insect species collected from piglet carrion accorded with those collected from human bodies in local forensic casework (Archer 2002). Further experimental work with guinea pig carcasses has since been conducted to examine within-region variation in annual and seasonal succession patterns between a bushland and an agricultural habitat near Perth, Western Australia (Voss et al. 2009a). Minimal variation in succession patterns ($n = 10$ per month at each site) was found between habitats and between years, although a significant seasonal effect on succession rate and species assemblage was detected.

The first forensic succession studies to feature adult-sized pig carcasses were conducted in Western Australia. Comparison has been made between succession on clothed and unclothed carcasses, and showed that clothing produced very few differences (Voss et al. 2011). Succession patterns within three vehicles have also been described for carbon monoxide poisoned pigs, and compared with pigs killed both with carbon monoxide and with a captive bolt gun and exposed outdoors. There was a delay in colonization of approximately 1 day for the pigs in vehicles, although succession preceded faster because of the higher temperatures recorded inside vehicles (Voss et al. 2008). Further work with adult-sized pigs in the Victorian summer evaluated the accuracy of preceding mean ambient temperature models for estimating mPMI, and compared them with the use of simple summary statistics of carrion taxon arrival and departure times. This work is the first to test the accuracy of succession data on real forensic cases with known death time, and concluded that summary statistics of carrion taxon arrival and departure days appeared to be the superior method for the data set under investigation (Archer 2014).

Succession patterns on pigs in open field, coastal dune, and native bush habitats around Auckland, New Zealand, have been described, and habitat differences in decomposition rates demonstrated between autumn and winter (Eberhardt and Elliot 2008). Succession work using adult-sized pig carcasses has also recently commenced in the Darling Downs of Queensland, and aims to describe succession from this region throughout the year, as well as compare taxa with those found in local Coronial cases (Farrell et al. 2012).

Remaining forensic entomology ecology work has described annual activity times of Victorian carrion taxa (Archer and Elgar 2003a) and the host associations, activity times, and development of hympenopteran parasites in Western Australia (Voss et al. 2009b, 2010). Species distribution data across Australia, and over bioclimatic and urbanization gradients, have been provided for a number of forensically important blow flies (Morris 1993, Johnson et al. 2009, Kavazos and Wallman 2012).

6.4 BEHAVIOR AND PHYSIOLOGY

Behavior and physiology research of forensic interest from Australia and New Zealand covers diverse topics, although carrion taxa other than blow flies have been largely neglected to date. A variety of topics have been investigated, and use of pure science and cross-disciplinary approaches to applied problems have long been popular in this region. For example, studies have been conducted in behavioral physiology (as in Barton Browne [1962]) and behavioral ecology (as in Archer and Elgar [2003b]). The major focus of work on forensically important taxa was previously the sheep blow fly problem, but dedicated forensic studies are more recently increasing in number.

Blow fly oviposition and larviposition behavior are particularly important in relation to understanding the genesis of delays between death and insect colonization of the dead body. Attraction to resources and factors affecting corpse colonization by forensically important flies have been investigated in the field with whole piglet carcasses (Archer and Elgar 2003b,c), meat baits (Vogt and Woodburn 1994; George et al. 2012), artificial baits (Morris et al. 1998a,b), and sheep carcasses (Woodburn and Vogt 1982). Work has also begun on identifying volatile organic compounds (VOCs) from decomposing guinea pigs, and on linking them to the attendance times of certain carrion fly and beetle taxa in Western Australia (Hung 2012).

Laboratory work has focused on factors influencing female *L. cuprina* oviposition choice, including the requirement for moisture at oviposition sites, availability of cavities in which to place eggs, and preference for places where other females have already oviposited (Barton Browne 1958, 1962; Barton Browne et al. 1969; Emmens and Murray 1982). Early behavioral work identified that gravid female *L. cuprina* were attracted to kairomones associated with feeding larvae, bacteria, and fleece (Eisemann and Rice 1987; Eisemann 1995). There has also been recent interest in the use of electroantenography to assess Australian blow fly responses to VOCs (Leitch et al. 2012). Laboratory studies have additionally been made of the larviposition behavior of the common Western Australian early colonizing blow fly, *C. dubia* (Cook and Dadour 2011).

Environmental factors affecting fly activity and corpse colonization is another focus of research in the region. Selected agricultural studies have made useful observations on the effect of trap height above ground on *L. cuprina* catches (Vogt et al. 1995), effect of temperature on blow fly activity in the laboratory (Nicholson 1934; Waterhouse 1939), and effects of combined features such as temperature, wind speed, solar radiation, humidity, and trap height on blow fly activity under field conditions (Vogt and Woodburn 1983; Vogt et al. 1985; Vogt 1988; Dymock et al. 1991). Extensive field investigation has also been undertaken in Victoria on the combined effects of abiotic factors, such as temperature, humidity, and wind speed, on blow fly colonization times. It was confirmed that the estimate of mPMI, rather than PMI, is essential because there can be a variable delay between exposure and colonization that is difficult to predict, even with statistical models that incorporate the influential factors (George et al. 2013b).

Nocturnal colonization by blow flies has been shown to be unlikely in southeastern Australia, but is nonetheless possible under certain circumstances. Nocturnal colonization of baits by several species of forensically important blow flies occurred in the field in Victoria only when flies were caged with the bait and able to walk to it, but not when forced to fly to baits surrounded by a water moat (George et al. 2013a). Nocturnal colonization was not demonstrated on whole piglet carcasses elevated 50 cm with several of the same species under study in New South Wales (Kavazos et al. 2010). Preliminary novel Victorian research shows that there is a molecular circadian rhythm to *C. vicina* oviposition, which is probably why the majority of oviposition occurs during daylight. However, a minority of females had arrhythmic oviposition patterns, and it was postulated that at least some nocturnal colonization could result from such arrhythmic females (George et al. 2014).

Initial studies on the physiology and natural history of important sheep blow flies generated data describing the mechanisms of larval feeding. This work also provided data on adult fecundity, ovarian maturation, and basic nutritional requirements (Mackerras 1933; Mackerras and Freney 1933). Juvenile carrion fly growth has also been the focus of much attention; however, this area is still noticeably lacking in a comprehensive regional database. Unfortunately, many of the existing data originated from agricultural research, rendering them unsuitable or impractical for application in the forensic setting.

The importance of increasing temperature and humidity on the egg hatch percentage and development rate has been shown for *L. cuprina*, and it was suggested that aggregation of egg batches, a behavior described for this species, does not actually increase survival under low humidity (Vogt and Woodburn 1980). Fecundity, juvenile growth rates, thermal death points, energetics, and diapause data have also been collected in varying detail for certain stages of the species listed in Table 6.1.

The influence of several extraneous factors on larval growth and development have also been investigated, such as food type (Day and Wallman 2006b), substrates thawed from frozen versus non-frozen (Day and Wallman 2006c), and food shortage or early cessation of feeding (Webber 1955; Levot et al. 1979; Williams and Richardson 1983). The effects of drug ingestion on larval growth have also been examined by several workers. The rate of amitriptyline uptake by *L. cuprina* larvae increased as the constant temperature of their surroundings increased (15°C–25°C), while fluctuating temperature appeared to reduce the rate of uptake (Kenny et al. 2011). However, it is not known whether this substance affects growth rates. Surprisingly, morphine at concentrations as high as 20 µg/g of feeding substrate was not found to affect *C. stygia* growth rates (George et al. 2009), which may be because the species rapidly excretes morphine via the Malphigian tubules (Parry et al. 2010). Conversely, the ingestion of methamphetamine and its metabolite *p*-hydroxymethamphetamine significantly increased growth rates and size in all life history stages of the same species at doses as low as 0.1 mg/kg methamphetamine and 0.015 mg/kg of *p*-hydroxymethamphetamine (Mullany et al. 2014).

Researchers worldwide are increasingly recognizing the role that behavior has on larval growth rates. Larval massing behavior is of particular relevance, especially in response to the often substantial thermogenesis produced within carcasses. Recent work in New South Wales and Canberra with *Ch. rufifacies* and *C. vicina* concentrated on quantifying the effect of larval mass volume on temperature production and larval growth, the heat produced by masses versus bacteria, and assessing

Table 6.1   Fly Species of Forensic Importance Whose Fecundity and Development Have Been Investigated in Australia

Calliphoridae

Calliphora augur	Mackerras (1933), Levot et al. (1979), Callinan (1980), O'Flynn (1983a), and Day (2006)
Calliphora dubia	Levot et al. (1979), Morris (1993), Dadour et al. (2001a), Wallman (1999), and Cook and Dadour (2011)
Calliphora hilli	Williams and Richardson (1984)
Calliphora ochracea	Fuller (1931)
Calliphora stygia	Mackerras (1933), Levot et al. (1979), O'Flynn (1983b), and Williams and Richardson (1984)
Calliphora varifrons Malloch	Cook et al. (2012)
Calliphora vicina	Wallman (1999)
Chrysomya megacephala	O'Flynn (1983a) and Nelson et al. (2009)
Chrysomya nigripes Aubertin	O'Flynn (1983b)
Chrysomya rufifacies	Mackerras (1933), Levot et al. (1979), O'Flynn (1983a), Wallman (1999), and Nelson et al. (2009)
Chrysomya saffranea (Bigot)	O'Flynn (1983a) and Nelson et al. (2009)
Chrysomya varipes (Macquart)	Levot et al. (1979) and O'Flynn (1983a)
Lucilia cuprina	Mackerras (1933), Levot et al. (1979), O'Flynn (1983a), Dallwitz (1984), Dallwitz and Wardaugh (1984), Williams and Richardson (1984), Morris (1993), and Day (2006)
Lucilia sericata	Mackerras (1933)

Muscidae

Australophyra rostrata	Mackerras (1933) and Dadour et al. (2001b)

Sarcophagidae

Sarcophaga (*Liopygia*) *crassipalpis* Macquart	Levot et al. (1979)
Sarcophaga impatiens	Roberts (1976) and Meiklejohn (2008)

Sciadoceridae

Sciadocera rufomaculata	Fuller (1934b)

an individual larva's heat exposure and temperature preference by following its activity within a mass (Johnson 2013; Johnson and Wallman 2014; Johnson et al. 2013b, 2014). Wallman (1998, 1999) laid earlier groundwork for exploring carcass heating and the associated role of bacteria.

Species interactions are an additional aspect of larval behavior work that is gaining interest worldwide. Early Australian research on the predator/prey system of *Ch. rufifacies* and *C. stygia* revealed that secretions of *Ch. rufifacies* larvae appear to actively repel *C. stygia* larvae (D'Alberto 2004). Further work in the area revealed a complex set of interactions between larvae of different species feeding together in two seasonal cohorts of three species each, and showed that larval size and/or development rate was significantly different for most species when compared with pure cultures of the same species (Melbourne et al. 2010).

6.5 FUTURE

Working in forensic entomology in Australia and New Zealand presents some special challenges. Australia comprises 7.5 million km² of land (Geoscience Australia 2013) with conditions ranging from tropical to desert to cool temperate and alpine. New Zealand, although much smaller at

267,710 km^2 (Central Intelligence Agency 2013), also has a diverse range of habitats and climates. Both countries also have a low population density in comparison to most others (The World Bank 2013), but there is a greatly increased coastal concentration of people in Australia because of the dry interior (Australian Bureau of Statistics 2013). These geographic factors combine with additional problems associated with working in the region, namely small numbers of researchers and practitioners, persistent lack of research coverage in remote areas, relatively low levels of research funding, insecure employment, and weather stations that may be spaced far apart in the more remote regions. There are also relatively low forensic case rates per annum for each practitioner (usually in the order of 5–20 medicolegal cases per year for the major cities, although restaurant fast-food cases can number significantly more). Nevertheless, Australian and New Zealand entomologists have "batted above their weight" when the contributions of the region described in this chapter are reviewed. Much of this is due to a head start from copious agricultural industry funding aimed at solving sheep strike. Therefore, part of our challenge includes making up the shortfall now that this money is being directed more toward molecular solutions that are not as readily and directly applicable to forensic work.

REFERENCES

Anderson, P.J., E. Shipp, J.M.E. Anderson, and W. Dobbie. 1988. Population maintenance of *Lucilia cuprina* (Wiedemann) in the arid zone. *Australian Journal of Zoology* 36: 241–249.

Archer, M.S. 2000. Natural history observations of the native carrion beetle, *Ptomaphila lacrymosa* Schreibers (Coleoptera: Silphidae). *Proceedings of the Royal Society of Victoria* 112: 133–136.

Archer, M.S. 2002. The ecology of invertebrate associations with vertebrate carrion in Victoria, with reference to forensic entomology. PhD Dissertation, The University of Melbourne, Australia.

Archer, M.S. 2003. Annual variation in arrival and departure times of carrion insects at carcasses: Implications for succession studies in forensic entomology. *Australian Journal of Zoology* 51: 569–576.

Archer, M.S. 2004a. The effect of time after body discovery on the accuracy of retrospective ambient temperature corrections in forensic entomology. *Journal of Forensic Sciences* 49: 553–559.

Archer, M.S. 2004b. Rainfall and temperature effects on the decomposition rate of exposed neonatal remains. *Science and Justice* 44: 35–41.

Archer, M.S. 2014. Comparative analysis of insect succession data from Victoria (Australia) using summary statistics vs preceding mean ambient temperature models. *Journal of Forensic Sciences* 59: 404–412.

Archer, M.S., R.B. Bassed, C.A. Briggs, and M.J. Lynch. 2005. Social isolation and delayed discovery of bodies in houses: The value of forensic pathology, anthropology, odontology and entomology in the medico-legal investigation. *Forensic Science International* 151: 259–265.

Archer, M.S. and M.A. Elgar. 2003a. Yearly activity patterns in southern Victoria (Australia) of seasonally active carrion insects. *Forensic Science International* 132: 173–176.

Archer, M.S. and M.A. Elgar. 2003b. Female breeding-site preferences and larval feeding strategies of carrion-breeding Calliphoridae and Sarcophagidae (Diptera): A quantitative analysis. *Australian Journal of Zoology* 51: 165–174.

Archer, M.S. and M.A. Elgar. 2003c. Effects of decomposition on carcass attendance in a guild of carrion-breeding flies. *Medical and Veterinary Entomology* 17: 263–271.

Archer, M.S., M.A. Elgar, C.A. Briggs, and D.L. Ranson. 2006. Fly pupae and puparia as potential contaminants of forensic entomology samples from sites of body discovery. *International Journal of Legal Medicine* 120: 364–368.

Archer, M.S. and D.L. Ranson. 2005. Potential contamination of forensic entomology samples collected in the mortuary: A case report. *Medicine, Science, and the Law* 45: 89–91.

Archer, M.S. and J.F. Wallman. 2002. A new distribution record from Victoria for the blowfly, *Chrysomya incisuralis* (Macquart) (Diptera: Calliphoridae). *Proceedings of the Royal Society of Victoria* 114: 59–60.

Australian Bureau of Statistics. 2013. Population. In *Measures of Australia's Progress 2010: Is life in Australia getting better?* http://www.abs.gov.au/ausstats/abs@.nsf/Lookup/by%20Subject/1370.0~2010~Chapter~Population%20distribution%20(3.3) (accessed April 4, 2013).

Barton, P.S., S.A. Cunningham, B.C.T. Macdonald, S. McIntyre, D.B. Lindenmayer, and A.D. Manning. 2013. Species traits predict assemblage dynamics at ephemeral resource patches created by carrion. *PLoS One* e53961.

Barton Browne, L. 1958. The choice of communal oviposition sites by the Australian sheep blowfly *Lucilia cuprina*. *Australian Journal of Zoology* 6: 241–247.

Barton Browne, L. 1962. The relationship between oviposition in the blowfly *Lucilia cuprina* and the presence of water. *Journal of Insect Physiology* 8: 383–390.

Barton Browne, L., R.J. Bartell, and H.H. Shorey. 1969. Pheromone-mediated behaviour leading to group oviposition in the blowfly *Lucilia cuprina*. *Journal of Insect Physiology* 15: 1003–1014.

Bezzi, M. 1927. Some Calliphoridae (Dipt.) from the South Pacific Islands and Australia. *Bulletin of Entomological Research* 17: 231–247.

Bishop, D.M. 1991. Variations in numbers of occipital setae for two species of *Lucilia* (Diptera: Calliphoridae) in New Zealand. *New Zealand Entomologist* 14: 28–31.

Bishop, D.M. 1993. Early records (1984–1987) of the Australian green blowfly (*Lucilia cuprina*) in New Zealand. *New Zealand Entomologist* 16: 22–24.

Bishop, D.M., A.C. Heath, and N.A. Haack. 1996. Distribution, prevalence and host associations of Hymenoptera parasitic on Calliphoridae occurring in flystrike in New Zealand. *Medical and Veterinary Entomology* 10: 365–370.

Blackburn, T. 1903. Further notes on Australian Coleoptera with descriptions of new genera and species. *Transactions of the Royal Society of South Australia* 32: 91–182.

Bornemissza, G.F. 1957. An analysis of arthropod succession in carrion and the effect of its decomposition on the soil fauna. *Australian Journal of Zoology* 5: 1–12.

Callan, E.Mc. 1974. *Hermetia illucens* (L.) (Dipt., Stratiomyidae), a cosmopolitan American species long established in Australia and New Zealand. *Entomologist's Monthly Magazine* 109: 232–234.

Callinan, A.P.L. 1980. Aspects of the ecology of *Calliphora augur* (Fabricius) (Diptera: Calliphoridae), a native Australian blowfly. *Australian Journal of Zoology* 28: 679–684.

Cantrell, B.K. 1981. The immature stages of some Australian Sarcophaginae (Diptera: Sarcophagidae). *Journal of the Australian Entomological Society* 20: 237–248.

Carvalho, F., I.R. Dadour, D.M. Groth, and M.J. Harvey. 2005. Isolation and detection of ingested DNA from the immature stages of *Calliphora dubia* (Diptera: Calliphoridae). *Forensic Science, Medicine and Pathology* 1: 261–265.

Central Intelligence Agency. 2013. New Zealand. In *The World Factbook* https://www.cia.gov/library/publications/the-world-factbook/geos/nz.html (accessed April 4, 2013).

Clarke, A.F. 1962. Scientific aspects of the Graeme Thorne kidnapping and murder in July 1960. *Australian Police Journal* 16: 181–237.

Cook, D.F. and I.R. Dadour. 2011. Larviposition in the ovoviviparous blowfly *Calliphora dubia. Medical and Veterinary Entomology* 25: 53–57.

Cook, D.F., S.C. Voss and I.R. Dadour. 2012. Larviposition in the ovoviviparous blowfly. *Calliphora varifrons Forensic Science International* 223: 44–46.

Crosby, T.K., J.C. Watt, A.C. Kistemaker, and P.E. Nelson. 1985. Entomological identification of the origin of imported cannabis. *Journal of the Forensic Science Society* 26: 35–44.

Dadour, I.R., D.F. Cook, J.N. Fissioloi, and W.J. Bailey. 2001a. Forensic entomology: Application, education and research in Western Australia. *Forensic Science International* 120: 48–52.

Dadour, I.R., D.F. Cook, and N. Wirth. 2001b. Rate of development of *Hydrotaea rostrata* under summer and winter (cyclic and constant) temperature regimes. *Medical and Veterinary Entomology* 15: 177–182.

D'Alberto, C. 2004. Predator-prey interactions between maggots of the necrophagous flies *Chrysomya rufifacies* Macquart and *Calliphora stygia* Fabricius. BSc (Hons) Dissertation, University of Melbourne, Australia.

Dallwitz, R. 1984. The influence of constant and fluctuating temperatures on development rate and survival of pupae of the Australian sheep blowfly *Lucilia cuprina. Entomologica Experimentalis et Applicata* 36: 89–95.

Dallwitz, R. and K.G. Wardaugh. 1984. Overwintering of prepupae of *Lucilia cuprina* (Diptera: Calliphoridae) in the Canberra region. *Journal of the Australian. Entomological Society* 23: 307–312.

Davidson, J. 1933. The species of blowflies in the Adelaide district of South Australia and their seasonal occurrence. *Journal of Agriculture of South Australia* 36: 1148–1153.

Day, D.M. 2006. Development of immature blowflies and their application to forensic science. MSc Dissertation, University of Wollongong, Australia.

Day, D.M. and J.F. Wallman. 2006a. Width as an alternative measurement to length for post-mortem interval estimations using *Calliphora augur* (Diptera: Calliphoridae) larvae. *Forensic Science International* 159: 158–167.

Day, D.M. and J.F. Wallman 2006b. Influence of substrate tissue type on larval growth in *Calliphora augur* and *Lucilia cuprina* (Diptera: Calliphoridae). *Journal of Forensic Sciences* 51: 657–663.

Day, D.M. and J.F. Wallman. 2006c. A comparison of frozen/thawed and fresh food substrates in development of *Calliphora augur* (Diptera: Calliphoridae) larvae. *International Journal of Legal Medicine* 120: 391–394.

Day, D.M. and J.F. Wallman 2008. Effect of preservative solutions on preservation of *Calliphora augur* and *Lucilia cuprina* larvae (Diptera: Calliphoridae) with implications for post-mortem interval estimates. *Forensic Science International* 179: 1–10.

Dear, J.P. 1985. Calliphoridae (Insecta: Diptera). *Fauna of New Zealand* 8: 1–86.

Disney, R.H.L. 2003. Tasmanian Phoridae (Diptera) and some additional Australasian species. *Journal of Natural History* 37: 505–639.

Durdle, A, R.A. van Oorschot, and R.J. Mitchell. 2013a. The human DNA content in artifacts deposited by the blowfly *Lucilia cuprina* fed human blood, semen and saliva. *Forensic Science International* 233: 212–219.

Durdle, A, R.A. van Oorschot and R.J. Mitchell. 2013b. The morphology of fecal and regurgitation artifacts deposited by the blow fly *Lucilia cuprina* fed a diet of human blood. *Journal of Forensic Sciences* 58: 897–903.

Durdle, A, R.A.H. van Oorschot, and R.J. Mitchell. 2009. The transfer of human DNA by *Lucilia cuprina* (Meigen) (Diptera: Calliphoridae). *Forensic Science International Genetics Supplement Series* 2: 180–182.

Dymock, J.J. and S.A. Forgie. 1993. Habitat preferences and carcase colonization by sheep blowflies in the northern North Island of New Zealand. *Medical and Veterinary Entomology* 7: 155–160.

Dymock, J.J. and S.A. Forgie. 1995. Large-scale trapping of sheep blowflies in the northern North Island of New Zealand using insecticide-free traps. *Australian Journal of Experimental Agriculture* 35: 699–704.

Dymock, J.J., M.O.E. Peters, T.J.B. Herman, and S.A. Forgie. 1991. A study of sheep blowflies at Limestone Downs sheep station in the northern Waikato, New Zealand, over two summers. *New Zealand Journal of Agricultural Research* 34: 311–316.

Eberhardt, T.L. and D.A. Elliot. 2008. A preliminary investigation of insect colonisation and succession on remains in New Zealand. *Forensic Science International* 176: 217–223.

Eisemann, C.H. 1995. Orientation by gravid Australian sheep blowflies, *Lucilia cuprina* (Diptera: Calliphoridae), to fleece and synthetic chemical attractants in laboratory bioassays. *Bulletin of Entomological Research* 85: 473–477.

Eisemann, C.H. and M.J. Rice. 1987. The origin of sheep blowfly, *Lucilia cuprina* (Wiedemann) (Diptera: Calliphoridae), attractants in media infested with larvae. *Bulletin of Entomological Research* 77: 287–294.

Emmens, R.L. and M.D. Murray. 1982. The role of bacterial odours in oviposition by *Lucilia cuprina* (Wiedemann) (Diptera: Calliphoridae), the Australian sheep blowfly. *Bulletin of Entomological Research* 72: 367–375.

Evans, A.C. 1933. Comparative observations on the morphology and biology of some hymenopterous parasites of carrion-infesting Diptera. *Bulletin of Entomological Research* 24: 385–405.

Fabricius, J.C. 1775. Systema entomologiae, sistens insectorum classes, ordines, genera, species adiectis synonymis, locis, descriptionibus, observationibus. Flensburg & Leipzig: Kortii.

Farrell, J., A. Whittington, and M. Zalucki. 2012. Carrion café—using insects to determine post-mortem intervals in Queensland. Paper presented at the 21st International Symposium on the Forensic Sciences, Hobart, Australia.

Froggatt, J.L. 1918. A study of the external breathing-apparatus of the larvae of some muscoid flies. *Proceedings of the Linnaean Society of New South Wales* 43: 658–667.

Fuller, M.E. 1931. The life history of *Calliphora ochracea* Schiner (Diptera, Calliphoridae). *Proceedings of the Linnaean Society of New South Wales* 56: 172–181.

Fuller, M.E. 1932a. The blowfly problem—notes on the effect of carcass burial. *Journal of the Council for Scientific and Industrial Research* 5: 162–164.

Fuller, M.E. 1932b. The larvae of the Australian sheep blowflies. *Proceedings of the Linnaean Society of New South Wales* 57: 77–91.

Fuller, M.E. 1934a. Observations on the flies responsible for striking sheep in Western Australia. *Journal of the Council for Scientific and Industrial Research* 7: 150–152.

Fuller, M.E. 1934b. The early stages of *Sciadocera rufomaculata* White (Diptera: Phoridae). *Proceedings of the Linnaean Society of New South Wales* 59: 9–15.

Fuller, M.E. 1934c. The insect inhabitants of carrion: A study in animal ecology. *Bulletin Council for Scientific and Industrial Research (Australia)* 82: 1–63.

George, K.A., M.A. Archer, L.M. Green, X.A. Conlan, and T. Toop. 2009. Effect of morphine on the growth rate of *Calliphora stygia* (Fabricius) (Diptera: Calliphoridae) and possible implications for forensic entomology. *Forensic Science International* 193: 21–25.

George, K.A., M.S. Archer, and T. Toop. 2012. Effects of bait age, larval chemical cues and nutrient depletion on colonization by forensically important Calliphorid and Sarcophagid flies. *Medical and Veterinary Entomology* 26: 188–193.

George, K.A., M.S. Archer, and T. Toop. 2013a. Nocturnal colonization behaviour of blowflies (Diptera: Calliphoridae) in southeastern Australia. *Journal of Forensic Sciences* 58: 112–116.

George, K.A., M.S. Archer, and T. Toop. 2014. Correlation of molecular expression with diel rhythm of oviposition in *Calliphora vicina* (Robineau-Desvoidy) (Diptera: Calliphoridae) and implications for forensic entomology. *Journal of Forensic Sciences.* doi: 10.1111/1556-4029.12585.

George, K.A., T. Toop, and M.S. Archer. 2013b. Abiotic environmental factors influencing blowfly colonisation patterns in the field. *Forensic Science International* 229: 100–107.

Geoscience Australia. 2013. Australia's size compared. http://www.ga.gov.au/education/geoscience-basics /dimensions/australias-size-compared.html (accessed April 4, 2013).

Gilmour, D., D.F. Waterhouse, and G.A. McIntyre. 1946. An account of experiments undertaken to determine the natural population density of the sheep blowfly, *Lucilia cuprina* Wied. *Bulletin of the Council for Scientific and Industrial Research (Australia)* 195: 1–39.

Gleeson, D.M. 1995. The effects on genetic variability following a recent colonization event: The Australian sheep blowfly, *Lucilia cuprina* arrives in New Zealand. *Molecular Ecology* 4: 699–707.

Gunn, J.A., C. Shelley, S.W. Lewis, T. Toop, and M. Archer. 2006. The determination of morphine in the larvae of *Calliphora stygia* using flw injection analysis and HPLC with chemiluminescence detection. *Journal of Analytical Toxicology* 30: 519–523.

Hardy, G.H. 1926. Notes on Australian flies of the genus *Calliphora*. *Proceedings of the Royal Society of Queensland* 37: 168–173.

Hardy, G.H. 1927. Notes on Australian and exotic sarcophagid flies. *Proceedings of the Linnaean Society of New South Wales* 52: 447–459.

Hardy, G.H. 1930. The Queensland species of *Calliphora* subgenus *Neopollenia*. *Bulletin of Entomological Research* 21: 441–448.

Hardy, G.H. 1932. Some Australian species of *Calliphora* (subgenera *Neopollenia* and *Proekon*). *Bulletin of Entomological Research* 23: 549–558.

Hardy, G.H. 1934. Notes on Australian Muscoidea (Calyptrata). *Proceedings of the Royal Society of Queensland* 45: 30–37.

Hardy, G.H. 1937. Notes on the genus *Calliphora* (Diptera). *Proceedings of the Linnaean Society of New South Wales* 62: 17–26.

Hardy, G. H. 1940. Notes on the Australian Muscoidea, V. Calliphoridae. *Proceedings of the Royal Society of Queensland* 51: 133–146.

Hardy, G.H. 1943. The Sarcophaginae of Australia and New Zealand. *Proceedings of the Linnaean Society of New South Wales* 68: 17–32.

Harvey, M.L., I.R. Dadour, and S. Gaudieri. 2003a. Mitochondrial DNA cytochrome oxidase I gene: Potential for distinction between immature stages of some forensically important fly species (Diptera) in Western Australia. *Forensic Science International* 131: 134–139.

Harvey, M.L., S. Gaudieri, M.H. Villet, and I.R. Dadour. 2008. A global study of forensically significant calliphorids: Implications for identification. *Forensic Science International* 177: 66–76.

Harvey, M.L., M.W. Mansell, M.H. Villet, and I.R. Dadour. 2003b. Molecular identification of some forensically important blowflies of southern Africa and Australia. *Medical and Veterinary Entomology* 17: 363–369.

Hawkeswood, T.J. and J.R. Turner. 2008. Record of the Australian burying beetle, *Ptomaphila perlata* Kraatz, 1876 (Coleoptera: Silphidae) feeding and breeding in the dead carcass of a Swamp Wallaby (*Wallabia bicolor,* Mammalia: Macropodidae), with a review of its biology and habitat. *Calodema Supplementary Paper* 79: 1–5.

Holloway, B.A. 1985. Immature stages of New Zealand Calliphoridae. *Fauna of New Zealand* 8: 12–14.

Holloway, B.A. 1991. Identification of third-instar larvae of flystrike and carrion-associated blowflies in New Zealand (Diptera: Calliphoridae). *New Zealand Entomologist* 14: 24–28.

Hung, W.-F. 2012. Synchronisation of necrophagous insects to the odours of decomposing remains. Paper presented at the 21st International Symposium on the Forensic Sciences, Hobart, Australia.

Johns, L., B. Reedy, and J.F. Wallman. 2010. Detection of ignitable liquids in entomological samples: A field-work study in Holsworthy, New South Wales. In *20th International Symposium on the Forensic Sciences: Abstracts of Oral Presentations*, eds. J.F. Wallman, C.P. Roux, and C.J. Lennard. *Forensic Science, Medicine, and Pathology* 7: 75–138.

Johnson, A.P. 2013. Novel approaches to the investigation of thermogenesis in the maggot masses of blow flies (Diptera: Calliphoridae). PhD Dissertation, University of Wollongong, Australia.

Johnson, A.P., M.S. Archer, L. Leigh-Shaw, M. Brown, C. O'Donnell, and J.F. Wallman. 2013a. Non-invasive visualisation and volume estimation of maggot masses using computed tomography scanning. *International Journal of Legal Medicine* 127: 185–194.

Johnson, A.P., M.S. Archer, L. Leigh-Shaw, M. Pais, C. O'Donnell, and J.F. Wallman. 2012a. Examination of forensic entomology evidence using computed tomography scanning: Case studies and refinement of techniques for estimating maggot mass volumes in bodies. *International Journal of Legal Medicine* 126: 693–702.

Johnson, A.P., K. Mikac, and J.F. Wallman. 2013b. Thermogenesis in decomposing carcasses. *Forensic Science International* 231: 271–277.

Johnson, A.P., C. Ryan, M.S. Archer, and J.F. Wallman. 2009. Distribution and dry stress tolerance in the Australian blowflies, *Calliphora augur* and *Calliphora dubia* (Diptera: Calliphoridae). Paper presented at Darwin 200: Evolution & Biodiversity, Darwin, Australia.

Johnson, A.P. and J.F. Wallman 2013. Infrared imaging as a non-invasive tool for documenting maggot mass temperatures. *Australian Journal of Forensic Sciences* 46: 73–79.

Johnson, A.P. and J.F. Wallman. 2014. Effect of massing on larval growth rate. *Forensic Science International* 241: 141–149.

Johnson, A.P., J.F. Wallman, and M.S. Archer. 2012b. Experimental and casework validation of ambient temperature corrections in forensic entomology. *Journal of Forensic Sciences* 57: 215–221.

Johnson A.P., S.J. Wighton, and J.F. Wallman. 2014. Tracking movement and temperature selection of larvae of two forensically important blow fly species within a "maggot mass." *Journal of Forensic Sciences.* doi: 10.1111/1556-4029.12472 [Epub ahead of print].

Johnston, T.H. and G.H. Hardy. 1923. A revision of the Australian Diptera belonging to the genus *Sarcophaga*. *Proceedings of the Linnaean Society of New South Wales* 48: 94–129.

Johnston, T.H. and O.W. Tiegs. 1921a. New and little-known sarcophagid flies from south-eastern Queensland. *Proceedings of the Royal Society of Queensland* 33: 46–90.

Johnston, T.H. and O.W. Tiegs. 1921b. On the biology and economic significance of the chalcid parasites of Australian sheep maggot flies. *Proceedings of the Royal Society of Queensland* 33: 99–128.

Johnston, T.H. and O.W. Tiegs. 1922. Sarcophagid flies in the Australian Museum collection. *Records of the Australian Museum* 13: 175–188.

Kavazos, C.R.J., B.D. Lessard, and J.F. Wallman. 2010. Nocturnal activity in carrion-breeding blowflies. In *20th International Symposium on the Forensic Sciences: Abstracts of Oral Presentations*, eds. J.F. Wallman, C.P. Roux, and C.J. Lennard. *Forensic Science, Medicine, and Pathology* 7: 75–138.

Kavazos, C.R.J., K.A. Meiklejohn, M.A. Archer, and J.F. Wallman. 2012. Carrion flies of Australia: A LUCID key for the online identification of the Australian flies known or suspected to breed in carrion. University of Wollongong, Australia.

Kavazos, C.R.J. and Wallman, J.F. 2012. Community composition of carrion-breeding blowflies (Diptera: Calliphoridae) along an urban gradient in south-eastern Australia. *Landscape and Urban Planning* 106: 183–190.

Kenny, D., J.F. Wallman, and P. Doble. 2011. Factors affecting larval drug uptake. In *20th International Symposium on the Forensic Sciences: Abstracts of Oral Presentations*, eds. J.F. Wallman, C.P. Roux, and C.J. Lennard. *Forensic Science, Medicine, and Pathology* 7: 75–138.

Kitching, R.L. 1976. The immature stages of the Old-World screw-worm fly, *Chrysomya bezziana* villeneuve, with comparative notes on other Australasian species of *Chrysomya* (Diptera, Calliphoridae). *Bulletin of Entomological Research* 66: 195–203.

Kitching, R.L. 1981. The sheep blowfly: A resource-limited specialist species. In *The Ecology of Pests: Some Australian Case Histories*, eds. R.E. Jones and R.L. Kitching. Canberra, Australia: CSIRO.

Kitching, R.L. and R. Voeten. 1977. The larvae of *Chrysomya incisuralis* (Macquart) and *Ch. Eucompsomyia semi-metallica* (Malloch) (Diptera: Calliphoridae). *Journal of the Australian Entomological Society* 16: 185–190.

Kurahashi, H. 1971. The tribe Calliphorini from Australian and Oriental regions. II. *Calliphora*-group (Diptera: Calliphoridae). *Pacific Insects* 13: 141–204.

Lang, M.D., G.R. Allen, and B.J. Horton. 2006. Blowfly succession from possum (*Trichosurus vulpecula*) carrion in a sheep-farming zone. *Medical and Veterinary Entomology* 20: 445–452.

Lea, A.M. 1925. On Australian Histeridae (Coleoptera). *Transactions of the Entomological Society of London* 1924: 239–263.

Leitch, O., C. Lennard, P. Kirkbride, J. Wallman, and A. Anderson. 2012. Electrophysiological and behavioural responses of the blowfly *Calliphora stygia* to volatile organic compounds liberated during mammalian decomposition. Paper presented at the 21st International Symposium on the Forensic Sciences, Hobart, Australia.

Lessard, B.D., J.F. Wallman, and M. Dowton 2009. Incorrect report of cryptic species within *Chrysomya rufifacies* (Diptera: Calliphoridae). *Invertebrate Systematics* 23: 507–514.

Levot, G.W. 2003. Insect fauna used to estimate the post-mortem interval of deceased persons. *General and Applied Entomology* 32: 31–39.

Levot, G.W., K.R. Brown, and E. Shipp. 1979. Larval growth of some calliphorid and sarcophagid Diptera. *Bulletin of Entomological Research* 69: 469–475.

Lopes, H.S. 1954. Contribution to the knowledge of the Australian sarcophagid flies belonging to the genus "Tricholioproctia" Baranov, 1938 (Diptera). *Anais da Academia Brasileira de Ciências* 26: 235–276.

Lopes, H.S. 1959. A revision of Australian Sarcophagidae (Diptera). *Studia Entomologica* 2: 33–67.

Lopes, H.S. and R. Kano. 1979. Notes on *Sarcorohdendorfia* with key of the species (Diptera, Sarcophagidae). *Revista Brasileira de Biologica* 39: 657–70.

Mackerras, M.J. 1933. Observations on the life-histories, nutritional requirements and fecundity of blowflies. *Bulletin of Entomological Research* 24: 353–362.

Mackerras, M.J. and M.R. Freney. 1933. Observations on the nutrition of maggots of Australian blow-flies. *Journal of Experimental Biology* 10: 237.

Macquart, J. 1851. Diptères exotiques nouveaux ou peu connus. Suite du 4.ᵉ supplement publié dans les memoires dᵉ 1849. *Mémoires de la Société (Royale) des Sciences, de l'Agriculture et des Arts de Lille* 1850: 134–294.

Malloch, J.R. 1927. Notes on Australian Diptera. No. XI. *Proceedings of the Linnaean Society of New South Wales* 52: 299–335.

Malloch, J.R. 1932. Notes on Australian Diptera. No. XXX. *Proceedings of the Linnaean Society of New South Wales* 57: 64–68.

Marks, E.N. 1991. Biographical history. In *The Insects of Australia*, eds. I.D. Naumann, P.B. Carne, J.F. Lawrence, E.S. Nielsen, J.P. Spradbury, R.W. Taylor, M.J. Whitten, and M. J. Littlejohn, pp. 198–220. Carlton, Australia: Melbourne University Press.

May, B.M. 1961. The occurrence in New Zealand and the life-history of the soldier fly *Hermetia illucens (L.)* (Diptera: Stratiomyiidae). *New Zealand Journal of Science* 4: 55–65.

Meiklejohn, K.A. 2008. DNA-based identification, phylogenetic inference and thermo-development of the forensically important Australian Sarcophagidae (Diptera). Honours Dissertation, University of Wollongong, Australia.

Meiklejohn, K.A., M. Dowton, T. Pape, and J.F. Wallman. 2013a. A key to the Australian Sarcophagidae (Diptera) with special emphasis on *Sarcophaga* (*sensu lato*). *Zootaxa* 3680: 148–189.

Meiklejohn, K.A., M. Dowton, and J.F. Wallman. 2012a. Notes on the distribution of 31 species of Sarcophagidae (Diptera) in Australia, including new records in Australia for eight species. *Transactions of the Royal Society of South Australia* 136: 56–64.

Meiklejohn, K.A., C.R.J. Kavazos, M. Dowton, T. Pape, and J.F. Wallman. 2012b. Australian *Sarcophaga sensu lato* (Diptera: Sarcophagidae): A LUCID key for the online identification of the Australian flesh flies. University of Wollongong, Australia.

Meiklejohn, K.M., J.F. Wallman, S.L. Cameron, and M. Dowton. 2012c. Comprehensive evaluation of DNA barcoding for the molecular species identification of forensically important Australian Sarcophagidae (Diptera). *Invertebrate Systematics* 26: 515–525.

Meiklejohn, K.A., J.F. Wallman, and M. Dowton. 2011. DNA-based identification of forensically important Australian Sarcophagidae (Diptera). *International Journal of Legal Medicine* 125: 27–32.

Meiklejohn, K.A., J.F. Wallman, and M. Dowton. 2013b. DNA barcoding identifies all immature life stages of a forensically important flesh fly (Diptera: Sarcophagidae). *Journal of Forensic Sciences* 58: 184–187.

Meiklejohn, K.A., J.F. Wallman and T. Pape. 2013c. Updates on the taxonomy and nomenclature of Australian *Sarcophaga* (*sensu lato*) (Diptera: Sarcophagidae), with descriptions of two new species. *Zootaxa* 3680: 139–147.

Melbourne, K.A., M.S. Archer, and J.F. Wallman. 2010. Interspecific interactions between larvae of carrion-breeding blowflies. Paper presented at the 7th International Congress of Dipterology, San Juan, Costa Rica.

Miller, D. 1939. Blow-flies (Calliphoridae) and their associates in New Zealand. *Cawthron Institute Monographs* 2: 1–68.

Miller, H. 1991. *Indelible Evidence.* Crows Nest, Australia: ABC Enterprises.

Mingari, L., P.J. Maynard, K.R. Brown, and J.F. Wallman. 2006. A novel method for the extraction of accelerants from fly larvae using solid-phase microextraction followed by gas chromatography-mass spectrometry. Paper presented at the 18th International Symposium on the Forensic Sciences, Fremantle, Australia.

Monzu, N. 1977. Coexistence of carrion-breeding Calliphoridae (Diptera) in Western Australia. PhD Dissertation, University of Western Australia, Australia.

Morgan, K. 2012. *Detective Piggott's Casebook: True Tales of Murder, Madness and the Rise of Forensic Science.* Richmond, Australia: Hardie Grant Books.

Morris, B. 1990. Description of the life history stages of *Calliphora nociva* Hardy (Diptera: Calliphoridae). *Journal of the Australian Entomological Society* 30: 79–82.

Morris, B. 1993. Physiology and taxonomy of blowflies. MAgSc Dissertation, The University of Adelaide, Australia.

Morris, B. and I. Dadour. 2011. Forensic entomology: The use of insects in legal cases. In *Expert Evidence*, eds. I. Freckleton and H. Selby, 91A. Sydney, Australia: Law Book.

Morris, M.C., L. Morrison, M.A. Joyce, and B. Rabel. 1998a. Trapping sheep blowflies with lures based on bacterial cultures. *Australian Journal of Experimental Agriculture* 38: 125–130.

Morris, M.C., A.D. Woolhouse, B. Rabel, and M.A. Joyce. 1998b. Orientation stimulants from substances attractive to *Lucilia cuprina* (Diptera, Calliphoridae). *Australian Journal of Experimental Agriculture* 38: 461–468.

Mullany, C., P.A. Keller, A. Nugraha, and J.F. Wallman. 2014. Effects of methamphetamine and its primary human metabolite, *p*-hydroxymethamphetamine, on the metabolism of the Australian blowfly, *Calliphora stygia*. *Forensic Science International* 241: 102–111.

Nelson, L.A., M. Dowton, and J.F. Wallman. 2009. Thermal attributes of *Chrysomya* species. *Entomologia Experimentalis et Applicata* 133: 260–275.

Nelson, L.A., C.L. Lambkin, P. Batterham et al. 2012. Beyond barcoding: A mitochondrial genomics approach to molecular phylogenetics and diagnostics of blowflies (Diptera: Calliphoridae). *Gene* 511: 131–142.

Nelson L.A., J.F. Wallman, and M. Dowton. 2007. Using COI barcodes to identify forensically and medically important blowflies. *Medical and Veterinary Entomology* 21: 44–52.

Nelson L.A., J.F. Wallman, and M. Dowton. 2008. Identification of forensically important *Chrysomya* (Diptera: Calliphoridae) species using the second ribosomal internal transcribed spacer (ITS2). *Forensic Science International* 177: 238–247.

Nicholson, A.J. 1934. The influence of temperature on the activity of sheep-blowflies. *Bulletin of Entomological Research* 15: 85–99.

Nicholson, A.J. 1950. Population oscillations caused by competition for food. *Nature* 165: 476–477.

Norris, K.R. 1959. The ecology of sheep blowflies in Australia. *Monographiae Biologicae* 8: 514–544.

Norris, K.R. 1965. The bionomics of blowflies. *Annual Review of Entomology* 10: 47–68.

Norris, K.R. 1966. Daily patterns of flight activity of blowflies (Calliphoridae: Diptera) in the Canberra district as indicated by trap catches. *Australian Journal of Zoology* 14: 835–853.

Norris, K.R. 1990. Evidence of the multiple exotic origin of Australian populations of the sheep blowfly, *Lucilia cuprina* (Wiedemann) (Diptera: Calliphoridae). *Australian Journal of Zoology* 38: 635–648.

Norris, K.R. 1994. Three new species of Australian 'Golden Blowflies' (Diptera: Calliphoridae: *Calliphora*), with a key to described species. *Invertebrate Taxonomy* 8: 1343–1366.

Norris, K.R. 1997. Supposed Australian record of the New Zealand blowfly *Calliphora quadrimaculata* (Swederus) (Diptera: Calliphoridae). *The Entomologist* 116: 37–39.

O'Flynn, M.A. 1980. Studies on blowflies and related flies from Queensland. PhD Dissertation, Department of Parasitology, University of Queensland, Australia.

O'Flynn, M.A. 1983a. The succession and rate of development of blowflies in southern Queensland and the application of these data to forensic entomology. *Journal of the Australian Entomological Society* 22: 137–148.

O'Flynn, M.A. 1983b. Notes on the biology of *Chrysomya nigripes* Aubertin (Diptera: Calliphoridae). *Journal of the Australian Entomological Society* 22: 341–342.

O'Flynn, M.A. and D.E. Moorhouse. 1979. Species of *Chrysomya* as primary flies in carrion. *Journal of the Australian Entomological Society* 18: 31–32.

O'Flynn, M.A. and D.E. Moorhouse. 1980. Identification of early immature stages of some common Queensland carrion flies. *Journal of the Australian Entomological Society* 19: 53–61.

Olliff, A.S. 1886. A revision of the Staphylinidae of Australia. *Proceedings of the Linnaean Society of New South Wales* 1: 403–473.

Palmer, D.H. 1975. A quantitative survey of the Calliphoridae (Insecta: Diptera) on carrion. BSc (Hons) Dissertation, Australian National University, Australia.

Palmer, D.H. 1980. Partitioning of the carrion resource by sympatric Calliphoridae (Diptera) near Melbourne. PhD Dissertation, La Trobe University, Australia.

Parry, S., S.M. Linton, P.S. Francis, M.J. O'Donnell, and T. Toop. 2010. Accumulation and excretion of morphine by *Calliphora stygia*, an Australian blow fly species of forensic importance. *Journal of Insect Physiology* 57: 62–73.

Patton, W.S. 1925. Diptera of medical and veterinary importance, II. The more important blowflies, Calliphorinae. *The Philippine Journal of Science* 27: 397–411.

Peck, S.B. 2001. Review of the carrion beetles of Australia and New Guinea (Coleoptera: Silphidae). *Australian Journal of Entomology* 40: 93–101.

Pont, A. 1973. Studies on Australian Muscidae (Diptera). IV. A revision of the subfamilies Muscinae and Stomoxyinae. *Australian Journal of Zoology Supplementary Series* 21: 129–296.

Pont, A. 1977. A revision of the Australian Fanniidae (Diptera: Calyptrata). *Australian Journal of Zoology Supplementary Series* 51: 1–60.

Porter, G. 2012. Zak coronial inquest and the interpretation of photographic evidence. *Current Issues in Criminal Justice* 24: 39–49.

Roberts, B. 1976. Larval development in the Australian flesh fly *Tricholioproctia impatiens Annals of the Entomological Society of America* 69: 158–164.

Roberts, B. and M. Warren. 1975. Diapause in the Australian flesh fly *Tricholioproctia impatiens.* (Diptera: Sarcophagidae). *Australian Journal of Zoology* 23: 563–567.

Roeterdink, E.M., I.R. Dadour, and R.J. Watling. 2003. Extraction of gunshot residues from the larvae of the forensically important blowfly *Calliphora dubia* (Macquart) (Diptera: Calliphoridae). *International Journal of Legal Medicine* 118: 63–70.

Sackett, P., D. Holmes, K. Abbott, S. Jephcott, and M. Barber. 2006. Assessing the economic cost of endemic disease on the profitability of Australian beef cattle and sheep producers. Sydney, Australia: Meat & Livestock Australia.

Smeeton, W.M.I., T.D. Koelmeyer, B.A. Holloway, and P. Singh. 1984. Insects associated with exposed human corpses in Auckland, New Zealand. *Medicine, Science and the Law* 24: 167–174.

Smith, P.H., R. Dallwitz, K.G. Wardhaugh, W.G. Vogt, and T.L. Woodburn. 1981. Timing of larval exodus from sheep and carrion in the sheep blowfly, *Lucilia cuprina. Entomologia Experimentalis et Applicata* 30: 157–162.

Steel, W.O. 1949. On the Australian species of *Creophilus* (Coleoptera: Staphylinidae). *Proceedings of the Linnaean Society of New South Wales*, 74: 57–61.

Taylor, F.H. 1917. *Sarcophagi froggatti*, sp. n.—A new sheep-maggot fly. *Bulletin of Entomological Research* 7: 265.

Vogt, W.G. 1988. Influence of weather on trap catches of *Chrysomya rufifacies* (Macquart) (Diptera: Calliphoridae). *Journal of the Australian Entomological Society* 27: 99–103.

Vogt, W.G., A.C.M Van Gerwen, and R. Morton. 1995. Influence of trap height on catches of *Lucilia cuprina* (Wiedemann) (Diptera: Calliphoridae) in wind-oriented fly traps. *Journal of the Australian Entomological Society* 34: 225–227.

Vogt, W.G. and T.L. Woodburn. 1980. The influence of temperature and moisture on the survival and duration of the egg stage of the Australian sheep blowfly, *Lucilia cuprina* (Wiedemann) (Diptera: Calliphoridae). *Bulletin of Entomological Research* 70: 665–671.

Vogt, W.G. and T.L. Woodburn. 1983. The analysis and standardisation of trap catches of *Lucilia cuprina* (Wiedemann) (Diptera: Calliphoridae). *Bulletin of Entomological Research* 73: 609–617.

Vogt, W.G. and T.L. Woodburn. 1994. Effects of bait age on the number, sex, and age composition of *Lucilia cuprina* (Wiedemann) (Diptera: Calliphoridae) in Western Australia blowfly traps. *Australian Journal of Experimental Agriculture* 34: 595–600.

Vogt, W.G., T.L. Woodburn, R. Morton, and B.A. Ellem. 1985. The influence of weather and time of day on trap catches of males and females of *Lucilia cuprina* (Wiedemann) (Diptera: Calliphoridae). *Bulletin of Entomological Research* 75: 315–319.

Voss, S.C., D.F. Cook, and I.R. Dadour. 2011. Decomposition and insect succession of clothed and unclothed carcasses in Western Australia. *Forensic Science International* 211: 67–75.

Voss, S.C., S.L. Forbes, and I.R. Dadour. 2008. Decomposition and insect succession on cadavers inside a vehicle environment. *Forensic Science, Medicine and Pathology* 4: 22–32.

Voss, S.C., H. Spafford, and I.R. Dadour. 2009a. Annual and seasonal patterns of insect succession on decomposing remains at two locations in Western Australia. *Forensic Science International* 193: 26–36.

Voss, S.C., H. Spafford, and I.R. Dadour. 2009b. Hymenopteran parasitoids of forensic importance: Host associations, seasonality, and prevalence of parasitoids of carrion flies in Western Australia. *Journal of Medical Entomology* 46: 1210–1219.

Voss, S.C., H. Spafford, and I.R. Dadour. 2010. Temperature-dependent development of the parasitoid *Tachinaephagus zealandicus* on five forensically important carrion fly species. *Medical and Veterinary Entomology* 24: 189–198.

Wallman, J.F. 1993. First South Australian record of the carrion-breeding blowfly *Calliphora nigrithorax* Malloch (Diptera: Calliphoridae). *Transactions of the Royal Society of South Australia* 117: 193.

Wallman, J.F. 1997. First record of the Oriental Latrine fly, *Chrysomya megacephala* (Fabricius) (Diptera: Calliphoridae), from South Australia. *Transactions of the Royal Society of South Australia* 121: 163–164.

Wallman, J.F. 1998. Seasonal influences on larval thermogenesis in blowflies (Calliphoridae). Paper presented at the 4th International Congress of Dipterology, Oxford, United Kingdom.

Wallman, J.F. 1999. Systematics and thermobiology of carrion-breeding blowflies (Diptera: Calliphoridae). PhD dissertation, The University of Adelaide, Australia.

Wallman, J.F. 2001a. A key to the adults of species of blowflies in southern Australia known or suspected to breed in carrion [corrigendum in *Medical and Veterinary Entomology* 16, 223]. *Medical and Veterinary Entomology* 15: 433–437.

Wallman, J.F. 2001b. Third-instar larvae of common carrion-breeding blowflies of the genus *Calliphora* (Diptera: Calliphoridae) in South Australia. *Invertebrate Taxonomy* 15: 37–51.

Wallman, J.F. 2002a. Key to adults of carrion-breeding Diptera of forensic importance in south-eastern Australia. In *Entomology and the Law: Flies as Forensic Indicators*, eds. B. Greenberg and J.C. Kunich, pp. 147–152. Cambridge: Cambridge University Press.

Wallman, J.F. 2002b. Key to third-instar larvae of carrion-breeding Diptera of forensic importance in south-eastern Australia. In *Entomology and the Law: Flies as Forensic Indicators*, eds. B. Greenberg and J.C. Kunich, pp. 143–147. Cambridge: Cambridge University Press.

Wallman, J.F. and M. Adams. 1997. Molecular systematics of Australian carrion-breeding blowflies of the genus *Calliphora* (Diptera: Calliphoridae). *Australian Journal of Zoology* 45: 337–356.

Wallman, J.F. and M. Adams. 2001. The forensic application of allozyme electrophoresis to the identification of blowfly larvae (Diptera: Calliphoridae) in southern Australia. *Journal of Forensic Sciences* 46: 681–684.

Wallman, J.F. and S.C. Donnellan. 2001. The utility of mitochondrial DNA sequences for the identification of forensically important blowflies (Diptera: Calliphoridae) in southeastern Australia. *Forensic Science International* 120: 60–67.

Wallman, J.F., R. Leys, and K. Hogendoorn. 2005. Molecular systematics of Australian carrion-breeding blow-flies (Diptera: Calliphoridae) based on mitochondrial DNA. *Invertebrate Systematics* 19: 1–15.

Waterhouse, D.F. 1939. Temperature preference in the Australian sheep blowfly, *Lucilia cuprina* Wied. *The Australian Journal of Science* 2: 31–32.

Waterhouse, D.F. 1947. The relative importance of live sheep and carrion as breeding grounds for the Australian sheep blowfly *Lucilia cuprina*. *Bulletin of the Council for Scientific and Industrial Research* 217: 1–31.

Waterhouse, D.F. and S.J. Paramonov. 1950. The status of the two species of *Lucilia* (Diptera, Calliphoridae) attacking sheep in Australia. *Australian Journal of Scientific Research Series B* 3: 310–336.

Webber, L.G. 1955. The relationship between larval and adult size of the Australian sheep blowfly *Lucilia cuprina* (Wied.). *Australian Journal of Zoology* 3: 346–353.

Williams, G. 1981. Records of the carrion beetle *Diamesus osculans* Vigor (Silphidae: Coleoptera) from New South Wales. *Australian Entomological Magazine* 8: 47.

Williams, H. 1984. A model for the aging of fly larvae in forensic entomology. *Forensic Science International* 25: 191–199.

Williams, H. and A.M.M. Richardson. 1983. Life history responses to larval food shortages in four species of necrophagous flies (Diptera: Calliphoridae). *Australian Journal of Ecology* 8: 257–263.

Williams, H. and A.M.M. Richardson. 1984. Growth energetics in relation to temperature for larvae of four species of necrophagous flies (Diptera: Calliphoridae). *Australian Journal of Ecology* 9: 141–152.

Woodburn, T.L. and W.G. Vogt. 1982. Attractiveness of merino sheep before and after death to adults of *Lucilia cuprina* (Wiedemann) (Diptera: Calliphoridae). *Journal of the Australian Entomological Society* 21: 131–134.

The World Bank. 2013. Population density (people per sq. km of land area). http://data.worldbank.org/indicator /EN.POP.DNST (accessed April 4, 2013).

Zwick, P. 1979. Contributions to the knowledge of Australian Cholevidae (Catopidae auct.: Coleoptera). *Australian Journal of Zoology, Supplement* 70: 1–56.

PART II

History, Accomplishments, and Challenges of Forensic Entomology in Europe

CHAPTER 7

Introduction to European Chapters

Martin J. R. Hall

Europe has a long and rich history of forensic entomology (Benecke 2001). In many countries, publications from the mid-nineteenth to early twentieth centuries introduced readers to the potential of insect evidence in criminal investigations, for example, in Austria (Von Hofmann 1886), Poland (Horoszkiewicz 1902; Niezabitowski 1902), France (Bergeret 1855; Mégnin 1894), Germany (Reinhard 1882), Spain (Graëlls 1886), and Sweden (Schöyen 1895).

However, there was a general lack of uptake of forensic entomology in European judicial systems and the discipline was not really considered a serious subject for entomological research until the landmark work of Kenneth Smith in 1986, *A Manual of Forensic Entomology* (Smith 1986). This work pulled together many strands of knowledge from a widely dispersed literature and acted as a catalyst to researchers from across Europe and further afield. It alerted a new generation to the potential for using insect evidence in criminal investigations and highlighted the work of scientists, such as Reiter (1984), who, with tools such as the iso-megalendiagram and iso-morphendiagram, were moving forensic entomology from a somewhat anecdotal art to a quantitative science. Since that time, manuals have been published in the German (Amendt et al. 2013; Grassberger and Schmid 2013), French (Charabidze and Gosselin 2014), Italian (Introna and Campobasso 1998; Magni et al. 2008), and Polish (Kaczorowska and Draber-Mońko 2009) languages.

A second catalyst to forensic entomology in Europe was the foundation of the European Association for Forensic Entomology (EAFE) in 2002 (Figure 7.1), following the First European Forensic Entomology Seminar, May 28–30, 2002, at the home of the Institut de Recherche Criminelle de le Gendarmerie Nationale (IRCGN) in Rosny Sous Bois, just outside Paris, France (Gendarmerie National 2002). The meeting was introduced by Marcel Leclercq, seminar president, and was in many ways a testament to his pioneering work in the field of forensic entomology, the casework aspects of which have been recently reviewed (Dekeirsschieter et al. 2013). Since that meeting there have been 11 annual meetings of the EAFE, with the 11th held once again in France, at Lille in April 2014. The 10 intervening meetings were held in: Frankfurt, Germany (2003); London, United Kingdom (2004); Lausanne, Switzerland (2005); Bari, Italy (2006); Brussels, Belgium (2007); Kolymbari, Greece (2008); Uppsala, Sweden (2009); Murcia, Spain (2010); Toruń, Poland (2012); and Coimbra, Portugal (2013). The diversity of locations demonstrates the broad Europe-wide participation in the EAFE. The 12th EAFE meeting in 2015 moves back to the United Kingdom, to the University of Huddersfield, which introduced the first MSc course in forensic entomology in 2014.

European Association for Forensic Entomology

Figure 7.1   The official logo of the European Association for Forensic Entomology, designed by Aïda Gomez and Luisa Diaz Aranda from the Universitad de Alcala de Henares, Madrid, Spain, and adopted in 2005. EAFE©.

The wide membership of the EAFE meant that their approval of the protocols paper authored by the then Board of the EAFE (Amendt et al. 2007) was a robust endorsement with credibility. The small numbers of forensic entomologists in each country make international communication and collaboration through media such as the EAFE of great importance. Few countries have their own forensic entomology groups, an exception being Italy, with the Gruppo Italiano di Entomologia Forense (GIEF) (Chapter 14).

The majority of European participants of the EAFE come from the 11 countries that have hosted meetings, but there are regular contributions from scientists where forensic entomology is in a more fledgling state, such as the Czech Republic, Bulgaria, Hungary, Romania, and Turkey. Emphasis for the future should be on supporting development of forensic entomology in these countries and others in a similar position, such as some of the Baltic states where awareness raising presentations were given to the 6th International Congress of the Baltic Medico-Legal Association, Vilnius, Lithuania in June 2007 (Amendt and Hall 2007; Hall and Amendt 2007). In addition to European members, the EAFE has a flourishing membership from non-European countries on the continents of America, Africa, Asia, and Australia.

Problems facing the growth of forensic entomology in Europe include limited awareness of its value in criminal investigations, even in countries where forensic entomology is practiced routinely such as in the United Kingdom. Although there might be a general awareness of the topic, there is often little awareness of specific issues, such as how to recognize and collect specimens in a manner that can lead to their eventual use as evidence. Modern digital methods to raise awareness and to disseminate information and make it widely available are to be applauded. However, they bring with them a responsibility to be used wisely. For example, software to enable estimation of egg-laying times (*ForenSeek*) (see Chapter 12) is a significant step forward, but users should be careful not to be seduced by statistics and computers, relying on programs to do the thinking and interpretation that a human must do. Investigators must keep their brains engaged to maximize the value of such aids, for example, checking the original sources of developmental data is important to appreciate how the various data sets were gathered.

Another issue, which several of the following chapters mention, is accreditation of forensic entomology, both of facilities and practitioners. The first European forensic entomology laboratory to gain ISO 17025 accreditation was that of the IRCGN in 2007 (Chapter 12). The IRCGN laboratory handles more forensic entomology cases than any other in Europe and continues to strive toward improving quality assurance plans, standard operating procedures, and laboratory best practices. Several other countries are exploring the requirements for accreditation of forensic entomology laboratories (e.g., Belgium, Germany, Spain, and United Kingdom) and for certifying individual practitioners. In the United Kingdom, any forensic entomology expert witness who intends to present their reports in court is recommended to undertake expert witness training, which should cover report writing, presenting evidence in court, and an introduction to criminal law and procedure.

In some countries, for example, those of Scandinavia (Chapter 11) and Austria (Chapter 13), the relatively small number of cases in which a forensic entomology input is required can create a situation where there are few personnel with the necessary expertise and experience to analyze insect evidence. The situation is compounded to an extent by the decline in entomology and taxonomy as core subjects in higher education. One of the first questions in any forensic entomology investigation is "which species have been collected." Differences in the developmental rates of even closely related species (Richards et al. 2009) means that an incorrect identification can send an investigation down the wrong path at an early stage. It is ironic that this situation is highlighted in Sweden (Chapter 11), the birthplace in 1707 of Carl Linnaeus (Carl von Linné after ennoblement) who developed the binomial system of species classification. Linnaeus himself named some species of forensic importance, for example, *Calliphora vomitoria* (Linnaeus, 1758) (Diptera: Calliphoridae) although he originally placed it in the genus *Musca*.

The publicity around high-profile cases that involve entomology can help raise awareness of the subject and stimulate research, such as was recorded in Germany in the 1990s (Chapter 13) and Sweden in 2004 (Chapter 11). A good example of raising public awareness of the roles of insects in decomposition is the "Dood doet leven" (Death brings life) project of Belgium (Chapter 10). However, raising awareness at a public level is not new; back in 1907 Jean-Henri Fabre (see Chapter 12) wrote in a captivating style about "The Greenbottles," "The Grey Flesh-Flies," and "The Bluebottle: The Maggot" in his book translated into English as *The Life of the Fly* (1913).

I am certain that the following chapters on the development of forensic entomology in Europe will show that it is not only the past history of forensic entomology that is rich on the continent, but the present too, with a wide variety of exciting and stimulating research and practice and a continuing expansion of its use within established centers and to new regions. I mentioned the importance of public awareness earlier—*The Life of the Fly* (Fabre 1913) was a significant stimulus to developing my own fascination in insects, especially blow flies, and so it was a genuine thrill and honor to be asked to portray both a blow fly adult and pupa in *A Consilience*, an arts/science video installation for the public created by Jean-Henri Fabre's great-grandson, Jan Fabre (2000) (Figure 7.2). Although seemingly tangential, this is a highly appropriate theme on which to conclude, because forensic entomology is itself a true consilience—a convergence of evidence—linking together principles from different disciplines to form a comprehensive theory.

Figure 7.2 **(See color insert.)** Belgian artist Jan Fabre (a beetle, right) and Natural History Museum, London, entomologist Martin Hall (a fly, left) appearing in *A Consilience*. (From Fabre, J., *A Consilience*, Film Installation by Jan Fabre at The Natural History Museum, London, U.K. 13/01/2000-29/02/2000. Commissioned by the Arts Catalyst, United Kingdom, 2000 [©The Trustees of the Natural History Museum and Angelos bvba / Jan Fabre.]).

REFERENCES

Amendt, J., C. Campobasso, E. Gaudry, C. Beiter, H. Le-Blanc, and M. Hall. 2007. Best practice in forensic entomology: Standards and guidelines. *International Journal of Legal Medicine* 121: 90–104.

Amendt, J. and M.J.R. Hall. 2007. Forensic entomology—Standards and guidelines. *Forensic Science International* 169: S27.

Amendt, J., R. Krettek, G. Nießen, and R. Zehner. 2013. *Forensische Entomologie: Ein Handbuch*. Germany: Verlag für Polizeiwissenschaft.

Benecke, M. 2001. A brief history of forensic entomology. *Forensic Science International* 120: 2–14.

Bergeret, M. 1855. Infanticide: Momification naturelle du cadavre. *Annales d'Hygiène Publique et de Médecine Légale* 4: 442–452.

Charabidze, D. and M. Gosselin. 2014. Insectes, cadavresetscènesdecrime, Prinipes et applications de l'entomologie medico-légale. Belgium: De Boeck.

Dekeirsschieter, J., C. Frederickx, F.J. Verheggen, P. Boxho, and E. Haubruge. 2013. Forensic entomology investigations from Doctor Marcel Leclercq (1924–2008): A review of cases from 1969 to 2005. *Journal of Medical Entomology* 50: 935–954.

Fabre, J. 2000. *A Consilience*. A multi-channel film installation by Jan Fabre with Jan Fabre, Martin Brendell, Ian Gauld, Martin Hall, Rory Post and Dick Vane-Wright at The Natural History Museum, London, U.K. 13/01/2000-29/02/2000. Commissioned by the Arts Catalyst, United Kingdom.

Fabre, J.-H. 1913. *The Life of the Fly*. Translated from the *Souvenirs Entomologiques* by A. Teixeira de Mattos. London: Hodder and Stoughton.

Gendarmerie National. 2002. *Proceedings of the First European Forensic Entomology Seminar*, May 28–30, 2002, p. 167. Rosny sous Bois, France.

Graëlls, M. 1886. Entomología Judicial. *Revista de la Real Academia de Ciencias Exactas, Físicas y Naturales* 1886: 458–447.

Grassberger, M. and H. Schmid. 2013. *Todesermittlung—Befundaufnahme und Spurensicherung. Ein praktischer Leitfaden für Polizei, Juristen und Ärzte*. 2nd Edition. Stuttgart, Germany: Wissenschaftliche Verlagsgesellschaft.

Hall, M.J.R. and J. Amendt. 2007. Forensic entomology—Scientific foundations and applications. *Forensic Science International* 169: S27.

Horoszkiewicz, von S. 1902. Casuistischer Beitrag zur Lehre von der Benagung der Leichen durch Insecten. *Vierteljahrsschrift für gerichtliche Medizin und öffentliches Sanitätswesen* 23: 235–239.

Introna, F. and C.P. Campobasso. 1998. *Entomologia Forense: il ruolo dei ditteri nelle indagini medico-legali*. SBM editore, Noceto (PR), Italy.

Kaczorowska, E. and A. Draber-Mońko. 2009. *Wprowadzenie do entomologii sądowej*. Gdańsk, Poland: Wydawnictwo Uniwersytetu Gdańskiego.

Magni, P., M. Massimelli, R. Messina, P. Mazzucco, and E. Di Luise. 2008. *Entomologia forense. Gli insetti nelle indagini giudiziarie e medico-legali*. Italy: Edizioni Minerva Medica.

Mégnin, P. 1894. La Faune des Cadavres: Application de L'entomologie à la Médecine Légale. Paris, France: Masson.

Niezabitowski, von E. 1902. Experimentelle Beiträge zur Lehre von der Leichenfauna. *Vierteljahrsschrift für gerichtliche Medizin und öffentliches Sanitätswesen* 1: 2–8.

Reinhard, H. 1882. BeitrÄge zur GrÄberfauna. *Verhandlungen der Kaiserlich-KÖniglichen Zoologisch-Botanischen Gesellschaft in Wien* 31: 207–210.

Reiter, C. 1984. Zum Wachstumsverhalten der Maden der blauen Schmeißfliege Calliphora vicina. *Zeitschrift für Rechtsmedizin* 91: 295–308.

Richards C.S., K.L. Crous, and M.H Villet. 2009. Models of development for blowfly sister species *Chrysomya chloropyga. Chrysomya putoria. Medical and Veterinary Entomology* 23: 56–61.

Schöyen, W.M. 1895. Et bidrag til gravenes fauna. *Entomologisk Tidskrift* 5: 121–124.

Smith, K.G.V. 1986. *A Manual of Forensic Entomology*. London: British Museum (Natural History).

Von Hofmann, E. 1886. Observation de larves de Diptéres sur des cadavres exhumés. *Bulletin ou comptes rendus des séances de la Société entomologique de Belgique* 74: 131–132.

CHAPTER 8

Poland

Daria Bajerlein, Szymon Konwerski, Szymon Matuszewski, and Krzysztof Szpila

CONTENTS

8.1 TURN OF THE NINETEENTH AND TWENTIETH CENTURY

The history of forensic entomology in Poland begins in the nineteenth century. The first published information on necrophilous fauna and their role in the process of carrion decomposition was in 1889. A Lvov professor, Leopold Wajgel, in his article entitled *O faunie grobów ludzkich* ("On the fauna of human graves"), presented research conducted by the German medical doctor Hermann Reinhard and the French pathologist Jean-Pierre Mégnin and emphasized the need for further studies on entomofauna of graves and its application in medicolegal investigations (Wajgel 1889). However, the true pioneers of forensic entomology in Poland were two medical examiners from the Institute of Forensic Medicine at the Jagiellonian University in Cracow, Stefan von Horoszkiewicz and Eduard von Niezabitowski, who published the first case report and experimental studies on insects attracted to carrion (Von Horoszkiewicz 1902; Von Niezabitowski 1902).

In 1899, Stefan von Horoszkiewicz, while performing an autopsy of a child, became interested in numerous abrasions that were visible on a substantial portion of the body. The mother of the child informed Horoszkiewicz that she had not noticed any abrasions earlier, only that the body was covered by many cockroaches. Horoszkiewicz decided to test experimentally whether these insects were the cause of the abrasions. For the study, he used small pieces of human tissue and put them in glass containers with cockroaches. It turned out that the insects were feeding on the tissue and, only after the skin had dried, a lot of abrasions were seen on the skin (Von Horoszkiewicz 1902).

Eduard von Niezabitowski contributed to the development of forensic entomology as the first Polish scientist to design an experiment detailing necrophilous entomofauna attracted to carcasses of various types. The aim of his study was to answer the following questions: (1) How much time is necessary for insects to decompose a human corpse or animal carrion outdoors? (2) What insects and in what sequence visit the carrion outdoors? (3) Are human corpses and animal carrion visited by different species of insects? (4) Is the carrion fauna specific to the season? (5) Are there any differences in carrion fauna between corpses decomposing in human settlements and in natural habitats? In his study, he used cat, fox, mole, rat, and calf carcasses. He also placed aborted human fetuses on the windowsill of the Institute and in the nearby garden (Niezabitowski 1902). Most importantly, he concluded that human corpses and animal carcasses are colonized by similar fauna. Moreover, he noticed that the process of carrion decomposition proceeds faster in summer as compared to spring and autumn and that this process is strongly influenced by insect activity, ambient temperature, and humidity. He found only slight seasonal differences in insect composition, but noticed that various habitat types (anthropogenic and natural) are characterized by specific entomofauna. Niezabitowski observed also that large and small carrion differ in the number of necrophilous insects that infest them.

8.2 TWENTIETH CENTURY

In the twentieth century, most studies related to necrophilous insects were focused on more general ecological questions. Some research was concerned directly with the role of invertebrates in the process of carrion decomposition (Kopyłówna 1935; Mroczkowski 1949; Błażejewski 1956; Nabagło 1973), but none emphasized forensics. Maciej Mroczkowski, one of the most famous Polish entomologists, a specialist in the taxonomy of skin beetles (Dermestidae) and carrion beetles (Silphidae), made a series of observations in 1948 concerning colonization of cat and squirrel carcasses by various species of the genus *Nicrophorus* Fabricius and *Neonicrophorus* Hatch (Coleoptera: Silphidae). He noticed that particular species differed in their numbers and appearance time on the carcasses (Mroczkowski 1949).

Natalja Kopyłówna was the first scientist to conduct more detailed experiments on beetle succession and carrion decomposition under various environmental conditions (Kopyłówna 1935). The main part of her study was conducted in the summer of 1930 and complementary observations were made in the spring and summer of 1931. In the study, carcasses of various types and size (mice, moles, frogs, fishes, birds, piglets, cats, and dogs) were exposed on 89 sites in four types of habitat: (1) mixed forest, (2) meadows and fields, (3) parks and scrubs by the water, and (4) sandy hills and dunes. As a result, she distinguished eight stages of carrion decomposition named as fresh carrion, slightly decomposed carrion, slightly deformed carrion, completely deformed carrion, drying carrion, completely dried carrion, bones with a small amount of dried tissue, and bones completely cleaned of tissue. She noticed that the process of carrion decomposition is continuous and differences between particular stages of decomposition are often difficult to observe. Moreover, she noticed that the rate of carrion decomposition depends on ambient temperature and rainfall. Kopyłówna provided ecological characteristics of beetle species from the families Silphidae, Ptiliidae, Staphylinidae, Histeridae, Cleridae, Dermestidae, Nitidulidae, and Scarabaeidae, with an emphasis on their preferences toward food, carrion size and type, stage of decomposition, and type of habitat. Furthermore, Kopyłówna discussed possibilities of using beetles for postmortem interval (PMI) determination.

Another entomologist who observed beetle succession and carrion decomposition, with special attention to factors affecting these processes, was Franciszek Błażejewski (1956). Błażejewski conducted his study from July 1950 to June 1951 in 63 sites partitioned between open meadow and forests. To study necrophilous beetles, he used carcasses of various types (fishes, birds, mammals, molluscs, and reptiles) as well as pieces of wild boar skin and the internal organs of cows. Błażejewski collected 2735 beetles belonging to 65 species and representing 17 families. The most abundant were carrion beetles (Silphidae), dung beetles (Aphodiidae, Geotrupidae), rove beetles

(Staphylinidae), and clown beetles (Histeridae). Błażejewski showed that the abundance and species composition of necrophilous beetles varied between seasons, habitats, and stage of decomposition. He recognized five stages of decomposition as follows: fresh carrion, slightly decomposed carrion, slightly deformed carrion, entirely deformed carrion, and drying carrion. In addition, he described beetle preferences toward a particular stage of decomposition and made observations on the time of their appearance on carrion. He did not record any relationships between carcass weight and the appearance of a given beetle species or any preferences of beetles toward particular carrion type.

Another study on this subject was set up by Leszek Nabagło, an entomologist from the Polish Academy of Sciences in Cracow. In the spring and summer of 1971 and winter of 1971/1972, he conducted field experiments designed to provide information on insect succession and decomposition of exposed and buried bank vole, *Myodes glareolus* Schreber carcasses in a forest habitat (Nabagło 1973). Nabagło distinguished three stages of carrion decomposition—preparatory, active decomposition, and residual—and observed a similar pattern of decomposition for all carcasses examined. He determined that the rate of carrion decomposition was strongly influenced by season, carcass location (buried/exposed), temperature, and humidity. Nabagło concluded that the duration of the decomposition process was strongly influenced by the duration of the preparatory stage, which was dependent on temperature. According to his observations, the rate of decomposition of buried carcasses was more influenced by humidity than temperature. Lower temperatures were associated with reduced insect activity, but high humidity accelerated the decomposition. As for the insect succession, Nabagło showed that the composition and number of insect species varied over the decompositional process and according to the carrion treatment. The greatest number of species was recorded in the active decomposition stage and the species diversity decreased with the loss of carrion mass. Nabagło also provided information on the temporal patterns of species presence/absence on carrion.

In summary, the studies carried out by Polish entomologists in the twentieth century, although not designed with forensic entomology in mind, represented the first experiments to reveal information on insect succession, carrion decomposition, and factors affecting these processes. However, it should be stressed that these studies focused mainly on necrophilous beetles with flies receiving little attention. Apart from successional studies, a few review articles concerning entomological methods of PMI estimation and carrion fauna succession were published in the second half of the twentieth century (Żółtowski 1953; Piotrowski 1981, 1990).

8.3 TWENTY-FIRST CENTURY

A renewed interest in forensic entomology was noted in Poland at the beginning of the twenty-first century. At that time, review papers (Kaczorowska et al. 2002; Matuszewski 2004a,b; Matuszewski et al. 2008a; Skowronek and Chowaniec 2010), case reports (Kaczorowska et al. 2003, 2004; Barzdo et al. 2007; Żydek et al. 2007), and methods for collecting and rearing insects (Kaczorowska 2002) were published.

In the beginning of the twenty-first century, the first forensically oriented experiments on insect succession and pig carrion decomposition were conducted (Matuszewski et al. 2008b, 2010b,c, 2011). The aim of the research was to study decomposition and insect succession on exposed pig carcasses according to season (spring, summer, and autumn) and forest type (pine-oak forest, hornbeam-oak forest, and alder forest), and eventually to provide entomological tools applicable to Polish cases that required PMI estimates.

It was found that season and forest type significantly affected decomposition of medium-sized exposed carcasses (mean mass: 25.8 kg). Decomposition, which was mosaic as a rule, manifested with putrefaction, active and advanced decay. Moreover, it was demonstrated that active decay was not always driven by blow fly (Diptera: Calliphoridae) larvae; instead in some conditions, it was associated with larvae of the carrion beetle, *Necrodes littoralis* (Linnaeus)

(Coleoptera: Silphidae). A later onset of all decompositional processes was found in spring with the rate of active decay being much higher in summer than spring or autumn. Pig carcasses decomposed faster in alder forest than in pine-oak forest or hornbeam-oak forest (Matuszewski et al. 2010b). The study revealed that carcasses may be visited by a high number of taxa; however, only a subset of these species is important from a forensic perspective. It was found that the amount of time that adult dipterans were present was shorter than for adult coleopterans. Residencies of the adult stages were fragmented (with absences clumped in the final part of the presence period or evenly distributed within the presence period), whereas residencies of the larval stages were continuous. The appearance of many taxa was closely related to the onset of bloating. Based on the length of the presence period and the regularity of the appearance on carrion, taxa have been divided into four groups of usefulness (low, moderate, high, and unknown) for the succession-based estimation of PMI (Matuszewski et al. 2010c).

The previously discussed research showed significant differences in insect occurrence on carrion between seasons, forest types, and years of the study. As for residence time, only season significantly affected most insect taxa. Furthermore, particular taxa appeared on carrion in similar sequence in different seasons, forest types, and years. The study eventually gave rise to general seasonal models of insect succession on medium-sized, exposed pig carcasses in forests of Central Europe (Matuszewski et al. 2011).

As demonstrated earlier in this chapter, insect succession studies on carrion provide unique opportunities to address ecologically based questions associated with necrophilous fauna. For example, seasonality, habitat preferences, and residency of histerid beetles on exposed pig carrion were examined in various forest types (Bajerlein et al. 2011). In this particular study, the abundance of adults of *Margarinotus* Marseul (Coleoptera: Histeridae) was influenced by both season and forest type, whereas abundance of *Saprinus* Erichson (Coleoptera: Histeridae) only by season. Most *Margarinotus* species were found in a hornbeam-oak forest and their abundance was relatively low and constant for the whole decomposition process. In contrast, a clear peak in abundance of *Saprinus* was found during an early stage of decomposition.

One of the newest achievements of Polish forensic entomology is the development of a novel approach for estimating the preappearance interval (PAI) of carrion insects (Matuszewski 2011, 2012). It was found for *N. littoralis* (Matuszewski 2011), *Creophilus maxillosus* (Linnaeus, 1758) (Coleoptera: Staphylinidae) (Matuszewski 2012), and many other species of carrion beetles (Matuszewski and Szafałowicz 2013) that adult and larval PAI is strongly, inversely related to daily temperatures averaged for the duration of PAI. Consequently, it was proposed that PAI could be estimated from temperature and several methods useful for that purpose were validated (Matuszewski 2011, 2012). Interestingly, such relationships appeared to be only partially true in the case of carrion flies (Matuszewski et al. 2013a).

Entomological methods for PMI estimation can be used in legal practice only if forensic entomologists, crime scene technicians, and lawyers cooperate. It is essential to train non-entomologists how to recognize, collect, and preserve entomological evidence to protect the integrity of the evidence. Therefore, in 2010, a three-part catalogue of insects useful for PMI estimation of remains located in Polish forests was published (Matuszewski 2010; Matuszewski and Szpila 2010; Matuszewski et al. 2010a). The first part of the catalogue introduces the reader to the entomological methods for PMI estimation, lists beetle and fly species of forensic importance, and discusses the general content of the catalogue (Matuszewski 2010). The second and third part are dedicated to fly and beetle species with regard to when developmental stages appear on carrion based on seasonality and reoccurrence, location on remains, and proper techniques for collecting and preserving specimens (Matuszewski and Szpila 2010; Matuszewski et al. 2010a).

Proper species identification is the key first step in any forensic entomology investigation; therefore, studies on the morphology of forensically important insects are of special importance. In the last few years, many papers concerning the morphology of forensically important Diptera were published by Polish entomologists. Of special importance are a large monograph summarizing the morphology, taxonomy, and distribution of Polish blow flies (Draber-Mońko 2004) and articles on morphology and keys for identification of European, Mediterranean, and African blow flies of forensic importance (Szpila 2010, 2012; Szpila and Villet 2011; Szpila et al. 2013a,b, 2014). Morphological studies of forensically important dipterans focused also on latrine flies (Fannidae) (Grzywacz et al. 2012), houseflies (Muscidae) (Velásquez et al. 2013), and flesh flies (Sarcophagidae) (Draber-Mońko et al. 2009).

There has also been an increase in the use of molecular methods to identify insect species of forensic importance. Scientists from the Polish Academy of Sciences in Warsaw succeeded in using DNA barcoding for the identification of *Sarcophaga argyrostoma* (Robineau-Desvoidy) (Diptera: Sarcophagidae), a species of high forensic importance (Draber-Mońko et al. 2009). Moreover, they successfully developed a new method for the identification of blow flies by coupling the real-time polymerase chain reaction and high-resolution melting analysis (Malewski et al. 2010).

In 2009, forensic entomology was popularized in Poland through the publication of the first Polish manual of forensic entomology (Kaczorowska and Draber-Mońko 2009). The book, titled *Wprowadzenie do entomologii sądowej* ("An introduction to forensic entomology"), covers a broad range of topics from the history of forensic entomology to the use of DNA analysis in forensic entomology. The book also provides checklists of insects recorded on particular types of carcasses in Poland.

In 2010, a surprising phenomenon of buried vertebrate carrion colonized by flesh flies representing the sarcophagid subfamily Miltogramminae was described (Szpila et al. 2010). Descriptions of the unique abilities of the first instars of these flies to penetrate dry soil in desert or sand dune habitats were given in a series of field and laboratory experiments.

Insects may sometimes indicate the postmortem relocation of a corpse. Recently, several carrion insects were found to demonstrate a strong preference toward open habitats as opposed to forest habitats (Matuszewski et al. 2013b). It was found that *Lucilia sericata* (Meigen) (Diptera: Calliphoridae); *Dermestes frischii* Kugelann; *Dermestes laniarius* (Illiger) (Coleoptera: Dermestidae); *Omosita colon* (Linnaeus); some species of *Nitidula* Fabricius (Coleoptera: Nitidulidae); and *Necrobia rufipes* (De Geer) (Coleoptera: Cleridae) breed exclusively in carrion exposed in open habitats. Consequently, these species are perfect candidates for indicators of corpse relocation from rural open to rural forest habitats of Central Europe.

8.4 TRAINING COURSES AND CONFERENCES IN FORENSIC ENTOMOLOGY

Recently, forensic entomology has gained greater appeal to investigators and students alike. Currently, courses in forensic entomology are offered to students at the Adam Mickiewicz University in Poznań, Jagiellonian University in Cracow, Nicolaus Copernicus University in Toruń, University of Gdańsk, and University of Lodz. Moreover, between the years 2010 and 2013, Krzysztof Szpila and Andrzej Grzywacz (Nicolaus Copernicus University) organized six international identification workshops on forensically important Diptera.

8.5 ONGOING PROJECTS

Field experiments dealing with insect succession during carrion decomposition under various conditions are currently being conducted by researchers from the Adam Mickiewicz University in Poznań and Nicolaus Copernicus University in Toruń. Initial projects were concerned with the effect of carcass mass and clothing on decomposition, residency of insects, and development of immature stages of selected species useful for PMI estimation. Other projects examined carrion decomposition and insect succession on exposed pig carcasses in blackthorn hedges in different seasons. Research projects concerned with carrion decomposition and insect succession on hanging pig carcasses in a forest habitat in different seasons are also underway. Experiments on mite (Acari) succession on exposed carcasses are being conducted as well. In addition to field projects, other forensically oriented studies are conducted in Poland, for example, a new project on the estimation of PAI (Adam Mickiewicz University) or a project on the morphology of preimaginal stages of European Calliphoridae and Muscidae of forensic importance (Nicolaus Copernicus University).

8.6 USE OF FORENSIC ENTOMOLOGY IN LEGAL INVESTIGATIONS

Although, a number of courses are offered, workshops are conducted for biology students, lawyers, and police personnel, and literature in forensic entomology is being published, the use of entomological methods for PMI estimation is still not widespread in Poland. It is worth noting that currently in Poland there is no institution responsible for the accreditation of expert witnesses in forensic entomology. Many entomologists are neither keen to be expert witnesses nor deal with the practical aspects of collecting evidence from corpses. As a result, entomology experts do not need to be licensed to testify in court but, at present, only a few individuals do so in Poland. Moreover, up to now, only eight cases have been published (Kaczorowska et al. 2003, 2004; Barzdo et al. 2007; Żydek et al. 2007).

It seems that presently the greatest hurdle preventing the application of forensic entomology in Poland is limited cooperation and communication between entomologists, police forces, and practitioners of forensic medicine. Crime scene technicians and medical examiners are not trained, or are trained only occasionally, how to recognize, collect, and preserve entomological evidence. In routine forensic casework, insect specimens are usually not collected and as a result forensic entomology is rather infrequently used. However, due to the increasing interest in the field, globally and in Poland, the future is bright for the discipline and its application in criminal investigations.

REFERENCES

Bajerlein, D., S. Matuszewski, and S. Konwerski. 2011. Insect succession on carrion: Seasonality, habitat preference and residency of histerid beetles (Coleoptera: Histeridae) visiting pig carrion exposed in various forests (Western Poland). *Polish Journal of Ecology* 59: 787–97.

Barzdo, M., L. Żydek, M. Michalski, E. Meissner, and J. Berent. 2007. Część II. Wykorzystanie metod entomologicznych do oceny czasu zgonu—opis przypadków (Part II. The use of entomological methods in determination of the time of death—case presentations). *Archiwum Medycyny Sądowej i Kryminologii* 57: 351–54.

Błażejewski, F. 1956. Chrząszcze trupożerne rezerwatu cisowego Wierzchlas (Necrophagous beetles of the Wierzchlas yew-tree preserve). *Zeszyty Naukowe Uniwersytetu Mikołaja Kopernika w Toruniu, Biologia* 1: 63–90.

Draber-Mońko, A. 2004. Calliphoridae—Plujki (Insecta: Diptera) (Calliphoridae—blowflies (Insecta: Diptera)). Fauna Polski, 23, Warszawa, Poland: MiIZ PAN.

Draber-Mońko, A., T. Malewski, J. Pomorski, M. Łoś, and P. Ślipiński. 2009. On the morphology and mitochondrial DNA barcoding of the flesh fly *Sarcophaga (Liopygia) argyrostoma* (Robineau-Desvoidy, 1830) (Diptera: Sarcophagidae)—An important species in forensic entomology. *Annales Zoologici* 59: 465–93.

Grzywacz, A., T. Pape, and K. Szpila. 2012. Larval morphology of the lesser housefly, *Fannia canicularis. Medical and Veterinary Entomology* 26: 70–82.

Kaczorowska, E. 2002. Zbieranie i hodowanie owadów nekrofagicznych, istotnych w odtwarzaniu daty śmierci metodą entomologiczną (Collecting and rearing necrophagous insects, important in determining date of death, basic on the entomological method). *Archiwum Medycyny Sądowej i Kryminologii* 52: 343–50.

Kaczorowska, E. and A. Draber-Mońko. 2009. *Wprowadzenie do entomologii sądowej (An introduction to forensic entomology)*. Gdańsk: Wydawnictwo Uniwersytetu Gdańskiego.

Kaczorowska, E., D. Pieśniak, and Z. Szczerkowska. 2002. Entomologiczne metody określania czasu śmierci (Entomological methods of determining time of death). *Archiwum Medycyny Sądowej i Kryminologii* 52: 305–12.

Kaczorowska, E., D. Pieśniak, and Z. Szczerkowska. 2003. The use of blowfly larvae (Diptera: Calliphoridae) in attempt at determination of the human body drowning time—a case history from Poland. *Polish Journal of Entomology* 72: 343–48.

Kaczorowska, E., D. Pieśniak, and Z. Szczerkowska. 2004. Wykorzystanie metod entomologicznych w próbach określania daty zgonu—opis przypadków (The use of entomological methods in attempts at determination of the time of death—case studies). *Archiwum Medycyny Sądowej i Kryminologii* 54: 169–76.

Kopyłówna, N. 1935. Z badań nad chrząszczami nekrotycznemi pow. dziśnieńskiego (Aus den Untersuchugen an nekrotischen Käfern im Bezirk Disna). *Prace Towarzystwa Przyjaciół Nauk w Wilnie* 9: 1–37.

Malewski, T., A. Draber-Mońko, J. Pomorski, M. Łoś, and W. Bogdanowicz. 2010. Identification of forensically important blowfly species (Diptera: Calliphoridae) by high-resolution melting PCR analysis. *International Journal of Legal Medicine* 124: 277–85.

Matuszewski, S. 2004a. Entomoskopia (Entomoscopy). In *Doctrina multiplex, veritas una. Księga jubileuszowa ofiarowana Profesorowi Mariuszowi Kulickiemu*, eds A. Bulsiewicz, A. Marek, and V. Kwiatkowska-Darul, 271–80. Toruń, Poland: Wydawnictwo UMK.

Matuszewski, S. 2004b. Przedmiot ekspertyzy entomoskopijnej (The matter of an expertise in entomoscopy). In *Czynności procesowo-kryminalistyczne w polskich procedurach. Materiały z konferencji naukowej i IV Zjazdu Katedr Kryminalistyki*, ed. V. Kwiatkowska-Darul, 109–116. Toruń, 5–7 maja 2004. Toruń, Poland: Wydawnictwo UMK.

Matuszewski, S. 2010. Katalog owadów przydatnych do ustalania czasu śmierci w lasach Polski. Część 1: Wprowadzenie (Catalogue of insects useful in time of death estimation in forests of Poland. Part I: Introduction). *Problemy Kryminalistyki* 267: 5–17.

Matuszewski, S. 2011. Estimating the pre-appearance interval from temperature in *Necrodes littoralis* L. (Coleoptera: Silphidae). *Forensic Science International* 212: 180–88.

Matuszewski, S. 2012. Estimating the preappearance interval from temperature in *Creophilus maxillosus* L. (Coleoptera: Staphylinidae). *Journal of Forensic Sciences* 57: 136–45.

Matuszewski, S., D. Bajerlein, and S. Konwerski. 2010a. Katalog owadów przydatnych do ustalania czasu śmierci w lasach Polski. Część 3: Chrząszcze (Insecta: Coleoptera) (Catalogue of insects useful in time of death estimation in forests of Poland. Part 3: Insecta (Coleoptera)). *Problemy Kryminalistyki* 269: 5–21.

Matuszewski, S., D. Bajerlein, S. Konwerski, and K. Szpila. 2008a. Entomologia sądowa w Polsce (Forensic entomology in Poland). *Wiadomości Entomologiczne* 27: 49–52.

Matuszewski, S., D. Bajerlein, S. Konwerski, and K. Szpila. 2008b. An initial study of insect succession and carrion decomposition in various forest habitats of Central Europe. *Forensic Science International* 180: 61–69.

Matuszewski, S., D. Bajerlein, S. Konwerski, and K. Szpila. 2010b. Insect succession and carrion decomposition in selected forests of Central Europe. Part 1: Pattern and rate of decomposition. *Forensic Science International* 194: 85–93.

Matuszewski, S., D. Bajerlein, S. Konwerski, and K. Szpila. 2010c. Insect succession and carrion decomposition in selected forests of Central Europe. Part 2: Composition and residency patterns of carrion fauna. *Forensic Science International* 195: 42–51.

Matuszewski, S., D. Bajerlein, S. Konwerski, and K. Szpila. 2011. Insect succession and carrion decomposition in selected forests of Central Europe. Part 3: Succession of carrion fauna. *Forensic Science International* 207: 150–63.

Matuszewski, S. and M. Szafałowicz. 2013. Temperature-dependent appearance of forensically useful beetles on carcasses. *Forensic Science International* 229: 92–99.

Matuszewski, S., M. Szafałowicz, and A. Grzywacz. 2013a. Temperature-dependent appearance of forensically useful flies on carcasses. *International Journal of Legal Medicine*. doi: 10.1007/s00414-013-0921-9.

Matuszewski, S., M. Szafałowicz, and M. Jarmusz. 2013b. Insects colonising carcasses in open and forest habitats of Central Europe: Search for indicators of corpse relocation. *Forensic Science International* 231: 234–39.

Matuszewski, S. and K. Szpila. 2010. Katalog owadów przydatnych do ustalania czasu śmierci w lasach Polski. Część 2: Muchówki (Insecta: Diptera) (Catalogue of insects useful in time of death estimation in forests of Poland. Part 2: Insecta: Diptera). *Problemy Kryminalistyki* 268: 26–38.

Mroczkowski, M. 1949. Uwagi o kolejnym pojawianiu się gatunków z rodzaju *Nicrophorus* Fabr. i *Neonicrophorus* Hetch. (*Col. Silphidae*) (Comments on the succession of the species from the genus *Nicrophorus* Hetch. (Col. Silphidae).) *Polskie Pismo Entomologiczne* 19: 196–99.

Nabagło, L. 1973. Participation of invertebrates in decomposition of rodent carcasses in forest ecosystems. *Ekologia Polska* 21: 251–70.

Piotrowski, F. 1981. Metoda entomologiczna w określaniu czasu zgonu (Entomological method in determination of the time of death). *Problemy Kryminalistyki* 150: 203–05.

Piotrowski, F. 1990. *Zarys entomologii parazytologicznej*. Warszawa, Poland: PWN.

Skowronek, R. and C. Chowaniec. 2010. Polska entomologia sądowa—rys historyczny, stan obecny i perspektywy na przyszłość (Polish forensic entomology—the past, present and future perspectives). *Archiwum Medycyny Sądowej i Kryminologii* 60: 55–58.

Szpila, K. 2010. Key for the identification of third instars of European blowflies (Diptera: Calliphoridae) of forensic importance. In *Current Concepts in Forensic Entomology*, eds J. Amendt, C.P. Campobasso, M.L. Goff, and M. Grassberger M, 43–56. Dordrecht/Heidelberg/London/New York: Springer.

Szpila, K. 2012. Key for identification of European and Mediterranean blowflies (Diptera, Calliphoridae) of medical and veterinary importance—adult flies. In *Forensic Entomology, An Introduction*, 2nd edition, ed. D. Gennard, 77–81 ss + plates 5.1–5.9. West Sussex, United Kingdom: Willey-Blackwell.

Szpila, K., M.J.R. Hall, T. Pape, and A. Grzywacz. 2013b. Morphology and identification of first instars of the European and Mediterranean blowflies of forensic importance. Part II: Luciliinae. *Medical and Veterinary Entomology* 27: 349–66.

Szpila, K., M.J.R. Hall, K.L. Sukontason, and T.I. Tantawi. 2013a. Morphology and identification of first instars of the European and Mediterranean blowflies of forensic importance. Part I: Chrysomyinae. *Medical and Veterinary Entomology* 27: 181–93.

Szpila, K., T. Pape, M.J.R. Hall, and A. Mądra. 2014. Morphology and identification of first instars of European and Mediterranean blowflies of forensic importance. Part III: Calliphorinae. *Medical and Veterinary Entomology* 28(2): 133–42. doi: 10.1111/mve.12021.

Szpila, K. and M. Villet. 2011. Morphology and identification of first instar larvae of African blowflies (Diptera: Calliphoridae) commonly of forensic importance. *Journal of Medical Entomology* 48: 738–52.

Szpila, K., J.G. Voss, and T. Pape. 2010. A new dipteran forensic indicator in buried bodies. *Medical and Veterinary Entomology* 24: 278–83.

Velásquez, Y., T. Ivorra, A. Grzywacz, A. Martínez-Sánchez, C. Magaña, A. García-Rojo, and S. Rojo. 2013. Larval morphology, development and forensic importance of *Synthesiomyia nudiseta* (Diptera: Muscidae) in Europe: A rare species or just overlooked? *Bulletin of Entomological Research* 103: 98–110.

Von Horoszkiewicz, S. 1902. Casuistischer Beitrag zur Lehre von der Benagung der Leichen durch Insecten (A case report concerning the feeding of insects on human corpses). *Vierteljahrsschrift für gerichtliche Medizin und öffentliches Sanitätswesen* 23: 235–39.

Von Niezabitowski, E. 1902. Experimentelle Beiträge zur Lehre von der Leichenfauna (Experimental studies on the cadaver fauna). *Vierteljahrsschrift für gerichtliche Medizin und öffentliches Sanitätswesen* 1: 2–8.

Wajgel, L. 1889. O faunie grobów ludzkich (On the fauna of human graves). *Kosmos* 14: 344–46.

Żółtowski, Z. 1953. *Entomologia sanitarna*, Tom I. Warszawa, Poland: Wydawnictwo Ministerstwa Obrony Narodowej.

Żydek, L., M. Barzdo, M. Michalski, E. Meissner, and J. Berent. 2007. Część I. Wykorzystanie metod entomologicznych do oceny czasu zgonu—opis przypadków (Part I. The use of entomological methods in determination of the time of death—case presentations). *Archiwum Medycyny Sądowej i Kryminologii* 57: 347–50.

CHAPTER 9

United Kingdom

Martin J.R. Hall, Amoret P. Whitaker, and Andrew J. Hart

CONTENTS

9.1 HISTORICAL PERSPECTIVES

On September 29, 1935, a young woman enjoying a holiday walk near the village of Moffat, Scotland, about 80 km southeast of Glasgow, was shocked to see a human arm protruding from a bundle on the steep banks of a small river flowing through a ravine known locally as the Devil's Beef Tub. The discovery sparked a major criminal investigation in which, for the first time in the United Kingdom, a significant contribution to the casework was made by the recovery of insect evidence and its interpretation through the science of forensic entomology. Blow fly (Diptera: Calliphoridae) larvae were collected from many of the 70 plus human body parts extracted from several bundles found on the riverbank. The larvae were identified as *Calliphora vicina* Robineau-Desvoidy by Dr. Alexander Mearns of the Institute of Hygiene, University of Glasgow (Glaister and Brash 1937; Mearns 1939). His estimate of the larval age corroborated other evidence of the likely date of deposition, September 16, 1935, of the dismembered bodies of the common-law wife and maid of Dr. Buck Ruxton, a doctor based in Lancaster, over 160 km to the south. Although not used in the trial, the larval aging supported the jigsaw of evidence that led to the conviction of Ruxton for double murder and to his execution in 1936. The case has rightly become famous for the insect evidence, among entomologists at least (Smith 1986; Erzinçlioglu 2000), and some of the larval specimens are retained at the Natural History Museum (NHM), London. More general books on the contribution of forensic science to solving cases discuss the Ruxton case in terms of the use of evidence such as fingerprints, odontology, and photographic superimposition (Owen 2000; Goodman 2005) and overlook the insect evidence. This is counterbalanced by the inclusion of this evidence in a more recent television documentary, *Forensic Firsts* (*Catching Killers: Insect Evidence*, Smithsonian Channel 2012).

Despite the successful use of forensic entomology in the Ruxton case in the United Kingdom as early as 1935, the next major milestone did not appear until more than 50 years later when Ken Smith published the first *Manual of Forensic Entomology* (Smith 1986). Smith's book brought together a previously scattered literature and gave scientific credibility to the fledgling field of forensic entomology. The somewhat anecdotal manner in which forensic entomology had been applied in the intervening years is exemplified by the "The Lydney Murder—a Riddle of Maggots," a chapter in the autobiography of the eminent pathologist Professor Keith Simpson (1980) who wrote of his identifying and aging fly larvae at a crime scene without the aid of microscope or any temperature measurements, although he did recover samples for verification.

Overlapping the period of activity of Ken Smith was that of Dr. Zakaria Erzinçlioğlu, based at Cambridge University, who received funding for his research and casework on forensic entomology from the Home Office, the U.K. government department responsible for justice. He not only produced some fine scientific papers, especially on the taxonomy of blow flies (Erzinçlioğlu 1985a), but also wrote several book chapters and a book that synthesized the available information (Erzinçlioğlu 1996a,b, 2000). After withdrawal of government funding, he continued in the forensic entomology field and more widely, looking into miscarriages of justice, until his untimely death in 2002.

The growth of university undergraduate courses involving some aspect of forensic sciences was dramatic toward the end of the twentieth century in the United Kingdom, and seeing an evident gap in the literature available for that level Dr. Dorothy Gennard wrote a popular introduction to forensic entomology (Gennard 2007). At the same time, Dr. Alan Gunn published an undergraduate textbook covering the broader subject of forensic biology, including a chapter on invertebrates, which covered forensic entomology (Gunn 2007). Ken Smith's manual was written while he was employed at the NHM, a natural home for forensic science because every investigation of objects recovered from a crime scene starts with their identification. With a collection of approximately 30 million insect specimens and a broad range of staff expertise, the NHM is well placed to provide accurate insect identifications and engage in casework supported by research. Many forensic science providers (FSPs) package entomology under the heading of forensic ecology, and a first textbook to combine the many elements of that discipline includes a chapter on forensic entomology authored by NHM staff (Hall et al. 2012). This takes the reader from crime scene to court and follows an earlier publication on practical field protocols (Hall et al. 2008).

9.2 RESEARCH

9.2.1 Introduction

Before 1986, research on forensic entomology in the United Kingdom was limited to work on improving the identification of all stages of insects of forensic significance (Erzinçlioğlu 1985a) and to general ecological studies of carrion fauna (Chapman and Sankey 1955; Lane 1975; Smith 1975). Academic studies on general blow fly biology (Cragg 1955) including imaginal ecdysis (Cottrell 1962) also had specific forensic applications. Ken Smith's landmark publication was a significant stimulus to research on forensic entomology, not only in the United Kingdom but also worldwide. It also stimulated reviews from other U.K. entomologists to raise general awareness of the subject (Erzinçlioğlu 1989b,c; Turner 1987, 1991). U.K. research on forensic entomology since 1986 is summarized below, in sections 9.2.2 through 9.2.4.

9.2.2 Taxonomy

Taxonomic studies have concentrated on morphology (Erzinçlioğlu 1985a, 1989a) and, increasingly, the use of molecular methods to identify insects of forensic importance (Stevens and Wall 2001; Smith and Godfrey 2009) including single-nucleotide polymorphisms (SNPs) in a 6-plex SNaPshot™ assay (Smith and Godfrey 2011). Henry Disney (1994) established himself as the world authority on the taxonomy of Phoridae (Diptera) and has been involved in many studies of their application to forensic entomology (Disney 2005, 2006), especially in winter periods when Calliphoridae is largely absent and Phoridae could have an important role (Disney and Manlove 2005; Manlove and Disney 2008; Cuttiford and Disney 2010). Even when parasitized, the value of phorids and other insects as forensic evidence should not be overlooked (Disney and Munk 2004).

The study of mites (Acari) as forensic indicators has been limited because of the difficult taxonomy of the group and the difficulty in assigning a correct identification to such fauna, but the potential for forensic acarology was reassessed in a special edition of *Experimental and Applied Acarology* (Braig and Perotti 2009; Perotti et al. 2009; Perotti and Braig 2009; Turner 2009). The value of mites was demonstrated by a reanalysis of one of the landmark cases of Jean Pierre Mégnin, from 1878, taking into account new studies of the biology of the *Tyrophagus* species (Astigmata: Acaridae) involved (Perotti 2009). However, forensic acarology is likely to remain a specialist area, not commonly applied in casework.

9.2.3 Ecology

Studies on the ecology of carrion have ranged from descriptive studies of which species occurred in different sites and different degrees of shade (Isiche et al. 1992), to more analytical studies relating species distribution to habitat (Smith and Wall 1999; Hwang and Turner 2005, 2009), to studies of which species could exploit carrion in different concealed environments, for example, buried carrion (Gunn and Bird 2011) and carrion within the very specific, casework-focused location of zipped suitcases (Bhadra et al. 2014). The effects of surface exposure or burial, before and after insect colonization, on the decomposition rate of carrion confirmed the overwhelming importance of insect activity in this process (Bachmann and Simmons 2010; Simmons et al. 2010a,b). Fly seasonality and adult attraction to carrion baits has been extensively studied in relation to the blow fly *Lucilia sericata* (Meigen) (Wall et al. 1992, 1993) and could be usefully applied to other species of forensic importance, to know when and where any species is likely to colonize a human body or, indeed, a live human or animal in cases of neglect. The attraction of myiasis-causing flies such as *L. sericata* to livestock overlaps with the attraction of carrion flies to dead bodies and has been reviewed (Hall 1995). The rarity of nocturnal flight activity of blow flies was supported by laboratory studies (Wooldridge et al. 2007). LeBlanc conducted a PhD study to identify the specific odors that attract adults of *Calliphora vicina* Robineau-Desvoidy (Diptera: Calliphoridae) to pig carrion (LeBlanc 2008), and the value of pig models as proxies for humans has been confirmed in studies by Turner and Wiltshire (1999) and with a pig–human comparison carried out in the United States by Whitaker (2014).

9.2.4 Physiology

The most frequent use of forensic entomology in the United Kingdom is in the estimation of minimum postmortem interval (PMI) (in 72% of >160 cases undertaken by the NHM) (M.J.R. Hall and A.P. Whitaker, unpublished data). Therefore, studies on improving estimation of insect age are of immense value. Davies and Ratcliffe (1994) undertook a series of rearing experiments at low temperatures that tend, even now, to be overlooked in developmental studies. Donovan et al. (2006) developed a nonlinear model for the larval development of the blow fly *C. vicina*, which is the most common species encountered on human cadavers in the United Kingdom. The model was based on entire populations of larvae and also just on the upper three quartiles, to reflect the practical situation at a scene, where evidence collection concentrates on the largest blow fly larvae. The model, derived from laboratory studies at constant temperatures, was validated using data from larvae developing under fluctuating field conditions. Similar developmental studies are under way on several other species to increase the applicability of forensic entomology in U.K. casework (Richards et al. 2013a). The use of accumulated degree hours (ADHs) is commonly applied in developmental models with a lower development threshold (LDT), below which larval development is arrested. However, the notion of a fixed LDT has been questioned (Ames and Turner 2003) along with constant ADH requirements regardless of temperature (Amendt et al. 2007). Clearly, the aging of fly larvae generally assumes that the oldest stages were deposited as eggs (except for members of the family Sarcophagidae that deposit larvae)—this might not always be the case because of the deposition of precocious larvae (Erzinçlioğlu 1990). Although the proportion of these is low (<3%) (Davies and Harvey 2012), their potential for impacting PMI estimates should be considered.

Many factors have been shown to subtly influence the rate of development of fly larvae, including the origin of the tissues consumed, even different organs from the same carrion species (Kaneshrajah and Turner 2004; Clark et al. 2006; Ireland and Turner 2006) and the stage of decomposition of the food source (Richards et al. 2013c). The potential effects of variable dispersal periods of postfeeding larvae on calculating PMIs were considered by Arnott and Turner (2008). Westgarth-Smith and Hereward (1989) devised an "automatic maggot counter" for studying the timing of dispersal of larvae from carrion, but it was not used further.

Although entomotoxicology has not been a major area for U.K. studies, research has demonstrated the effects of drugs such as paracetamol on larval development (O'Brien and Turner 2004). The value of fly larvae in providing evidence of drugs from the tissues they feed on was studied by Sadler et al. (1995, 1997). They showed how drug concentrations fell significantly on cessation of feeding and at pupariation and recommended sampling actively feeding larvae (Sadler et al. 1995). However, they also showed that drug concentrations in larvae were significantly lower than those in their food source and, therefore, concluded that the absence of a drug even from feeding larvae would not necessarily imply its absence from the food source (Sadler et al. 1997). Many other more academic studies have potential forensic impact, including the exhaustive investigation of the control of diapause led by Saunders (2001).

The application of aging techniques to fly larvae frequently requires studies of preserved specimens. Preservation techniques have been shown to have a significant effect on the size of larvae, and Adams and Hall (2003) recommended a method for best preservation, that is, kill larvae in hot water (>80°C) for 30 seconds and then preserve in 80% ethanol. Richards et al. (2013b) extended this research by studying changes in the length of all larval instars up to 1 year after larval preservation. Brown et al. (2012) made a similar analysis of the effects of preservation techniques on blow fly pupae.

The pupal stage occupies about 50% of the period from egg laying to adult emergence and, therefore, knowledge of pupal development could be critical in estimating minimum PMI. Several studies have addressed pupal aging, from observation of external and internal morphological changes using standard microscopical and histological techniques (Davies and Harvey 2013) to the use of micro–computed tomography scanning, a technique that mimics the virtual autopsy of a human (Dirnhofer et al. 2006) but applies it to a fly pupa (Richards et al. 2012). The potential for using cuticular hydrocarbons in the identification and aging of insect evidence is an area of current interest (Moore et al. 2013), with the application of artificial neural network analysis (Butcher et al. 2013). This will be of particular value if it can be applied to empty fly puparia as a tool to determine when the adult fly emerged.

The potential effects of increased larval mass temperature on larval development were first raised by Turner and Howard (1992), who noted temperatures in masses up to 27°C above ambient in a temperate U.K. environment. The effect of these on larval development is still being researched, using noninvasive thermal imaging techniques to measure heat (Hall and Brandt 2006a,b) (Figure 9.1).

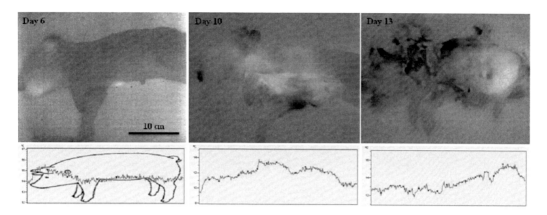

Figure 9.1   (See color insert.) Decomposition of a 1-kg piglet by approximately 730 larvae of *Calliphora vicina* recorded by thermal imaging. The infestation started in the oral cavity, with raised temperatures first becoming visible on Day 6. The larval mass gradually moved toward the rear of the piglet, with larval exodus being completed on Day 14. The temperature across a line from the snout (left) to the rump (right) of the piglet is shown in the graph below each thermal image.

9.3 UNITED KINGDOM CASEWORK

Although some early case reports discuss insect evidence (Holden and Camps 1952; Easton and Smith 1970; Erzinçlioğlu 1985b), there is a relative scarcity of publications on U.K. forensic entomology casework (Erzinçlioğlu 2000). Hart et al. (2008) discussed three case examples of how forensic entomology can be applied. Two of the cases demonstrated how insects can aid in minimum PMI estimation when normal pathology methods are insufficient. For instance, when low temperatures dramatically slow down the rate of decomposition or when fire damage to a body interferes with the normal processes of decomposition. As elsewhere, the role of blow flies as agents of myiasis on live bodies has enabled them to be used as indicators of neglect of humans and animals in the United Kingdom (Hart et al. 2008; Amendt et al. 2011; Hall et al. 2012). Although the analysis of insect evidence on bodies is the most common application of forensic entomology, the ubiquity of insects in nature, hence at crime scenes, means that there are many other types of investigation to which insect evidence can contribute. For example, U.K. entomologists assisted with the identification of insects found in cannabis plant material seized in New Zealand (Crosby et al. 1986). Knowledge of the natural distributions of these insects helped to identify the origin of the consignment to the Tenasserim region of Myanmar (Burma), and the defendants in the case changed their plea from not guilty to guilty of importation (Crosby et al. 1986). The broader value of forensic entomology is shown by its application to archaeological situations, to explore hypotheses concerning the internment of remains (Adams 1990; Panagiotakopulu and Buckland 2012).

9.4 FORENSIC ENTOMOLOGY SERVICES IN THE UNITED KINGDOM

Forensic entomology in the United Kingdom is offered by independent experts, subcontractors working for FSPs, or "in-house" by private companies. If a body is found that is likely to require the services of an entomologist, a crime scene manager will contact a forensic entomologist either directly or via an FSP (Buckley and Langley 2012). The crime scene managers will usually maintain their own list of contacts or access a database of experts held by the National Crime Agency (NCA). To be listed on the database requires a formal application from experts with references and examples of past casework. Forensic entomology has been combined with other "ecological" services (Márquez-Grant and Roberts 2012), including, for example, palynology, anthropology, diatom and soil analysis, and archaeology, so that an integrated and coordinated response can be offered by the provider. This increases their ease of use for the police as they become a "one-stop shop" of ecological services, which can work in synergy with more mainstream forensic services like DNA profiling and blood pattern analysis. A prime example of this was the "Natural Justice Unit" developed over a period of years at the former Forensic Science Service, where forensic entomology formed the backbone of a range of ecological specialisms that could be utilized by the police in their investigations. The majority of the cases investigated during the lifetime of the Natural Justice Unit involved an aspect of entomology. Currently in the United Kingdom, there are integrated ecological services that include forensic entomology on offer from several independent providers and one public body, which can undertake casework for the prosecution and the defense. However, given the relative rarity of this evidence type, it is unusual for a forensic scientist to be employed by the police or by an FSP solely to practice entomology casework; instead, the majority subcontracts the analysis of insect evidence to entomologists working independently or in other institutes such as universities and museums.

9.5 FUTURE

Forensic entomology evidence in the United Kingdom may eventually form part of the Streamlined Forensic Reporting (SFR) process, which was designed by the Metropolitan Police Service as a response to the Criminal Justice initiative on more effective case management. Its purpose is to deliver forensic evidence proportionate to the needs of the Criminal Justice System with its primary aims being the reduction of costs and delay associated with forensic evidence where such evidence adds no value to the administration of justice, compliance with the regularly revised Criminal Procedure Rules of the U.K.'s Ministry of Justice, and for more effective case management with the early identification of issues (Templeton 2012). One of the key benefits of SFR is as a tool enabling the courts, and specifically the judge, to clearly identify any forensic issues that need to be addressed before a case goes to trial. Similarly, SFR gives the defendant an opportunity to comment on any forensic evidence at the earliest opportunity. Disagreements involving forensic entomology, for example, in a case where the minimum time since death estimate differed from a defendant's account as to when he last saw the victim alive, could be addressed by the interested parties at the pretrial stage using the SFR approach. SFR is currently being implemented within the Criminal Justice System on a national basis.

In 1997, Erzinçlioğlu announced that he would only carry out future forensic work if paid by the judiciary. Explaining his decision, he claimed that under the adversarial system, "incompetent and dishonest forensic scientists" were undermining Britain's Criminal Justice System (Erzinçlioğlu 1998). Similarly, Disney (2002, 2011) shared concerns over low standards in forensic entomology. Practices in forensic science in the United Kingdom are now being standardized to raise confidence in the validity of expert evidence in the Criminal Justice System (Kershaw 2009). Quality assurance will be provided by audits and accreditation, the latter involving third-party assessment of competence to carry out specific tasks, with competence being defined as the evidence of skills, knowledge, and understanding to carry out the role over time. The United Kingdom Accreditation Service (UKAS) accredits and assesses work in the forensic science sector. In 2008, a U.K. government Home Office Forensic Science Regulator (2011) was appointed to oversee quality, and the "Codes of Practice and Conduct" have been developed to assist with this process. These codes are based on the International Standard ISO/IEC 17025:2005, which details the general requirements for the competence of testing and calibration laboratories. FSPs in the United Kingdom are required to meet these standards, and their practices are regularly audited by UKAS. However, historically, niche services such as forensic entomology have tended to fall outside this remit because of their specialist nature and relatively small number of practitioners. Attempts are being made to address this, and eventually guidelines relating to forensic entomology may be included within the codes of the forensic regulator and accreditation could include input from professional bodies such as the Forensic Science Society and the Royal Entomological Society (Hart and Hall 2012). There will be potential challenges around the accreditation of forensic entomology, which include validating the techniques used by each practitioner or organization; achieving, demonstrating, and maintaining competence; and evaluating the strength of the entomological evidence. Further pressure on forensic scientists, including entomologists, to provide sound expert evidence was provided by the United Kingdom's Supreme Court in 2011, when it overturned the long-standing rule that expert witnesses are immune from liability for damages to parties that have engaged them and to which they owe a duty of care.

Forensic entomology is an excellent, practical example of the societal value of scientific studies and, as such, has been used in the school curriculum of the United Kingdom (Hall et al. 2006) and in outreach to schools through the educational media (Hall and Brandt 2006a; Whitaker and Hall 2008). The general raising of awareness of the value of forensic entomology both for professionals (Hall 2000; Hart and Whitaker 2006; Hart et al. 2008; Hall et al. 2012) and for the general public can only benefit the development of forensic entomology and its contribution to the Criminal Justice System of the United Kingdom.

REFERENCES

Adams, R.G. 1990. *Dermestes leechi* Kalik (Coleoptera: Dermestidae) from an Egyptian mummy. *Entomologist's Gazette* 41: 119–120.

Adams, Z.J.O. and M.J.R. Hall. 2003. Methods used for the killing and preservation of blowfly larvae, and their effect on post-mortem larval length. *Forensic Science International* 138: 50–61.

Amendt, J., C. Campobasso, E. Gaudry, C. Reiter, H. LeBlanc, and M. Hall. 2007. Best practice in forensic entomology—standards and guidelines. *International Journal of Legal Medicine* 121: 90–104.

Amendt, J., C.C. Richards, C.P. Campobasso, R. Zehner, and M.J.R. Hall. 2011. Forensic entomology: Applications and limitations. *Forensic Science, Medicine and Pathology* 7: 379–392.

Ames, C. and B. Turner. 2003. Low temperature episodes in development of blowflies: Implications for post-mortem interval estimation. *Medical and Veterinary Entomology* 17: 178–186.

Arnott, S. and B. Turner. 2008. Post-feeding larval behaviour in the blowfly, *Calliphora vicina*: effects on post-mortem interval estimates. *Forensic Science International* 177: 162–167.

Bachmann, J. and T. Simmons. 2010. The influence of preburial insect access on the decomposition rate. *Journal of Forensic Sciences* 55: 893–900.

Bhadra, P., A. Hart, and M.J.R. Hall. 2014. Factors affecting accessibility of bodies disposed in suitcases to blowflies. *Forensic Science International* 239: 62–72.

Braig, H.R. and M.A. Perotti. 2009. Carcasses and mites. *Experimental and Applied Acarology* 49: 45–84.

Brown, K., A. Thorne, and M. Harvey. 2012. Preservation of *Calliphora vicina* (Diptera: Calliphoridae) pupae for use in post-mortem interval estimation. *Forensic Science International* 223: 176–183.

Buckley, R. and A. Langley. 2012. Aspects of Crime Scene Management. In *Forensic Ecology Handbook: From Crime Scene to Court*, eds. N. Márquez-Grant and J. Roberts, pp. 7–21. Chichester, United Kingdom: Wiley.

Butcher J., H. Moore, C. Day, C. Adam, and F. Drijfhout. (2013). Artificial neural network analysis of hydrocarbon profiles for the ageing of *Lucilia sericata* for post mortem interval estimation. *Forensic Science International* 232: 25–31.

Chapman R.F. and J.H.P. Sankey. 1955. The larger invertebrate fauna of three rabbit carcasses. *Journal of Animal Ecology* 24: 395–402.

Clark, K., L. Evans, and R. Wall. 2006. Growth rates of the blowfly, *Lucilia sericata*, on different body tissues. *Forensic Science International* 156: 145–149.

Cottrell, C.B. 1962. General observations on the imaginal ecdysis of blowflies. *Transactions of the Royal Entomological Society of London* 114: 317–333.

Cragg, J.B. 1955. The natural history of sheep blowflies in Britain. *Annals of Applied Biology* 42: 197–207.

Crosby, T.K., J.C. Watt, A.C. Kistemaker, and P.E. Nelson. 1986. Entomological identification of the origin of imported cannabis. *Journal of the Forensic Science Society* 26: 35–44.

Cuttiford, L.A., and R.H.L. Disney. 2010. Colonisation of pig carrion by *Triphleba* Rondani (Dipt., Phoridae) during a very cold British winter. *Entomologist's Monthly Magazine* 146: 1754–1759.

Davies, K. and M. Harvey. 2012. Precocious egg development in the blowfly *Calliphora vicina*: implications for developmental studies and post-mortem interval estimation. *Medical and Veterinary Entomology* 26: 300–306.

Davies, K. and M. L. Harvey. 2013. Internal morphological analysis for age estimation of blow fly pupae (Diptera: Calliphoridae) in postmortem interval estimation. *Journal of Forensic Sciences* 58: 79–84.

Davies, L. and G.G. Ratcliffe. 1994. Development rates of some pre-adult stages in blowflies with reference to low temperatures. *Medical and Veterinary Entomology* 8: 245–254.

Dirnhofer, R., C. Jackowski, P. Vock, K. Potter, and M.J. Thali. 2006. VIRTOPSY: Minimally invasive, imaging-guided virtual autopsy. *Radiographics* 26: 1305–1333.

Disney, R.H.L. 1994. *Scuttle Flies: The Phoridae*. London: Chapman and Hall.

Disney, R.H.L. 2002. Fraudulent forensic scientists. *Journal de Médecine Légale droit Médical* 45: 225–230.

Disney, R.H.L. 2005. Duration of development of two species of carrion-breeding scuttle flies and forensic implications. *Medical and Veterinary Entomology* 19: 229–235.

Disney, R.H.L. 2006. Duration of development of some Phoridae (Dipt.) of forensic significance. *Entomologist's Monthly Magazine* 142: 129–138.

Disney, R.H.L. 2011. Forensic science is not a game. *Pest Technology* 5: 16–22.

Disney, R.H.L. and J.D. Manlove. 2005. First occurrence of the phorid, *Megaselia abdita* in forensic cases in Britain. *Medical and Veterinary Entomology* 19: 489–491.

Disney, R.H.L. and T. Munk. 2004. Potential use of Braconidae (Hymenoptera) in forensic cases. *Medical and Veterinary Entomology* 18: 442–444.

Donovan, S.E., M.J.R. Hall, B.D. Turner, and C.B. Moncrieff. 2006. Larval growth rates of the blowfly, *Calliphora vicina*, over a range of temperatures. *Medical and Veterinary Entomology* 20: 106–114.

Easton, A.M. and K.G.V. Smith. 1970. The entomology of the cadaver. *Medicine, Science and the Law* 10: 208–215.

Erzinçlioğlu, Y.Z. 1985a. Immature stages of British *Calliphora* and *Cynomya*, with a re-evaluation of the taxonomic characters of larval Calliphoridae (Diptera). *Journal of Natural History* 19: 69–96.

Erzinçlioğlu, Y.Z. 1985b. The entomological investigation of a concealed corpse. *Medicine, Science and the Law* 25: 228–230.

Erzinçlioğlu, Y.Z. 1989a. The value of chorionic structure and size in the diagnosis of blowfly eggs. *Medical and Veterinary Entomology* 3: 281–285.

Erzinçlioğlu, Y.Z. 1989b. Entomology and the forensic scientist: How insects can solve crimes. *Journal of Biological Education* 23: 300–302.

Erzinçlioğlu, Y.Z. 1989c. Entomology, zoology and forensic science: The need for expansion. *Forensic Science International* 43: 209–213.

Erzinçlioğlu, Y.Z. 1990. On the interpretation of maggot evidence in forensic cases. *Medicine, Science and the Law* 30: 65–66.

Erzinçlioğlu, Y.Z. 1996a. Entomological investigation of the scene. In *Suspicious Death Scene Investigation*, eds. P. Vanezis and A. Busuttil, pp. 89–101. London: Arnold.

Erzinçlioğlu, Y.Z. 1996b. *Blowflies*. Slough, United Kingdom: Richmond.

Erzinçlioğlu, Y.Z. 1998. British forensic science in the dock. *Nature* 392: 859–860.

Erzinçlioğlu, Y.Z. 2000. *Maggots, Murder and Men*. Colchester, United Kingdom: Harley Books.

Forensic Science Regulator. 2011. Codes of Practice and Conduct for Forensic Science Providers and Practitioners in the Criminal Justice System. Birmingham, United Kingdom: Forensic Science Regulator.

Glaister, J. and J.C. Brash. 1937. *Medico-Legal Aspects of the Ruxton Case*. Edinburgh, United Kingdom: E. and S. Livingstone.

Goodman, J. 2005. The Jigsaw murder case (Dr Buck Ruxton, UK 1935). In *The Giant Book of Murder*, ed. R. Wilkes, pp. 432–447. London: Magpie Books.

Gunn, A. 2007. *Essential Forensic Biology*. Chichester, United Kingdom: Wiley.

Gunn, A. and J. Bird. 2011. The ability of the blowflies *Calliphora vomitoria* (Linnaeus), *Calliphora vicina* (Rob-Desvoidy) and *Lucilia sericata* (Meigen) (Diptera: Calliphoridae) and the muscid flies *Muscina stabulans* (Fallén) and *Muscina prolapsa* (Harris) (Diptera: Muscidae) to colonise buried remains. *Forensic Science International* 207: 198–204.

Hall, A., M. Reiss, C. Rowell, and A. Scott (Eds.). 2006. *Salters-Nuffield Advanced Biology A2 Student Book*. Oxford: Heinemann Educational Publishers.

Hall, M.J.R. 1995. Trapping the flies that cause myiasis: Their responses to host-stimuli. *Annals of Tropical Medicine and Parasitology* 89: 333–357.

Hall, M.J.R. 2000. On maggots and murders – forensic entomology. *Spotlight: The Journal of the National Crime Faculty* 3: 5–8.

Hall, M.J.R. and A.P. Brandt. 2006a. Forensic entomology. *Science in School* 2: 49–53.

Hall, M.J.R. and A.P. Brandt. 2006b. The use of thermal imaging to study the effect of larval masses on the development of blowfly larvae. [Abstract] Page 33 in *Program and Abstracts of the Fourth Meeting of the European Association for Forensic Entomology*, 26th–29th April 2006, Bari, Italy.

Hall, M.J.R., T. Brown, P. Jones, and D. Clark. 2008. Forensic sciences. In *The Scientific Investigation of Mass Graves: Towards Protocols and Standard Operating Procedures*, eds. M. Cox, A. Flavel, I. Hanson, and J. Laver, pp. 463–497. USA: Cambridge University Press.

Hall, M., A. Whitaker, and C. Richards. 2012. Forensic entomology. In *Forensic Ecology Handbook: From Crime Scene to Court*, eds. N. Márquez-Grant and J. Roberts, pp. 111–140. Chichester, United Kingdom: Wiley.

Hart, A.J. and M.J.R. Hall. 2012. Towards accreditation of forensic entomology in the United Kingdom [abstract]. In *Proceedings of the 10th Annual Meeting of the North American Forensic Entomology Association*; 2012 Jul 17–19; Las Vegas, Nevada: NAFEA, 2012. Session 1: #3.

Hart, A.J. and A.P. Whitaker. 2006. Forensic entomology. *Antenna* 30: 159–164.

Hart, A.J., A.P. Whitaker, and M.J.R. Hall. 2008. The use of forensic entomology in criminal investigations: How it can be of benefit to SIOs. *The Journal of Homicide and Major Incident Investigation* 4: 37–47.

Holden, H.S. and F.E. Camps. 1952. The investigation of some human remains found at Fingringhoe. *The Police Journal* July-September: 8.

Hwang, C.C. and B.D. Turner. 2005. Spatial and temporal variability of necrophagous Diptera from urban to rural areas. *Medical and Veterinary Entomology* 19: 379–391.

Hwang, C.C. and B.D. Turner. 2009. Small-scaled geographical variation in life-history traits of the blowfly *Calliphora vicina* between rural and urban populations. *Entomologia Experimentalis et Applicata* 132: 218–224.

Ireland, S. and B. Turner. 2006. The effects of larval crowding and food type on the size and development of the blowfly, *Calliphora vomitoria Forensic Science International* 159: 175–181.

Isiche, J., J.E. Hillerton, and F. Nowell. 1992. Colonization of the mouse cadaver by flies in southern England. *Medical and Veterinary Entomology* 6: 168–170.

Kaneshrajah, G. and B. Turner. 2004. *Calliphora vicina* larvae grow at different rates on different body tissues. *International Journal of Legal Medicine* 118: 242–244.

Kershaw, A. 2009. Professional standards, public protection and the administration of justice. In *Handbook of Forensic Science*, eds. J. Fraser and R. Williams, pp. 546–571. Cullompton, Devon, United Kingdom: Willan Publishing.

Lane, R.P. 1975. An investigation into blowfly (Diptera: Calliphoridae) succession on corpses. *Journal of Natural History* 9: 581–588.

LeBlanc, H.N. 2008. Olfactory stimuli associated with the different stages of vertebrate decomposition and their role in the attraction of the blowfly *Calliphora vomitoria* (Diptera: Calliphoridae) to carcasses. PhD Thesis, University of Derby, United Kingdom.

Manlove, J.D. and R.H.L. Disney. 2008. The use of *Megaselia abdita* (Diptera: Phoridae) in forensic entomology. *Forensic Science International* 175: 83–84.

Márquez-Grant, N. and J. Roberts (Eds.). 2012. *Forensic Ecology Handbook: From Crime Scene to Court*. Chichester, United Kingdom: Wiley.

Mearns, A.G. 1939. Larval infestation and putrefaction. In *Recent Advances in Forensic Medicine*, 2nd Edition, ed. S. Smith, pp. 250–255. London: Churchill.

Moore, H.E., C.D. Adam, and F.P. Drijfhout. 2013. Potential use of hydrocarbons for aging *Lucilia sericata* blowfly larvae to establish the postmortem interval. *Journal of Forensic Sciences* 58: 404–412.

O'Brien, C. and B. Turner. 2004. Impact of paracetamol on *Calliphora vicina* larval development. *International Journal of Legal Medicine* 118: 188–189.

Owen, D. 2000. *Hidden Evidence*. London: Quintet Publishing.

Panagiotakopulu, E. and P.C. Buckland. 2012. Forensic archaeoentomology: An insect fauna from a burial in York Minster. *Forensic Science International* 221: 125–130.

Perotti, M.A. 2009. Mégnin re-analysed: The case of the newborn baby girl, Paris, 1878. *Experimental and Applied Acarology* 49: 37–44.

Perotti, M.A. and H.R. Braig. 2009. Phoretic mites associated with animal and human decomposition. *Experimental and Applied Acarology* 49: 85–124.

Perotti, M.A., M.L. Goff, A.S. Baker, B.D. Turner, and H.R. Braig. 2009. Forensic acarology: An introduction. *Experimental and Applied Acarology* 49: 3–13.

Richards, C., C. Rowlinson, and M. Hall. 2013a. Unlocking the secrets of *Lucilia caesar*—A new instrument in the forensic entomology toolkit. [Abstract] Page 22 in 10th Meeting of the European Association for Forensic Entomology, April 10–13, 2013, Coimbra, Portugal.

Richards, C.S., C.C. Rowlinson, L. Cuttiford, R. Grimsley, and M.J.R. Hall. 2013c. Decomposed liver has a significantly adverse affect on the development rate of the blowfly *Calliphora vicina International Journal of Legal Medicine* 127: 259–262.

Richards, C.S., C.C. Rowlinson, and M.J.R. Hall. 2013b. Effects of storage temperature on the change in size of *Calliphora vicina* larvae, during preservation in 80% ethanol. *International Journal of Legal Medicine* 127: 231–241.

Richards, C.S., T.J. Simonsen, R.L. Abel, M.J.R. Hall, D.A. Schywn, and M. Wicklein. 2012. Virtual Forensic Entomology: Improving estimates of minimum post-mortem interval with 3D micro-computed tomography. *Forensic Science International* 220: 251–264.

Sadler, D.W., C. Fuke, F. Court, and D.J. Pounder. 1995. Drug accumulation and elimination in *Calliphora vicina* larvae. *Forensic Science International* 71: 191–197.

Sadler, D.W., L. Robertson, G. Brown, C. Fuke, and D.J. Pounder. 1997. Barbiturates and analgesics in *Calliphora vicina* larvae. *Journal of Forensic Sciences* 42: 481–485.

Saunders, D.S. 2001. The blow fly *Calliphora vicina*: A "clock-work" insect. In *Insect Timing: Circadian Rhythmicity to Seasonality*, eds. D.L. Denlinger, J.M. Giebultowicz, and D.S. Saunders, pp. 1–14. Amsterdam, the Netherlands: Elsevier.

Simmons, T., R.E. Adlam, and C. Moffatt. 2010a. Debugging decomposition data: Comparative taphonomic studies and the influence of insects and carcass size on decomposition rate. *Journal of Forensic Sciences* 55: 8–13.

Simmons, T., P.A. Cross, R.E. Adlam, and C. Moffatt. 2010b. The influence of insects on decomposition rate in buried and surface remains. *Journal of Forensic Sciences* 55: 889–892.

Simpson, K. 1980. *Forty Years of Murder*. London, United Kingdom: Granada Publishing Limited [First published by George G. Harrap and Company Limited, 1978].

Smith, J. and H. Godfrey. 2009. The application of genetics in forensic entomology. *Antenna* 33: 81–86.

Smith, J. and H. Godfrey. 2011. A SNaPshot™ assay for the identification of forensically important blowflies. *Forensic Science International: Genetics Supplement Series* 3: e479–e480.

Smith, K.E. and R. Wall. 1997. The use of carrion as breeding sites by the blowfly *Lucilia sericata* and other Calliphoridae. *Medical and Veterinary Entomology* 11: 38–44.

Smith, K.G.V. 1975. The faunal succession of insects and other invertebrates on a dead fox. *Entomologist's Gazette* 26: 277–287.

Smith, K.G.V. 1986. *A Manual of Forensic Entomology*. London, United Kingdom: British Museum (Natural History).

Stevens, J. and R. Wall. 2001. Genetic relationships between blowflies (Calliphoridae) of forensic importance. *Forensic Science International* 120: 116–123.

Templeton, H. 2012. *SFR Briefing Note*. London: Metropolitan Police Service.

Turner, B. 2009. Forensic entomology: A template for forensic acarology? *Experimental and Applied Acarology* 49: 15–20.

Turner, B. and T. Howard. 1992. Metabolic heat generation in dipteran larval aggregations: A consideration for forensic entomology. *Medical and Veterinary Entomology* 6: 179–181.

Turner, B. and P. Wiltshire. 1999. Experimental validation of forensic evidence: A study of the decomposition of buried pigs in a heavy clay soil. *Forensic Science International* 101: 113–122.

Turner, B.D. 1987. Forensic entomology: Insects against crime. *Science Progress* 71: 133–144.

Turner, B.D. 1991. Forensic entomology. *Forensic Science Progress* 5: 129–151.

Wall, R., N. French, and K.L. Morgan. 1992. Effects of temperature on the development and abundance of the sheep blowfly *Lucilia sericata* (Diptera: Calliphoridae). *Bulletin of Entomological Research* 82: 125–131.

Wall, R., N. French, and K.L. Morgan. 1993. Predicting the abundance of the blowfly *Lucilia sericata* (Diptera: Calliphoridae). *Bulletin of Entomological Research* 83: 431–436.

Westgarth-Smith, A.R. and C. Hereward. 1989. An automatic maggot counter for use in blowfly development studies. *Medical and Veterinary Entomology* 3: 323–324.

Whitaker, A.P. 2014. Development of blowflies (Diptera: Calliphoridae) on pig and human cadavers: Implications for forensic entomology casework. PhD Thesis. King's College, University of London.

Whitaker, A.P. and M.J.R. Hall. 2008. Forensic entomology: Using insects to solve crime. *Catalyst* 19: 1–3.

Wooldridge, J., L. Scrase, and R. Wall. 2007. Flight activity of the blowflies, *Calliphora vomitoria Lucilia sericata*, in the dark. *Forensic Science International* 172: 94–97.

CHAPTER 10

Belgium, the Netherlands, and Luxembourg

Luc Bourguignon, Yves Braet, Françoise Hubrecht, and Sofie Vanpoucke

CONTENTS

10.1 HISTORY AND GEOGRAPHY

Belgium, the Netherlands, and Luxembourg are three parliamentary monarchies located in the middle of Europe. The history of these rather small territories has been intertwined for several centuries. Historically known as the Low Countries, they were alternatively under the domination of several other countries such as France, Austria, and Spain. The borders we know today were progressively fixed during the course of the nineteenth century.

In 1921, Belgium and Luxembourg signed an economic treaty, often considered the first step toward the construction of Benelux—an economic convention between Belgium, the Netherlands, and Luxembourg facilitating trade by abolishing internal custom taxes and harmonizing custom taxes for importations (1944). As a logical consequence of these first economic treaties, came the desire to establish deeper judicial cooperation. In June of 1962, a treaty was signed between the King of Belgium, the Queen of the Netherlands, and the Great Duchess of Luxembourg, establishing a common set of rules for extradition and judicial cooperation. This friendship is shown today by the good relationship these three countries have regarding the field of forensics, including forensic entomology.

10.2 FORENSIC ENTOMOLOGY IN BELGIUM

10.2.1 History

In Belgium, during the first half of the twentieth century, virtually no legal doctors or police officers understood the potential importance of carrion flies (Diptera). Nevertheless, for decades the French-speaking scientific world had been very active in forensic entomology in France with scientists such as Bergeret, Mégnin, Lacassagne, and Balthazar (Cherix and Wyss 2006).

The modern forensic entomology had its beginning in Belgium in 1947 when a medical student named Marcel Leclercq (1924–2008) displayed his interest in dipterology and was given attention by his professors. When he was asked to give an opinion on a blow fly (Diptera: Calliphoridae) sample recovered from the corpse of a child (Leclercq 2009), based on experiments conducted fortuitously a few years before with his brother Jean Leclercq, he concluded that the eggs of these flies were probably deposited on the body soon after the child was reported missing. The police investigation later confirmed this hypothesis (Leclercq J. 2008). From then on, Marcel Leclercq balanced his job as a general physician (until 1994, at the age of 70 years), his own entomological studies (he was a renowned specialist of hematophagous Diptera, particularly the family Tabanidae), and the demands of the judicial world. Charles Verstraeten from the University of Gembloux collaborated with Leclercq for almost 20 years assisting when expertise on nondipteran insects of forensic importance was important to a case. In 1948, Leclercq regularly published scientific articles and case reports on forensic entomology. A complete list of his articles can be found in the study by Leclercq J. (2008, 2009), and Dekeirsschieter et al. (2013) reviewed several cases in which Leclercq was involved between 1969 and 2005.

Leclercq's influence can still be witnessed in forensic entomology. Starting in 1992, he acted as a consultant and taught his knowledge of forensic entomology to two young scientists from the French Gendarmerie (see Chapter 12 on France for description of the agency). Around the year 2002, under the influence of the French Gendarmerie, the first European forensic entomology seminar took place in Paris. Many scientists from all over Europe and overseas attended this conference. This conference was one of the last public appearances of Marcel Leclercq before his death in 2008.

10.2.2 Current Trends

The laboratory Microtraces and Entomology, located in Brussels, which is hosted by the National Institute for Criminalistics and Criminology (NICC), offers expertise and scientific research in a variety of forensic fields to the Belgian judicial community. The laboratory studies and investigates trace evidence of natural origin, including human and animal hair, plants, soil, diatoms, and insects.

In 1993, the laboratory established contact and exchanged information with Marcel Leclercq. Due to the pressure of casework at the time, namely the infamous Dutroux case, the laboratory effectively began to work in forensic entomology on a regular basis in 2003. The laboratory participated in the entomology seminar in Paris, and it is one of the founding members of the European Association for Forensic Entomology (EAFE), which held the fifth meeting of the association in Brussels in May 2007.

The laboratory is actively involved in both research and casework, and as of 2013 it has examined 150+ human cases and a several animal cases.

10.2.2.1 Casework

Belgium is divided into several judicial districts. Each judicial district has a scientific police team that performs some acts and observations of scientific order, seizures on crime scenes, and sampling of trace evidence. Regarding entomology, a sampling kit designed by the Microtraces and Entomology Laboratory has been provided to each judicial district with instructions to return evidence to the laboratory within 24 hours of the sample collection. The members of the Microtraces and Entomology Laboratory are available on call for 24 × 7, and the police laboratories often call the lab directly for instructions on how to sample and what to look for at the body recovery site. Many legal doctors are also aware of the potential and usefulness of entomological studies and will suggest to prosecutors associated with cases involving such evidence. These close collaborations are made possible by relatively small size of the country, which can be crossed within approximately 2 hours. The laboratory examines annually 15–20 entomology cases.

The vast majority of entomological casework being performed today involves minimum postmortem interval estimations on human corpses, as well as animal carcasses (game, cattle). The minimum postmortem interval (PMI) estimation is based on the accumulated degree days (ADD) concept, with a particular emphasis on climatic parameters. Correlations are made between temperature measured at the death scene and a dataset from a local weather station. The date of oviposition is calculated on this basis, and the minimum PMI is then estimated and discussed, taking into account other climatic parameters (e.g., rain, wind) or circumstances (e.g. access to the body made difficult for the insects, presence/absence of wounds) to adjust the time interval during which the death may have occurred (Bourguignon 2011; Bourguignon et al. 2013).

Particular emphasis is also placed on collecting all the species present on the corpse, which often results in extensive sampling and massive rearing. This method allows for the capture of specimens present in low numbers, which is often beneficial as the estimation of the minimum PMI can be realized simultaneously with several species, reinforcing the accuracy of the final result.

10.2.2.2 Research and Development

10.2.2.2.1 Measures of ADD

During the early years (2003–2006), most of the research conducted had been focusing on succession studies of arthropods on carrion. Many experiments with pig remains were conducted over a 3-year period, covering a wide variety of realistic situations (e.g., body left in the open air, wrapped in packaging, hanged, buried) (Bourguignon et al. 2006). These experiments resulted in extensive trapping over several months and a large variety of observations, unfortunately most of them were unpublished.

To calculate the moment of adult fly oviposition and estimate the minimum time of death, the laboratory mainly uses the ADD method. For this reason, the Microtraces and Entomology Laboratory has been performing rearing experiments to compare the different data available in the literature with the local biological reality of the species found in Belgium. Blow flies (Diptera: Calliphoridae) *Lucilia sericata* (Meigen), *Calliphora vomitoria* Linnaeus, and *Calliphora vicina* Robineau-Desvoidy have historically been the first species examined because of their prevalence in association with casework (Karapetian 2003). Others species followed, such as *Sarcophaga (Liopygia) argyrostoma* (Robineau-Desvoidy) (Diptera: Sarcophagidae), *Muscina prolapsa* (Harris) (Diptera: Muscidae), *Cynomya mortuorum* (Linnaeus) (Diptera: Calliphoridae), and *Phormia regina* Meigen (Diptera: Calliphoridae), as they are also present regularly on human remains in this geographical region. The list continues to increase each time new species are collected from human remains (Braet et al. 2010). ADD and other thermal parameters are measured during experiments performed both under constant and variable temperature conditions.

Rearing protocols have been developed for each species, and the uniqueness of the research lies in the fact that the experimental rearing and casework rearing are almost always done with 24-hour cyclic temperatures and lighting conditions (nights cooler than days). Indeed, it is often under these conditions—rather than constant conditions—measurements and observations appear the most reproducible for casework. Coincidentally, overall mortality rates of larvae are lower under cyclic conditions, potentially reflecting a minimum level of stress on insects (unpublished observations).

10.2.2.2.2 Mathematical and Meteorological Aspects

The Microtraces and Entomology Laboratory places great importance on understanding the impact of climatic conditions on the accuracy of calculations regarding the insect colonization of human and animal remains. Temperature is one of the main parameters affecting the development of most insects, and the daily mean temperature is the primary factor considered in the ADD modeling of insect development. Relative humidity is also considered an important parameter for some species, such as *M. prolapsa* (Braet et al. 2010), but this parameter is not yet taken mathematically into account.

The manner in which weather conditions are used for interpreting forensic entomology evidence in Belgium is unique compared with methods used in other countries. The Royal Meteorological Institute of Belgium provides the laboratory with a large amount of meteorological data daily (e.g., atmospheric temperature, ground temperature, underground temperature [−10 cm], sunshine, and rain duration). These datasets are measured by a web of automatic stations scattered throughout the country. The Microtraces and Entomology Laboratory has developed methods to compare and select the most reliable dataset to be used for the reconstruction of the preexisting climate, allowing for the assessment of the mean daily temperature of the previous weeks with increased accuracy. Tests made in the laboratory show that this method gives a preexisting climate that may be an approximation, but that the mean daily temperature is accurate (unpublished results).

The NICC is involved in a quality assurance process (ISO 17025). The Microtraces and Entomology Laboratory has thus had the opportunity to devote relatively large budgets to the automatic control of temperature and humidity. In practice, from the moment of sampling from human or animal remains, each sample is permanently accompanied by one or more data (temperature) loggers, and breeding chambers are under constant surveillance with an early warning system to provide warning when the temperature or relative humidity level exceeds the preset level.

10.2.2.3 Collaborations

10.2.2.3.1 Molecular Biology: Joint Experimental Molecular Unit (JEMU)

Many forensically relevant dipteran species have not been studied in great detail and remain largely unknown in Belgium. Besides individuals in the Microtraces and Entomology Laboratory, no calliphorid specialist is known in the country. However, modern molecular tools allow for refined taxonomic work to be accomplished. In collaboration with the JEMU project, the Microtraces and Entomology Laboratory continues to study the taxonomy of many Sarcophagidae, including some species of forensic interest, as well as the blow flies *Lucilia caesar* Linnaeus (Diptera: Calliphoridae) and *Lucilia illustris* Meigen (Diptera: Calliphoridae) (Jordaens et al. 2013a; Sonnet et al. 2012; Sonnet et al. 2013).

This project is a collaboration between the Royal Museum for Central Africa (MRAC —Musée Royal de l'Afrique Centrale) and the Royal Belgian Institute of Natural Sciences (IRSNB—Institut Royal des Sciences Naturelles de Belgique) where different species have been examined within the broader framework of the Barcoding of Life project (BOL). The Microtraces and Entomology Laboratory supplies study samples, which are identified using morphological taxonomic methods. This combined approach allows for validation of identifications made as well as an estimate of the accuracy of the laboratory to identify such specimens.

At the NICC, research has been conducted in collaboration with the DNA laboratory located in the agency to compare *L. caesar* and *L. illustris* and determine the genetic variation between geographically different strains of the same species (Desmyter and Gosselin 2009). Similar studies are performed along with the JEMU, between North American and West European strains of *P. regina* (Jordaens et al. 2013b).

10.2.2.3.2 Project "Dood doet leven"

The project *"Dood doet leven"* ("Death brings life") was primarily initiated in the Netherlands around 2005 by various organizations for the conservation of nature. The underlying premise was to promote the placement of animal carcasses in the field to allow nature to do its work of degradation. The aim of this project was to raise awareness about the ecological importance of different animals and plants that are involved in the decomposition of corpses. This project has generated interest in Belgium, and Dirk Raes, a ranger working in the great forest south of Brussels (Agentschap Natuur en Bos), has also created a similar initiative. Because of this initiative, the Microtraces and Entomology Laboratory naturally began to collaborate with research associated with these programs. This type of project helps to educate the public (whose reactions have been surprisingly positive) on ecological and applied aspects of decomposition ecology (Vanpoucke et al. 2010). On the very practical side, it has allowed us to use animals to examine decomposition processes at minimal costs and to have the opportunity to communicate about our research and expertise with the general public. Unfortunately, the Brussels *Dood doet leven* project was officially discontinued at the beginning of 2013; however, the research continues due to the diligent efforts of researchers previously associated with the project.

10.2.2.3.3 Forensic Entomology, Wildlife, and Livestock

For various reasons, it may be necessary to estimate the date of death of animal carcasses, such as poached wildlife or livestock found dead in a meadow. Methods used for human cadavers are transferable to dead animals. We regularly assist in such cases, in particular to support the UAB (Unité Anti-Braconnage—anti-poaching unit). Poaching of male deer occurs in the great forests of the south of Belgium, which are killed for their antlers, while the rest of the carcass is left to decay. An estimation of minimum PMI is of great interest to contradict suspect alibis.

Cattle killed by accidental or meteorological events are usually paid for by insurance, while this is not the case if death occurs due to disease. Therefore, there are attempts to attribute those events to lightning strikes during storm-related occurrences. An estimation of PMI gives the ability to assess the likelihood of such allegations.

Estimations of animal minimum PMI represents up to 10% of the annual casework.

10.2.2.3.4 Universities

Marcel Leclercq had close contacts with the University of Gembloux, where his brother Jean was a professor and Charles Verstraeten managed the arthropod collections. Some degree of collaboration between the Microtraces and Entomology Laboratory and entomologists associated with this university was bound to occur. To date, several masters theses have been produced, two of which have been conducted in close collaboration with the laboratory. One thesis studied the applicability of the method of Marchenko for Belgium (Karapetian 2003), and the second focused its efforts on the volatile organic components (VOC) emitted by decaying pig carcasses (Dekeirsschieter 2007).

10.2.2.4 Education

The institute as a whole, and among others the Microtraces and Entomology Laboratory, provides many courses to police, prosecutors, and more rarely, workshops specifically oriented for scientific police laboratories.

Due to public interest being aroused by the profusion of television series, the laboratory is regularly asked to participate in broadcasts, interviews, conferences, and exhibits in museums and universities.

10.2.3 Future Challenges

The Microtraces and Entomology Laboratory is mostly focused on applied science and therefore devotes its scientific efforts to very practical issues, which can lead to better minimum PMI estimations. The ADD values of numerous species encountered in Belgium are unknown and will be measured each time one of these species can be reared in the laboratory. Some coleopterans, such as *Necrodes littoralis* Linnaeus (Coleoptera: Silphidae), exhibit ecological and behavioral characteristics that are of interest for minimum PMI estimation and could become a new element in the forensic entomologist's toolbox. Genetic studies can shed new light on many taxonomic issues, particularly with sarcophagids or staphylinids.

Although many of these insect species associated with human remains are relatively common, little is known about their biology and ethology. Even distribution maps are relatively sparse. A minimum PMI estimation precise to the hour level will probably remain an ambitious goal due to the inherent variability of the living world. Nevertheless, a better understanding of the very subtleties of insect behavior and development is the necessary path to keep forensic entomology as a modern and reliable science, meeting all the quality standards.

10.3 FORENSIC ENTOMOLOGY IN THE NETHERLANDS

Between the mid-1980s and the dawn of the twenty-first century, the Dutch police used forensic entomology more than 80 times (Huijbregts 2008). As with Belgium, the Netherlands have a national forensic institute (Nationaal Forensisch Instituut [NFI]), which provides research and expertise in a wide variety of fields. Forensic entomology has only recently entered the list of services they provide. Previously, the main forensic entomologists available in the Netherlands were found in the Dutch natural history museum in Leiden (Naturalis) (Krikken and Huijbregts 2001).

When insects or other arthropods are present, scenes of crime police officers collect insect samples on the corpse and at the crime scene following a list of instructions. Occasionally, the prosecutor in a criminal case wishes a minimum PMI estimation to be made and hires a forensic entomologist for that purpose. The police provide the expert with samples, pictures, sampling locations, meteorological data, and any useful information. In the lab, the entomologist decides what to preserve or what to rear to the adult stage. A standardized report is finally written in which, taking into account all the data and samples available, a minimum PMI estimation or a reconstruction of forensically relevant circumstances is proposed (Krikken and Huijbregts 2001).

In the department Non-Human Biologic Traces, the scientists of the NFI develop molecular approaches to identify or date specimens or parts of specimens (de Graaf and Uitdehaag 2012; de Graaf et al. 2013).

Problems are similar to those experienced in Belgium and Luxembourg: the country is too small to allow for an expert to live full time from solely working on forensic entomology. Indeed, the limited geographical area and the relatively high population density cause many bodies to be found very quickly, if not too quickly, as cynical as this statement may seem. The experts in this field must then have a broader range of activities in entomology or even in other domains of biology (e.g., microbiology, botany, palynology, diatomology, or soil analysis, among others). Biological data on the fly species present in the Benelux were mainly measured in neighboring countries. Here also existed and still exists the need to check the data for the territory concerned.

10.4 FORENSIC ENTOMOLOGY IN LUXEMBOURG

Despite legal agreements that allow the Luxembourg courts to use the NICC, currently no entomological expertise has been demanded from Luxembourg.

10.5 CONCLUSION

Belgium, the Netherlands, and Luxembourg are three relatively small countries located in the center of Europe. They share many similar characteristics, including judicial organization, insect fauna, and the way forensic entomology is used. These are all reasons to support a reinforcement of trans-national cooperation.

REFERENCES

Bourguignon, L. 2011. Entomologie forensique. In: *Manuel de l'enquête forensique,* Edition Politeia SA, Brussels, Belgium.

Bourguignon, L., M. Gosselin, Y. Braet, T. Boonen, K. Vits, and F. Hubrecht. 2006. Insect colonization of a hanged body: an explorative study. *Proceedings of the 2006 EAFE meeting,* Bari, Italy.

Bourguignon, L., S. Van Poucke, Y. Braet, and F. Hubrecht. 2013. No need to be dead yet...: a case report. *Proceeding of the 2013 EAFE meeting,* Coimbra, Portugal.

Braet, Y., L. Bourguignon, E. Dupont, S. Vanpoucke, and F. Hubrecht. 2010. New ecological data on *Muscina prolapsa* (Harris, 1780) (Diptera: Muscidae) collected in Belgium. *Proceeding of the 2010 EAFE meeting,* Murcia, Spain.

de Graaf, M. L., and S. Uitdehaag. 2012. Preliminary results of a DNA-study of the inter, intra and regional variation of flies using the COI marker. *Proceedings of the 2012 EAFE meeting,* Toruń, Poland.

de Graaf, M L., M. Wesselink, S. Uitdehaag, and I. Kuiper. 2013. Preliminary results: inter- and intraspecies variation of marker COI in necrophagous flies. *Proceedings of the 2013 APST meeting,* the Hague, the Netherlands.

Dekeirsschieter, J. 2007. Etude des odeurs émises par des carcasses de porc (*Sus domesticus* L.) en décomposition et suivi de la colonisation postmortem par les insectes nécrophages. Unpublished master thesis, Gembloux Agro-Bio Tech, Gembloux, Belgium. http://hdl.handle.net/2268/27148 (consulted October 30, 2013).

Dekeirsschieter, J. R., F. J. Verheggen, C. Frederickx, P. Boxho, and E. Haubruge. 2013. Forensic entomology investigations from Doctor Marcel Leclercq (1924–2008): a review of cases from 1969 to 2005. *Journal of Medical Entomology* 50(5): 935–54.

Desmyter, S., and M. Gosselin. 2009. COI sequence variability between *Chrysomyinae* of forensic interest. *Forensic Science International. Genetics* 3(2): 89–95.

Huijbregts, H. 2008. Bromvliegen als laatste getuigen. In *Passie voor kleine beestjes,* EIS-Nederland publishers, Leiden. pp. 30–31.

Jordaens, K., G. Sonet, Y. Braet, M. De Meyer, T. Backeljau, F. Goovaerts, L. Bourguignon, and S. Desmyter. 2013b. DNA barcoding and the differentiation between North American and West European *Phormia regina* (Diptera: Calliphoridae: Chrysomyinae). *Zookeys* 365: 149–174.

Jordaens, K., G. Sonet, R. Richet, E. Dupont, Y. Braet, and S. Desmyter. 2013a. Identification of forensically important *Sarcophaga* species (Diptera: Sarcophagidae) using the mitochondrial COI gene. *International Journal of Legal Medicine.* 127(2): 491–504.

Karapetian, J. 2003. Evaluation des tables de Marchenko pour la détermination de la date de décès, dans différents biotopes, grâce à deux espèces de Diptères: *Calliphora vicina* Robineau-Desvoidy (1830) et *Lucilia sericata* Meigen (1826). Unpublished master thesis, Gembloux Agro-Bio Tech, Gembloux, Belgium.

Krikken, J., and H. Huijbregts. 2001. Insects as forensic informants: the Dutch experience and procedure: medical and forensic entomology. *Proceedings of the section experimental and applied entomology of the Netherlands Entomological Society (NEV),* Amsterdam, volume 12, the Netherlands pp. 159–163.

Leclercq, J. 2009. Marcel Leclercq (1924–2008), médecin, diptériste, parasitologue et pionnier de l'entomologie forensique. *Faunistic Entomology–Entomologie faunistique* 61 (4): 129–150.

Leclercq, M. 1993. Private correspondence 14 08 1993.

Sonet, G., K. Jordaens, Y. Braet, L. Bourguignon, A. Dupont, T. Backeljau, M. De Meyer, and S. Desmyter. 2013. Utility of GenBank and the Barcode of Life Data Systems (BOLD) for the identification of forensically important Diptera from Belgium and France. *Zookeys* 365: 307–328.

Sonet, G., K. Jordaens, Y. Braet, and S. Desmyter. 2012. Why is the molecular identification of the forensically important blowfly species *Lucilia caesar* and *L. illustris* (family Calliphoridae) so problematic? *Forensic Science International* 223: 1–3.

Vanpoucke, S., L. Bourguignon, Y. Braet, E. Dupont, D. Raes, and F. Hubrecht. 2010. When death is life: ecology, nature conservation and forensic entomology. *Proceedings of the 2010 EAFE meeting,* Murcia, Spain.

Wyss, C., and D. Cherix. 2006. *Traité d'entomologie forensique – les insectes sur la scène de crime.* Presses polytechniques et universitaires romandes, Lausanne, Switzerland. p. 317.

INTERNET RESOURCES

www.eafe.org: EAFE—European Association for Forensic Entomology.

www.incc.fgov.be: NICC—National Institute for Criminalistics and Criminology (Belgium).

www.forensicinstitute.nl: NFI—National Forensic Institute (the Netherlands).

www.naturalis.nl: Natural History Museum (the Netherlands).

www.dooddoetleven.nl: "Dood doet leven" (the Netherlands).

CHAPTER 11

Sweden, Finland, Norway, and Denmark

Anders Lindström

CONTENTS

11.1 HISTORICAL PERSPECTIVES

The history of forensic entomology casework in Sweden is brief although there have been historical contributions to the knowledge of carrion fauna in the literature. In notes (Vahl 1770) taken by the Danish citizen Martin Vahl from lectures held by Carl Linnaeus, he writes that "2 fluer er i stand at at aede et aadsel op som en ulv allene ved deres eg." This statement means that Linnaeus indicated that two flies are capable of devouring a carcass, just like a wolf, through the larvae hatched from their eggs. In 1895, W. M. Schöyen wrote a paper called *Et bidrag til gravenes fauna* (A contribution to the fauna of the grave) (Schöyen 1895) and is a comment and rejoinder to Mégnin's La Faune des Tombeaux (Mégnin 1987). In the paper, he reports the first observation of *Ophyra anthrax* Meig., today known as *Hydrotaea capensis* (Diptera: Muscidae), from human remains unearthed in an exhumation in Kristiania, today known as Oslo, the capital of Norway, which at the time belonged to Sweden.

In 1946, a book titled *Rättsmedicin* (Forensic medicine) was published (Sjövall 1946) describing the development of an undetermined "fly"; eggs hatch to larvae within 1–2 days, larvae grow for 10–12 days before they pupate, and the pupal stage lasts another 10–14 days. There are no instructions on collection or preservation of these larvae or any hints on how postmortem interval can be calculated, only a comment that they might be useful in determining time of death.

In 1949, a book titled *Brottsplatsundersökning* (Crime scene investigation) was published (Wendel and Svensson 1949). This book has a section devoted to the damages inflicted on corpses by insects and other animals. Here it is clearly stated that the study of insects is of great importance to forensic medicine but does not say that it has been used in Sweden. The same approximation of a fly life cycle taken from the handbook in forensic medicine published in 1946 is repeated again. However, this book also gives a description on how adult and larval specimens should be sampled and preserved. The author states that a minimum time of death can be estimated based on the insects on the corpse and describes that this has to be done by an entomologist. The chapter ends with four references to other publications. The third edition from 1959 (Wendel and Svensson 1959) also mentions the development of the same undetermined fly and sampling advice but contains 20 references to other publications, including among them papers by Leclerq and Bequaert (Bequaert 1942; Leclercq and Leclercq 1948; Leclercq 1949; Leclercq and Quinet 1949).

Pekka Nourteva (Nuorteva et al. 1967) gives a reference to two Finnish forensic pathology textbooks from 1901 (Löfström 1901) and 1961 (Uotila 1961), where the possibility to use insects to determine the time of death is mentioned, but he also states that in practice forensic entomology was rarely used and any conclusions would be uncertain due to lack of knowledge of phenology, distribution, development rate, synanthropy, and other bionomic factors under Finnish conditions.

In 1953, Paul Ardö wrote his paper *Likflugan, Conicera tibialis, i Sverige (Dipt., Phoridae)* (The coffin fly, *Conicera tibialis*, in Sweden) in which he gives a detailed description of the life cycle of the coffin fly (Ardö 1953). He had obtained the flies from an exhumation he attended and kept them in a large glass jar where they completed several generations. The study ended abruptly when his housemaid put the jar in the window on a sunny day and all the flies died. This paper is still considered one of the most detailed descriptions of this enigmatic fly.

A nonscientific account of the graveyard beetle, *Rhizophagus parallelocollis* (Coleoptera: Monotomidae) was published by the journalist and amateur entomologist Anton Jansson in his book *Dagsländan och dödgrävaren* (May fly and carrion beetle) from 1947 (Jansson 1947). One of the chapters begins with a description of how the author was sitting in a graveyard on a hot damp evening near the end of May and watched the beetles swarm. He proceeds to give anecdotes about the beetle from his and a friend's beetle-collecting careers. But being an entomologist, he also provides valuable information and observations on the life cycle of the beetle.

In 1972, Roy Danielsson from Lund University in southern Sweden wrote his MSc thesis on Coleoptera visiting carcasses of small animals (Danielsson 1972). He compared the arthropod communities associated with bird, mammal, and fish carcasses. In total, he collected 8514 specimens represented by 168 species and concluded, among other things, that spruce forest holds the most diverse carrion fauna in southern Sweden.

In 2000, Anders Lindström wrote a BSc thesis on the possibilities of using insects as postmortem indicators of human remains in Sweden (Lindström 2000). During the summer of 1998, insects were collected from human remains that were transported to the Department of Forensic Pathology in Uppsala. The methods that were used to age the larvae were taken from literature. A paper on four of the cases was published in 2003 (Lindström et al. 2003). This study was the beginning of modern forensic entomology in Sweden. Since 1998, Lindström has collaborated on cases with the police and forensic pathology departments, averaging two to three homicide or suspected homicide cases per year. Apart from casework, his collaboration with the police involves supervision of crime scene technicians during their education and lectures at meetings. In 2001, Lindström accepted a PhD position in honeybee (Hymenoptera: Apidae) pathology and performed no research in forensic entomology during the following years although he assisted with some investigations. In Sweden, the TV series CSI began broadcasting in 2001 with the character Gil Grissom portrayed as a forensic entomologist. Although the "CSI-effect" is a problem within many areas of forensic science and the forensic entomology analyses in the series leave a lot to be desired, it has only been positive for forensic entomology in Sweden. This show brought an understanding of the scientific field of forensics to a broader audience and the knowledge that analysis of insects from dead humans could be useful during certain investigations.

A working group on forensic archeology was formed in 1991 due to an initiative from the crime scene investigation unit in Stockholm. The intent of forming this working group was to develop an interdisciplinary scientific resource for the forensic community. Originally, this working group consisted of police officers, an archaeologist, a geologist, an osteologist, and a forensic pathologist. In 1996, the group published an overview of their disciplines including a chapter with a brief overview of forensic entomology written by forensic osteologist Rita Larje (Anonymous 1996). However, the activity in the group dwindled, and the group more or less ceased to exist by the end of the 1990s. In 2011, the group was reformed with police, archaeologists and osteologists, forensic pathologists, and Lindström as a forensic entomologist. The group assists the police in cases with buried or badly decomposed remains. In 2013, the working group organized a theoretical and practical field course for crime scene technicians with exhumation of pig carcasses. In the latest edition of *Grundläggande kriminalteknik* (Basic Forensic Science), a brief description on how insects should be collected and instructions to contact Anders Lindström for further guidance was provided (Olsson and Kupper 2013).

In August 2004, there was a high-profile case when a female psychologist disappeared with her car from central Stockholm. Her car was later found in the ferry harbor of Nynäshamn where ships leave for the Baltic countries. Because she had worked as a prison psychologist, the police feared that a former client had kidnapped her. Her body was found in a forested area northwest of Stockholm 11 days after her disappearance. She had been stabbed in the back multiple times. One of the questions was if she had been held captive for a period of time before being killed? Forensic pathologists had a difficult time estimating the time of death and asked for assistance from a forensic entomologist. A number of *Calliphora vicina* Robineau-Desvoidy (Diptera: Calliphoridae) larvae were collected at the autopsy and results of the analyses suggested that they were approximately 11 days old. The entomological analysis led to the conclusion that she had been killed shortly after the abduction. A few months later, her phone was tracked to a police officer in Lithuania who had bought the phone second hand. When approached, she told investigators the name of the man she had bought it from. It turned out that the man and his girlfriend had travelled to Norway and on their way back did not have money to pay for the boat trip back to Lithuania. They decided to rob the psychologist and steal her car. The man stabbed the psychologist 11 times, and then the couple fled to the ferry harbor and took the ferry to Lithuania. The woman was sentenced to 4 years in prison for the abduction and the man was sentenced to 20 years in prison for murder. Although forensic entomology played a minor role in the investigation of this homicide, it was a milestone for the discipline because it demonstrated its usefulness in a real case to police officers and forensic pathologists. Since that case, there has been a marked increase in the annual casework assistance and request for lectures and training.

11.2 RESEARCH

In Sweden, the coleopteran fauna of a dead cat (Ferrer et al. 2004), badger (Ferrer et al. 2006), and blackbird were studied (Ferrer et al. 2009). There has been some international collaboration on the descriptions of some Diptera observed on human remains. Together with Fremdt et al. (2012), Lindström described the first European observations of *Lucilia silvarum* Meigen (Diptera: Calliphoridae) on human remains. In another forthcoming paper, the first finding of *Hydrotaea similis* Meade (Diptera: Muscidae) is reported from a human cadaver in a homicide case (Grzywacz et al. 2014), as well as occurrence of the phorid *Dohrniphora cornuta* Bigot (Diptera: Phoridae) from human remains in Sweden (Disney et al. 2014). Recently, Lindström's position has allowed him to collaborate closely with a statistician, which has led to publications on the use of Bayesian networks in forensic entomology (Andersson et al. 2013; Lindström and Andersson 2015). The use of Bayesian networks for time of death analysis has been extended to include forensic taphonomy as well in collaboration with the Department of Forensic Pathology in Uppsala. A forthcoming paper describes how these statistical methods can be applied in this field to yield additional insight into the decomposition of human remains and how to calculate the time of death (also see Chapter 20).

11.3 THE OTHER NORDIC COUNTRIES

11.3.1 Finland

A great asset to forensic entomology in Sweden and the rest of the Nordic countries is the thorough dipterological work by Professor Pekka Nuorteva in Finland. In 1955, he accepted a position as an assistant professor at Helsinki University, and in 1958, he became a curator at the Zoological Museum in Helsinki. From 1959 to 1977, he published more than 25 papers on different aspects of blow fly distribution (Nuorteva 1959a, 1959c, 1960, 1964a, 1966; Nuorteva and Räsänen 1968; Nuorteva and Hedström 1971), ecology (Nuorteva 1959b, 1964b, 1965, 1970, 1972; Nuorteva and Skarén 1960), synanthropy (Nuorteva 1963, 1967; Nuorteva et al. 1964; Nuorteva and Laurikainen 1964; Nuorteva and Vesikari 1966), and forensic entomology (Nuorteva 1974, 1977, 1987; Nuorteva et al. 1967, 1974). The blow fly studies were initiated on the suspicion that these insects might be involved in the transmission of polio virus. In 1977, he wrote a chapter on forensic entomology in a forensic pathology textbook compiling and summarizing the information that was available at that time and what he had collected during the previous decades (Nuorteva 1977). In 1974, he became the first Finnish professor in environmental protection and his efforts became increasingly directed toward environmental protection leaving less time for forensic and entomological work.

Another prominent Finnish researcher is Professor Ilkka Hanski from the University of Helsinki. He has mainly studied community ecology and conservation biology. Although never involved in forensic entomology casework (Ilkka Hanski, pers. comm.), he has published several papers on topics with relevance to forensic entomology. He has authored several papers on the colonization of carrion and the competition between species on a limited resource such as a carcass (Hanski 1976a, 1987a; Hanski and Kuusela 1977). Further, other papers deal with the assimilation of nutrients by *Lucilia illustris* Meigen (Diptera: Calliphoridae), which is of forensic interest. He has also contributed to the studies of carrion fly communities in different seasons and different types of carrion. (Hanski 1976b, 1977, 1987b; Hanski and Kuusela 1980; Kuusela and Hanski 1982; Hanski and Prinkkilä 1995; Kouki and Hanski 1995).

In 2007, Ilari Sääksjärvi from Turku began accepting forensic cases under the supervision of Professor Nuorteva. Sääksjärvi wrote his PhD dissertation on the diversity of Peruvian Amazonian insects. His present research themes include the diversity of parasitoid wasps and tropical entomology. His efforts with regard to forensic entomology have been done in collaboration with, amongst others, Professor Pekka Saukko from the Turku University, the editor-in-chief of the journal *Forensic Science International*. Dr. Sääksjärvi has annual introductory lectures at the Finnish Police Academy and organized an international course on forensic entomology at the University of Turku in 2007. As in the other Nordic countries, the case rate in Finland is low, around one to two cases per year. A study on indoor forensic cases was published in 2010 (Pohjoismäki et al. 2010).

11.3.2 Norway

Another cornerstone of dipterology, and thus forensic entomology in the Nordic countries is Professor Knut Rognes from Stavanger in Norway. He has recently moved to Oslo and is now professor emeritus although he continues to work on blow fly systematics and taxonomy. His key to the Calliphoridae of Scandinavia and Denmark (Rognes 1991) has been invaluable to anyone interested in this group of flies in the Nordic countries and he briefly mentions forensic entomology in the biology chapter. Professor Rognes did work on some cases between 1988 and 1995 and also cooperated with Dr. Zakaria Erzinçlioğlu (Knut Rognes, pers. comm.).

Morten Staerkeby was another Norwegian entomologist who turned to forensic entomology in the 1990s. His website "International Forensic entomology pages" (http://web.archive.org /web/20070814053432/http:/www.forensicentomology.info/forens_ent/forensic_entomology.html) was at the time one of the best online information sources on the topic. He initiated the website in 1994 while he was studying entomology at the University of Oslo, and it was maintained until 2006. Staerkeby was involved in a forensic entomology project at the Forensic Pathology Institute in Oslo (Staerkeby 1996, 1997, 2001). He now works as a science teacher at a gymnasium in Oslo. Staerkeby still accepts cases if the police requests his services, although the number of cases is low (Morten Staerkeby, pers. comm.).

11.3.3 Denmark

In Denmark, Boy Overgaard Nielsen, professor emeritus at Aarhus University, has been consulting on forensic cases for many years. In the last 14 years, he has assisted on nine cases (Boy Overgaard Nielsen, pers. comm.). He has been working mainly with the ecology of Diptera of medical importance such as biting midges (Diptera: Ceratopogonidae) and mosquitoes (Diptera: Culicidae), but has also published on Calliphoridae and Muscidae. Other work has involved insect–plant interactions, and Overgaard has published some popular accounts of forensic entomology that familiarizes the public with this growing science (Nielsen 2006, 2008).

Thomas Pape, associate professor at the Natural History Museum in Denmark, studies mainly on taxonomy, systematics, phylogeny, and biogeography of Calyptratae. He worked for 10 years between 1994 and 2004 in Stockholm at the Swedish National History Museum. A majority of his research has focused on flesh flies (Diptera: Sarcophagidae), and he has published several papers on the forensic use of Sarcophagidae and Calliphoridae (Szpila et al. 2008, 2010, 2013; Cherix et al. 2012). He was also the PhD supervisor of Professor Krzysztof Spzila from Torún, Poland.

11.4 FUTURE PERSPECTIVES

In all Nordic countries, the continuation of the discipline of forensic entomology is depending on one or very few persons. This limited number of individuals makes forensic entomology a vulnerable discipline. Almost all scientists involved in forensic entomology in the Nordic countries have another primary job within entomology, and several are dependent on research grants to fund their daily jobs. If, for some reason, grants are not funded, the future of forensic entomology in that country could be seriously threatened.

Another problem is the general decline of entomology as a discipline. There are very few pure entomology courses, if any, taught for example in Sweden in recent years. This makes it difficult for people interested in pursuing a career in forensic entomology to get the necessary basic entomological training. A solid background in entomology is essential for developing trust with the police. This is hardly a problem that can be addressed by the forensic entomologists, but remains a political issue with science in Sweden in general. Traditional morphological taxonomy finds disinterested students who prefer to do species identification by molecular methods such as barcoding. While molecular methods certainly have a place within all biological disciplines, today there is commonly a lack of understanding of the ecology and taxonomy of important forensic species due to people having a background in molecular biology and not in entomology.

Refreshingly, Nordic countries have a rather low murder rate. Often the police are not interested in paying for a forensic entomology analysis unless it is a homicide case. In Sweden, there are approximately 100 murders per year, with between 68 and 111 homicides annually during the last 10-year period (Anonymous 2012). The trend is a declining number of homicides. While this is of course good for society, remaining active as a professional forensic entomologist can be quite difficult due to a lack of work. However, by expanding this field of expertise to include also forensic palynology and forensic taphonomy, Lindström has made an effort to increase his participation in forensic investigations. However, these methods are often applicable to the same cases and the increase in case work is so far negligible. During 2015, there will be a major reorganization of the Swedish police force on a national level. The effect of this reorganization on the future of forensic entomology is difficult to foresee but will certainly affect this relevance of this discipline in criminal investigations.

REFERENCES

Andersson, G., A. Lindström, and A. Sundström. 2013. Bayesian networks for evaluation of evidence from forensic entomology. *Biosecurity and Bioterrorism: Biodefense Strategy, Practice, and Science*. 11: S64–S77.

Anonymous. 1996. *Forensisk arkeologi. Arbetsgruppen för forensisk arkeologi* [Forensic archaeology. Working group for Forensic Archaeology]. RPS Rapport, Stockholm, 1996: 5.

Anonymous. 2012. *Brottsutvecklingen i Sverige 2008-2011* [Crimes in Sweden 2008–2011]. Brottsförebyggande rådet Rapport 2012: 13, Stockholm (https://www.bra.se/download/18.22a7170813a0d141d2180007794/1363861007580/2012_13_Brottsutvecklingen_i_Sverige_2008_2011.pdf).

Ardö, P. 1953. Likflugan, *Conicera tibialis* Schmitz, i Sverige (Dipt. Phoridae). *Opuscula Entomologica* 18: 33–36.

Bequaert, J. 1942. Some observations on the fauna of putrefaction and its potential value in establishing the time of death. *New England Journal of Medicine*, 227–856.

Cherix, D., C. Wyss, and T. Pape. 2012. Occurrences of flesh flies (Diptera: Sarcophagidae) on human cadavers in Switzerland, and their importance as forensic indicators. *Forensic Science International* 220: 158–163.

Danielsson, R. 1972. En jämförande studie över skalbaggsfaunan på olika typer av kadaver i olika biotoper [A comparative study of the coleopteran fauna on different types of carrion in different biotopes]. *Entomologen* 1/2: 5–17.

Disney, R.H.L., A. Garcia-Rojo, A. Lindström, and J.D. Manlove. 2014. Further occurrences of *Dohrniphora cornuta* (Bigot) (Diptera, Phoridae) in forensic cases. *Revista Espanola de Medicina Legal* 241: e20–22. doi: 10.1016/j.forsciint.2014.05.010.

Ferrer, J., Y. Gomy, S. Snäll, and P. Whitehead. 2004. Zoosaprophagous coleoptera from a dead domestic cat in a Swedish forest: A comparative perspective. *Entomologist´s Gazette* 55: 185–200.

Ferrer, J., F. Whitehead, and J. Bonet. 2009. Coleoptera, Diptera and other invertebrates at a dead blackbird (*Turdus merula* L.). *Entomologist´s Gazette* 60: 259–268.

Ferrer, J., F. Whitehead, C. Collingwood, Y. Gomy, and S. Snäll. 2006. Zoosaprophagous coleoptera from a dead badger (*Meles meles* L.), in a Swedish forest. *Entomologist's Gazette* 57: 237–248.

Fremdt, H., K. Szpila, J. Huijbregts, A. Lindström, R. Zehner, and J. Amendt. 2012. *Lucilia silvarum* Meigen, 1826 (Diptera: Calliphoridae)—a new species of interest for forensic entomology in Europe. *Forensic Science International* 222: 335–339.

Grzywacz, A., A. Lindström, and M. Hall. 2014. *Hydrotaea similis* Meade (Diptera: Muscidae) newly reported from a human cadaver. *Forensic Science International* doi: 10.1016/j.forsciint.2014.07.014.

Hanski, I. 1976a. Breeding experiments with carrion flies (Diptera) in natural conditions. *Annales Entomologici Fennici* 42: 113–121.

Hanski, I. 1976b. Assimilation by *Lucilia illustris* (Diptera) larvae in constant and changing temperatures. *Oikos* 27: 288–299.

Hanski, I. 1977. An interpolation model of assimilation by larvae of the blowfly *Lucilia illustris* Calliphoridae in changing temperatures. *Oikos* 28: 187–195.

Hanski, I. 1987a. Nutritional ecology of dung- and carrion-feeding insects. In: Slansky, F. Jr. and J.G. Rodriguez (Eds.) *Nutritional Ecology of Insects, Mites, and Spiders*, John Wiley and Sons, New York, pp. 837–884.

Hanski, I. 1987b. Carrion fly community dynamics: Patchiness, seasonality and coexistence. *Ecological Entomology* 12: 257–266.

Hanski, I. and S. Kuusela. 1977. An experiment on competition and diversity in the carrion fly community. *Annales Entomologici Fennici* 43: 108–115.

Hanski, I. and S. Kuusela. 1980. The structure of carrion fly communities: Differences in breeding seasons. *Annales Zoologici Fennici* 17: 185–190.

Hanski, I. and M.L. Prinkkilä. 1995. Complex competitive interactions in four species of *Lucilia* blowflies. *Ecological Entomology* 20: 261–272.

Jansson, A. 1947. *Dagsländan och dödgrävaren. Biologiska skisser* [Mayfly and carrion beetle. Biological sketches]. Svensk Natur, Göteborg, Sweden.

Kouki, J. and I. Hanski. 1995. Population aggregation facilitates coexistence of many competing carrion fly species. *Oikos* 72: 223–227.

Kuusela, S. and I. Hanski. 1982. The structure of carrion fly communities: The size and the type of carrion. *Holarctic Ecology* 5, 337–348.

Leclercq, J. and M. Leclercq. 1948. Données bionomiques pour Calliphora erythrocephala (Meigen) et cas d'application à la médecine légale. *Bulletin de la Société entomologique de France* 53(7), 101–103.

Leclerq, M. 1949. Entomologie et médecine légale. *Acta Med Leg et Soc* 2: 179–202.

Leclercq, M. and L. Quinet. 1949. Quelques cas d'application de l'Entomologie à la détermination de l'époque de la mort. *Annales de Médecine légale* 29, p. 324–326.

Lindström, A. 2000. *Rättsentomologi: tio svenska fallstudier.* BSc Thesis, Uppsala University, Sweden.

Lindström, A. and G. Andersson. 2015. Bayesian statistics and predictive modeling.In: Tomberlin, J.K. and M.E. Benbow (Eds.) Forensic Entomology: International Dimensions and Frontiers. CRC Press, Boca Raton, USA.

Lindstrom, A., T. Jaenson, and O. Lindquist 2003. Forensic entomology: First Swedish case studies. *Canadian Society of Forensic Science* 36: 207–210.

Löfström, T. 1901. *Oikeuslääketieteellinen käsikirja* [Forensic Pathology Handbook]. Duodecim-seura, Finland.

Mégnin, P. 1887. *La faune des tombeaux.* [The Fauna of the Grave] *C. R. Comptes rendus hebdomadaires des séances de l'Académie des Sciences 105: 948–951.*

Nielsen, B.O. 2006. Spyefluer [Blow flies]. *Natur og Museum.* 1: 1–35.

Nielsen, B.O. 2008. Fluen og liget [The fly and the corpse]. *Kaskelot.* 170: 6–9.

Nuorteva, P. 1959a. Studies on the significance of flies in the transmission of poliomyelitis I. The occurence of *Lucilia* species (Dipt., Calliphoridae) in relation to the occurrence of poliomyelitis in Finland. *Annales Entomologica Fennici* 25: 124.

Nuorteva, P. 1959b. Studies on the significance of flies in the transmission of poliomyelitis III. The composition of blowfly fauna and the activity of flies in relation to the weather during the epidemic season of poliomyelitis in South Finland. *Annales Entomologica Fennici* 25: 121–136.

Nuorteva, P. 1959c. Studies on the significance of flies in the transmission of poliomyelitis IV. The composition of the blowfly fauna in different parts of Finland during the year 1958. *Annales Entomologica Fennici* 25: 137–162.

Nuorteva, P. 1960. Förekomsten av asflugan *Phormia terraenovae* R. D. i Finland (Areal and seasonal occurrence and synanthropy of *Phormia terraenovae* in Finland). *Notulae Entomol.* 40: 35–45.

Nuorteva, P. 1963. Synanthropy of blowflies (Dipt., Calliphoridae) in Finland. *Annales Entomlogica Fennici* 29: 149.

Nuorteva, P. 1964a. The zonal distribution of blowflies (Dipt., Calliphoridae) on the arctic hill Ailigas in Finland. *Annales Entomologica Fennici* 30: 218–226.

Nuorteva, P. 1964b. Differences in the ecology of *Lucilia caesar* (L.) and *Lucilia illustris* (Meig.) (Dipt., Calliphoridae) in Finland. *Wiadomosci. Parazytol.* Wroclaw 10: 583–587.

Nuorteva, P. 1965. The flying activity of blowflies (Dipt., Calliphoridae) in subarctic conditions. *Annales Entomologica Fennici* 31: 242–245.

Nuorteva, P. 1966. Local distribution of blowflies in relation to human settlement in an area around the town of Forssa in SouthFinland. *Annales Entomologica Fennici* 32: 128–137.

Nuorteva, P. 1967. The synanthropy and bionomics of blowflies in subarctic Nothern Finland. *Wiadomosci Parazytol.* 13: 603–607.

Nuorteva, P. 1970. Histerid beetles as predators of blowflies (Diptera, Calliphoridae) in Finland. *Annales Zoologica Fennici* 7: 195–198.

Nuorteva, P. 1972. A three year survey of the development of *Cynomyia mortuorum* (L.) (Diptera,Calliphoridae) in the conditions of a subarctic fell. *Annales Entomologica Fennici* 38: 65–74.

Nuorteva, P. 1974. Age determination of blood stain in a decaying shirt by entomological means. *Forensic Science.* 3: 89–94.

Nuorteva, P. 1977. Sarcosaprophagous insects as forensic indicators. In: C.G. Tedeshi, W.G. Eckert, and L.G. Tedeshi (Eds.) *Forensic Medicine II,* W. B. Saunders Company, Philadelphia, PA, pp. 1072–1095.

Nuorteva, P. 1987. Empty puparia of *Phormia terraenovae* R.-D. (Diptera, Calliphoridae) as forensic indicators. *Annales Entomologica Fennici* 53: 53–56.

Nuorteva, P. and L. Hedström. 1971. Zonal distribution of flies on the hill Ailigas in subarctic northern Finland. *Annales Entomologica Fennici* 37: 121–125.

Nuorteva, P., M. Isokoski, and K. Laiho. 1967. Studies on the possibilities of using blowflies (Dipt.) as medicolegal indicators in Finland I. Report of four indoor cases from the city of Helsinki. *Annales Entomologica Fennici* 33: 217–227.

Nuorteva, P., M. Isokoski, K. Laiho, and H. Schumann. 1974. Studies on the possibilities of using blowflies (Dipt., Calliphoridae) as medicolegal indicators in Finland 2. Four cases where species identification was performed from larvae. *Annales Entomologica Fennici* 40: 70–74.

Nuorteva, P., T. Kotimaa, L. Pohjolainen, and T. Räsänen. 1964. Blowflies (Dipt., Calliphoridae) on the refuse depot of the city of Kuopio in Central Finland. *Annales Entomologica Fennici* 30: 94–104.

Nuorteva, P. and E. Laurikainen. 1964. Synanthropy of blowflies (Dipt., Calliphoridae) on the island of Gotland, Sweden. *Annales Entomologica Fennici* 30: 187–190.

Nuorteva, P. and T. Räsänen. 1968. The occurrence of blowflies (Dipt., Calliphoridae) in the archipelago of lake Kallavesi, Central Finland. *Annales Entomologica Fennici* 5: 188–193.

Nuorteva, P. and U. Skarén. 1960. Studies on the significance of flies in the transmission of poliomyelitis V. Observations on the attraction of blowflies to carcasses of micromammals in the commune of Kuhmo, East Finland. *Annales Entomologica Fennici* 26: 221–226.

Nuorteva, P. and T. Vesikari. 1966. The synanthropy of blowflies (Diptera, Calliphoridae) on the coast of the Arctic Ocean. *Annales Medicinae Experimentalis et Biologiae Fenniae* 44: 544–548.

Olsson, J. and T. Kupper. 2013. *Grundläggande kriminalteknik.* 2nd ed. Jure Förlag, Stockholm, Sweden.

Pohjoismäki, J.L., P.J. Karhunen, S. Goebeler, P. Saukko, and I.E. Sääksjärvi. 2010. Indoors forensic entomology: Colonization of human remains in closed environments by specific species of sarcosaprophagous flies. *Forensic Science International* 199: 38–42.

Rognes, K. 1991. *Blowflies (Diptera, Calliphoridae) of Fennoscandia and Denmark Fauna Entomologica Scandinavica Vol 24.* E. J. Brill/Scandinavian Science Press, Leiden, The Netherlands.

Schöyen, W.M. 1895. Et bidrag til gravenes fauna. *Entomologisk Tidskrift* 5: 121–124.

Sjövall, E. 1946. *Rättsmedicin. Kriminologisk handbok [Forensic pathology. Criminological handbook].* Wahlström & Widstrand, Stockholm, Sweden.

Staerkeby, M. 1996. Bestemmelse av dødstidspunkt ved hjelp av rettsentomologi [Determination of the time of death by forensic entomology]. *Nordisk Rettsmedisin* 2: 3–7.

Stærkeby, M. 1997. Beregning av det postmortale intervall på grunnlag av insektstudier [Calculation of the postmortem interval from insect data]. In: Torleiv Ole Rognum (Ed.) *Lundevalls Rettsmedisin,* Universitetsforlaget (In Norwegian).

Staerkeby, M. 2001. Dead larvae of *Cynomya mortuorum* (L.) (Diptera, Calliphoridae) as indicators of the postmortem interval: A case history from Norway. *Forensic Science International* 120: 77–78.

Szpila K., M.J.R. Hall, T. Pape, and A. Grzywacz. 2013. Morphology and identification of first instars of the European and Mediterranean blowflies of forensic importance. Part II. Luciliinae. *Medical Veterinary Entomology* 27: 349–366.

Szpila, K., T. Pape, and A. Rusinek. 2008. Morphology of the first instar larva of *Calliphora vicina, Phormia regina Lucilia illustris* (Diptera, Calliphoridae). *Medical Veterinary Entomology* 22: 16–25.

Szpila, K., T. Pape, and J.G. Voss. 2010. A new forensic indicator for buried bodies (Diptera, Sarcophagidae, Miltogramminae). *Medical and Veterinary Entomology* 24: 278–283.

Uotila, A. (Ed.) 1961. *Oikeuslääketiede* [Forensic Pathology]. WSOY, Porvoo, Finland.

Vahl, M. 1770. *Praelectiones privatissimae.* [Private lectures] Handwritten document.

Wendel, O. and A. Svensson. 1949. Handbok i brottsplatsundersökning. 1st ed. Bröderna Siösteens boktryckeri AB, Stockholm, Sweden.

Wendel, O. and A. Svensson. 1959. Brottsplatsundersökning. 3rd ed. AB Ragnar Lagerblads boktryckeri, Karlshamn.

CHAPTER 12

France

Damien Charabidze and Emmanuel Gaudry

CONTENTS

12.1 INTRODUCTION

The first documented case that considered entomological evidence in France was published in 1855 (Bergeret 1855). A physician, Dr. Bergeret (1814–1893), reported his estimation of the postmortem interval (PMI) based on the analysis of necrophagous insects associated with the body of a newborn child. He identified Lepidoptera species, which were collected from the mummified remains of an infant found behind a bricked up fireplace of a house (in Jura region, Eastern part of France). He estimated 2 years had passed since the death of the child, resulting in the owners of the property suspected as the parents of the newborn and having been responsible for its death. This first application of forensic entomology (FE) in France was somewhat errone-ous, as Bergeret based his estimate on the incorrect assumption that insects only undergo one life cycle each year. Other works followed by Hippolyte Camille Brouardel in 1880 (Lacassagne 1906), who was a member of the French Academy of Medicine and Pierre Mégnin in 1894

(Mégnin 1894), who first described the chronological succession of insect taxa through eight stages of decomposition. The French naturalist, Jean-Henri Fabre (1823–1915), also referred to the legions of "enterprising butchers" colonizing a Linnet (a sparrow) and the behavior of *Necrophorus* (Coleoptera), which bury small rodent carcasses to feed their offspring (Fabre 1907). He also studied the egg-laying behavior of blow flies (Diptera: Calliphoridae) under various "controlled" conditions (Fabre 1907).

There are currently two research groups carrying out FE research in France. One group of researchers belongs to the Forensic Science Institute of The French Gendarmerie (IRCGN[*]) state laboratory. The other belongs to the Institute of Legal Medicine (Lille University).[†] These two research groups have been carrying out research relating to FE in France for the past two decades.

12.2 FORENSIC ENTOMOLOGY DEPARTMENT: IRCGN

12.2.1 Brief History

The *Section Technique d'Investigation Criminelle (STIC)* was formed in 1987. In 1992, the STIC became the *Criminal Research Institute of the National Gendarmerie (IRCGN)* with the FE Department also being established. The FE Department was initially staffed by two individuals, but because of the lack of infrastructure, the investigators relied on voucher specimens loaned by the Natural History Museum in Paris (Muséum National d'Histoire Naturelle de Paris) to confirm their identifications. The FE Department has since expanded and now is composed of five individuals, either gendarmes (policemen with a military status) or civilians (Gaudry et al. 2001), and is now considered one of the largest FE groups in Europe. To date, the FE Department has assessed over 1300 cases for the French justice system. Furthermore, the FE Department has also occasionally provided a second opinion on entomological evidence to overseas judicial systems. It independently and presently handles nearly 50% of all entomological analysis for Europe (unpublished data). Entomologists from the IRCGN can be deployed to a crime scene to support field units and can also appear as expert witnesses in court to respond to the requests filed by various jurisdictions for expert testimony in criminal trials as well.

The FE Department was accredited in 2007 (COFRAC: Comité français d'accréditation—French Committee of Accreditation) (Gaudry and Dourel 2013) and is actively involved in the promotion of the field of FE along with its standards and practices on collection and analysis. The FE Department predominately assists with investigations; however, it also provides technical assistance and is involved in the education and training of death scene technicians (from both police forces: gendarmerie and police). It participates in workshops for investigators and judges with the goal of educating them about the techniques employed in FE and the benefit to law enforcement investigations. It also offers courses in forensic medicine and forensic science. The FE Department assists the National Office of Hunting and Wildlife (ONCFS) with training courses for investigators concerned with environmental crime, public health, and wildlife protection. Members of the FE Department have been actively involved with the European Association for Forensic Entomology (EAFE, www.eafe.org), an organization that serves as the professional body for individuals active in FE research, training, and casework in Europe.

---

[*] Institut de recherche criminelle de la gendarmerie nationale. Emmanuel GAUDRY MSc, Laurent DOUREL MSc, Benoit VINCENT MSc, Thierry PASQUERAULT MSc, Laëtitia CERVANTES MSc.
[†] Professor Valery HEDOUIN, Dr. Damien CHARABIDZE, Julien BOULAY MSc, Cindy AUBERNON MSc.

12.2.2 Entomological Expertise in the Forensic Entomology Department

The FE Department is not often required to attend crime scenes. However, given that the IRCGN is a single national laboratory for the Gendarmerie, it often relies on crime scene units located throughout the country to collect insect evidence, which is then sent to the FE Department for analysis. Because of the reliance on crime scene units to collect evidence, the FE Department provides training to these personnel with special emphasis on collecting evidence at crime scenes, deposition sites (Figure 12.1), and during postmortems. To assist these units, a collection kit has been designed and made available to crime scene units to facilitate the sampling, preservation, and transportation of entomological evidence to the laboratory. It contains all the equipment required for collecting and handling entomological evidence and includes a practical guide that outlines the methodology, packaging, and transportation. The protocols provided by the FE Department also emphasize the importance of chain of custody as the material is "legal evidence," and must comply with the French Penal Code (www.legifrance.fr).

12.2.3 Laboratory Examination of Insect Evidence

All materials received are photographed prior to opening the packaging and before examination, helping to maintain the chain of custody documentation. Containers are labeled following a standard operating procedure: IRCGN registration number, date of rearing, technician's name, and seal reference, and then stored in a secure room. In France, during criminal proceedings, the seal must be returned to the applicant (judge, prosecutor, and investigator) together with the report. Specimen identification criteria are photographed and appended to the peer-reviewed report by other members of the FE Department (Gaudry et al. 2009).

The FE Department predominately uses accumulated degree days (ADD) as the method for estimating the age of immature insects collected from human remains (Smith 1986; Marchenko 2001), thus inferring an associated minimum postmortem interval (minPMI). However, Lefebvre and Gaudry (2009) used insect succession as a method for estimating the minimum PMIs of the remains over a much longer period of time than this minPMI (weeks or months) (Figure 12.2).

While the FE Department is predominately concerned with the practical application of FE, the IRCGN also conducts research in areas such as DNA typing for identification (Malgorn and Coquoz

Figure 12.1 Crime scene technicians (gendarmerie) collecting samples using an entomological sampling kit. (P. Latrubesse ©.)

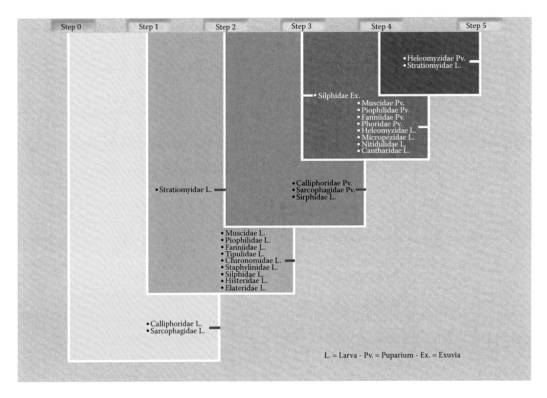

Figure 12.2   Most parsimonious tree showing family succession. (From Lefebvre F. and E. Gaudry. *Annales de la Société Entomologique de France* 45, 377–392, 2009.)

1999), impact of cold temperature on PMI estimations (Myskowiak and Doums 2002), development time of insects of forensic interest (Lefebvre and Pasquerault 2003), insect succession on carrion and human remains (Lefebvre and Gaudry 2009), and temperature reconstruction (Dourel et al. 2010).

12.2.4 Forensic Entomology and Quality

In France, FE is still considered a young science, or even a "nonscience" by some. In 1997, in compliance with the European Network of Forensic Science Institute policy (2003), the IRCGN imposed a principle of quality assurance (QA) practice for all associated forensic departments. The goal for implementing such steps was to eventually obtain accreditation with the International Organization for Standardization (ISO): NF EN ISO/CEI 17025 standard (2005). Until 2007, despite existing references such as the Guidelines for Forensic Science Laboratories (2002), very few guidelines were published on FE (Catts and Haskell 1990; Greenberg and Kunich 2002) in France. The implementation of QA in FE has been problematic due to many practitioners coming from a variety of different institutions and backgrounds (universities, museums, forensic science institutes, and institutes of legal medicine) with each having a different set of specific procedures.

12.2.5 Forensic Entomology Department (IRCGN) Quality Plan

During the past decade, casework in the FE Department of the IRCGN has increased by approximately 80% (Emmanuel Gaudry, pers. comm.). Because of increased reliance on entomological evidence, greater emphasis is now being placed on QA, specifically chain of traceability reducing

Figure 12.3 Data logger used at the scene. (P. Latrubesse ©.)

the risk of errors occurring in PMI estimates. Attempts are therefore being made to achieve three goals for FE based on ISO NF EN ISO/CEI 17025.

- Develop a QA plan: The current document is composed of nine chapters that address the following: (1) personnel, (2) accommodation and environmental conditions, (3) test and calibration methods and method validation, (4) equipment, (5) measurement traceability, (6) sampling, (7) handling of test and calibration items, (8) assuring the quality of testing and calibration results, and (9) reporting the results (including annual survey on results).
- Develop standard operating procedures: (1) sampling method performed by the forensic entomologist (use of sampling kit and data logger (Figure 12.3); (2) preserving, packaging, and storing insect samples; (3) preparing insect samples (larvae and adults) before macroscopic and microscopic observation; (4) rearing dipterans at immature stages; and (5) estimating the PMI, including the existing methods to estimate PMI and criteria justifying the choice of method used by department staff.
- Guidelines for laboratory best practice: (1) use of critical equipment (climatic chambers, data loggers, and temperature data acquisition station), (2) retrieval of weather data, (3) good housekeeping in the laboratory, and (4) management of chemicals and disposable products/disposal of organic materials.

The accreditation of the Forensic Entomology Department of the IRCGN (EN ISO/IEC 17025 norm) in 2007 was a major achievement. However, efforts are still being made at a global level to raise the confidence of judges, prosecutors, investigators, and the general public.

12.3 LILLE FORENSIC TAPHONOMY UNIT ENTOMOLOGY DEPARTMENT

12.3.1 Framework

The cooperation between entomologists and pathologists was initiated in 1995 with the formation of the Lille Forensic Entomology Laboratory. Currently, forensic entomologists at the Lille unit average working on 15–20 cases per year. While initially established to address the application of entomology in criminal investigations, Lille laboratory has since expanded to include research activities. This research team is now part of the Forensic Taphonomy Unit, which also includes anthropology, odontology, and forensic medicine. Some of their current research priorities are outlined in the following sections of this chapter.

12.3.2 Research on Larval Behavior

Blow fly larvae typically aggregate in larval masses (Serra et al. 2011). This gregarious behavior results in an increase in local temperature (larval mass-effect or thermogenesis) that can be extreme (Slone and Gruner 2007; Johnson et al. 2012). To analyze this larval-mass effect, Lille researchers performed mass–temperature measurements using different numbers of *Lucilia sericata* (Meigen) (Diptera: Calliphoridae) larvae under various ambient temperatures (Charabidze et al. 2011). Results demonstrated that larval-mass effect was strongly related to food availability, weight of the larval mass, and ambient temperature. Furthermore, the observation of larval masses revealed complex self-organized structures. Such organization can be linked to a discontinuous gregarious feeding behavior of the larvae. To test this hypothesis, the researchers dissected and measured the crops of *L. sericata* larvae raised under controlled conditions (Charabidze et al. 2013). Crop surfaces of third instar just removed from the food source ranged from 0 to 16.6 mm², indicating a continuous variation of satiation/feeding activity in the population. Starving experiments showed a rather long digestive process: after 150 minutes without food at 25°C, only half of the population had an empty crop. Finally, they used starved larvae to measure the kinetics of food absorption and demonstrated that larvae can ingest faster than they digest. These results demonstrate that larvae do not feed continuously. Such discontinuous foraging behavior of the larvae likely creates a permanent turnover inside the larval masses. This turnover may affect the temperature experienced by the larvae during their development and lowering the impact of larval-mass temperature on development time.

Thus, the larval-mass effect appears as a complex self-regulated mechanism. Currently, the mechanisms supporting the aggregative behavior of necrophagous larvae are mostly unknown. In a 2013 study, we determined that *L. sericata* larvae that are randomly spread in a homogeneous environment will rapidly form an aggregation (0.5 hours) (Boulay et al. 2013). Furthermore, larvae were able to detect a signal (currently unknown) on the area where conspecifics had previously been. This larval signal had a movement-retentive effect, and thus increased the time spent by larvae on a previously visited area. This result demonstrates for the first time the implication of a signal deposited by larvae in their gregariousness. The larval masses thus result from dynamic processes mediated by attractive and retentive signals. Together, these results provide insight into the individual and collective behavior of blow fly larvae. Although some of these topics started from very fundamental questions, the results are relevant in forensic cases (e.g., larvae development time in larval masses and drug metabolism for entomotoxicology studies).

12.3.3 ForenSeek: A Software to Calculate Egg-Laying Times

12.3.3.1 Development Time Calculation

The age of a larva can be calculated using several measurements such as size, weight, instar, and development duration. The most widely used method in France, and in most of Europe, is based on time required for larvae to reach a given stage of development at a known temperature (Amendt et al. 2006). This method starts from a final observed state (i.e., the developmental stage of the sample at time *t*) and reverses through the timeline to determine the time of fly oviposition (i.e., the time when the development of the insect begins external of the adult female).

Development time is primarily dependent on temperature. Unfortunately, data produced have an intrinsic variability due to differences between individuals (Wells and Lamotte 2001; Richards and Villet 2008; Gosselin et al. 2010), and this variability can lead to decreased accuracy when making predictions of a PMI with such data. To further complicate the matter, the comparison of different data sources (studies) for any given species shows interexperimental variability. In other words, the data obtained by a researcher in a laboratory are never exactly the same as those of another researcher in another laboratory (Figure 12.4). Several factors can be attributed to this variation, such as genetic

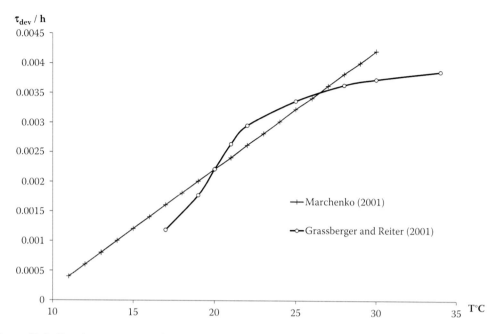

Figure 12.4   Development rate per hour (τ_{dev} = 1/development time) to emergence dependent on temperature (°C) for *Lucilia sericata*. Data of these two different publications show clear differences in values and shape of fitting curves between Grassberger and Reiter data (2001) and Marchenko values (2001).

background, breeding conditions, and reliability of measurements, to name a few. Consequently, the entomologist has to be prepared to explain the sources of such variability. Why use one data set over another? The choice of a data source may thus depend on any criteria that the expert considers relevant. Another solution is to pool several published sources to estimate average developmental times.

Once the developmental data set has been selected, it is necessary to model the data and determine the mathematical relationship between temperature and developmental time. The easiest method postulates a linear relationship, easily modeled by a linear equation (e.g., ADD). However, for extreme temperatures, this linearity assumption cannot be made and it is necessary to refer to more complex, nonlinear models (e.g., sigmoid) (Stinner et al. 1974; Byrd and Castner 2009). Thus, given the wealth of data and mathematical models available, it is important to compare resulting data. To expedite the process of making these comparisons, ForenSeek was developed.

12.3.3.2 ForenSeek: The Forensic Entomology Software

ForenSeek software was developed in partnership with computer scientists. It consists of two components: a collaborative development-time database and an expert tool. The project has also received financial and technical support from the Technology Transfer Office SATT Nord and is available on www.forenseek.org.

12.3.3.2.1 Development-Time Database

The ForenSeek database includes the development times of Diptera of forensic importance that have been published over the last 30 years. The user selects the species and instar and accesses the corresponding developmental data. When selecting different sources (e.g., studies), the user visualizes it on a temperature to development rate graph. This representation allows the user to quickly see discrepancies between sources (Figure 12.4). Once the data are selected, they can be fitted with the appropriate model (e.g., linear, asymmetric sigmoid, or Spline). The user selects the models to be applied, and the results are displayed directly on the graph. This step allows the user to highlight the nonlinearity of the data and thus to choose the most suitable mathematical model.

More than a compilation of published data, this tool is also a mechanism for sharing new development data. Indeed, each user can add and store experimental data on its workspace. By default though, these data are private and therefore not usable by others. However, it is also possible to set the data as "public" and thus to make them available for all other users. An alert system allows the user to report data that appear erroneous. It is therefore a truly collaborative and interactive tool designed to facilitate the sharing and dissemination of development-time data.

12.3.3.2.2 Calculation of a Minimum Postmortem Interval

To estimate the age of fly samples, the first step is to describe their thermal history. For each step (e.g., development of larvae on the body, sampling and transport, storage, breeding), the user can define a constant or variable temperature. In the latter case, the user must provide the corresponding file (e.g., data from a weather station). Once the thermal history is fully described, it is possible to visualize it graphically and to save it.

The user then accesses the samples input screen. Each sample is identified by a; (1) species, (2) stage, and (3) date of observation (e.g., empty pupal cases of *L. sericata* the 5th of May). The user is free to add as many samples as he or she wishes, including the observation of eggs or empty pupal cases. For each sample, the user selects the developmental data and fitting method to be used (see Section 12.3.3.2.1). When all samples are reported and corresponding developmental data are selected, the calculations are initiated. For each sample, the program will determine, based on user-selected development data and fittings, the corresponding oviposition events. If the sample observation refers to the emergence, the corresponding oviposition date will be a single point. If the observation refers to a development stage (e.g., pupal), ForenSeek will report all the oviposition dates enabling the samples to reach this stage of development at time *t*. In other words, a time interval will be produced. Results are provided as both text and graphics. The graph shows on a timescale the oviposition dates calculated with the selected model (Figure 12.5). This view clearly shows the time of first oviposition, and is an efficient way to compare differences resulting from the choice of developmental data and fittings. All of these results can be saved, exported, and integrated into a report.

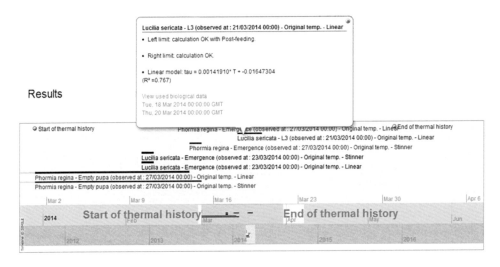

Figure 12.5 Screenshot of the ForenSeek results timeline. Times of flies oviposition are reported according to samples and calculation methods used.

12.4 FUTURE OF FORENSIC ENTOMOLOGY IN FRANCE

FE is well established in France. The vitality of this success has been due to a number of factors. Pioneering scientists played a major role, while efforts in terms of communication, training, and research also made major contributions. However, this is not where the story ends, because forensic entomologists still have many hurdles to overcome.

One of these factors is the taphonomic process associated with human remains. Arthropod species associations, larval behaviors, and competition and predation may have a significant influence on the development time of necrophagous species, and thereby influence PMI estimations. Furthermore, the impact of environment on the decomposition process still remains relatively understudied in France (Gaudry 2010). Since FE is based on both fundamental knowledge and field data, research should reflect and support this dual need. Publication of forensic case reports, development of new forensic tools such as computer software, and the study of important environmental parameters all contribute to improving the accuracy of PMI estimations (Myskowiak and Doums 2002; Charabidze et al. 2009; Dourel et al. 2010; Gosselin et al. 2010).

Such research, along with molecular biology, entomotoxicology, and other fields, has gradually led to the professionalization of FE in France. As a part of this dynamic, a 1999 decision by the European Council placed forensic entomologists within the fields of DNA and fingerprint analyses. Because these laboratories have strong QA standards, European FE laboratories should take steps to comply with these processes as the implementation of this decision moves forward. Certifying individual experts via general requirements, education standards, professional experiences, examinations, and renewal and revocation of certifications is not currently required, but perhaps should be considered in the future.

ACKNOWLEDGMENTS

Emmanuel Gaudry is very grateful to Philippe Latrubesse (photographs) and Frédéric Thomas. He also thanks Pina Grodi for her very precious help. Damien Charabidze would like to thank Professor D. Gosset and Professor V. Hedouin for their constant will to develop and promote forensic entomology. He also thanks A. Veremme for his involvement in the ForenSeek software. The authors would also like to thank anonymous reviewer comments on an earlier draft of this chapter.

REFERENCES

Amendt, J., C.P. Campobasso, E. Gaudry, C. Reiter, H.N. LeBlanc, and M. Hall. 2006. Best practice in forensic entomology—standards and guidelines. *International Journal of Legal Medicine* 121: 90–104.

Bergeret, M. 1855. Infanticide Momification naturelle du cadavre. Annals of public hygiene and legal medicine 4: 442–452.

Boulay, J., C. Devigne, D. Gosset, and D. Charabidze. 2013. Evidence of active aggregation behaviour in *Lucilia sericata* larvae and possible implication of a conspecific mark. *Animal Behaviour* 85: 1191–1197.

Byrd, J.H. and J.L. Castner. 2009. *Forensic Entomology: The Utility of Arthropods in Legal Investigations.* 2nd revised edition. CRC Press, Boca Raton, Florida.

Catts, E.P. and N.H. Haskell. 1990. *Entomology & Death: A Procedural Guide.* Joyce's Print Shop, Inc., Clemson, SC.

Charabidze, D., B. Bourel, and D. Gosset. 2011. Larval-mass effect: Characterization of heat emission by necrophageous blowflies (Diptera: Calliphoridae) larval aggregates. *Forensic Science International* 211: 61–66.

Charabidze, D., B. Bourel, V. Hedouin, and D. Gosset. 2009. Repellent effect of some household products on fly attraction to cadavers. *Forensic Science International* 189: 28–33.

Charabidze, D., V. Hedouin, and D. Gosset. 2013. Discontinuous foraging behavior of necrophagous *Lucilia sericata* (Meigen 1826) (Diptera Calliphoridae) larvae. *Journal of Insect Physiology* 59: 325–331.

Dourel, L., T. Pasquerault, E. Gaudry, and B. Vincent. 2010. Using estimated on-site ambient temperature has uncertain benefit when estimating postmortem interval. *Psyche Journal of Entomology* 2010: 1–7.

Fabre, J.H. 1907. *Souvenirs Entomologiques*, Ed. Delagrave, Paris, France. 10, p. 350.

Gaudry, E. 2010. The insects' colonization on buried remains. In: J. Amendt, M. Lee Goff, C.P. Campobasso, and Grassberger M. (Eds.) *Current Concepts in Forensic Entomology*. 1st edition. Springer, Dordrecht, The Netherlands, 273–312.

Gaudry, E. and L. Dourel. 2013. Forensic entomology: Implementing quality assurance for expertise work. *International Journal of Legal Medicine* 127: 1031–1037. doi: 10.1007/s00414-013-08 92-x.

Gaudry, E., J.B. Myskowiak, B. Chauvet, T. Pasquerault, F. Lefebvre, and Y. Malgorn. 2001. Activity of forensic entomology department of the French Gendarmerie. *Forensic Science International* 120: 68–71.

Gaudry, E., T. Pasquerault, L. Dourel, B. Chauvet, and B. Vincent. 2009. L'entomologie légale: une identification ciblée pour une reponse adaptée. Mémoires de la Société Entomologique de France 8: 85–92.

Gosselin, M., D. Charabidze, C. Frippiat, B. Bourel, and D. Gosset. 2010. Development time variability: Adaptation of Régnière's method to the intrinsic variability of belgian *Lucilia sericata* (Diptera, Calliphoridae) population. *Journal of Forensic Research* 1: 1–3. Available from: http://www.omicsonline.org/2157-7145/2157-7145-1-109.digital/2157-7145-1-109.html.

Grassberger, M. and C. Reiter. 2001. Effect of temperature on *Lucilia sericata* (Diptera: Calliphoridae) development with special reference to the isomegalen and isomorphen diagram. *Forensic Science International* 120: 32–36.

Greenberg B. and J.C. Kunich. 2002. *Entomology and the Law*. Cambridge University Press, Cambridge, United Kingdom.

Johnson, A., M. Archer, L. Leigh-Shaw, M. Pais, C. O'Donnell, and J. Wallman. 2012. Examination of forensic entomology evidence using computed tomography scanning: Case studies and refinement of techniques for estimating maggot mass volumes in bodies. *International Journal of Legal Medicine* 126: 693–702.

Lacassagne, A. 1906. Paul Brouardel: L'homme, le professeur, l'expert. Archives D'anthropologie Criminelle, de Médecine Légale et de Psychologie Normale et Pathologique 21: 759–764.

Lefebvre F. and E. Gaudry. 2009. Forensic entomology: A new hypothesis for the chronological succession pattern of necrophagous insect on human corpses. Annales de la Société Entomologique de France 45: 377–392.

Lefebvre, F. and T. Pasquerault. 2003. Temperature-dependent development of *Ophyra aenescens* (Wiedemann, 1830) and *Ophyra capensis* (Wiedemann, 1818) (Diptera, Muscidae). *Forensic Science International* 139: 75–79.

Malgorn, Y. and R. Coquoz. 1999. DNA typing for identification of some species of Calliphoridae: An interest in forensic entomology. *Forensic Science International* 102: 111–119.

Marchenko, M.I. 2001. Medico-legal relevance of cadaver entomofauna for the determination of the time since death. *Acta Medical and Legal Society* 120: 89–109.

Mégnin, P. 1894. *La Faune des Cadavres: application de L'entomologie à la Médecine Légale*, Paris, Masson.

Myskowiak, J.B. and C. Doums. 2002. Effects of refrigeration on the biometry and development of *Protophormia terraenovae* (Robineau-Desvoidy) (Diptera: Calliphoridae) and its consequences in estimating postmortem interval in forensic investigations. *Forensic Science International* 125: 254–261.

Norm EN ISO/IEC 17025. 2005. General requirements for the competence of testing and calibration laboratories (ISO/IEC 17025:2005),© 2005 CEN/CENELEC Brussels.

Richards, C.S. and M.H. Villet. 2008. Factors affecting accuracy and precision of thermal summation models of insect development used to estimate post-mortem intervals. *International Journal of Legal Medicine* 122: 401–408.

Serra, H., M. Costa, and W. Godoy. 2011. Allee effect in exotic and introduced blowflies. *Neotropical Entomology* 40: 519–528.

Slone, D.H. and S.V. Gruner. 2007. Thermoregulation in larval aggregations of carrion-feeding blow flies (Diptera: Calliphoridae). *Journal of Medical Entomology* 44: 516–523.

Smith, K.J.V. 1986. *A Manual of Forensic Entomology*. Trustees of the British Museum, Natural History and Cornell University Press, London, United Kingdom.

Stinner, R.E., A.P. Gutierrez, and G.D. Butler. 1974. An algorithm for temperature-dependent growth rate simulation. *Canadian Entomology* 106: 519–524.

Wells, J.D. and L.R. Lamotte. 2001. Estimating the Postmortem Interval. In: Byrd, J. H., Castner, J.L., (Eds.) *Forensic Entomology*. CRC Press, Boca Raton, FL: 263–285.

Wyss, C. and D. Chérix. 2006. *Traité D'entomologie Forensique. Les Insectes sur la Scène de Crime*. Lausanne, Switzerland, Presses Polytechniques et Universitaires Romandes.

CHAPTER 13

Austria, Switzerland, and Germany

Jens Amendt, Daniel Cherix, and Martin Grassberger

CONTENTS

13.1 AUSTRIA

The first record regarding forensic entomology research in Austria was in the late nineteenth century. It consisted of a systematic analysis of the grave fauna by the German medical doctor Hermann Reinhard, together with the renowned Viennese entomologist Friedrich Moritz Brauer. Brauer was custodian of the Diptera collection at the Royal Museum of Natural History in Vienna, and later became the director of the museum. He was also elected an Honorary Fellow of the Entomological Society of London in 1900. The results of this early interdisciplinary cooperation were first presented at the annual meeting of the Royal Zoological-Botanical Society in Vienna on April 6, 1881 and were subsequently published in the *Transactions of the Royal Zoological-Botanical Society* (Reinhard 1882). The work titled *Beiträge zur Gräberfauna* (Contributions on the Fauna of Graves) dealt predominantly with scuttle flies (Diptera: Phoridae) and can be credited as an early classic work in the field of forensic entomology.

At the Institute of Forensic Medicine University of Vienna, which was founded in 1805, insect specimens were frequently collected from corpses throughout the nineteenth and early twentieth centuries, which is evident from an old (undated) entomology exhibit display box at the Institute's own museum (MG personal observation). In the pioneering and influential text, *Lehrbuch der gerichtlichen Medizin* (Forensic Medicine) by renowned Viennese forensic pathologist Eduard Ritter von Hofmann (together with Von Hofmann and Haberda, 1923), postmortem changes due to insect activity and observations regarding insect development were repetitively reported. After having published his first observations regarding Diptera larvae on exhumed corpses in 1886, von Hofmann notes in his textbook that oviposition by blow flies (Diptera: Calliphoridae) commences at the eyes and mouth within the first 12 hours after death during summer months and larvae emerge from their eggs on the next day (von Hofmann 1898a). In the 10th edition, published in 1923, several timeframes for body decomposition due to fly larval activity are given (e.g., almost complete skeletonization of an adult body during summer within 8 days) and several influential authors are cited, among them Mégnin, with regard to postmortem interval (PMI) estimation (Von Hofmann and Haberda 1923). In the 1927 edition of the book, it is reported that timing and speed of decomposition due to larval activity is dependent on the cause of death, in that open wounds facilitate rapid disintegration and larval presence can provide possible suggestion of preexisting injuries (von Hofmann and Haberda 1927). Fritz Reuter from Graz (Styria, Austria) also reported in his textbook on forensic medicine on the time-dependent variability of body decomposition as a result of fly larvae and beetle (Coleoptera) activity (Reuter 1933). However, as with von Hofmann, Reuter does not go into detail regarding the particular species involved.

Fifty years later, the Vienna-based forensic pathologist Christian Reiter began work on identification keys for forensically important Calliphoridae (Diptera) larvae and puparia (Reiter and Wollenek 1983a,b), building extensively on the morphological work of Hubert Schumann from the German Democratic Republic (Schumann 1954). In addition, Reiter published some experimental work on the growth behavior of the blue blow fly *Calliphora vicina* Robineau-Desvoidy larvae at various temperatures (Reiter 1984).

In the beginning of the new millennium, forensic entomology once again gained momentum at the Institute of Forensic Medicine in Vienna when Grassberger, a forensic pathologist and biologist, together with Reiter published further experimental work on the development of various necrophagous insect species of forensic importance along with some case reports with focus on Calliphoridae and Sarcophagidae (Diptera) (Grassberger and Reiter 2001, 2002a,b; Grassberger et al. 2003). Based on earlier work by Reiter (1984), they proposed the isomegalen and isomorphen diagrams for the blow fly *Lucilia sericata* (Meigen) in which each line represents identical larval length or a certain morphological change at various temperatures (Grassberger and Reiter 2001). Having compared their developmental data to those of investigators from other regions, they suggested a different response pattern of the same fly species in various zoogeographic regions. In 2003, Grassberger et al. reported an apparent recent northward expansion of the tropical and subtropical blow fly *Chrysomya albiceps* (Wiedemann) beyond southern Europe. It was also noted that the aggressive feeding behavior of *C. albiceps* larvae could reset the postmortem insect clock by clearing a corpse of all earlier arrivers. To include the parasitic wasp *Nasonia vitripennis* (Walker) (Hymenoptera: Pteromalidae) into estimations of the extended PMI, Grassberger and Frank (2003a) studied the wasp development times at various temperatures. Unique to this day was their effort to analyze succession and composition of the carrion-visiting fauna within a central European urban habitat (Vienna) using medium-sized (33–45 kg), clothed domestic pig carcasses for their decomposition studies (Grassberger and Frank 2004).

During the past decade, there was relatively little demand for forensic entomology expert reports in Austrian courts, despite regular education and professional training. During the same period, there was, however, increasing interest in forensic-entomology-related methods from the Austrian and German archaeological community, offering stimulating research possibilities in archaeoentomology (Scharrer-Liška and Grassberger 2005; Grassberger and Scharrer-Liška 2006).

Being a small country with approximately 8.5 million inhabitants and an explicitly low crime rate, there are only a few cases each year in Austria, where insect evidence is of relevance for subsequent legal proceedings. Consequently, there is rather little demand for entomological expertise by a trained full-time forensic entomologist. Moreover, as for example with forensic anthropology issues, entomological casework is mainly performed by forensic pathologists and medical examiners with a degree in biology or special training and knowledge in this field of forensic science. However, since entomology-related questions (e.g., time since death) regularly arise after the recovery of a body, initial crime scene investigation, and autopsy, there is regular training for police and crime scene investigators and basic insect evidence collection techniques in our death investigation handbook (Grassberger and Schmid 2013).

13.2 SWITZERLAND

For a long time, Swiss entomologists were not very interested in blow flies and flesh flies (Sarcophagidae). The only interesting insect found in an entomological collection was a specimen of the blow fly *C. albiceps* collected in 1966 at Col de Bretolet (altitude 1920 m, at the border between Switzerland and France). This record was followed by a second one found in 1995 in a flat in Zurich where the police discovered a dead man (B. Merz, private collection).

In 1993, Claude Wyss, police inspector at the Criminal Police of Canton de Vaud (Switzerland), learned about the Belgian entomologist Marcel Leclerq, one of the pioneers of European forensic entomology, demonstrating how to collect insects found at a crime scene. Wyss was very much impressed and decided to collect flies and larvae at the next crime scene. Although he collected fly larvae associated with death investigations and reared them to adults in his lab, he realized rapidly that identification of this material was not easy. Therefore, he contacted the entomologist Daniel Cherix who was the curator at the Museum of Zoology in Lausanne and professor at the University of Lausanne. They worked together for more than 15 years, analyzing very special cases (Faucherre et al. 1999). Cherix was able to get numerous students interested in rearing different species at different temperatures, collect necrophagous species with baited traps in the field, and test the influence of drugs on the growth of blow flies such as *L. sericata* (Kharbouche et al. 2008). During this time, they analyzed evidence associated with more than 160 cases. However, to obtain more reliable data for use in their cases, they also conducted research with dead pigs, observing the colonization and succession of necrophagous insect species on the remains (Wyss and Cherix 2006, 2013). Recently, they published their findings with flesh fly species found on corpses in Switzerland (Cherix et al. 2012).

Today both Wyss and Cherix are retired and no one has taken their lead. Wyss still works on two to three cases per year, but only in the Lausanne area west of Switzerland. The knowledge and awareness about this method among the investigative authorities and forensic pathologists is still very poor and it is not clear whether this will change in the next decade.

13.3 GERMANY

History always describes events of negligence and superficiality. The biologist August Weismann had already published developmental data of forensically important species in 1864 (Weismann 1864). But at that time, there was no interdisciplinary or sufficient cooperation between entomologists and (forensic) pathologists. Consequently, Weismann's work and data were not used before 1940, when Waldemar Weimann (1940) referred to this publication in his own review of the fauna of cadavers. Nevertheless, the idea of using insect infestations of dead bodies for forensic purposes arose in Germany in the late nineteenth century. Exemplarily, the works by von Hofmann can be mentioned, where he provided a description of the colonization and development of fly larvae on human corpses and, based on developmental data of blow flies, even published some entomology-based estimations of the PMI in 1898 (von Hofmann 1898a,b). Beginning with the twentieth century, various case reports were published concerning the necrophagous species composition on human cadavers (Merkel 1930), duration of development (Meixner 1922), oviposition (Merkel 1925), and feeding behavior (Meixner 1922; Walcher 1933). In 1940, Weimann referred to the temperature-dependent development of necrophagous insects and discussed the question of nocturnal oviposition, which is still of current interest on a global scale. Eventually, Schumann began publishing articles dealing with the taxonomy and ecology of blow flies (Schumann 1954, 1959, 1971). In the 1960s, Schumann began writing entomological reports for the police forces of the German Democratic

Republic (Hubert Schumann, pers. comm.) and his activities in this field are highlighted by his cooperation with the Finnish forensic entomologist Pekka Nuorteva (Nuorteva et al. 1974).

It became generally accepted that the correct identification of the species and the knowledge of the ambient temperature are crucial for an accurate calculation of the time of colonization and to infer a PMI. But despite all the efforts put forth through research and case work, a regular use of insect evidence in forensic investigations was not established in Germany until the 1980s, when the Austrian forensic pathologist Reiter published his works on forensically important blow flies. He assisted with cases in Germany and was the only contact for such special requests. This changed in the 1990s when a high-profile murder case required forensic expertise (Benecke and Seifert 1999). Ants (Hymenoptera: Formicidae) were collected as crime scene evidence and were used to link the suspect to the incident. Moreover, the body was infested with blow fly larvae and a German biologist with some knowledge in the field of entomology was asked to estimate the PMI using the entomological evidence. Because the case was such a public matter, it served as a springboard, launching the application of forensic entomology in Germany and initiating several research projects.

In Cologne, the biologist Mark Benecke reported on several examinations of human corpses colonized by necrophagous insects, specifically with cases of neglect of children and the elderly (Benecke and Lessig 2001). Since the beginning of the twenty-first century, the constant application of forensic entomology eventually led to a surge in publications by Schroeder et al. (2002, 2003), Amendt and colleagues (Zehner et al. 2004a,b; Amendt et al. 2008; Böhme et al. 2010, 2013; Fremdt et al. 2012; Mai and Amendt 2012; Baqué and Amendt 2013), Reibe and Madea (2010a,b), and Niederegger et al. (2010), dealing with genetic and morphological identification of relevant insects, their development, and their biology and ecology. Amendt et al. (2007) published the first best practice paper about forensic entomology in a legal medicine journal, which illustrates that forensic entomology had been received and utilized in the autopsy room, and it seems that the mandatory fusion between entomology, forensic pathology and criminal investigation has slowly taken place. This is encouraging for the increased use of forensic entomology in the future.

Despite these positive advancements in the field, the possible benefit of this discipline still suffers from insufficient sampling at the crime scene. In some cases, this lack of attention to forensic entomology at a crime scene is due to investigators simply not being aware of the value of the field and, therefore, they just do not collect the evidence. In other cases, attempts are made to collect the entomological evidence, but due to a lack of training, the samples are of poor quality. For example, there may not be enough specimens, they may not be preserved correctly, or there was too much time between sampling and forwarding these samples to an expert.

Moreover, the court and legal officials still have difficulty understanding the value of entomological evidence. This is highlighted by a recent example where a prosecutor sent two different samples (from the same homicide) to two different forensic entomologists; the expectation was to get identical and more confident results at a possible trial later on. They were not aware of the possibility that two different samples could lead to two different estimates regarding the PMI. This case demonstrates that the next necessary step is the accreditation of the methods as well as the experts. There is also the task of the forensic entomology community to educate police, judges, and lawyers to give them a better understanding of how the PMI is derived so that they are able to ask specific questions and more thoroughly scrutinize an entomological report for possible flaws. Lastly, there are first steps toward the accreditation of this discipline, which will improve not just the acceptance of forensic entomology in court, but also the quality of the reports and experts.

REFERENCES

Amendt, J., C.P. Campobasso, E. Gaudry, C. Reiter, H.N. LeBlanc, and M.J.R. Hall. 2007. Best practice in forensic entomology: Standards and guidelines. *International Journal of Legal Medicine* 121: 90–104.

Amendt, J., R. Zehner, and F. Reckel. 2008. The nocturnal oviposition behaviour of blowflies (Diptera: Calliphoridae) in Central Europe and its forensic implications. *Forensic Science International* 175: 61–64.

Baqué, M. and J. Amendt. 2013. Strengthen forensic entomology in court: The need for data exploration and the validation of a generalised additive mixed model. *International Journal of Legal Medicine* 127: 213–223.

Benecke, M. and R. Lessig. 2001. Child neglect and forensic entomology. *Forensic Science International* 120: 155–159.

Benecke, M. and B. Seifert. 1999. Forensische Entomologie am Beispiel eines Tötungsdeliktes. Eine kombinierte Spuren und Liegezeitanalyse [Forensic entomology exemplified by a homicide. A combined stain and postmortem time analysis]. *Archiv für Kriminologie* 204: 52–60 (in German).

Böhme, P., J. Amendt, R.H.L. Disney, and R. Zehner. 2010. Molecular identification of carrion-breeding scuttle flies (Diptera: Phoridae) using COI barcodes. *International Journal of Legal Medicine* 124: 577–581.

Böhme, P., P. Spahn, J. Amendt, and R. Zehner. 2013 Differential gene expression during metamorphosis: A promising approach for age estimation of forensically important *Calliphora vicina* pupae (Diptera: Calliphoridae). *International Journal of Legal Medicine* 127: 243–249.

Cherix, D., C. Wyss, and T. Pape. 2012. Occurrences of flesh flies (Diptera:Sarcophagidae) on human cadavers in Switzerland, and their importance as forensic indicators. *Forensic Science International* 220: 158–163.

Faucherre, J., D. Cherix, and C. Wyss. 1999. Behavior of *Calliphora vicina* (Diptera, Calliphoridae) under extreme conditions. *Journal of Insect Behaviour* 12: 687–690.

Fremdt, H., K. Szpila, J. Huijbregts, A. Lindström, R. Zehner, and J. Amendt. 2012. *Lucilia silvarum* (Diptera: Calliphoridae), Meigen 1826: A new species of interest for forensic entomology in Europe. *Forensic Science International* 222: 335–339.

Grassberger, M. and C. Frank. 2003a. Temperature-related development of the parasitoid wasp *Nasonia vitripennis* as forensic indicator. *Medical and Veterinary Entomology* 17: 257–262.

Grassberger, M. and C. Frank. 2004. Initial study of arthropod succession on pig carrion in a central European urban habitat. *Journal of Medical Entomology* 41: 511–523.

Grassberger, M., E. Friedrich, and C. Reiter. 2003. The blowfly *Chrysomya albiceps* (Wiedemann) (Diptera: Calliphoridae) as a new forensic indicator in Central Europe. *International Journal of Legal Medicine* 117: 75–81.

Grassberger, M. and C. Reiter. 2001. Effect of temperature on *Lucilia sericata* (Diptera: Calliphoridae) development with special reference to the isomegalen- and isomorphen-diagram. *Forensic Science International* 120: 32–36.

Grassberger, M. and C. Reiter. 2002a. Effect of temperature on development of *Liopygia* (=*Sarcophaga*) *argyrostoma* (Robineau-Desvoidy) (Diptera: Sarcophagidae) and its forensic implications. *Journal of Forensic Sciences* 47: 1332–1336.

Grassberger, M. and C. Reiter. 2002b. Effect of temperature on development of the forensically important holarctic blow fly *Protophormia terraenovae* (Robineau-Desvoidy) (Diptera: Calliphoridae). *Forensic Science International* 128: 177–182.

Grassberger, M. and G. Scharrer-Liška. 2006. Archaeoentomology: Examination and interpretation of entomological specimens from a 1200-year-old burial site in Austria. Abstract presented at the *4th Annual Meeting of the European Association of Forensic Entomology*, Bari, Italy.

Grassberger, M. and H. Schmid. 2013. Todesermittlung: Befundaufnahme und Spurensicherung. Ein praktischer Leitfaden für Polizei, Juristen und Ärzte. Second Edition. Wissenschaftliche Verlagsgesellschaft Stuttgart (in German).

Kharbouche, H., M. Augsburger, D. Cherix, F. Sporkert, C. Giroud, C. Wyss, C. Champod, and P. Mangin. 2008. Codeine accumulation and elimination in larvae, pupae, and imago of the blowfly *Lucilia sericata* and effects on its development. *International Journal of Legal Medicine* 122: 205–211.

Mai, M. and J. Amendt. 2012. Effect of different post-feeding periods on the total time of development of the blowfly *Lucilia sericata* (Diptera: Calliphoridae). *Forensic Science International* 221: 65–69.

Meixner, K. 1922. Leichenzerstörung durch Fliegenmaden (Destruction of corpses caused by blowfly maggots), *Zeitschrift für Medizinalbeamte und Krankenhausärzte* 35: 407–413 (in German).

Merkel, H. 1925. Die Bedeutung der Art der Tötung für die Leichenzerstörung durch Madenfraß, *Deutsche Zeitschrift für die gesamte gerichtliche Medizin* 5: 34–44 (in German).

Merkel, H. 1930. Todeszeitbestimmung an menschlichen Leichen (Post mortem interval estimation on human corpses), *Deutsche Zeitschrift für die gesamte gerichtliche Medizin* 15: 285–288 (in German).

Niederegger, S., J. Pastuschek, and G. Mall. 2010. Preliminary studies of the influence of fluctuating temperatures on the development of various forensically relevant flies. *Forensic Science International* 199: 72–78.

Nuorteva, P., H. Schumann, M. Isokoski, and K. Laiho. 1974. Studies on the possibilities of using blowflies (Dipt., Calliphoridae) as medicolegal indicators in Finland. *Annales entomologici Fennici* 40: 70–74.

Reibe, S. and B. Madea. 2010a. How promptly do blowflies colonise fresh carcasses? A study comparing indoor with outdoor locations. *Forensic Science International* 195: 52–57.

Reibe, S. and B. Madea 2010b. Use of *Megaselia scalaris* (Diptera: Phoridae) for post-mortem interval estimation indoors. *Parasitology Research* 106: 637–640.

Reinhard, H. 1882. Beiträge zur Gräberfauna. Verhandlungen der Kaiserlich-KÖniglichen Zoologisch-Botanischen Gesellschaft in Wien 31: 207–210 (in German).

Reiter, C. 1984. Growth behavior of the blue blowfly *Calliphora vicina* maggots. *Zeitschrift für Rechtsmedizin* 91: 295–308 (in German).

Reiter, C. and G. Wollenek. 1983a. Determination of maggot species of forensically significant blow flies. *Zeitschrift für Rechtsmedizin* 90: 309–316 (in German).

Reiter C. and G. Wollenek. 1983b. Species determination of puparia of forensically important blowflies. *Zeitschrift für Rechtsmedizin* 91: 61–69 (in German).

Reuter, F. 1933. *Lehrbuch der gerichtlichen Medizin mit gleichmäßiger Berücksichtigung der deutschen und österreichischen Gesetzgebung und des gemeinsamen Entwurfes 1927*. Urban und Schwarzenberg, Berlin, Germany (in German).

Scharrer-Liška, G. and M. Grassberger. 2005. Archaeoentomology on grave no. 34 from the Avar cemetery at Frohsdorf, Lower Austria. *Archäologisches Korrespondenzblatt* 35: 531–544. (in German).

Schroeder, H., H. Klotzbach, L. Oesterhelweg, and K. Püschel. 2002. Larder beetles (Coleoptera, Dermestidae) as an accelerating factor for decomposition of a human corpse. *Forensic Science International* 127: 231–236.

Schroeder, H., H. Klotzbach, and K. Püschel. 2003. Insects' colonization of human corpses in warm and cold season. *Legal Medicine* 5, Suppl 1: S372–374.

Schumann, H. 1954. Morphologisch-Systematische Studien an Larven von hygienisch wichtigen mitteleuropäischen Dipteren der Familie Calliphoridae-Muscidae (Formal morphological investigations on mid-European Diptera of hygienic relevance (Calliphoridae, Muscidae), *Wissenschaftliche Zeitschrift Universität Greifswald, math.-naturwiss, Reihe III* 4/5: 245–274 (in German).

Schumann, H. 1959. Zur Larvalsystematik synanthroper und symboviner Fliegen. *Zeitschrift für angewandte Zoologie* 46: 382–386 (in German).

Schumann, H. 1971. Die Gattung *Lucilia* (Goldfliegen). Merkblätter über angewandte Parasitenkunde und Schädlingsbekämpfung. *Angewandte Parasitologie* 12: 1–20 (in German).

von Hofmann, E. 1886. Observation de larves de Diptéres sur des cadavres exhumés. *Bulletin ou comptes rendus des séances de la Société entomologique de Belgique* 74: 131–132 (in French).

von Hofmann, E. 1898a. Fliegeneier in den Augen- und Mundwinkeln [*Fly eggs in eyes and corners of the mouth*], Gerichtliche Medicin, Verlag J.F. Lehmann, München, Germany (in German).

von Hofmann, E. 1898b. Hochgradige faule, zum grossen Teil von Fliegenmaden aufgefressene Leiche eines alten Mannes, die erst 16 Tage nach dem Tode aufgefunden wurde (Corpse of an old man found in the stage of advanced putrefaction with marked feeding defects of fly maggots 16 days post mortem), Gerichtliche Medicin, Verlag J. F. Lehmann, München, Germany. pp. 192–193 (in German).

von Hofmann, E. and A. Haberda. 1923. *Lehrbuch der gerichtlichen Medizin mit gleichmäßiger Berücksichtigung der deutschen und österreichischen Gesetzgebung*. Zehnte Auflage. Urban and Schwarzenberg, Berlin, Germany (in German).

von Hofmann, E. and A. Haberda. 1927. *Lehrbuch der gerichtlichen Medizin mit gleichmäßiger Berücksichtigung der deutschen und österreichischen Gesetzgebung*. Elfte Auflage. Urban and Schwarzenberg, Berlin, Germany (in German).

Walcher, K. 1933. Eindringen von Maden in die Spongiosa der großen Röhrenknochen (Maggots penetrating the spongiosa of long bones). *Deutsche Zeitschrift für die gesamte gerichtliche Medizin* 20: 469–471 (in German).

Weimann, W. 1940. Leichenfauna (Fauna of corpses), In: F.V. Neureiter, F. Pietrusky, E. Schütt (eds.) *Handwörterbuch der gerichtlichen Medizin und medizinischen Kriminalistik*, Verlag J. Springer, Berlin, Germany. pp. 441–444 (in German).

Weismann, A. 1864. Die nachembryonale Entwicklung der Musciden nach Beobachtungen an *Musca vomica* und *Sarcophaga carnaria* (On the development of *Musca vomica and Sarcophaga carnaria*) *Z. wissensch. Zool* 14:188–336 (in German).

Wyss C. and D. Cherix. 2006. *Traité d'entomologie forensique. Les insectes sur la scène de crime.* Presses polytechniques et universitaires romandes, Lausanne, Switzerland (Suisse).

Wyss C. and D. Cherix. 2013. *Traité d'entomologie forensique. Les insectes sur la scène de crime.* Deuxième édition revue et augmentée. Presses polytechniques et universitaires romandes, Lausanne, Switzerland (Suisse).

Zehner, R., J. Amendt, and R. Krettek 2004b. STR typing of human DNA from fly larvae fed on decomposing bodies. *Journal of Forensic Sciences* 49: 337–340.

Zehner, R., J. Amendt, S. Schütt, J. Sauer, R. Krettek, and D. Povolny. 2004a. Genetic identification of forensically important flesh flies (Diptera: Sarcophagidae). *International Journal of Legal Medicine* 118: 245–247.

CHAPTER 14

Italy

Simonetta Lambiase and M. Denise Gemmellaro

CONTENTS

14.1 FIRST STEPS IN MEDICOLEGAL ENTOMOLOGY: THE ORIGINS AND THE SCHOOL OF BARI

The famous Italian zoologist, Francesco Redi, in the 1600s, is recognized worldwide for his research disputing the theory of spontaneous generation, showing how decomposing meat would not be infested by fly larvae if adults were denied access to the resource (Roncalli Amici 2001). Because of his observations, in the world of entomology, he is considered the first Italian linked to the discipline of forensic entomology in its most basic form, decomposition ecology. However, historically, Giuseppe Müller (1880–1967), an entomologist and director of the Natural History Museum of Trieste (Mezzena 1964; Invrea 1966; Alberti 1995), was one of the first entomologists to collect and preserve entomological specimens from a body that was found hanging. These specimens can still be found in the entomological collections of the Museum of Trieste. After Müller, no other entomologists were engaged in collecting entomological specimens from criminal investigations, as all other Italian entomologists were primarily interested in systematic and phylogenetic questions (as Müller himself had been).

In the nineteenth century, Italy was the theater of one of the first criminal cases in Europe, solved in part with the use of entomological evidence; in 1874, in the city of Padua, the remains of a mummified corpse were found in an attic. Because of the advanced state of decomposition, the entomologist, Lazzaretti, helped to calculate the time of death using the puparia that were recovered on the scene (Porta 1929).

Italian forensic pathologists became aware of the use of arthropods as evidence and began applying it as part of their investigations. Bruno Altamura (1943–2001), a forensic pathologist and director of the Institute of Forensic and Insurance Medicine, University of Trieste, is often credited as being the first forensic pathologist to appreciate the relationship between Diptera colonizing a dead body and time of death estimates (Altamura and Introna 1981, 1982). Francesco Introna Jr., another forensic pathologist, and professor of Legal Medicine at the "Aldo Moro" University of Bari, influenced by Altamura's teachings, and guided by the first European publications in the field of forensic entomology, also began utilizing entomological evidence to contribute to forensic cases (Introna et al. 1989). During his more than 30 years of practice, he specialized in criminal sciences often involving the identification of human remains. He also made significant contributions to the global community of forensic entomology. He documented considerable information regarding the roles of Diptera and other arthropods in the decomposition of human remains (Campobasso et al. 2001) and emphasized the necessity for pathologists and entomologists to collaborate when investigating human death (Campobasso and Introna 2001).

Introna is most known for his pioneering work on entomotoxicology, working with Carlo Campobasso, a forensic pathologist and associate professor in the Department of Medicine and Health Sciences at the University of Molise, Italy, and Lee Goff, former professor in the Department of Entomology at the University of Hawai'i at Manoa and professor emeritus at Chaminade University, Honolulu, Hawaii. Collaboratively, they initiated research examining the detection of toxins and drugs in entomological evidence recovered from decomposing human remains (Introna et al. 2001).

Campobasso is very active in forensic pathology, forensic anthropology, and forensic entomology with much of his work on the latter focusing on the genetic analyses of empty puparia (Mazzanti et al. 2010) and insect gut contents recovered from human remains (Campobasso et al. 2005). Further, Campobasso has evaluated the presence of postmortem artifacts on human remains made by ants (Hymenoptera: Formicidae), and their role in the rate of decomposition (previous Campobasso et al. 2005: Wells et al. 2001; Campobasso et al. 2009). He also played an important role in the delineation of standards and guidelines for forensic entomology in Europe (Amendt et al. 2007, 2011) and published a text with others entitled "Current Concepts in Forensic Entomology" (Amendt et al. 2010). Moreover, Campobasso continued the entomotoxicology research initiated by Introna (Campobasso et al. 2004a). Together, Campobasso and Introna published an Italian book on the role of Diptera in medicolegal investigations (Introna and Campobasso 1998) and contributed to the *Manual of Palaearctic Diptera* (Introna and Campobasso 2000).

Other individuals have made significant contributions to the field of forensic entomology in Italy. Francesco Porcelli, an agricultural entomologist from Bari, has examined the succession of arthropods on swine carcasses in Italy (Introna et al. 2007; Bonacci et al. 2011) as well as myiasis and the use of entomological evidence to estimate a postmortem interval (PMI) (Gherardi et al. 2009). He has also assisted with many medicolegal investigations in Italy (Gherardi et al. 2009: Gherardi and Lambiase 2006; Introna et al. 2011).

The first international seminar on forensic entomology in Italy was organized in 1998 by the Medico-Legal School of Bari. This seminar, which attracted the involvement of about 30 participants, was intended to serve as a platform for scientists from all over the world to discuss many topics including the development and application of entomology as a forensic tool (Di Vella and Campobasso 1999). The School has published research describing the life cycles of forensically relevant Diptera and studies of entomotoxicology (Introna et al. 1990, 2007), and has provided opportunities to collaborate with other specialists from abroad (Introna et al. 1991). Furthermore, the School of Bari attracted new scientists supportive of forensic entomology (Introna et al. 1998; Campobasso and Introna 2001; Campobasso et al. 2001).

The involvement of Italy and of the School of Bari in the forensic entomology community continued to evolve throughout the years; the European Association of Forensic Entomology (EAFE), which was initiated in 2002, indeed, held its fourth meeting in Bari in 2006.

14.2 FIRST UNIVERSITY LAB OF FORENSIC ENTOMOLOGY IN ITALY

In the 1980s, the University of Bari began conducting research in forensic entomology using growth cabinets to follow the life cycle of different fly species (Introna et al. 1989). In 2005, the Department of Legal Medicine and the Department of Animal Biology in Pavia, Italy, initiated a collaboration that resulted in the first academic forensic entomology laboratory under the direction of Simonetta Lambiase in 2006. The aim of this lab was to integrate and support pathologists working in the morgue of the university with the analyses of entomological evidence collected during autopsies.

This collaboration has evolved with many new forensic entomology projects being created, including the development of educational courses and training in forensic entomology. Presently, the Forensic Entomology Lab is involved in training undergraduate and graduate students in the practices of forensic entomology; they offer lectures, lab activities, and fieldwork in forensic entomology. The lab is engaged in several research projects including an assessment of the biodiversity and seasonality of the necrophilous entomofauna associated with vertebrate carrion, insect morphology, and development studies of necrophilous species under controlled conditions and in the natural environment.

In 2008, the Forensic Entomology Lab of Pavia also organized an important conference, L'entomologia forense, un valore aggiunto per le indagini criminalistiche—translated as Forensic Entomology—an added value for criminal investigations. This event was intended for police, magistrates, and pathologists with the goal to introduce the background and practice of forensic entomology in criminal investigations. The Deputy Chief of the Italian Police was invited to give a talk and attended this meeting. Considering that the only way forensic entomology analyses may be implemented in routine crime scene investigation is through the sensitization of police forces to this discipline, the fact that an event focused only on this subject drew the interest and active participation of the highest rank of the Italian police was of significant importance. During that event, the first national forensic entomology course for law enforcement was presented; the course, which focused on the collection of insect evidence from a crime scene, was realized after 1 year and involved Campobasso as a lecturer. After the first course, the Forensic Entomology Lab of Pavia, in collaboration with the Ministry of Internal Affairs, organized other courses for police officers. Since then, the relationship with the police has become progressively stronger. The collaboration between the Forensic Entomology Lab of Pavia and some sections of the Italian Police is based on a formally stipulated agreement between the two parties that involves teaching activities, research projects, and death scene investigations. This collaboration promotes a deeper integration between the law enforcement and the academic environment that facilitates joint contributions to the scientific community (Lambiase and Fredella 2012).

14.3 GROWTH OF FORENSIC ENTOMOLOGY

Forensic entomology in Italy, although founded on the work of a few individuals, has been able to diversify with research in areas such as entomotoxicology (Campobasso et al. 2004a) and entomogenetics (Campobasso et al. 2005). More recently, researchers in forensic entomology, including some in these references (Turchetto and Vanin 2004; Gherardi and Lambiase 2006), developed a working group entitled the Italian Group of Forensic Entomology (Gruppo Italiano di Entomologia Forense [GIEF]).

The GIEF was created in 2007 to increase the visibility of forensic entomology in Italy as well as to facilitate collaborations between different disciplines. Carlo Campobasso, Mirella Gherardi, Francesco Introna, Simonetta Lambiase, Giorgio Nuzzaci, Francesco Porcelli, Margherita Turchetto, Stefano Vanin, and Augusto Vigna Taglianti worked together to institute the GIEF that was quickly strengthened with the addition of new members, including both pathologists and entomologists. Francesco Introna was the first president of the GIEF, from 2007 to 2010. Together, the members of the GIEF worked to disseminate information regarding forensic entomology and its potential in medicolegal investigations throughout Italy. In addition, the GIEF created a website (www.giefitalia.org), which increased the presence of forensic entomology in Italy and offers references to those who become interested in the discipline; the purpose and past and current activities of the GIEF can be found at this website. The GIEF continues to encourage the exchange of information and experiences within the Italian community of forensic entomologists and pathologists, organizes workshops and meetings, and promotes the definition of common standards of forensic entomology. Moreover, the GIEF periodically organizes forensic entomology courses, which are open to anyone who has an interest in the subject. One of the founders of the GIEF is Stefano Vanin, an Italian biologist currently teaching forensic entomology in England at the University of Huddersfield, where he also coordinates a Master's of science in forensic entomology. Since 1998, while working in the lab of Turchetto, he has been involved in forensic research and investigations, taking part in several autopsies and offering consultancy services for various institutes of legal medicine in Italy. He has also worked criminal cases that have received much media attention.

The interests of Vanin in the field of forensic entomology are numerous. He has performed consistent research in the composition of the entomofauna from various areas in Italy (Vanin et al. 2009, 2011; Vanin 2010), genetic analyses on bacteria and insects of forensic interest for the purpose of identifying the species (Adler et al. 2011), as well as on entomological evidence with the purpose of extracting human DNA (Marchetti et al. 2013). Vanin has also collaborated with several forensic anthropologists, with whom he has worked on the analysis of human remains from a mass grave of World War I and the entomological evidence potentially associated with them (Vanin et al. 2009; Gaudio et al. 2013). In 2012, Vanin and other colleagues published the results of their study examining circadian rhythm in *Drosophila* in *Nature* (Vanin et al. 2012), demonstrating his broad research interest in the basic and applied sciences.

The number of researchers in forensic entomology has continued to increase with the creation of the Italian Group of Forensic Entomology Laboratory (F. E. Lab) in Turin. This lab, located in the city morgue, was created with the support of the national health-care service and was directed by Dr. Paola Magni. She obtained a PhD in evolutionary biology and biodiversity conservation with an emphasis in forensic entomology from the University of Turin, in collaboration with the Centre of Forensic Science of the University of Western Australia; she is one of the first Italian scientists to have been formally trained in forensic entomology. In the past, Magni coordinated the Forensic Entomology Lab realized by the Local Health Unit of Turin and the Laboratory of Forensic Entomology and Entomotoxicology in the Chemistry Department at the University of Turin. Magni has worked extensively in the training of Carabinieri in the practices and use of forensic entomology and has been a consultant for the popular Italian TV show, *RIS, Delitti Imperfetti*. Magni, in collaboration with the Carabinieri and with several pathologists, published an Italian textbook on forensic entomology (Magni et al. 2008); she has also performed research in various fields of forensic entomology (Magni et al. 2012, 2013a,b). In 2013, Magni, together with Ian Dadour of the University of Western Australia and in collaboration with Italian Law Enforcement, designed the first Forensic Entomology App, named SMARTINSECTS Forensic Entomology. This app is intended to act as guideline for the collection and preservation of entomological evidence from a death scene and offers quick, easy, and simple information about forensic entomology practices and methods. The app also includes photos of insects of forensic interest as well as information on recording environmental data. This is an example of how Italian entomologists are moving the field of forensic entomology into the future.

The Italian Zooprophylactic Institutes have also begun to focus their attention on the applications of forensic entomology to veterinary medicine; Francesco Defilippo, who works for one of the institutes, has been involved in research and investigations on veterinary forensic entomology.

Currently, in Sicily, one of the major regions of Italy, great attention is being paid to forensic entomology due to the efforts of Denise Gemmellaro, PhD student at Rutgers University, and Claudia Sollami. In collaboration with local administration, judiciary system, police forces as well as foreign institutes and experts in the field, several research projects are being conducted.

14.4 TRAINING AND EDUCATION IN FORENSIC ENTOMOLOGY

Presently, no universities in Italy offer an official degree or certification in forensic entomology, which is, in part, why forensic pathologists have held a defining role in the discipline in Italy. However, students are able to attend postgraduate courses in biological sciences offered by various universities and taught by faculty that are considered experts in the field. Thus, they receive training that allows them to practice in the field. Moreover, the GIEF periodically organizes workshops and educational courses throughout the territory.

14.4.1 University Activities (University Class for Undergraduate Science Majors, Level I Master's Program)

Academically, forensic entomology in Italy is exclusively taught within master's programs (postgraduate courses, which differently from the Anglo-Saxon master's of science degrees, do not confer a University Degree) in Forensic Sciences. Among these, the European master's degree program organized by the University of Parma, with the collaboration of the Carabinieri and with other three European Universities, allows students to have a general knowledge of forensic entomology. In 2009, the Department of Legal Medicine of the University of Pavia organized a level I master's program in the forensic sciences. The program covers a variety of disciplines applicable to criminal investigations, such as ballistics, pathology, blood pattern analysis, and forensic palynology. However, the three major modules of the program are forensic genetics, forensic toxicology, and forensic entomology, along with other areas of the natural sciences (e.g., botany, pedology, acarology, and zoology) that can be used in crime investigations.

Moreover, students at the University of Pavia have the opportunity to attend a class titled *Forensic Methodologies*, which is centered on forensic genetics, toxicology, and entomology. Students attending this course are trained to collect entomological evidence from a mock crime scene, identify specimens, and determine the developmental stage of specimens and correlate this information with weather and temperature data for the estimation of a PMI.

14.4.2 Training of Law Enforcement (Collecting Evidence)

With the previously described agreements between the police and forensic entomologists, police officers are now able to attend various educational courses on forensic entomology. Furthermore, lectures are periodically provided to the Italian Police Academy in Nettuno where those in attendance receive basic information on the life cycle and ethology of insects and how and why insects can be important indicators of PMI. Further, an operative protocol has been developed and distributed to all the scientific police offices. All the officers of the scientific police that have attended the courses organized by the Forensic Entomology Lab of Pavia are authorized by the Ministry of Internal Affairs to collect entomological evidence from the crime scene and can make it available to prosecutors.

14.5 JUDICIAL SYSTEM AND FORENSIC ENTOMOLOGY

Currently, the judicial system in Italy has a limited understanding of forensic entomology and its benefit to death investigations. The education of most judges is in the humanities indicating limited appreciation for the natural sciences in general within the judicial system. This lack of training results in the magistrates not always appreciating the potential of the natural sciences and its range of applications in forensics; moreover, due to the lack of a scientific background, there is the risk that people in the judiciary system may not have the technical background and training to appropriately and critically evaluate a technical (and therefore also a forensic entomological) report (Mori 2011).

After a criminal act is reported or after a body is found, the police forces process the scene. These investigators are part of the State Police and the Carabinieri; the latter are a security body belonging to the Italian Army. Their organization is illustrated in Figures 14.1 and 14.2.

The space limitation of this chapter precludes a more detailed description of all the scientific labs that are part of law enforcement agencies; Figures 14.1 and 14.2 are intended to provide a better understanding of the two police forces that operate in Italy. More information about the scientific departments of the two agencies can be found at the following websites:

http://www.poliziadistato.it/articolo/23433/
http://www.carabinieri.it/Internet/Arma/Oggi/RACIS/

After a crime is reported, the squads of the State Police or Carabinieri go to the scene. Who goes to the scene is dependent on its location: the presence of the Carabinieri is more common throughout Italy as there are many of their offices spread across the country. The police, on the other hand, are present only in larger cities and towns. Paramedics arrive on the scene along with law enforcement when a body is found to assess the situation and, if necessary, pronounce the death of the individual. The law enforcement staff involved in the first intervention on the scene will also notify the prosecutor of the facts and will suggest the possibility of involving the Scientific Police, or for the Carabinieri, the RIS (*Reparto Investigazioni Scientifiche*, Scientific Investigation Department) in the analysis of the scene. These teams will perform analyses and collect evidence from the crime scene and may ask the prosecutor to request expert consultants when needed. The magistrate is not always present at the crime scene, and therefore the ability of the investigative team is fully responsible for the investigation and transmitting all relevant scientific information to the magistrate. However, this approach does not apply to forensic toxicological and genetic analyses, which have now become routine in the normal investigative procedures. But for other types of evidence that are infrequently encountered (i.e., entomology), and are not always well known by the prosecutor, the information supplied by the first intervention squads is essential.

14.6 CASES AND FUTURE CHALLENGES

In the past, entomological evidence from investigations was rarely utilized (Magni et al. 2012); however, with increased education and training of law enforcement and the magistrates of Italy, it is expected that forensic entomology will take a more important role as evidence in the future. A very peculiar crime scene investigation that required assistance from a forensic entomologist was conducted with the Carabinieri of Lecco in 2007 (Lambiase and Camerini 2012) and with the Carabinieri of Lodi in 2011, when a dismembered corpse was recovered near a river and a forensic entomologist used egg development to provide an entomology-based PMI.

Further, there are two cases of homicide that are still being analyzed and represent some of the first cases in Italy to include the police officers of a Scientific Police Department directly in evidence collection and rearing forensically important insects under lab conditions, in collaboration with academic forensic entomologists, while waiting for the evaluation of the prosecutor. This unique experience shows the potential future of such police activity in collecting entomological evidence and working with expert entomologists using established forensic entomology protocols. This is just an individual example of the growing importance of forensic entomologist within a crime scene investigation. Thanks to the efforts of the pathologists, the labs, the universities, and the GIEF law enforcement agencies are starting to require consultation with a forensic entomologist more frequently, making it possible for this discipline to assist with even the most complicated crimes.

Without the entomological and biogeographical data supplied through research, in the event of a case, there would be no database to be used as a reference when comparing the evidence collected

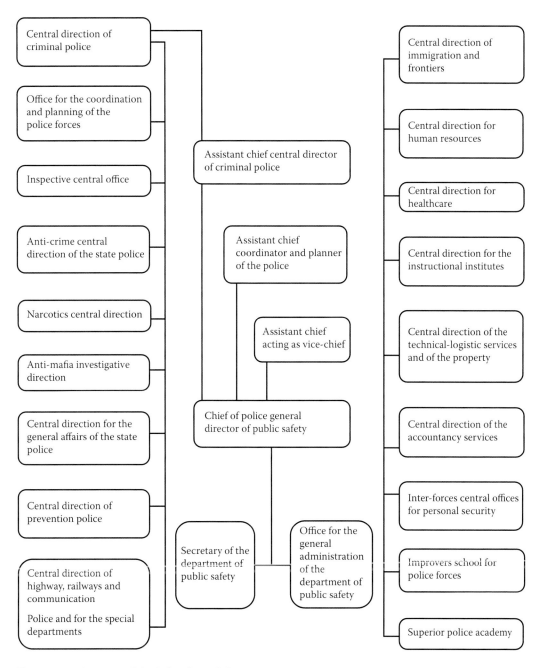

Figure 14.1 Structure of the Italian State Police.

on the scene to what it is normally observed in that area. This means that more resources and time spent on research of forensic entomologists would facilitate the development and refinement of forensically useful databases. When founding the GIEF, it was requested that the members put as much effort as possible toward research projects that included broad collaborations throughout the nation. This would greatly improve communication within the community and eventually will prove fundamental for the growth of forensic entomology in Italy. We are all still committed to that principle, and surely hope to see its realization at all levels as soon as possible.

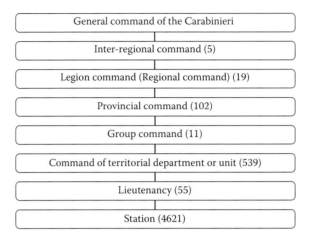

Figure 14.2 Structure of the Carabinieri.

REFERENCES

Adler, B., C. Thèves, E. Crubézy, A. Senescau, S. Vanin, C. Keyser, F. Ricaut et al. 2011. Molecular identification of bacteria by total sequence screening: Determining the cause of death in ancient human subjects. *PLOS ONE* 6, p. e21733.

Alberti, G. 1995. Uomini e Insetti. Le collezioni entomologiche del Museo Civico di Storia Naturale di Trieste e la loro storia. Museo Civico Storia Naturale Trieste: 1.

Altamura, B.M. and F. Introna. 1981. Ditteri cadaverici ed epoca della morte. *La Tipografia Nazionale. Bari.*

Altamura, B.M. and F. Introna. 1982. New possibility of applying the entomological method in forensic medicine: Age determination in post mortem mutilation. Medicina Legale—Quaderni Camerti 4: 127–130.

Amendt, J., C.P. Campobasso, E. Gaudry, C. Reiter, H.N. LeBlanc, and M.J.R. Hall. 2007. Best practice in forensic entomology: Standards and guidelines. *International Journal of Legal Medicine* 121: 90–104.

Amendt, J., C. P. Campobasso, M. L. Goff, and M. Grassberger. 2010. *Current Concepts in Forensic Entomology.* Springer Science, Dordrecht, The Netherlands, p. 376.

Amendt J., C.S. Richards, C.P. Campobasso, R. Zehner, and M.J. Hall. 2011. Forensic entomology: Applications and limitations. *Forensic Science, Medicine, and Pathology* 7: 379–392.

Bonacci, T., T. Zetto Brandmayr, P. Brandmayr, V. Vercillo, and F. Porcelli. 2011. Successional patterns of the insect fauna on a pig carcass in southern Italy and the role of *Crematogaster scutellaris* (Hymenoptera, Formicidae) as carrion invader. *Journal of Entomological Science* 14: 125–132.

Campobasso, C.P. 2003. Costituzione dell'European Association for Forensic Entomology. *Rivista Italiana di Medicina Legale* 25: 245–255.

Campobasso, C.P., R.H.L. Disney, and F. Introna. 2004b. A case of *Megaselia scalaris* (Loew) (Dipt., Phoridae) breeding in a human corpse. *Anil Aggrawal's Internet Journal of Forensic Medicine and Toxicology* 5: 3–5.

Campobasso, C.P., G. Di Vella, and F. Introna. 2001. Factors affecting decomposition and Diptera colonization. *Forensic Science International* 120: 18–27.

Campobasso, C.P., M. Gherardi, and M. Caligara. 2004a. Drug analysis in blowfly larvae and in human tissues: A comparative study. *International Journal of Legal Medicine* 118: 210–214.

Campobasso, C.P. and F. Introna. 2001. The forensic entomologist in the context of the forensic pathologist's role. *Forensic Science International* 120: 132–139.

Campobasso, C.P., J.G. Linville, and J.D. Wells. 2005. Forensic genetic analysis of insect gut contents. *American Journal of Forensic Medicine and Pathology* 26: 161–165.

Campobasso, C.P., D. Marchetti, F. Introna, and M. Colonna. 2009. Post-mortem artifacts made by ants and the effect of ant activity on decompositional rates. *American Journal of Forensic Medicine and Pathology* 30: 84–87.

Di Vella, G. and C.P. Campobasso. 1999. International seminar in forensic entomology. Rivista Internazionale di Medicina Legale 351–355.

Gaudio, D., A. Betto, S. Vanin, A. De Guio, A. Galassi, and C. Cattaneo. 2013. Excavation and study of skeletal remains from a World War I mass grave. *International Journal of Osteoarchaeology*. doi: 10.1002/ oa.2333.

Gherardi, M. and S. Lambiase. 2006. Miasi ed entomologia forense: segnalazione casistica. Rivista Italiana di Medicina Legale 28: 617–628.

Gherardi, M., E. Ragusa, S. Convertini, and F. Porcelli. 2009. Miasi e stima dell'intervallo di tempo trascorso dalla morte. In: *Atti XXII Congresso Nazionale Italiano di Entomologia*, Ancona 15–18 Giugno 2009, p. 263.

Introna, F., B. Altamura, and A. Dell'Erba. 1989. Time since death definition by experimental reproduction of *Lucilia sericata* cycles in growth cabinet. *Journal of Forensic Science* 34: 478–480.

Introna, F. and C.P. Campobasso. 1998. Entomologia Forense: il ruolo dei ditteri nelle indagini medico-legali. SBM editore, Noceto (PR).

Introna, F. and C.P. Campobasso. 2000. Forensic dipterology. In: Papp L. and Darvas B. (eds.), *Manual of Palaearctic Diptera*. Vol. I. Science Herald, Budapest, Hungary, pp. 793–846.

Introna, F., C.P. Campobasso, and A. Di Fazio. 1998. Three case studies in forensic entomology from southern Italy. *Journal of Forensic Science* 43: 210–214.

Introna, F., C.P. Campobasso, and M.L. Goff. 2001. Entomotoxicology. *Forensic Science International* 120: 42–47.

Introna, F., A. De Donno, V. Santoro, S. Corrado, V. Romano, F. Porcelli, and C.P. Campobasso. 2011. Two bodies in enclosed underground environment: The sad story of two missing children. *Forensic Science International* 207: e40–e47.

Introna, F., C. Lo Dico, and Y.H. Caplan. 1990. Opiate analysis in cadaveric blowfly larvae as an indicator of narcotic intoxication. *Journal of Forensic Science* 35: 118–122.

Introna, F., M. Mazzanti, D. Urso, A. Veneziani, F. Alessandrini, A. Tagliabracci, F. Porcelli, A. Caselli, and C.P. Campobasso. 2007. Succession of necrophagous insects over winter in an urban area of southern Italy. In: *Proceeding EAFE 2007 5th Meeting of the European Association for Forensic Entomology*. Brussels, Belgium, May 2–5, 2007.

Introna, F., T.W. Suman, and J.E. Smialek. 1991. Sarcosaprophagous fly activity in Maryland. *Journal of Forensic Science* 36: 238–243.

Invrea, F. 1966. Memorie della Società entomologica italiana 45: 135–148.

Lambiase, S. and G. Camerini. 2012. Spread and habitat selection of *Chrysomya albiceps* (Wiedemann) (Diptera Calliphoridae) in northern Italy: Forensic implications. *Journal of Forensic Science* 57: 799–801.

Lambiase, S. and L. Fredella. 2012. The recovery of a body found indoor which was precociously colonized by *Megaselia scalaris* (Diptera, Phoridae). Poster presented to the 2012 EAFE Meeting in Torun.

Magni, P., E. Di Luise, and S. Scolaro. 2013a. La scena criminis in ambiente acquatico. In Curtotti D. and Saravo L. (eds.). *Le investigazioni sulla scena del crimine*. Norme, tecniche, scienze. Giappichelli Ed., pp. 897–925.

Magni, P., M. Massimelli, R. Messina, P. Mazzucco, and E. Di Luise. 2008. Entomologia forense. Gli insetti nelle indagini giudiziarie e medico-legali. Edizioni Minerva Medica.

Magni, P.A., M.L. Harvey, L. Saravo, and I.R. Dadour. 2012. Entomological evidence: Lessons to be learnt from a cold case review. *Forensic Science International* 223: e31–e34.

Magni, P.A., C. Perez-Banon, M. Borrini, and I.R. Dadour. 2013b. *Syritta pipiens* (Diptera: Syrphidae), a new species associated with human cadavers. *Forensic Science International* 231: e19–e23.

Marchetti, D., E. Arena, I. Boschi, and S. Vanin. 2013. Human DNA extraction from empty puparia. *Forensic Science International* 229: e26–e29.

Mazzanti, M, F. Alessandrini, A. Tagliabracci, J.D. Wells, and C.P. Campobasso. 2010. DNA degradation and genetic analysis of empty puparia: Genetic identification limits in Forensic Entomology. *Forensic Science International* 195: 99–102.

Mezzena, R. 1964. Ricordi di Giuseppe Muller, illuminata figura de uomo e di scienziato. Natura Milano 55: 264–267.

Mori, E. 2011. La drammatica situazione delle scienze forensi in Italia. EARMI. Available at http://www.earmi.it/varie/scienze%20forensi.pdf; accessed on January 13, 2013.

Porta, C.F. 1929. Contributo allo studio dei fenomeni cadaverici: l'azione della microfauna cadaverica terrestre nella decomposizione del cadavere. *Archivio di Antropologia Criminale, Psichiatria e Medicina Legale* 49: 3–55.

Roncalli Amici, R. 2001. The history of Italian parasitology. *Veterinary Parasitology* 98: 3–30.

Turchetto, M. and S. Vanin. 2004. Forensic entomology and climatic change. *Forensic Science International* 146S: S207–S209.

Vanin, S. 2010. Two new species of snow fly Chionea from Italian Alps (Diptera: Limoniidae). *Lavoro Società Veneziana di Scienze Naturali* 35: 5–12.

Vanin, S., S. Bhutani, S. Montelli, P. Menegazzi, E. Green, M. Pegoraro, F. Sandrelli, R. Costa, and C. Kyriacou. 2012. Unexpected features of *Drosophila* circadian behavioral rhythms under natural conditions. *Nature* 484: 371–375.

Vanin, S., L. Caenazzo, A. Arseni, G. Cecchetto, C. Cattaneo, and M. Turchetto. 2009a. Records of *Chrysomya albiceps* in northern Italy: An ecological and forensic perspective. Memórias do Instituto Oswaldo Cruz 104: 555–557.

Vanin, S., M. Gherardi, V. Bugelli, and M. Di Paolo. 2011. Insects found on a human cadaver in central Italy including the blowfly *Calliphora loewi* (Diptera, Calliphoridae), a new species of forensic interest. *Forensic Science International* 207: e30–e33.

Vanin, S., M. Turchetto, A. Galassi, and C. Cattaneo. 2009b. Forensic entomology and the archaeology of war. *Journal of Conflict Archaeology* 5: 127–139.

Wells, J.D., F. Introna, G. Di Vella, C.P. Campobasso, J. Hayes, and F.A.H. Sperling. 2001. Human and insect mitochondrial DNA analysis from maggots. *Journal of Forensic Science* 46: 685–687.

CHAPTER 15

Spain

Marta I. Saloña-Bordas, Concepción Magaña-Loarte, and Ana M. García-Rojo

CONTENTS

15.1 BRIEF REVIEW OF THE HISTORY OF FORENSIC ENTOMOLOGY IN SPAIN

The usefulness of arthropods in forensic research in Spain was first reported by Mariano de la Paz Graëlls (1809–1898). This medical doctor and naturalist made a pleasant review of the entomological changes associated with the decomposition process, listing the insects and mites involved during the decay process (Graëlls 1886), and describing previous cases reported in France by Bergeret (1855), Lichtenstein et al. (1885), and Mégnin (1894), to support the usefulness of arthropods as applied tools in forensic research. Graëlls made a special effort to introduce his students to the importance of having an adequate background about the biology of the arthropods in their environment, especially those related to the decomposition process. He also encouraged magistrates to consider the value of entomological evidence in court. Nevertheless, the potentiality of forensic entomology in Spain remained disregarded for years.

More than a decade later, Teodoro Ríos (1856–1908), who was a professor of the University of Zaragoza, although reluctant about the application of entomology in forensic research, gave a detailed list of the arthropods associated with the putrefaction process of corpses (Ríos, 1902a,b). In his review *Los insectos y la putrefacción de los cadáveres* (Insects and corpses putrefaction), he provided a detailed compilation of the necrophagous fauna known at that time, together with the sequence of appearance of the different groups, even though he did not believe in "Mégnin squads" as clearly and closely defined groups (González Peña 1997).

Apart from some isolated reports (Lecha-Marzo 1924; Piga Pascual 1928) of Mégnin squads, there was no advance on the knowledge of arthropods of forensic interest beyond specific taxonomic keys essential for adequate knowledge of our fauna (González Mora 1988; González Mora and Peris 1989; Peris and González Mora 1991). However, the Iberian Peninsula is an extensive and broad territory with different bioclimatic regions (Mediterranean, continental, subdesert, oceanic, etc.). Therefore, a caution was always necessary when making any extrapolation from different geographic areas. The need to record our own observations and conduct our own research about the distribution and life cycle of necrophagous fauna was a matter pending during the whole twentieth century (cf. Castillo Miralbés 2002; Arnaldos et al. 2006b; Gómez-Gómez et al. 2007).

For decades, there was little mention of forensic entomology in the literature (González Mora et al. 1990). Contributions from Báguena (1952), Domínguez Fernández et al. 1957, 1963; González Mora et al. 1990. But still referred to Mégnin squads in their work. However, they represent some of the first references to the development of a necrophagous insect succession model in Spain. Pascual (1928) cautioned others regarding extrapolating information from one model in a given region to other regions and emphasized the need for greater cooperation between pathologists and entomologists.

On the occasion of the first centenary of the Spanish Royal Society of Natural History, the pathologist Pérez de Petinto y Bertomeu (1957) published an article in which he included a review of artistic representations of insects in association with death, including a description of several practical cases, the life cycle of several dipteran species, as well as the importance of coleopterans in human decomposition (Castillo Miralbés 2002). Nevertheless, the number of studies and publications remained very scarce and limited during the twentieth century (cf. Gómez-Gómez et al. 2007).

It was not until 2000, during the biannual Congress of the Iberian Association of Entomology, that forensic entomology became established in some regions of Spain. A paper given during that meeting introduced to the audience the importance of necrophagous fauna as evidence in criminal investigations (cf. Magaña and Hernández 2000). The seed had finally germinated, and the integration of our country in the recently established European Association for Forensic Entomology (EAFE) was a reality since the first meetings (García Arribas et al. 2002; Ribas Ozonas et al. 2002; Arnaldos et al. 2003a,b,c; Ordoñez and García 2003; Saloña 2003; García-Rojo and Honorato 2004; Saloña et al. 2004).

Since the integration of Spain into the forensic entomology research community, different research models have been developed to improve our knowledge of the dynamics of necrophagous fauna associated with vertebrate remains in terrestrial (Castillo Miralbés 2002; Arnaldos et al. 2003; García-Rojo 2004) and aquatic (Morales et al. 2013) environments, necrophagous arthropod development (Díaz Martín et al. 2013; Gobbi et al. 2013a,b), and molecular characterization of arthropod species (Ramos de Pablo et al. 2006; Pancorbo et al. 2006; GilArriortua et al. 2013). International collaboration, in particular with researchers in Portugal and Latin American countries, has contributed to the understanding and prestige of this discipline (Battán Horestein et al. 2005; Magaña et al. 2006; Cainé et al. 2009; Prado e Castro et al. 2010a,b, 2012). Moreover, soil fauna dynamics during the decomposition process have been analyzed in detail, and forensic acarology has an important background in our international contributions (Saloña et al. 2010; Gonzalez Medina et al. 2012; Mašán et al. 2013).

Continued efforts have resulted in new arthropod records in the Iberian Peninsula and Europe (Bahillo de la Puebla and López Colón 1999; Ortuño et al. 2010; Bahillo de la Puebla et al. 2004; Saloña and González Mora 2005; Martínez Sánchez et al. 2011; Martín-Vega 2011; Martín-Vega et al. 2010a,b, 2012, 2013; Díaz Martín and Saloña Bordas 2012; Martínez Ibáñez et al. 2012; Martín-Vega and Baz Ramos 2010, 2011, 2012, 2013; Prado e Castro et al. 2012; Velasquez et al. 2012; Ventura et al. 2012; Velasquez et al. 2013, among others). Of special mention is the rediscovery of a species that was considered extinct in Europe during the last century. Two different teams in different regions of Spain confirmed the presence of *Thyreophora cynophila* Panzer 1794 (Diptera: Piophilidae), a cheese skipper that was considered extinct in 2007 (Carles Tolrá et al. 2010; Martín-Vega et al. 2010c).

During this last decade, the number of reported forensic entomology cases (García-Rojo and Honorato 2004; Arnaldos et al. 2005; García-Rojo and Honorato 2005; García-Rojo et al. 2006; García-Rojo et al. 2008; García-Rojo et al. 2009; González Medina et al. 2011a,b; González Herrera et al. 2012; González Medina et al. 2012a,b, etc.), experimental works (Castillo 2001; Castillo Miralbés 2001, 2002; Arnaldos et al. 2003; García-Rojo 2004; Baz et al. 2007, 2010), and taxonomical reviews (Bahillo de la Puebla and López Colón 1999; Aznar Cervantes 2006; Gómez-Gómez et al. 2007; Martín-Vega et al. 2012), together with other publications, have increased considerably (cf. Gómez-Gómez et al. 2007), demonstrating the establishment and maturity of research groups at different institutions, including groups such as the National Institute of Toxicology and Forensic Sciences, Forensic Institutions (IAF), the National Police Force (General Commissariat of Scientific Police), and a number of universities. This culmination of expertise was demonstrated in a special issue on forensic entomology published in 2006, *Ciencia Forense Revista Aragonesa de Medicina Legal*. In this special issue, several aspects of forensic entomology were reviewed, including the training of experts (Luna Maldonado and García García 2006), best practices (Arnaldos et al. 2006; Pasquerault et al. 2006), field research (Arnaldos et al. 2006; Romero Palanco et al. 2006), molecular entomology (Pancorbo et al. 2006), and casework (García-Rojo and Honorato 2006).

15.2 CURRENT STATUS OF FORENSIC ENTOMOLOGY IN SPAIN

Although forensic entomology has attained recognition at the national and international levels, the use of arthropods as forensic evidence is still scarce in Spain. This failure to implement the science is partly due to not having adequate taxonomic reviews for most arthropod families of forensic interest. Recent efforts have been done to compile taxonomic data on arthropods distributed in the Iberian Peninsula (*Fauna Ibérica* project), and publications for select groups now exist (Martínez Sánchez et al. 2000; Prieto Piloña and Pérez Valcarcel 2000, 2002; González Mora and Peris 2004; Pérez Moreno et al. 2006; Velasquez et al. 2010; Martín-Vega 2011; Gómez-Gómez et al. 2010; Martín Vega et al. 2011, 2012; Ubero Pascual et al. 2012). The primary federal laboratories of the General Commissariat of the Scientific Police within the National Police Force and the Crime Unit laboratories of the Civil Guard are both associated with the Ministry of the Interior. The main task of these laboratories is to provide assistance with police investigations in a technical way. The National Police Force is composed of units located throughout Spanish territory.

The Institutes of Legal Medicine (IMLs) are directly related to the Administration of Justice, under the Ministry of Justice. However, some autonomous communities in Spain have assumed coordination of their IMLs (i.e., Basque Country and Catalonia). Regardless, it is the responsibility of these institutions (e.g., IML) to assist all aspects of investigations and their litigation. To summarize the judicial structure and departments involved in investigations and associated prosecutions, a diagram has been developed (Figure 15.1). The Chairs of Legal Medicine together with specific research groups specialized on forensic entomology in universities are reviewed in Section 15.4.

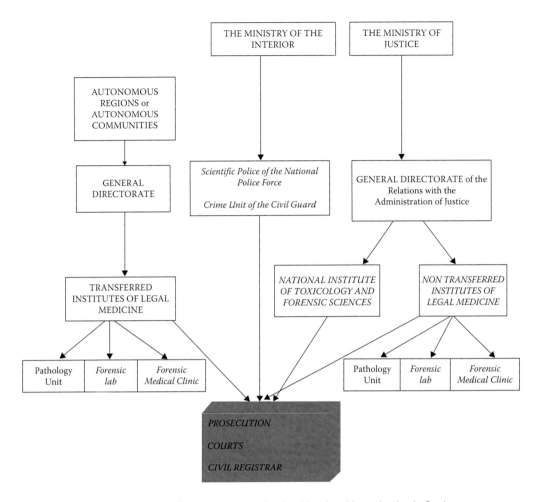

Figure 15.1 Hierarchic structure of the departments involved in a legal investigation in Spain.

15.3 PROCEDURES AND REGULATIONS FOR FORENSIC ENTOMOLOGY IN SPAIN

15.3.1 National Police Force

The National Police Force depends on the Ministry of the Interior. If a crime has been suspected, the police units verify whether the offense was committed and protect the scene to avoid the alteration of any physical evidence. Pursuant to title III of the Spanish Criminal Procedure Code (Ley de Enjuiciamiento Criminal 1882), when offenses are liable to public prosecution crime scene investigators have the power to gather any effects or instruments from the scene. Crime scene investigators (CSI officers) assigned to the Units of Scientific Police are responsible for collecting all identified evidence at scene, including entomology-related materials. A technical investigation utilizes a systematic procedure considered the Best Practice of the General Commissariat of Scientific Police (CGPC). All documentation is made available to assigned experts via the intranet associated with the police website.

CSI officers are highly trained through specialized courses. In relation to forensic entomology, CSI officers are trained to recognize the typical fauna of a site and to the best practices for gathering, preserving, and sending specimens to the Forensic Entomology Laboratory "to assure the chain of custody in all the stages" and to collect temperature data by placing "loggers" where human remains are located. Like many law enforcement agencies globally, evidence is only collected if it is considered to be useful for the investigation. The entomological evidence collected for each case is sent to the Forensic Entomology Laboratory, located in the CGPC (UNE-EN ISO/IEC 17025, Madrid 2000). Presently, entomological evidence is analyzed following the minimum standards and recommendations given by the European Association of Forensic Entomology (Amendt et al. 2007). Furthermore, the laboratory has made it a priority to be accredited (UNE-EN ISO/IEC 17025), in the same way as other laboratories inside the CGPC.

15.3.2 Institute of Legal Medicine

The Institute of Legal Medicine (ILM) depends on the Ministry of Justice. As it is shown in the diagram (Figure 15.1), each institute (ILM) has its own pathology service that is responsible for a given region. Pathologists are responsible for the medicolegal investigation during a criminal investigation together with the identification of human and nonhuman remains, following magistrate instructions. During the investigation, pathologists can order complementary analysis of any evidence collected during autopsy (e.g., toxicological, histopathological, radiological, and entomological). However, there is no Spanish regulation describing where the evidence should be analyzed. Currently, such evidence can be analyzed in the same institution if they have adequate resources or the institution can request the assistance of an external expert. In Madrid, there are some specialized researchers in forensic entomology who have the responsibility of processing entomological samples. They belong to the Institute of Anatomy (*Instituto Anatómico Forense*) and the National Institute of Toxicology and Forensic Sciences. Both agencies are under the Ministry of Justice, and the General Commissariat of Scientific Police (the National Police Force, the Ministry of the Interior). Currently, these agencies are developing collaborations with researchers, universities, and external institutes.

If forensic entomology is successfully integrated into the pathology office standard operating procedure due to such interactions, the resulting reports generated from investigations will require entomological analyses. Such reports will contain specific details about the case and should be attached to the samples including details about the circumstances of the death, developmental stage of the arthropods, and other complementary details (cf. Arnaldos et al. 2001; Magaña Loarte and Prieto Carrero 2009). To date, no formal datasheet has been developed and adapted; therefore, reports currently follow the international standards (Arnaldos et al. 2001) and recommendations of the EAFE (cf. Amendt et al. 2007). Unfortunately, entomological evidence is considered as complementary information in the autopsy report. Therefore, when entomological evidence is considered of interest to an investigator, the sample has to be sent to a specific service of criminology located in the Institute of Toxicology and Forensic Sciences in Madrid following legal regulations (see Section 15.6). Exceptionally, the Institute of Forensic Anatomy has its own expert working in the Laboratory of Forensic Anthropology and Dentistry that has systematically included forensic entomology in the reports for the last 18 years (Magaña 2001). Reports related to forensic cases should be referred to a Court of First Instance, which will process the case.

15.3.3 University

Punctual collaborations between the official institutions detailed previously and the academic research groups are being formalized with different universities and have recently reported fruitful results (cf. Arnaldos et al. 2005; Castillo-Miralbés et al. 2007; García-Rojo et al. 2009; Martínez Sánchez et al. 2011; Saloña et al. 2010; González Medina et al. 2011a,b, 2012a,b; González Herrera et al. 2012). The Ministry of the Interior signed a collaborative agreement with the University of Alcalá de Henares, Madrid, in 1999. Under that agreement, an Institute of Research in Police Sciences (IUICP) was created in 2007. Now, at the University of Alcalá de Henares postgraduate programs are taught, research is conducted, and specialized courses on different areas of forensic science are carried out. In a similar way, the Basque Police (*Ertzaintza*) and the University of the Basque Country (UPV/EHU) have developed personal agreements, and together with the Basque Institute of Criminology (IVAC) they have developed a specific masters on forensic analysis for the research and training of experts through postgraduate courses (Section 15.4).

15.4 WHERE TO STUDY FORENSIC ENTOMOLOGY IN SPAIN

With the acceptance of forensic entomology in crime laboratories, there are increased efforts by academia to train forensic entomologists. Models vary across regions ranging from short training courses specifically designed for specialists (pathologists, researchers, and/or police officers) to regulated disciplines involving specific postgraduate courses developed in universities where faculty are actively conducting forensic entomology research (Universidad de Alcalá [UA], Universidad de Alicante [UAL], Universidad Autónoma de Madrid [UAM], Universidad de Córdoba [UC], Universidad de Murcia [UM], and Universidad del País Vasco [UPV/EHU]). To summarize, the participating universities are reported in a detailed list with links to more specific information about the referred courses in Sections 15.4.1 through 15.4.6. There are several different options for forensic entomology training in Spain at the postgraduate level.

15.4.1 University of Alcalá

Through the IUICP, University of Alcalá has organized the *"Master Universitario en Ciencias Policiales"* (police sciences) and PhD in criminology.

http://www.uah.es/IUICP/
http://www.uah.es/estudios/postgrado/programa.asp?CdPlan = M076

Forensic entomology is offered as a course of specialization.

15.4.2 University of the Basque Country

Master universitario en análisis forense (forensic analysis) http://www.masteranalisisforense.ehu.es

- Introducción a la entomología forense
- Entomología forense II
- Técnicas de estudio en entomología forense

The first course is obligatory, and the other two are optional.

15.4.3 University of Cordoba and University of Lincoln

Master en Ciencias Forenses (forensic sciences) through the European program Erasmus Mundus

http://www.uco.es/estudios/idep/masteres/ciencias-forenses

15.4.4 University of Murcia

Master universitario en Ciencias Forenses (forensic sciences) http://www.um.es/web/biologia/contenido/estudios/masteres/ciencias-forenses

- Entomología forense

For contents, see Arnaldos Sanabria M.I., García García M.D. 2009. Entomología Médico-legal. Open Course War, U. Murcia http://ocw.um.es/ciencias/entomologia-medico-legal.

15.4.5 University of Zaragoza

The University of Zaragoza organizes a summer course on medical, veterinary entomology, where forensic entomology is offered in alternate years.

http://moncayo.unizar.es/cv%5Ccursosdeverano.nsf/CursosPorNum/10

15.4.6 Universitarias Lucía Botín

The university organizes a course on medical entomology titled "entomología sanitaria y control de vectores."

* Curso de introducción a la entomología forense

http://www.luciabotin.com/infocursos/ef/Index.html

For additional courses, see http://www.emagister.com/cursos-entomologia-forense-kwes-36120.htm.

15.5 SCIENTIFIC ASSOCIATIONS

Spain has a long tradition in entomology research. Insects have captivated many people, not limited to scientists, and particular collections have reported very interesting news regarding specific taxa including necrophagous fauna. Moreover, because many arthropods are vectors of various pathogens responsible for life-threatening illnesses in humans (e.g., malaria was endemic of the Iberian Peninsula until the twentieth century), entomology departments have been established in schools of medicine and veterinary science associated with several institutions, and they are summarized here. Furthermore, many amateur and scientific organizations contribute to understanding the roles (good and bad) of arthropods in Spain. Sections 15.5.1 through 15.5.4 discuss some of the most active organizations with interest in forensic entomology.

15.5.1 Asociación Española de Entomología
http://www.entomologica.es/

This is the first and more mature association of entomology in Spain. Members organize annual meetings and produce high-quality materials on general and applied entomology, including a textbook that is used in basic and applied entomology (Barrientos 2004).

15.5.2 Sociedad Entomológica Aragonesa
http://www.sea-entomologia.org/

Although the name refers to a specific region of Spain, it brings together members from different countries of the world, including non-Spanish-speaking members. The electronic list is extremely active, and members have produced specific monographs on forensic entomology included in the References section.

15.5.3 Asociación Andaluza de Entomología
http://www.sociedadandaluzadeentomologia.com

With their head office in Córdoba, they have worked since 1995 for the conservation and study of insects and the diffusion of entomology. They promote education campaigns to develop new consciousness in society and to ensure the preservation of our natural patrimony, with special emphasis on insects and their environment.

15.5.4 Gipuzkoako Entomologia Elkartea: Asociación Gipuzcoana de Entomología
http://www.heteropterus.org/c_presentacion.html

This young association was born in 1999 to promote the research on arthropods in our country. Members organize an annual meal to meet with the simple excuse of having some relaxation time talking about arthropods around a table.

15.6 REGULATION

Legal aspects are regulated through a civil code (Código Civil, BOE 1889) and a penal code (Código Penal and Ley Orgánica 1995) in Spain that describe how to proceed after a civil offense and a criminal act, respectively. These codes were subsequently transcribed into Royal Orders that promulgated laws to regulate all aspects related to the civil and criminal courts in Spain, professional competences, and so on. On September 14, 1882, a Royal Orders (*Real Decreto* RD *de 14 Septiembre de 1882*) defined the Spanish Criminal Procedure Law (*Ley de Enjuiciamiento Criminal Española*, LECr) to regulate procedures in criminal proceedings.

The Royal Order RD 296/1996 regulates the pathologist's role in forensics, and RD 386/1996 regulates the IML's functions (cf. Sánchez Sánchez 2010).

The powers, roles, and legal guarantees of the actions taken by the State Security Forces under the Ministry of Interior are regulated in the Spanish Constitution 1978, Judicial Organization Act 6/1985 (1985) (articles 547–550), Criminal Procedure Law (articles 282 et seq.), State Security Forces Organic Law 2/1986 (1986) (article 11.1.g), Royal Decree 767/1987 (1987) regulating the Criminal Investigation Department Police, and Jurisprudence of the Supreme Court (2005).

15.7 FUTURE TRENDS IN FORENSIC ENTOMOLOGY IN SPAIN

Major advances in forensic entomology have occurred in Spain during the last decade. Nevertheless, continued efforts are needed, especially with describing the arthropod fauna associated with vertebrate remains in different environments and regions of Spain. Doing so will allow for more focused efforts on examining the biology and behavior of those arthropods deemed most important in these different locations. Presently, efforts to develop multidisciplinary teams across the institutions previously mentioned are being developed. Coordinated efforts such as these will go a long way in identifying potential sources of errors or ways to remedy them.

REFERENCES

Amendt, J., C.P. Campobasso, E. Gaudry, C. Reiter, H.N. LeBlanc, and M.J. Hall. 2007. Best practice in forensic entomology—standards and guidelines. *International Journal of Legal Medicine* 121: 90–104.
Antón Barberá, F. 2000. La entomología forense en la investigación policial *Ciencia Policial: Revista del Instituto de Estudios de la Policía* 54: 21–30.

Arnaldos, M.I., M.D. García, E. Romera, J.J. Presa, and A. Luna. 2005. Estimation of postmortem interval in real cases based on experimentally obtained entomological evidence. *Forensic Science International* 149: 57–65.

Arnaldos, M.I., A. Luna, J.J. Presa Asensio, E. López-Gallego, and M.D. García. 2006a. Entomología Forense en España: hacia una buena práctica profesional. *Ciencia forense: Revista Aragonesa de Medicina Legal* 8: 17–38.

Arnaldos, M.I., C. Prado e Castro, J.J. Presa Asensio, E. López-Gallego, and M.D. García. 2006b. Importancia de los estudios regionales de fauna sarcosaprófaga: aplicación a la práctica forense. *Ciencia forense: Revista Aragonesa de Medicina Legal* 8: 63–82.

Arnaldos, M.I., E. Romera, M.D. García, and E. Baquero. 2003a. Nuevos datos sobre la fauna de Mymaridae (Hymenoptera, Chalcidoidea) de la Península Ibérica de una comunidad sarcosaprófaga. *Boletín de la Asociación española de Entomología* 27: 225–228.

Arnaldos, M.I., E. Romera, M.D. García, and A. Luna. 2001. Protocolo para la recogida, conservación y remisión de muestras entomológicas en casos forenses. *Cuadernos de Medicina Forense* 25: 65–73.

Arnaldos, M.I., E. Romera, J.J. Presa, A. Luna, and M.D. García. 2003b. Studies on seasonal arthropod succession on carrion in the southeastern Iberian Peninsula. *International Journal of Legal Medicine* 118: 197–205.

Arnaldos, M.I., E. Romera, J.J. Presa, A. Luna, and M.D. García. 2003c. Relationships within entomosarcosaprophagous fauna in a west mediterranean environment. *Proceedings of the First European meeting of the European Association for Forensic Entomology*, Frankfurt, p. 39.

Aznar Cervantes, S.D. 2006. Los Phoridae, todo un reto para la entomología forense. *Eubacteria* 17: 19–21.

Báguena, L. 1952. Algunas notas sobre entomología médico-legal. *Graellsia: Revista de Zoología* 10: 67–101.

Bahillo de la Puebla, P. and J.I. López Colón. 1999. Citas interesantes de cléridos de la Península Ibérica (Coleoptera, Cleridae). *Zoologica Baetica*, 10: 207–209.

Bahillo de la Puebla, P., M.I. Saloña Bordas, and J.I. López Colón. 2004. Confirmación de la presencia de *Omosita depressa* (Linnaeus, 17S8) en la Península Ibérica (Coleoptera, Nitidulídae). *Boletín de la SEA* 34: 161–162.

Barrientos, J.A. ed. 2004. *Curso práctico de Entomología*. Asociación española de Entomología, Universidad de Alicante CIBIO, Universitá Autònoma de Barcelona.

Battán Horestein, M., M.I. Arnaldos Sanabria, B. Rossode Ferradás, and M.D. García. 2005. Estudio preliminar de la comunidad sarcosaprófaga en Córdoba (Argentina): aplicación a la entomología. *Anales de Biología* 27: 91–202.

Baz, A., B. Cifrian, L.M. Diaz-Aranda, and D. Martin-Vega. 2007. The distribution of adult blowflies (Diptera: Calliphoridae) along an altitudinal gradient in central Spain. *Annales de la Societe Entomologique de France* 43: 289–296.

Baz, A., B. Cifrian, D. Martin-Vega, and M. Baena. 2010. Phytophagous insects captured in carrion-baited traps in central Spain. *Bulletin of Insectology* 63: 21–30.

Bergeret, M. 1855. Infanticide, Momification naturelle du cadavre. *Annales d'hygiène publique et de médecine légale* 4: 442–452.

Cainé, L.M., F. Corte Real, M.I. Saloña-Bordas, M.A. Martínez de Pancorbo, G. Lima, T. Magalhães, and F. Pinheiro. 2009. DNA typing of Diptera collected from human corpses in Portugal. *Forensic Science International* 184: e21–e23.

Carles-Tolrá, M., P.C. Rodríguez, and J. Verdú. 2010. *Thyreophora cynophila* (Panzer, 1794): collected in Spain 160 years after it was thought to be extinct (Diptera: Piopilidae: Thyreophorini). *Boletín de la SEA* 46: 1–7.

Castillo, M. 2001. El papel de las moscardas en la entomología forense: Dípteros necrófagos y colonización de carroñas. *Quercus* 186: 24–27.

Castillo Miralbés, M. 2001. Principales especies de Coleópteros necrófagos presentes en carroña de cerdos en la comarca de La Litera (Huesca). *Graellsia: revista de zoología* 57: 85–90.

Castillo Miralbés, M. 2002. *Estudio de la entomofauna asociada a cadáveres en el Alto Aragón (España)*. Monografías de la SEA. 94 pp.

Código Civil, 1889. Boletín Oficial del Estado; núm. 206, de 25 de julio de 1889, accessed, 14 November 2012.

Código Penal and Ley Orgánica 1995. *Boletín Oficial del Estado* BOE de 24 de Noviembre de 1995, 281: 1–610.

Díaz Martín B., A. López Rodríguez and M. I. Saloña Bordas. 2013. Primeros resultados sobre desarrollo de *Calliphora vicina* (Diptera, Calliphoridae) bajo condiciones controladas de temperatura. *Ciencia Forense* 11–12, in press.

Díaz Martín, B. and M.I. Saloña Bordas. 2012. Primera cita y nuevos datos sobre los hábitos necrófagos de *Trox scaber* (Linnaeus, 1767) (Coleoptera, Trogidae) en la Comunidad Autónoma del País Vasco (C. A. P. V.) *Boletín de la Asociación Española de Entomología* 36: 53–59.

Domínguez Fernández J. and L. Gómez Fernández. 1957. Observaciones en torno a la entomología tanatológica. Aportación experimental a los estudios de la fauna cadavérica. *Revista Ibérica de Parasitología* 17: 4–30.

Domínguez Fernández J. and L. Gómez Fernández. 1963. Momificación cadavérica particularmente rápida, operada bajo la acción de numerosas larvas de *Chrysomya albiceps*, Wiedemam, 1819. *Revista Ibérica de Parasitología* 23: 43–62.

García-Arribas, O., M. Pérez-Calvo, B. Ribas-Ozonas, and H. Labrousse. 2002. Diptera larvae biotest as alternative assay for detection of toxicity. *Proceedings of the First European Forensic Entomology* Seminar, Rosny, p. 100.

García-Rojo, A.M. 2004. Estudio de la Sucesión de Insectos en cadáveres en Alcalá de Henares (Comunidad Autónoma de Madrid) utilizando cerdos domésticos como modelos animales. *Boletín de la Sociedad Entomológica Aragonesa* 34: 263–269.

García-Rojo, A.M. and L. Honorato. 2004. A Case Description. *Proceedings of the Second meeting of the European Forensic Entomology,* London, p. 39.

García-Rojo, A.M. and L. Honorato. 2005. Three forensic cases involving *Hydrotaea capensis* (Diptera). *Proceedings of the Third meeting of the European Association for Forensic Entomology,* Lausanne, p. 34

García-Rojo, A.M. and L. Honorato. 2006. La Entomología Forense y la práctica policial en España: Estimación del intervalo post-mortem en un cadáver hallado en el interior de una arqueta en la Comunidad de Madrid. *Ciencia Forense* 8: 57–62.

García-Rojo, A.M., L. Honorato, and M.I. Saloña. 2008. The insect fauna associated with corpse in advance stage of decomposition found indoors in an abandoned building at Puerto de la Cruz (Tenerife, Canary Islands). *Proceedings of the the Sixth meeting of the European Association for Forensic Entomology,* Kolymbari, p. 73.

García-Rojo A.M, L. Honorato, M. González, and A. Téllez. 2009. Determinación del intervalo post-mortem mediante el estudio de la sucesión de insectos en dos cadáveres hallados en el interior de una finca rústica en Madrid. *Cuadernos de Medicina Forense,* 15 (56): 137–145.

GilArriortua, M., M.I. Saloña-Bordas, L.M. Cainé, F. Pinheiro, and M.M. de Pancorbo. 2013. Cytochrome b as a useful tool for the identification of blowflies of forensic interest (Diptera, Calliphoridae). *Forensic Science International* 228: 132–136.

Gobbi, P., A. Martínez-Sánchez, and S. Rojo. 2013a. The effects of larval diet on adult life-history traits of the Black Soldier Fly, *Hermetia illucens* (L.) (Diptera, Stratiomyidae). *European Journal of Entomology* 110: 461–468.

Gómez-Gómez, A., D. Martín-Vega, C. Botías Talamantes, A. Baz Ramos, and L.M. Díaz Aranda. 2007. La Entomología Forense en España: pasado, presente y perspectivas de futuro. *Cuadernos de medicina forense* 13: 21–31.

Gómez-Gómez, A., D. Martin-Vega, H.P. Tschorsnig, A. Baz, B. Cifrian, and L. Díaz-Aranda. 2010. Tachinids associated with carrion in the centre of the Iberian peninsula (Diptera: Tachinidae). *Entomologia Generalis* 32: 217–226.

González Herrera L, A. González Medina, G. Jiménez Ríos, and A. Valenzuela Garach. 2012. Estudio necrópsico de la muerte súbita cardíaca de origen isquémico, en un cuerpo en avanzado estado de descomposición. *Gaceta internacional de ciencia Forense* 3: 22–30.

González Medina A., L. González Herrera, and G. Jiménez Ríos. 2012a. Edaphic fauna and buried corpses: Arthropods found in coffins during exhumations in Granada (Spain). *Proceeding of 22nd Congress of the International Academy of Legal Medicine. Istanbul (Turkey).*

González Medina A., L. González Herrera, I. Martínez Téllez, F. Archilla Peña, J. de la Higuera Hidalgo, and G. Jiménez Ríos. 2011a. Estimación del intervalo post-emersión de un cadáver hallado en un embalse en Granada (España). *Cuadernos de medicina forense* 17: 137–144.

González Medina A., L. González Herrera, I. Martínez Téllez, F. Archilla Peña, and G. Jiménez Ríos. 2011b. Análisis patológico y entomológico de unos restos humanos hallados en una zanja en Granada (España). *Revista Española de Medicina Legal* 37: 113–116.

González Medina A., L. González Herrera, M.A. Perotti, and G. Jiménez Ríos. 2012b. Occurrence of *Poecilochirus austroasiaticus* (Acari: Parasitidae) in forensic autopsies and its application on postmortem interval estimation. *Experimental and Applied Acarology*, 59: 297–305.

González Mora, D. 1989. Los Calliphoridae de España, II: Calliphorini. *Eos*, 65: 39–59.

González Mora, D. and S.V. Peris. 1988. Los Calliphoridae de España, 1: Rhiniinae y Chrysomyinae. *Eos*, 64: 91–139.

González Mora, D. and S.V. Peris. 2004. Clave de identificación para los géneros de Calliphoridae del Mundo. Subfamilia con vena *remigium* desnuda y creación de una nueva subfamilia. *Boletín de la Real Sociedad Española de Historia Natural. Sección biológica* 99: 115–144.

González Mora D., S.V. Peris, and J.D. Sánchez Pérez. 1990. Un caso de entomología forense. *Revista Española de Medicina Legal* 17 (62–63/64–65): 19–21.

González Peña, C.F. 1997 Los insectos y la muerte. In: *Los Artrópodos y el Hombre*. Monografías de la SEA 20: 285–290.

Graells, M. 1886. Entomología Judicial. *Revista de la Real Academia de* Ciencias *Exactas, Físicas y Naturales* 1886: 458–447.

Judicial Organization Act 1985. *Boletín Oficial del Estado* BOE, 2 July 1985.

Jurisprudence of the Supreme Court. 2005. Sentencia nº 75/2005 de TS, Sala 2ª, de lo Penal, 25 de Enero de 2005.

Lecha Marzo, A. 1924. Tratado de autopsias y embalsamamientos. El diagnostico médico-legal en el cadáver. Ed. Plus Ultra.

Ley de Enjuiciamiento Criminal, 1882. *Boletín Oficial del Estado* BOE nº 260 de 17 de septiembre de 1882.

Lichtenstein J., A. Moitessier, and J. Alphonse. 1885. *Application de l'entomologie à la Médecine légale*. Ed. Montpellier.

Luna Maldonado, A. and M.D. García García. 2006. La enseñanza de entomología forense: la realidad española. *Ciencia forense: Revista Aragonesa de Medicina Legal* 8: 11–16.

Magaña, C. 2001. La Entomología Forense y su aplicación a la medicina legal: data de la muerte. *Boletín de la SEA* 28: 49–57.

Magaña, C., C. Andara, M.J. Contreras, A.J. Coronado Fonseca, E. Guerrero, D. Hernández, M. Herrera Millán, et al. 2006. Estudio preliminar de la fauna de insectos asociada a cadáveres en Maracay, Venezuela. *Entomotrópica: Revista Internacional Para El Estudio de la Entomología Tropical* 21: 53–59.

Magaña, C. and M. Hernández. 2000. Aplicaciones de la Entomología Forense en las investigaciones médico-legales. *Actas IX Congreso Ibérico de Entomología*, Zaragoza, España.

Magaña Loarte, C. and J.L. Prieto Carrero. 2009. Recogida de muestras para estudio entomológico forense. *Revista Española de Medicina Legal* 35: 39–43.

Martín-Vega, D. 2011. Skipping clues: Forensic importance of the family Piophilidae (Diptera). *Forensic Science International* 212: 1–5.

Martín-Vega, D. and A. Baz Ramos. 2010a. Datos sobre Dermaptera capturados en trampas cebadas con carroña en el centro de la Península Ibérica. *Boletín de la SEA* 46: 571–573.

Martín-Vega, D. and A. Baz Ramos. 2010b. *Prochyliza nigricornis* (Meigen, 1826) (Diptera: Piophilidae): Nueva especie para la fauna de la Península Ibérica. *Boletín de la Asociación Española de Entomología* 34: 249–251.

Martín-Vega, D. and A. Baz Ramos. 2011. Variation in the colour of the necrophagous fly, *Prochyliza nigrimana* (Diptera: Piophilidae): A case of seasonal polymorphism. *European Journal of Entomology* 108: 231–234.

Martín-Vega, D. and A. Baz Ramos. 2012. Spatiotemporal distribution of Necrophagous beetles (Coleoptera: Dermestidae, Silphidae) assemblages in natural habitats of Central Spain. *Annals of the Entomological Society of America* 105: 44–53.

Martín-Vega, D. and A. Baz Ramos. 2013. Sarcosaprophagous Diptera assemblages in natural habitats in central Spain: Spatial and seasonal changes in composition. *Medical and Veterinary Entomology* 27: 64–76.

Martín-Vega, D., A. Baz, and L.M. Díaz-Aranda. 2012. The immature stages of the necrophagous fly, *Prochyliza nigrimana*: Comparison with *Piophila casei* and medicolegal considerations (Diptera: Piophilidae). *Parasitology Research* 111: 1127–1135.

Martín-Vega, D., A. Baz, and V. Michelsen. 2010c. Back from the dead: *Thyreophora cynophila* (Panzer, 1798) (Diptera: Piophilidae) 'globally extinct' fugitive in Spain. *Systematic Entomology* 35: 607–613.

Martín-Vega, D., A. Gómez-Gómez, and A. Baz. 2011a. The "Coffin Fly" *Conicera tibialis* (Diptera: Phoridae). Breeding on buried human remains after a postmortem interval of 18 years. *Journal of Forensic Sciences* 56: 1654–1656.

Martín-Vega, D., A. Gómez-Gómez, A. Baz, and L.M. Díaz Aranda. 2011b. New piophilid in town: The first Palaearctic record of *Piophila megastigmata* and its coexistence with *Piophila casei* in central Spain. *Medical and Veterinary Entomology* 25: 64–69.

Martínez Ibáñez, M.D., E. López-Gallego, M.I. Arnaldos Sanabria, and M.D. García. 2012. New record of *Plagiolepis grassei* Le Masne, 1956 (Hymenoptera: Formicidae: Formicinae) in the Iberian Peninsula and its relation to the sarcosaprophagous fauna. *Boletín de la Asociación Española de Entomología* 36: 211–214.

Martínez Sánchez, A., C. Magaña, M. Saloña, and S. Rojo. 2011. First record of *Hermetia illucens* (Diptera:Stratiomyidae) on human corpses in Iberian Peninsula. *Forensic Science International* 206: e76–e78.

Martínez-Sánchez, A., S. Rojo, and M.A. Marcos García. 2000. Sarcofágidos necrófagos y coprófagos asociados a un agrosistema de dehesa (Diptera: Sarcophagidae). *Boletín de la Asociación Española de Entomología* 23 (1–2): 148–50.

Mašán, P., M.A. Perotti, M.I. Saloña-Bordas, and H.R. Braig. 2013. *Proctolaelaps euserratus*, an ecologically unusual melicharid mite (Acari, Mesostigmata) associated with animal and human decomposition. *Experimental and Applied Acarology* 2013: 1–15.

Mégnin, J. P. *La Faune des cadavres: application de l'entomologie à la médécine legale.* Enciclopedie Scientifique des Aide-Memories, Paris: Masson et Gautiers, 1894.

Morales Rayo, J., G. San Martín Peral, and M.I. Saloña Bordas. 2013. Primer estudio sobre la reducción cadavérica en condiciones sumergidas en la Península Ibérica, empleando un modelo de cerdo doméstico (Sus scrofa L., 1758) en el Río Manzanares (Comunidad Autónoma de Madrid). *Cienciaforense:Revista Aragonesa de Medicina Legal* 11–12, in press.

Ortuño, V.M., J. Fresneda, and A. Baz. 2010. New data on *Troglorites breuilli* Jeannel, 1919 (Coleoptera: Carabidae: Pterostichini) a hypogean Iberian species with description of a new subspecies. *Annales de la Societe Entomologique de France* 46: 537–549.

Pancorbo, M.M., R. Ramos, M. Saloña, and P. Sánchez. 2006. Entomología molecular forense. *Ciencia forense: Revista Aragonesa de Medicina Legal* 8: 107–132.

Pasquerault, T., B. Vincent, L. Dourel, B. Chauvet, and E. Gaudry. 2006. Los muestreos entomológicos: de la escena del crimen a la peritación. *Ciencia forense: Revista Aragonesa de Medicina Legal* 8: 39–56.

Pastor, B., H. Čičková, M. Kozánek, A. Martínez-Sánchez, and P. Takáč. 2011. Effect of the size of the pupae, adult diet, oviposition substrate and adult population density on egg production in *Musca domestica* (Diptera: Muscidae). *European Journal of Entomology* 108: 587–596.

Pérez de Petinto y Bertomeu, M. 1975. *La miasis cadavérica en la esqueletización.* Real Sociedad Española de Historia Natural. Consejo Superior de Investigaciones Científicas. Madrid.

Pérez-Moreno, S., M.A. Marcos-García, and S. Rojo. 2006. Comparative morphology of early stages of two Mediterranean *Sarcophaga* Meigen, 1826 (Diptera; Sarcophagidae) and a review of the feeding habits of Palaearctic species. *Micron* 37: 169–179.

Peris, S.V. and D. González Mora. 1991. Los Calliphoridae de España, III Luciliini (Diptera). *Boletín de la Real Sociedad Española de Historia Natural* 87: 187–207.

Piga Pascual, A. 1928. *Medicina legal de urgencia: la autopsia judicial.* Ed. Mercurio 719 pp.

Prado e Castro, C., E. Cunha, A. Serrano, and M.D. García. 2012. *Piophila megastigmata* (Diptera:Piophilidae): First records on human corpses. *Forensic Science International* 214: 23–26.

Prado e Castro, C., M.D. García, M.I. Arnaldos Sanabria, and D. González Mora. 2010a. Sarcophagidae (Diptera) atraídos a cadáveres de cochinillo, con nuevas citas para la fauna portuguesa. *Graellsia: Revista de Zoología* 66: 285–294.

Prado e Castro, C., M.D. García, A. Serrano, P. Gamarra Hidalgo, and R. Outerelo. 2010b. Staphylinid forensic communities from Lisbon with new records for Portugal (Coleoptera: Staphylinidae). *Boletín de la Asociación Española de Entomología* 34: 87–98.

Prieto Piloña, F. and J. Pérez Valcárcel. 2000. Coleoptera: Fam. 39. Silphidae. *Catalogus de la Entomofauna Aragonesa* 21: 3–10.

Prieto Piloña, F. and J. Pérez Valcárcel. 2002. Catálogo de los Silphidae (Coleoptera) de la Península Ibérica e Islas Baleares. *Boletín de la Sociedad Entomológica Aragonesa* 30: 1–32.

Ramos de Pablo, R., M. Saloña, E. Sarasola, S. Cardoso, and M.M. de Pancorbo. 2006. Molecular identification of *Stearibia nigriceps*: An example of the usefulness of Cytochrome b gene for the identification of entomofauna species. *International Congress Series* 1288: 864–866.

Ribas-Ozonas, B., O. García-Arribas, M. Pérez-Calvo, E.L.B. Novelli, and J.M. Escribano. 2002. Effects of magnetic fields 50 hz 2.7 mt on physiological parameters during ontogeny of diptera *P. argyrostoma*. *Proceedings of the First European Forensic Entomology Seminar*, 117–126.

Ríos, T. 1902a. Los insectos y la putrefacción de los cadáveres (I-II). La Clínica moderna. *Revista de Medicina y Cirugía* 1: 74–80 Royal Decree 767/1987. 1987. *Boletín Oficial del Estado* BOE, 22 June 1987.

Ríos, T. 1902b. Los insectos y la putrefacción de los cadáveres (III-IV). La Clínica moderna. *Revista de Medicina y Cirugía* 1: 171–180.

Romero Palanco, J.L., F. Munguía Girón, and J. Gamero Lucas. 2006. Entomología cadavérica en la provincia de Cádiz (S. de España). *Ciencia forense: Revista Aragonesa de Medicina Legal* 8: 83–106.

Saloña, M. 2003. Introducing forensic entomology in the Basque Country. *Proceedings of the Second meeting of the European Association for Forensic Entomology, Frankfurt*, p. 35.

Saloña, M. and D. González Mora. 2005. Primera cita de *Liosarcophaga aegyptica* Salem 1935 en la Península Ibérica, con descripción de sus fases larvarias II y III, puparios y adultos. *Boletín de la SEA* 36: 251–255.

Saloña, M.I., M. Carles-Tolrá, P. Bahillo, V. Iraola, and R. Alcaraz. 2004. Forensic Entomology Introduced in the Basque Country: A case study. *2nd European meeting of the European Association for Forensic Entomology*, London, p. 17.

Saloña, M.I, M.L. Moraza, M. Carles-Tolrá, V. Iraola, P. Bahillo, T. Yélamos, R. Outerelo, and R. Alcaraz. 2010. Searching the soil. Report about the importance of the edaphic fauna after the removal of a corpse. *Journal of Forensic Sciences* 55: 1652–1655.

Sánchez Sánchez, J.A. 2010. Organización de la medicina legal y forense en España. *Jano: Medicina y humanidades* 1766: 77–80.

Spanish Constitution, 1978. *Boletín Oficial del Estado* BOE 311, 29 December 1978, pp. 29313–29424.

State Security Forces Organic Law 2/1986, 1986. *Boletín Oficial del Estado* BOE 63, 14 March 1986, pp. 9604–9616.

Ubero-Pascal, N., R. López-Esclapez, M.D. García, and M.I. Arnaldos. 2012. Morphology of preimaginal stages of *Calliphora vicina* Robineau-Desvoidy, 1830 (Diptera, Calliphoridae): A comparative study. *Forensic Science International* 219: 228–243.

UNE-EN ISO/IEC 17025. Requisitos generales relativos a la competencia de los laboratorios de ensayo y calibración. AENOR, Madrid, 2000.

Velásquez Y., T. Ivorra, A. Grzywacz, A. Martínez-Sánchez, C. Magaña, A. García-Rojo, S. Rojo. 2012. Larval morphology, development and forensic importance of Synthesiomyia nudiseta (Diptera: Muscidae) in Europe: A rare species or just overlooked? *Bulletin of Entomological Research* 103: 98–110.

Velásquez, Y., C. Magaña, A. Martínez-Sánchez, and S. Rojo. 2010. Diptera of forensic importance in the Iberian Peninsula: Larval identification key. *Medical and Veterinary Entomology* 24: 293–308.

Velásquez Y, A. Martínez-Sánchez, and S. Rojo. 2013. First record of *Fannia leucosticta* (Meigen) (Diptera: Fanniidae) breeding in human corpses. *Forensic Science International* 229: e13–e15.

Ventura, D., B. Diaz, and M.I. Saloña. 2012. *Crossopalpus humilis* (Frey, 1913) en la Península Ibérica y la relación de la familia Hybotidae con cadáveres de vertebrados (Diptera, Empidoidea, Hybotidae). *Boletín de la SEA*, 50: 507–532.

PART III

History, Accomplishments, and Challenges of Forensic Entomology in Africa

CHAPTER 16

Africa

Martin H. Villet

CONTENTS

16.1 HISTORY

Forensic entomology is characteristically subdivided into medico-criminal, stored-product, urban, and environmental branches; only the first branch has been developed substantially in Africa. Fortunately, many of the insect species that commonly appear in stored-product and urban cases are synanthropic and therefore very widely distributed, so that these branches of forensic entomology can draw on studies of the same species conducted outside Africa. Organized medico-criminal forensic entomological research has a brief history in Africa, arguably starting with the publication of a series of papers on the carrion community (Prins 1979, 1982, 1983, 1984a,b), but drawing on a history of taxonomic, agricultural, medical, and ecological studies (Williams and Villet 2006a).

Taxonomic research on African carrion insects goes back to Linnaeus, Fabricius, Wiedemann, and Robineau-Desvoidy, and continues to this day through the activity of various international taxonomists, particularly in the area of molecular systematics.

The initial phase of biological research (1921–1950) was initiated by the South African wool industry because of the contemporary economic preeminence of this export, and terminated in George Ullyett's opus on the control of sheep strike (Ullyett 1950) and the advent of modern veterinary insecticides. Economic diversification into mining in South Africa lead to a period of research (1952–1965) on medical entomology related to migrant labor, primarily the studies around myiasis published by Fritz K.E. Zumpt. Subsequently (1968–1990), a series of ecological studies of the entomology of decomposition were published independently by André Prins, Leo Braack, and Ivan Meskin. Prins' work was a response to requests for forensic services from the South African Police and the Department of Health (Prins 1983), but the first explicitly forensic publication appeared only a decade later (Louw and van der Linde 1993) with subsequent accumulation of medico-criminal forensic entomology research in South Africa. All of these studies were reviewed by Williams and Villet (2006a) and several studies from across Africa have since been published. This review focuses on these developments.

16.2 ACCOMPLISHMENTS

The first requirement of all branches of forensic entomology is to identify the insects in question because this is an access point to the biological knowledge that will illuminate an investigation. Although the precise details are contingent on the particular case, the relevant biological information generally relates to the niche requirements of each species and to the development rates of its immature stages; other data, such as toxicology and taphonomy, may sometimes be relevant too.

16.2.1 Systematics and Identification

Many key members of the carrion insect community are synanthropic species that are readily transported by humans (Williams and Villet 2006b; Midgley et al. 2010), whereas others are sufficiently localized (Strümpher and Scholtz 2009; Midgley et al. 2012) to be at a risk of extinction. Carrion insect communities from different parts of the world will become more homogenous as these two processes continue, but presently these different biogeographical regions still retain distinctive carrion faunas. The cumulative catalogue of the African carrion insect community is now quite extensive, but it certainly remains incomplete because the indigenous faunas of many countries are thinly documented and because foreign species are still immigrating. The characteristic fauna of corpses has now received attention in several countries across the major faunistic zones of West, North, East, and southern Africa (reviewed by Villet [2011]). The principal community members (blow flies, flesh flies, larder beetles, and carrion beetles) are present throughout the region, whereas groups like the hide beetles, hister beetles, and spider beetles show more regional diversity.

Additions to the African list of carrion-inhabiting insects since 2006 include the parasitoid wasp, *Coelalysia nigriceps* (Szépligeti) (Hymenoptera: Braconidae) (Braet et al. 2012), the dung beetle, *Frankenbergerius forcipatus* (Harold) (Coleoptera: Scarabaeidae) (Midgley et al. 2012), and the sap beetle, *Omosita japonica* Reitter (Coleoptera: Nitidulidae) (Midgley et al. unpublished data). Although some *Omosita* species have been inadvertently dispersed globally by humans, this is the first record of the genus occurring in Africa. *Pachylister* (*Sulcignathos*) *luctuosus* (Marseul) (Coleoptera: Histeridae) has been recovered from cow dung and beneath a dead camel (Mazur 2010), and a group of small, flightless, locally endemic hide beetles (Coleoptera: Trogidae) have been described from South Africa (Strümpher and Scholtz 2009).

The numerous identification guides to the known African carrion insects have not been consolidated, but a review of relevant taxonomic resources is available (Villet 2011). Since that review, keys were published for the morphological identification of the first-instar larvae of most African carrion blow flies (Calliphoridae), including a critical review of literature on the identification of all blow fly instars (Szpila and Villet 2011), adults of the (sub)genera of larder beetles (Dermestidae) (Háva et al. 2013), and adults of the African species of *Phradonoma* Jacquelin Du Val (Dermestidae) (Háva et al. 2013). Work is in progress on the larvae of the African carrion beetles (Silphidae).

Molecular identification of most African carrion flies is now possible based on mitochondrial genes, primarily cytochrome oxidase I (*COI*) and 16S rRNA (*16S*) (Harvey et al. 2008;

Tourle et al. 2009; Marinho et al. 2012; Williams and Villet 2013). Parallel research on carcass beetles is well advanced (Midgley et al. 2012; Ridgeway et al. 2014; Strümpher et al. 2014; Collett et al. unpublished data). These mitochondrial markers generally perform well, but there are a few instances where nuclear markers are necessary. In particular, *Lucilia sericata* (Meigen) and *L. cuprina* (Wiedemann) (Calliphoridae) apparently hybridize naturally in southern Africa (Williams and Villet 2013) and a lineage of *L. cuprina* is present in Africa that carries mitochondrial DNA derived from *L. sericata*, apparently acquired by introgression about 300,000 years ago (Harvey et al. 2008; Tourle et al. 2009; Williams and Villet 2013). Nuclear genes that are reliable taxonomic markers in African carrion insects include the period protein (*per*), 28S rRNA (*28S*), 18S rRNA (*18S*), and the second internal transcribed spacer (*ITS2*) (Marinho et al. 2012; Williams and Villet 2013; Strümpher et al. 2014).

16.2.2 Ecology

All branches of forensic entomology rely on information about the ecological requirements and behavior of their focal insects. The bulk of this information is about the autecology of individual species, but the community ecology of carcasses and corpses is germane to medico-criminal forensic entomology (Tomberlin et al. 2011; Villet 2011).

At the ecophysiological level, the level of individual insects, the time that insects take to detect and colonize a corpse is partially dictated by the interaction of their thermophysiological thresholds, basking behavior, and the weather. The acute upper lethal temperature of mature larvae, preferred pupation temperature, and acute upper lethal temperature and thermal thresholds for uncoordinated and coordinated neuromuscular activity of adults have been established for seven widespread species of blow flies (Richards et al. 2009b). These physiological indices are related to these species' seasonal distributions and relative spatial distributions on carcasses and corpses (Richards et al. 2009b).

At the population level, the ecology of forensically relevant African insects has yet to be studied.

At the species level, insight into the ecological niche characteristics of African blow flies was generated by a comprehensive survey of the geographical distributions for blow flies in South Africa (Richards et al. 2009c). Some species, like *Chrysomya inclinata* Walker, had restricted spatial distributions, whereas others did not but showed seasonal changes in abundance. Using climatic data from localities where the flies were known to occur, a model of the potential distribution of each species was constructed and analyzed to identify the climatic variables that were most influential in the model. The presence of summer and winter rainfall regions in South Africa provides a natural control for the correlation between temperature and moisture. The general conclusion was that moisture was slightly more important than temperature in predicting geographical distributions. The high vagility of individual blow flies (Braack and Retief 1986; Braack and de Vos 1990) and their ability to thermoregulate by basking may partially explain the uncoupling of distribution from temperature. This insight allows a more nuanced interpretation of the occurrence of each species at a particular body.

Biogeographical models of this sort are useful for predicting the distribution of invasive species. Such species (e.g., *Chrysomya megacephala* Fabricius and *Calliphora vicina* Robineau-Desvoidy) are still colonizing Africa (Williams and Villet 2006b), and an African cheese skipper, *Piophila megastigmata* McAlpine (Diptera: Piophilidae), is now being found on corpses on other continents (Prado e Castro et al. 2011). *C. albiceps* (Wiedemann), which is ubiquitous in Africa, has become widespread in South America and Europe (do Prado and Guimarães 1982; Grassberger et al. 2003; Verves 2004), possibly due as much to changing climate as to direct human intervention (Verves 2004; cf. Rosati and VanLaerhoven 2007).

At the community level, the pattern of ecological succession on pig carcasses, was studied in Nigeria (Ekanem and Dike 2010), South Africa (Kelly et al. 2008, 2009, 2011), and Egypt (Aly et al.

2013), and fits the general pattern described from other continents (Villet 2011). Modifications of the succession process due to the presence of clothing, wrapping of the body, and wounds to the body were described (Kelly et al. 2008, 2009, 2011), and also followed patterns were reported from elsewhere.

The place of indigenous species in the succession process was predictable from their taxonomy. For example, African sexton beetles of the genus *Thanatophilus* Leach arrived in a time frame usually associated with flesh flies (Sarcophagidae), as do congeneric beetles on other continents (Ridgeway et al. 2014). This means that the succession pattern can be understood abstractly in terms of ecological guilds rather than particular species, and a framework for understanding the functional dynamics of carrion community succession has been proposed, based on a revised classification of guilds centered on diet (Villet 2011). This model gives focus to adaptations of life history traits to diet quality, competition, and risk of predation. It also suggests that there are only two distinctive stages in the succession, a "wet" phase dominated by flies and a "dry" phase dominated by beetles of various families (Schoenly and Reid 1987). The stages of physical decomposition of bodies that are commonly recounted in textbooks about forensic entomology should be treated as milestones that have descriptive rather than explanatory value for understanding the succession process in carrion communities (Schoenly and Reid 1987; Villet 2011).

Ecological interactions between species that alter the *course* of ecological succession appear to be uncommon in African carrion communities (Villet 2011). However, interguild predation (e.g., of eggs and larvae by ants) can affect the abundance of each species that is present, and may even extirpate a functionally important guild (at least in the laboratory) (Grassberger et al. 2003; Kheirallah et al. 2007), and so affect the course of succession. On the other hand, intraguild predation can change the relative abundances of guild members, but the course of succession alters little because the same functional guilds are still present. There is evidence that the presence of eggs or larvae of *C. albiceps*, a species with facultatively predatory larvae (Grassberger et al. 2003), discourages oviposition by other flies (Gião and Godoy 2007), and competitive or predatory interactions between *C. albiceps* and species such as or *L. sericata*, *C. megacephala*, and *Cochliomyia macellaria* (Fabricius) (since some of this research was done in Brazil, where *C. albiceps* has been introduced) affected survivorship under confined laboratory conditions (Grassberger et al. 2003; Rosa et al. 2004, 2006; Kheirallah et al. 2007). Factors that ameliorate these interactions in the field include the late arrival of *C. albiceps* at oviposition sites (Villet 2011) and the spatial partitioning of larvae on carcasses (Richards et al. 2009b; Ridgeway et al. 2014).

16.2.3 Developmental Biology

Qualitative accounts of the development of some carrion flies and beetles were given by Prins (1982), and quantitative developmental models have been parameterized for several common African blow flies, flesh flies, and carrion beetles (Table 16.1). The sparse data for flesh flies from around the globe were reviewed by Villet et al. (2006). An important insight from the work on beetle larvae is that, even in the laboratory, individual larvae may become unwell, cease normal growth, and eventually die (Ridgeway et al. 2014). These individuals are easily identified in laboratory studies of beetles, but in flies, which have to be raised in groups if they are to grow normally, these outliers can be difficult to detect, creating an additional source of imprecision in developmental models for these insects.

Comparisons of the development dynamics of sister species under the same laboratory and analytical conditions has provided the first explicit confirmation that developmental parameters cannot be transferred from one species to another (Richards et al. 2009a; Ridgeway et al. 2014). The quality assurance implications of standardizing conditions were illustrated by a review of 12 studies of *C. megacephala* (Richards and Villet 2009). Internationally applicable guidelines for statistically informative sampling of larvae for developmental studies were published (Richards and Villet 2008).

Table 16.1   Species of African Blow Flies, Flesh Flies, and Carrion Beetles for Which Quantitative Developmental Models Have Been Parameterized Using African Populations

Taxon	Source
Diptera: Flies	
Calliphoridae: Blow flies	
Chrysomya albiceps	Richards et al. (2008)
C. chloropyga	Richards and Villet (2008); Richards et al. (2009a)
C. megacephala	Richards and Villet (2009)
C. putoria	Richards et al. (2009a)
C. marginalis	N. Pegg, C. Richards, and M. Villet, unpublished data
C. inclinata	M. Villet, unpublished data
Lucilia sericata	N. Pegg and M. Villet unpublished data
L. cuprina	Z. Kotze, unpublished data
Sarcophagidae: Flesh flies	
Sarcophaga (Bercaea) Africa	Madubunyi (1986); Aspoas (1991)
S. (Liopygia) argyrostoma	Hafez (1940); Zohdy and Morsy (1982)
S. (Liopygia) nodosa	Aspoas (1991)
S. (Liosarcophaga) dux	Aspoas (1991)
S. (Liosarcophaga) tibialis	Aspoas (1991); Villet et al. (2006)
Wohlfahrtia pachytyli	Price and Brown (2006)
Coleoptera: Beetles	
Silphidae: Carrion beetles	
Thanatophilus micans	Midgley and Villet (2009a); Ridgeway et al. (2014)
T. mutilatus	Ridgeway et al. (2014)

Finally, there are strong indications that the developmental dynamics of populations of a species may be adapted to local conditions. A review of the development of *C. albiceps* indicated that there may be a latitudinal gradient in the values of the lower developmental thresholds and thermal summation constants of different populations (Richards et al. 2008). It would be useful to confirm this pattern, but until then this result suggests that developmental data should be generated for every species present at a locality in a climatic zone that is different from that of the benchmark populations used in published studies.

On the technical side, it has been shown (Midgley and Villet 2009b) that, because their exoskeletons are unevenly sclerotized, beetle larvae curl when killed using the standard method (immersion in near-boiling water) recommended for killing flies (Adams and Hall 2003; Amendt et al. 2007), which makes them difficult to measure. It was, therefore, recommended that beetle larvae relevant to an investigation should be drowned and preserved in 96% ethanol, which facilitates both measurement and DNA extraction. The ethanol should be drained after a few hours and replaced for long-term storage because the body fluids in the larvae dilute the ethanol (Martin 1977; King and Porter 2004), leading to deterioration of the specimens, particularly when the ethanol concentration falls below about 70%. The specimens preserved in this way by Midgley and Villet (2009b) were still in good condition after 5 years.

The design for an easily manufactured gauge for measuring live and preserved larvae has been published (Villet 2007).

16.2.4 Taphonomy and Toxicology

Taphonomic research involving insects was done on mummies and bones of archaeological ages. Rats were experimentally mummified using two ancient methods and attracted six carrion insects (Abdel-Maksoud et al. 2011). The methods are not representative of most medico-criminal cases, but have academic importance. On the other hand, termites (Isoptera) colonize carcasses, which is of taphonomic interest because they may damage bone in characteristic ways and affect the final stages of the decomposition process (Backwell et al. 2012). This damage includes subparallel striations, star-shaped marks, and etching produced by chewing, all of which might be mistaken for toolmarks by naïve observers.

African entomotoxicological research is very sparse, and research done in other regions and countries is not necessarily directly pertinent to African species. A catabolic glucocorticoid (hydro-cortisone) and a barbiturate stimulant (sodium methohexital) had detectable but erratic effects on the developmental rate of *Sarcophaga tibialis* Macquart (Musvasva et al. 2001), whereas two contraceptive steroids (medroxyprogesterone acetate and norethisterone enanthate) had no detectable effects on the developmental rate of *C. chloropyga* (Wiedemann) (da Silva and Villet 2006). The only theoretical contribution that has been made to this field is that, even when a drug is known to affect the rate of development of insects in a predictable way, it is practically impossible to work backward from a measurement of the drug concentrations in the insects' tissues or the body to a precisely corrected estimate of the insects' developmental rates. This is because (1) the amounts of drugs consumed by the person and the larvae are decoupled by the interaction between drug tropism in the body and migration of the larvae among body regions while feeding and (2) the amounts of drugs consumed by the larvae and subsequently detected in them by laboratory analysis are decoupled by time-dependent processes of metabolism, excretion, and sequestration (Musvasva et al. 2001; da Silva and Villet 2006).

16.2.5 Casework

The South African police have consulted local professional entomologists at various times, including M.J. Oosthuizen (1950s), Albert J. Hesse (1960s), Fredrik W. Gess (1960s), André Prins (1970s–1980s), Theunis C. van der Linde (1900s–2010), and Mervyn W. Mansell (mid-1990s–2000s), while Barry Blair (mid-1960s–1990s) filled a similar role in Zimbabwe (Williams and Villet 2006). An academic report on the entomology of 17 South African murder cases was published by Louw and van der Linde (1993) and another case was included in an assessment of the value of African *Thanatophilus* beetles in forensic entomology (Ridgeway et al. 2014).

Few of the early cases required court testimony, but in 2000 a high-profile conviction to draw on insect-related evidence occurred when the Johannesburg Supreme Court, South Africa, convicted Albert Myburgh of murder using entomological evidence provided by Dr. Mansell (du Plessis and Meintjes-van der Walt 2004). Mansell had assisted the South African police in the Moses Sithole serial murders in 1995 and subsequently in the Phoenix serial murders, the Pretoria Highwayman serial murders, the Pretoria Koppies serial murders, and the high-profile murder cases of Jacqueline Carter, Michael Bottrill, Marlene Konings, and Leigh Matthews. To transfer relevant skills to police members, Mansell also lectured to the South African Serious and Violent Crime Units and the Investigative Psychology Units, and in 2005, entomological evidence was presented in court in the Leigh Matthews case by Police Forensic Science Laboratory biologist Sergeant André Massyn.

Judging from the small volumes of research literature emanating from other African countries, there is a pronounced need for capacity development at the academic and applied levels throughout the region.

16.3 CHALLENGES

There is obvious scope for the development of entomotoxicology in Africa, but forensic entomology faces even bigger human, academic, and administrative challenges, some of which are universal.

The most pressing of these relate to personnel and the deployment of forensic entomology in casework. A steady stream of postgraduates has qualified in topics that explicitly involved forensic entomology (Lunt 2002; Williams 2003; Kelly 2006; Midgley 2008; Richards 2008; Kolver 2009; Brink 2012). This is impressive, given that the country has only 48 million inhabitants, most of whom lack tertiary education. Their recruitment into professional forensics has been less spectacular: three of these graduates are currently employed as museum curators, two are in industry, one is a mammal ecologist, and one now is a university lecturer specializing in parasitology. Three professional entomologists have provided forensic services to the South African Police Service in the last two decades (Williams and Villet 2006). Of these, Prof. Theunis van der Linde (who supervised several of the recent graduates) has recently retired and Prof. Mervyn Mansell (who investigated hundreds of South African cases) is now involved in the food biosecurity sector. The situation in the rest of Africa is unclear, but at least two students in Cameroon and Zimbabwe are actively conducting research toward higher degrees in forensic entomology.

Secondary to the challenge of human capacity is that of academic research capacity. Many African insects of forensic significance are widespread and the research already accomplished on them in South Africa will therefore be helpful in many other countries, but this assumption needs to be tested. A caveat has been raised by the geographical variation in developmental parameters that has been illustrated in *C. albiceps* (Richards et al. 2008), which implies that forensic entomological research programs should be instigated in the major biogeographical regions of Africa. Within Africa there is a growing interest in forensic entomological research, with recent publications coming from Egypt, Nigeria, and Cameroon (Adham et al. 2001; Ekanem and Umoetuk 2009; Ekanem and Dike 2010; Braet et al. 2012; Aly et al. 2013).

Besides personnel and benchmark data, the third major challenge facing African forensic entomology is the certification of techniques and the accreditation of service providers (Villet and Amendt 2011). These topics have barely been broached in Africa, where legal systems tend to follow developments elsewhere (du Plessis and Meintjes-van der Walt 2004). Unfortunately, the matters of quality assurance, certification, and accreditation of expert evidence in forensic entomology are far from being resolved anywhere in the world. On the other hand, peer-reviewed studies of African insects have now been published that go some way to meeting standards such as the Frye and Daubert standards of the United States for expert evidence (Villet and Amendt 2011), and the assumptions underlying the interpretation of such evidence are being made explicit (Richards and Villet 2008 2009; Villet et al. 2010; Villet 2011). The "CSI Effect" (Cole 2013; Hayes and Levitt 2013) has increased public enthusiasm for forensic science in general, and its negative aspects are less likely to affect civil law judicial systems that do not widely use a jury system, for example, the Islamic, French, Spanish, and Roman-Dutch laws that dominate Africa.

16.4 CONCLUSION

Forensic entomology in Africa has a good groundwork of baseline data to draw on and has been exercised in a few hundred investigations, some of which have even required expert testimony to be presented to court (Williams and Villet 2006a). The growing corps of relevantly qualified graduates exceeds the current demand for their forensic services in Africa, and there is more than ample opportunity for forensic entomology to be taken up throughout the continent. The insights that have come from African research in this field are internationally applicable, and are apparently most likely to bear fruit overseas at present (Villet 2011). A perception that entomological evidence is imprecise and that expertise is scarce (du Plessis and Meintjes-van der Walt 2004) has not been substantially eroded by recent high-profile cases, but educational initiatives are starting to generate more positive perceptions, and international citation of African research will help to establish its reliability with African judicial systems.

ACKNOWLEDGMENTS

This review is dedicated to Theunis van der Linde and Mervyn Mansell, who have greatly stimulated and promoted the formal development of forensic entomology in Africa. I thank Jeff Tomberlin and Eric Benbow for contributing to this chapter and Rhodes University for sustained logistical support in completing it.

REFERENCES

Abdel-Maksoud, G., E. Abed Al-Sameh Al-Shazly, and A.R. El Amin. 2011. Damage caused by insects during the mummification process: An experimental study. *Archaeological and Anthropological Sciences* 3: 291–308.

Adams, Z.J.O. and M.J.R. Hall. 2003. Methods used for the killing and preservation of blowfly larvae, and their effect on post-mortem larval length. *Forensic Science International* 138: 50–61.

Adham, F.K., M.A. Abdel, M.A. Tawfik, and R.M. El-Khateeb. 2001. Seasonal incidence of the carrion breeding blowflies *Lucilia sericata* (Meigen) and *Chrysomya albiceps* (Wied.) (Diptera: Calliphoridae) in Abu-Rawash Farm, Giza, Egypt. *Journal of Veterinary Medicine* 49: 377–383.

Aly, S.M., J. Wen, X. Wang, J. Cai, Q. Liu, and M. Zhong, 2013. Identification of forensically important arthropods on exposed remains during summer season in northeastern Egypt. *Journal of the Central South University (Medical Science)* 38: 1–6.

Amendt, J., C.P. Campobasso, E. Gaudry, C. Reiter, H.N. Leblanc, and M.J.R. Hall. 2007. Best practice in forensic entomology—Standards and guidelines. *International Journal of Legal Medicine* 121: 90–104.

Backwell, L., A.H. Parkinson, E.M.Roberts, F. D'Errico, and J-B. Huchet. 2012. Criteria for identifying bone modification by termites in the fossil record. *Palaeogeography, Palaeoclimatology, Palaeoecology* 337–338: 72–87.

Braack, L. and V. de Vos. 1990. Feeding habits and flight range of blow-flies (*Chrysomya* spp.) in relation to anthrax transmission in the Kruger National Park, South Africa. *Onderstepoort Journal of Veterinary Research* 57: 141–142.

Braack, L. and P.F. Retief. 1986. Dispersal, density and habitat preference of the blow-flies *Chrysomyia albiceps* (Wd.) and *Chrysomyia marginalis* (Wd.) (Diptera: Calliphoridae). *Onderstepoort Journal of Veterinary Research* 53: 13–18.

Braet, Y., C. van Achterberg, and F.D. Feugang Youmessi. 2012. *Coelalysia nigriceps* (Szépligeti, 1911) reared during a forensic study in Cameroon, with remarks on synonymy and biology (Hymenoptera, Braconidae, Alysiinae). *Bulletin de la Société Entomologique de France* 117: 381–389.

Brink, S.L. 2012. Key diagnostic characteristics of the developmental stages of forensically important Calliphoridae and Sarcophagidae in central South Africa, ix + 257 pp. PhD thesis, University of the Free State, Bloemfontein, South Africa.

Cole, S.A. 2013. A surfeit of science: The "CSI effect" and the media appropriation of the public understanding of science. *Public Understanding of Science*. DOI: 10.1177/0963662513481294.

da Silva, C. and M.H. Villet. 2006. Effects of prophylactic progesterone in decomposing tissues on the development of *Chrysomya chloropyga* (Wiedeman) (Diptera: Calliphoridae). *African Entomology*. 14: 199–202.

do Prado, A.P. and J.H. Guimarães. 1982. Estado atual de dispersao e distribuicao do gênero *Chrysomya* Robineau-Desvoidy na regiao neotropical [Current status of dispersal and distribution of the genus *Chrysomya* Robineau-Desvoidy in the Neotropical Region] (Diptera, Calliphoridae). *Revista Brasileira de Biologia* 26: 225–231.

du Plessis, M. and L. Meintjes-van der Walt. 2004. Forensic entomology: Relevant to legal dispute resolution? *Journal for Juridical Science* 29: 100–121.

Ekanem, M.S. and M.C. Dike. 2010. Arthropod succession on pig carcasses in southeastern Nigeria. *Papéis Avulsos de Zoologia (São Paulo)* 50: 561–570.

Ekanem M.S. and S. Umoetuk. 2009. The immature stages of three carrion breeding blowflies (Diptera: Calliphoridae) in south eastern Nigeria. *The Zoologist* 7: 152–161.

Gião, J.Z. and W. Godoy. 2007. Ovipositional behavior in predator and prey blowflies. *Journal of Insect Behavior* 20: 77–86.

Grassberger, M., E. Friedrich, and C. Reiter. 2003. The blowfly *Chrysomya albiceps* (Wiedemann) (Diptera: Calliphoridae) as a new forensic indicator in central Europe. *International Journal of Legal Medicine* 117: 75–81.

Harvey, M.L., S. Gaudieri, M.H. Villet, and I.R. Dadour. 2008. A global study of forensically significant calliphorids: Implications for identification. *Forensic Science International* 177: 66–76.

Háva, J., T. Lackner, and J. Mazancová. 2013. Description of *Phradonoma blabolili* sp. n. (Coleoptera, Dermestidae, Megatominae), with notes on the dermestid beetles from Angola. *ZooKeys* 293: 65–76.

Hayes, R.M. and L.M. Levett. 2013. Community members' perceptions of the CSI effect. *American Journal of Criminal Justice* 38: 216–235.

Kelly, J.A. 2006. The influence of clothing, wrapping and physical trauma on carcass decomposition and arthropod succession in central South Africa, 218 pp. PhD thesis, University of the Free State, Bloemfontein, South Africa.

Kelly, J.A., T.C. van der Linde, and G.S. Anderson. 2008. The influence of clothing and wrapping on carcass decomposition and arthropod succession: A winter study in central South Africa. *Journal of the Canadian Society of Forensic Science* 41: 135–147.

Kelly, J.A., T.C. van der Linde, and G.S. Anderson. 2009. The influence of clothing and wrapping on carcass decomposition and arthropod succession during the warmer seasons in central South Africa. *Journal of Forensic Science* 54: 1105–1112.

Kelly, J.A., T.C. van der Linde, and G.S. Anderson. 2011. The influence of wounds, severe trauma, and clothing, on carcass decomposition and arthropod succession in South Africa. *Canadian Society of Forensic Science Journal* 44: 144–157.

Kheirallah, A.M, T.I. Tantawi, A.H. Aly, Z.A. El-Moaoty. 2007. Competitive interaction between larvae of *Lucilia sericata* (Meigen) and *Chrysomya albiceps* (Wiedemann) (Diptera: Calliphoridae). *Pakistan Journal of Biological Science* 10: 1001–1010.

King, J.R. and S.D. Porter. 2004. Recommendations on the use of alcohols for preservation of ant specimens (Hymenoptera, Formicidae). *Insectes Sociaux* 51: 197–202.

Kolver, J.H. 2009. Forensic entomology: The influence of the burning of a body on insect succession and calculation of the postmortem interval, xiii + 302 pp. PhD thesis, University of the Free State, Bloemfontein, South Africa.

Louw, S. and T.C. van der Linde. 1993. Insects frequenting decomposing corpses in central South Africa. *African Entomology* 1: 265–269.

Lunt, N. 2002. Applied studies of some southern Africa blowflies (Diptera Calliphoridae) of forensic importance, xii + 224 pp. MSc thesis, Rhodes University, Grahamstown, South Africa.

Madubunyi, L.C. 1986. Laboratory life history parameters of the red-tailed fleshfly, *Sarcophaga haemorrhoidalis* Fallen (Diptera: Sarcophagidae). *Insect Science and Its Application* 7: 617–621.

Marinho, M.A.T., A.C.M. Junqueira, D.F. Paulo, M.C. Esposito, M.H. Villet, and A.M.L. Azeredo-Espin. 2012. Molecular phylogenetics of Oestroidea (Diptera: Calyptratae) with emphasis on Calliphoridae: Insights into the inter-familial relationships and additional evidence for paraphyly among blowflies. *Molecular Phylogenetics and Evolution* 65: 840–854.

Martin, J.E.H. 1977. *The Insects and Arachnids of Canada Part 1: Collecting, Preparing, and Preserving Insects, Mites, and Spiders*, 182 pp. Publication 1643. Hull, Québec, Canada: Research Branch, Canada Department of Agriculture.

Mazur, S. 2010. *Sulcignathos*, a new subgenus of *Pachylister* Lewis, 1904 (Coleoptera: Histeridae). *Annales Zoologici* 60: 209–214.

Midgley, J.M. 2008. Aspects of the thermal ecology of six species of carcass beetles in South Africa, xii + 68 pp. MSc thesis, Rhodes University, Grahamstown, South Africa.

Midgley, J.M., I.J. Collett, and M.H. Villet. 2012. The distribution, habitat, diet and forensic significance of the scarab *Frankenbergerius forcipatus* (Harold, 1881) (Coleoptera: Scarabaeidae). *African Invertebrates* 53: 745–749.

Midgley J.M., C.S. Richards, and M.H. Villet. 2010. The utility of Coleoptera in forensic investigations. In: Amendt J., Campobasso C.P., Goff M.L., and Grassberger M., eds. *Current Concepts in Forensic Entomology*, pp. 57–68. Heidelberg, Germany: Springer.

Midgley J.M. and M.H. Villet. 2009a. Development of *Thanatophilus micans* (Fabricius 1794) (Coleoptera: Silphidae) at constant temperatures. *International Journal of Legal Medicine* 123: 285–292.

Midgley J.M. and M.H. Villet. 2009b. Effect of the killing method on post-mortem change in length of larvae of *Thanatophilus micans* (Fabricius, 1794) (Coleoptera: Silphidae) stored in 70% ethanol. *International Journal of Legal Medicine* 123: 103–108.

Musvasva E, K.A. Williams, W.J. Muller, and M.H. Villet. 2001. Preliminary observations on the effects of hydrocortisone and sodium methohexital on development of *Sarcophaga (Curranea) tibialis* Macquart (Diptera: Sarcophagidae), and implications for estimating post mortem interval. *Forensic Science International* 120: 37–41.

Prado e Castro, C., E. Cunha, A. Serrano, and M.D. García. 2012. *Piophila megastigmata* (Diptera: Piophilidae): First records on human corpses. *Forensic Science International* 214: 23–26.

Prins, A.J. 1979. Discovery of the oriental latrine fly *Chrysomyia megacephala* (Fabricius) along the south-western coast of South Africa. *Annals of the South African Museum* 78: 39–47.

Prins, A.J. 1982. Morphological and biological notes on six South African blowflies (Diptera, Calliphoridae) and their immature stages. *Annals of the South African Museum* 90: 201–217.

Prins, A.J. 1983. Morphological and biological notes on some South African arthropods associated with decaying organic matter. I: Chilopoda, Diplopoda, Arachnida, Crustacea, and Insecta. *Annals of the South African Museum* 92: 53–112.

Prins, A.J. 1984a. Morphological and biological notes on some South African arthropods associated with decaying organic matter II: The predatory families Carabidae, Hydrophilidae, Histeridae, Staphylinidae and Silphidae (Coleoptera). *Annals of the South African Museum* 92: 295–356.

Prins, A.J. 1984b. Morphological and biological notes on some South African arthropods associated with decaying organic matter. III: The families Dermestidae, Cantharidae, Melyridae, Tenebrionidae, and Scarabaeidae (Coleoptera). *Annals of the South African Museum* 94: 203–304.

Richards, C.S. 2008. Effects of temperature on the development, behaviour and geography of blowflies in a forensic context, xvi + 144 pp. MSc thesis, Rhodes University, Grahamstown, South Africa.

Richards, C.S., K.L. Crous, and M.H. Villet. 2009a. Models of development for the blow fly sister species *Chrysomya chloropyga* and *C. putoria*. *Medical and Veterinary Entomology* 23: 56–61.

Richards, C.S., I.D. Paterson, and M.H. Villet. 2008. Estimating the age of immature *Chrysomya albiceps* (Diptera: Calliphoridae), correcting for temperature and geographical latitude. *International Journal of Legal Medicine* 122: 271–279.

Richards, C.S., B.W. Price, and M.H. Villet. 2009b. Thermal ecophysiology of seven carrion-feeding blowflies in Southern Africa. *Entomologia Experimentalis et Applicata* 131: 11–19.

Richards, C.S. and M.H. Villet. 2008. Factors affecting accuracy and precision of thermal summation models of insect development used to estimate postmortem intervals. *International Journal of Legal Medicine* 122: 401–408.

Richards, C.S. and M.H. Villet. 2009. Data quality in thermal summation models of development of forensically important blowflies (Diptera: Calliphoridae): A case study. *Medical and Veterinary Entomology* 23: 269–276.

Richards, C.S., K.A. Wiliams, and M.H. Villet. 2009c. Predicting geographic distribution of seven blowfly species (Diptera: Calliphoridae) in South Africa. *African Entomology* 17: 170–182.

Ridgeway, J., J.M. Midgley, I.J. Collett, and M.H. Villet. 2014. Advantages of using development models of the carrion beetles *Thanatophilus micans* (Fabricius) and *T. mutilatus* (Castelneau) (Coleoptera: Silphidae) for estimating minimum post mortem intervals, verified with case data. *International Journal of Legal Medicine*. 128: 207–220.

Rosa, G.S., L.R. de Carvalho, S.F. Dos Reis, and W. Godoy. 2006. The dynamics of intraguild predation in *Chrysomya albiceps* Wied. (Diptera: Calliphoridae): Interactions between instars and species under different abundances of food. *Neotropical Entomology* 35: 775–780.

Rosa, G.S., L.R. de Carvalho, and W. Godoy. 2004. Survival rate, body size and food abundance in pure and mixed blowfly cultures. *African Entomology* 12: 97–105.

Schoenly, K.G and W. Reid. 1987. Dynamics of heterotrophic succession in carrion arthropod assemblages: Discrete seres or a continuum of change? *Oecologia* 73: 192–202.

Strümpher, W. and C.H. Scholtz. 2009. New species and status changes of small flightless relictual *Trox* Fabricius from southern Africa (Coleoptera: Trogidae). *Insect Systematics and Evolution* 40: 71–84.

Strümpher, W., C.L. Sole, M.H. Villet, and C.H. Scholtz. 2014. Phylogeny of the family Trogidae (Coleoptera: Scarabaeoidea) inferred from mitochondrial and nuclear ribosomal DNA sequence data. *Systematic Entomology* 39: 548–562.

Szpila, K. and M.H. Villet. 2011. Morphology and identification of first instar larvae of African blowflies (Diptera: Calliphoridae) commonly of forensic importance. *Medical and Veterinary Entomology* 48: 738–752.

Tomberlin, J.K., R. Mohr, M.E. Benbow, A.M. Tarone, and S.L. Vanlaerhoven. 2011. A roadmap for bridging basic and applied research in forensic entomology. *Annual Review of Entomology* 56: 401–421.

Tourle, R.A., D.A. Downie, and M.H. Villet. 2009. Flies in the ointment: A morphological and molecular comparison of (*Lucilia cuprina)* and *L. sericata* (Diptera: Calliphoridae) in South Africa. *Medical and Veterinary Entomology* 23: 6–14.

Ullyett, G.C. 1950. Competition for food and allied phenomena in sheep-blowfly populations. *Philosophical Transactions of the Royal Society series B* 234: 77–174.

Verves, Yu. G. 2004. Records of *Chrysomya albiceps* in the Ukraine. *Medical and Veterinary Entomology* 18: 308–310.

Villet, M.H. 2007. An inexpensive geometrical micrometer for measuring small, live insects quickly without harming them. *Entomologia Experimentalis et Applicata* 122: 279–280.

Villet, M.H. 2011. African carrion ecosystems and their insect communities in relation to forensic entomology: A review. *Pest Technology* 5: 1–15.

Villet, M.H. and J. Amendt. 2011. Advances in entomological methods for estimating time of death. In: Turk, E.E., ed. *Forensic Pathology Reviews*. Heidelberg, Germany: Humana Press, 213–238.

Villet, M.H., B. Mackenzie, and W.J. Muller. 2006. Larval development of the carrion-breeding flesh fly *Sarcophaga Liosarcophaga tibialis* Macquart (Diptera: Sarcophagidae) at constant temperatures. *African Entomology* 14: 357–366.

Villet, M.H., C.S. Richards, and J.M. Midgley. 2010. Contemporary precision, bias and accuracy of minimum post-mortem intervals estimated using development of carrion-feeding insects. In: Amendt J., Campobasso C.P., Goff M.L., and Grassberger M., eds. *Current Concepts in Forensic Entomology*, pp. 109–137. Heidelberg, Germany: Springer.

Williams, K.A. 2003. Spatial and temporal occurrence of forensically important South African blowflies (Diptera: Calliphoridae), x + 64 pp. MSc thesis, Rhodes University, Grahamstown, South Africa.

Williams, K.A. and M.H. Villet. 2006a. A history of southern African research relevant to forensic entomology. *South African Journal of Science* 102: 59–65.

Williams, K.A. and M.H. Villet. 2006b. A new and earlier record of *Chrysomya megacephala* in South Africa, with notes on another exotic species, *Calliphora vicina* (Diptera: Calliphoridae). *African Invertebrates* 47: 347–350.

Williams, K.A. and M.H. Villet. 2013. Ancient and modern hybridization between *Lucilia sericata* and *Lucilia cuprina* (Diptera: Calliphoridae). *European Journal of Entomology* 110: 187–196.

PART IV

History, Accomplishments, and Challenges of Forensic Entomology in the Americas

CHAPTER 17

South America

Thiago C. Moretti and Wesley A. C. Godoy

CONTENTS

17.1 INTRODUCTION

Forensic entomology can be divided into the following three main areas: (1) urban entomology, (2) stored-products entomology, and (3) medicolegal or medico-criminal entomology (Lord and Stevenson 1986). This chapter focuses only on medicolegal entomology.

The chapter aims to provide a general overview of the history, accomplishments, and challenges of forensic entomology in South America. Numerous forensic entomology studies have been conducted in South America. In addition to Brazil, which has the longest established presence in the field of forensic entomology in South America and is therefore emphasized here, countries such as Argentina and Colombia have also made important contributions to the field. Rather than providing an exhaustive list of bibliographic references on forensic entomology published in South America, we present a brief overview of the pioneering works and milestones for these countries.

17.2 COLOMBIA

The first Colombian forensic entomology studies were carried out by Wolff et al. (2001) and Pérez et al. (2005) in Medellin, in the northwestern part of the country. Medellin is the second most populous city in Colombia. Wolff et al. (2001), despite using only one pig carcass (*Sus scrofa* L.) throughout the entire decompositional study (207 days), provided a comprehensive list of the arthropods attracted to each of the five decay stages (fresh, bloated, active decay, advanced decay, and dry remains). Arthropod specimens (n = 2314) were collected from the orders Diptera, Hymenoptera, Coleoptera, Dermaptera, Hemiptera, and Lepidoptera. The arthropods were also classified into ecological categories (necrophagous, predators, omnivorous, and incidental). Unfortunately, identification of the specimens at the species level was not possible for certain arthropod groups.

In the study by Pérez et al. (2005), three domestic pigs were left to decompose in the urban area of Medellin. A total of 11,937 arthropod specimens belonging to 12 orders were collected, and Diptera was the most abundant. Camacho (2005) performed the first successional study of carrion-related entomofauna in a savanna of Bogota, using a pig carcass. The insect sampling lasted 6 months. The most abundant families were Calliphoridae (Diptera), Muscidae (Diptera), Fanniidae (Diptera), and Silphidae (Coleoptera). The author included a table listing the probability of the appearance of various families of insects for each day of sampling.

The insect succession in the Colombian Páramo at 3035 m above sea level was described by Martinez et al. (2007) for the five decay stages used by Wolff et al. (2001). In the former study, the authors recognized indicator species for each of the stages and provided a succession table for the necrophagous arthropods of the region.

17.3 ARGENTINA

In Argentina, Mariluis and Schnack (1989, 1996), Mariluis et al. (1994), and Schnack et al. (1995) studied the ecology of blow flies (Diptera: Calliphoridae). However, studies by Oliva (1997) and Centeno et al. (2002) are among only a few conducted in the country on the possible legal application of studying carrion-related insects. Oliva (1997) presented the first checklist of insects of forensic interest in Buenos Aires Province and also provided bionomic data for some species. Centeno et al. (2002) studied the patterns of arthropod visitation on pig carcasses throughout the year, also in Buenos Aires Province; compiled a checklist of carrion-related arthropods; and investigated possible insect succession patterns. Battán-Horenstein et al. 2005, 2010, studied the species of arthropods associated with chicken and pig carrion, respectively, in Cordoba, in the central region of the country.

17.4 BRAZIL

Brazil has great potential for studies in forensic entomology because it encompasses a large territory with wide variation in temperature, altitude, rainfall patterns, and population densities (Moretti and Godoy 2013), which propitiate a great diversity of carrion-associated entomofauna. In addition, owing to high homicide rates, there are many corpses to which forensic entomology tools could be applied; an unfortunate phenomenon of a developing country.

17.5 HOMICIDES IN BRAZIL: THE MOST CHALLENGING OBSTACLE FACING A DEVELOPING COUNTRY

The number of homicides in Brazil rose from 13,910 in 1980 to 49,932 in 2010, an increase of 259%, equivalent to an increment of 4.4% per annum (Waiselfisz 2011). According to national surveys, the Brazilian population also grew during the same period but at a much lower rate. The population climbed from 119.0 to 190.7 million inhabitants, an increase of 60.3% (Waiselfisz 2011). The homicide rate increased from 11.7/100,000 inhabitants in 1980 to 26.2/100,000 inhabitants in 2010, an increase of 124% over the whole period or 2.7% per annum (Waiselfisz 2011). The Mortality Information System of the Brazilian Ministry of Health recorded 1.1 million homicides in this 30-year period (1980–2010). To provide perspective of the scale of this number, only 13 cities in Brazil reached this number of inhabitants in the annual survey in 2010 (Waiselfisz 2011).

Brazil has experienced a high level of internalization of violence, and the dynamic poles of violence have shifted from the capitals of the states and metropolitan regions (MRs) to country towns (Waiselfisz 2011). In 1995, there were 40.1 homicides/100,000 inhabitants in the capitals/MRs and 11.7 homicides/100,000 inhabitants in country towns. In 2010, the homicide rate in the capitals/MRs dropped to 33.6/100,000 inhabitants, while the rate increased to 22.1/100,000 inhabitants in country towns (Waiselfisz 2011).

Dissemination of violence also occurred. The 17 states with the lowest homicide rates at the turn of the century experienced significant increases in their average homicide rate (from 15.4/100,000 inhabitants in 2000 to 28.4 homicides/100,000 inhabitants in 2010), whereas the seven states that had the highest rates in the previous decade experienced a reduction in their average homicide rate (from 45.6/100,000 inhabitants in 2000 to 22.6 homicides/100,000 inhabitants in 2010) (Waiselfisz 2011).

In the first decade of this century, novel policies were implemented in Brazil to reduce crime levels—particularly homicides (UNODC 2011). In 2003, legislation introducing stricter control over firearms was enacted, and there were disarmament campaigns. At the national level, such measures contributed to a minor reduction in homicide rates after 2004, although their impact was greatest in São Paulo (the most populous state), where enforcement of these measures was particularly successful because of, in addition to these measures, preexisting efforts to restrain violent crimes through new policing methods (UNODC 2011). These two processes led to migration of the dynamic poles of violence from a limited number of capitals and or MRs, which were able to improve the efficiency of their security apparatus, to less protected areas in either country towns or other federal units (Waiselfisz 2011). Homicide continues to be a very large problem in Brazil, and medicolegal entomology may help in solving these crimes and consequently inhibit criminality in the country.

17.6 BIODIVERSITY IN BRAZIL: FLIES, BEETLES, AND MORE

Insects are one of the most numerous of all known groups of animals on Earth, comprising 58%–67% of all currently described eukaryotic species (Foottit and Adler 2009). Some authors have estimated that there are 30 million insect species worldwide. Therefore, the estimated number of insect species in Brazil—a little more than 400,000—is reasonable and even quite conservative (Rafael et al. 2012). The species richness of insects of forensic importance for both South America and Brazil in particular is also high. Diptera and Coleoptera, the most important orders in forensic entomology (Souza and Linhares 1997), are referred to as megadiverse because they are among the largest insect orders in terms of number of species (Rafael et al. 2009).

Identification keys for 23 families of Diptera, specifically species that feed/oviposit on vertebrate carrion or have larvae reared on this type of substrate, were developed by Smith (1986). In Brazil,

Carvalho et al. (2000) surveyed species from 22 families of Diptera associated with vertebrate car-rion, seven of which (Calliphoridae, Muscidae, Fanniidae, Phoridae, Piophilidae, Sarcophagidae, and Stratiomyidae) are of actual forensic utility. Carvalho and Mello-Patiu (2008) provided a key for the identification of adults of 12 dipteran families that are associated with human or other ani-mal remains in South America. Eight of these families are known to feed on human corpses, and four—Anthomyiidae, Sepsidae, Sphaeroceridae, and Ulidiidae (Otitidae)—feed on vertebrate car-rion. Oliveira-Costa (2011) listed the following 17 dipteran families as being of forensic interest in Brazil: Psychodidae, Culicidae, Stratiomyidae, Phoridae, Syrphidae, Sepsidae, Sphaeroceridae, Piophilidae, Drosophilidae, Chloropidae, Milichiidae, Heleomyzidae, Ulidiidae, Calliphoridae, Fanniidae, Muscidae, and Sarcophagidae.

Luederwaldt (1911) found 62 Coleoptera carcass-associated species of forensic importance in Brazil. Pessôa and Lane (1941) performed studies on Scarabaeinae fauna of medicolegal inter-est in southeast Brazil and surveyed 113 species in 26 genera. Working with beetles of forensic interest in South America, Almeida and Mise (2009) yielded a list of 221 species in 15 families. Scarabaeidae was the most diverse family, with 121 species, followed by Staphylinidae, with 68.

It has been estimated that the orders Diptera and Coleoptera account for approximately 60% of the arthropod fauna associated with vertebrate remains, and they are therefore very important in medicolegal entomology (Souza and Linhares 1997; Moretti et al. 2008a). However, other insect orders (e.g., Lepidoptera, Hymenoptera, Blattodea, Hemiptera, Isoptera, and Dermaptera) and even other arthropod groups, such as Araneae, Opiliones, Acari, Chilopoda, and Diplopoda, are frequently found with decomposing remains; however, these taxa are usually not included among cadaveric fauna because very little is known of their potential function in the decomposition process (Moretti et al. 2008a).

To exemplify these unusual taxa associated with vertebrate carrion in Brazil, stingless bees *Tetragonisca angustula* Latreille (Hymenoptera: Apidae) have been collected from rodent carcasses (Moretti et al. 2008a). This finding may be related to the bees' mineral or humidity needs rather than a protein requirement. They could also be attracted to fungi that grow on carcasses during the black putrefaction stage (Bornemissza 1957). Some mineral and protein-rich fungi that develop in microhabitats used by stingless bees play a key role in *T. angustula* nourishment (Gilliam et al. 1988). The report of *Mischonyx cuspidatus* (Roewer) (Opiliones: Gonyleptidae) found on rodent carrion (Moretti et al. 2008a) is equally unexpected because harvestmen are rarely referred to in medicolegal entomology. Nevertheless, further investigation of the feeding habits of this arthropod group is required to elucidate whether it benefits from the carcass or is a predator of the regular visiting fauna.

17.7 FORENSIC ENTOMOLOGY IN BRAZIL: 105 YEARS OF HISTORY

It is not possible to comment on the history of forensic entomology in Brazil without referring to the comprehensive article by Pujol-Luz et al. (2008a), which celebrated the 100th anniversary of the pioneering forensic entomology studies conducted in Brazil. Below is a brief analysis of the first reports on medicolegal entomology produced in the country that are presented in chronological order.

The beginning of forensic entomology in Brazil was associated with the work of Oscar Freire in 1908, only 14 years after the publication of *La faune des cadavres* by Mégnin (1894), the world's first book on the subject. Freire presented the first Brazilian collection of necrophagous insects, as well as the results of his research on arthropod fauna associated with human remains and carcasses of small animals, to the Medical Society of Bahia (Northeast Brazil) (Pujol-Luz et al. 2008a). In November 1908, Roquette-Pinto published an article entitled "Nota sobre a fauna cadaverica, no Rio de Janeiro" (translation, "Note on the necrophagous fauna in Rio de Janeiro"). In this article, the author, who had studied the fate of human remains found at the Tijuca forest (Rio de Janeiro, southeast Brazil), stated that estimating the date of death using the entomological succession on a corpse, as proposed by Mégnin (1894) would produce inaccurate outcomes in tropical zones such as Brazil. Roquette-Pinto (1908) called Mégnin's method a "zoological calendar" and attributed its inaccuracy to differences between Europe and tropical areas regarding weather and patterns of corpse decomposition. According to Pujol-Luz et al. (2008a), Freire and Roquette-Pinto established the baseline for the development of forensic entomology in the tropics.

In 1911, Hermann Luederwaldt was in charge of revising the collection of beetles of the Museu Paulista (currently the Museum of Zoology of the University of São Paulo, southeast Brazil). In his spare time, he collected insects of forensic importance. Not surprisingly, he concentrated on the coleopteran fauna associated with carcasses. Luederwaldt was mainly interested in the systematics of this group, but he also made important observations regarding the ecology of carrion-related beetles. In addition, he produced a list of insects of forensic importance in the state of São Paulo, which included, besides beetles identified to species level, hymenopterans, dipterans, orthopterans, and hemipterans, all of which had been insufficiently identified. According to Luederwaldt (1911), it was necessary to consider the differences between Europe and the tropics, with respect not only to climate regimes but also to faunal diversity patterns and ecology of the insects before applying Mégnin's (1894) method in Brazil. In summary, according to Luederwaldt (1911), all studies regarding both the zoological and the chronological succession would need to be recreated for Mégnin's calendar to have practical value in tropical regions of the world.

Luederwaldt also made important contributions to the field of myrmecology. As described in his 1926 work "Observações biologicas sobre formigas brasileiras especialmente do estado de São Paulo" (translation, "Biological observations on Brazilian ants, especially from the State of São Paulo"), he collected several species of ants (Hymenoptera: Formicidae) from a broad range of substrates, including meat, cold meats, plant material, human and animal excrement, and mushrooms. Some ants were collected from vertebrate carrion or fly larvae. The result was one of the first lists of ants of forensic potential in Brazil.

Freire (1914a,b) published "Algumas notas para o estudo da fauna cadaverica na Bahia" (translation, "Some notes for the study of the necrophagous fauna in Bahia"). According to Pujol-Luz et al. (2008a), Freire studied many dipteran species, mainly from the families Muscidae, Calliphoridae, and Sarcophagidae.

Belfort Mattos (1919) published a doctoral thesis under the supervision of Oscar Freire on the genus *Sarcophaga* (Diptera: Sarcophagidae) in the state of São Paulo. In this comprehensive work, the author described the external morphology of adult males, females, larvae, and pupae from several species of *Sarcophaga*. He also analyzed their breeding habits, their potential medicolegal application, and myiasis associated with them.

In 1923, Freire's article "Fauna cadaverica brasileira" (translation, "Brazilian necrophagous fauna") was posthumously published. In this article, he provided an inventory of what he called "workers of death," that is, insects collected from carrion. He included a list of species that he had recorded together with other researchers working with cadaverous fauna in Brazil (Luederwaldt, Roquette-Pinto, and Belfort Mattos). This list included species from the orders Diptera, Coleoptera, Lepidoptera, Orthoptera, Hymenoptera, and also some insufficiently identified mites. Freire organized another list comparing the fauna associated with vertebrate carcasses in North America, Europe, and Brazil (Freire 1923). In this article, he strongly criticized Mégnin's conclusions (1894), calling his work "excessively theoretical and schematic" (Pujol-Luz et al. 2008a). The author agreed with Mégnin on the existence of some pattern or order to the succession of insects at carcasses, but he claimed that such patterns would only be frequent, but not immutable. The following important points of the article by Freire (1923) were emphasized by Pujol-Luz et al. (2008a): (1) there is no exclusivity of species for each decomposition stage, (2) vital concurrence is very important among necrophagous insects, (3) there is no isochronism in the periods of cadaveric decomposition, and (4) an entirely precise chronology is impossible. As Roquette-Pinto (1908) had previously concluded, Freire (1923) determined that, because of the differences between Europe/North America and Brazil regarding weather and the entomofauna, no method developed in a foreign country could be put directly into practice in Brazil.

An article on necrophagous beetles was published by Pessôa and Lane (1941), giving special attention to the family Scarabaeidae (Coleoptera) and providing identification keys, illustrations, and taxonomical and biological data. The authors intended to enable forensic physicists and nonspecialist entomologists to identify at least the most common species from Scarabaeidae of medicolegal interest in the state of São Paulo and its vicinity.

After a lag of approximately four decades (1940–1980), during which virtually nothing was published on forensic entomology *sensu stricto* in Brazil, Monteiro-Filho and Penereiro (1987) wrote an article on the decomposition of, and insect succession on, small rodent carcasses at a secondary forest in the state of São Paulo.

Studies on the taxonomy, natural history, and ecology of biontophagous and necrophagous flies, which are the foundation for the development of forensic entomology, have been performed by Hugo de Souza Lopes (Sarcophagidae), Rubens Pinto de Mello, José Henrique Guimarães, and Nelson Papavero (Calliphoridae) (Pujol-Luz et al. 2008a).

Master's and PhD theses focusing on insect succession and the effects of several parameters (e.g., altitude, carcass size, type of vegetation, and drugs) on the entomofauna associated with vertebrate carrion began in 1991 under the supervision of Arício Xavier Linhares (University of Campinas, São Paulo). Salviano et al. (1996) published a list of calliphorids collected from human remains at the Medico-Legal Institute of Rio de Janeiro. Later, articles on diversity, ecology, taxonomy, and insect succession on carcasses were published by Arício Xavier Linhares, Cláudio José Barros de Carvalho (Federal University of Paraná, Paraná), and their respective teams (Moura et al. 1997; Souza and Linhares 1997; Carvalho et al. 2000).

17.8 CURRENTLY ACTIVE RESEARCH GROUPS IN BRAZIL

Caneparo et al. (2012) searched the Lattes Platform, a Brazilian public database that contains the *curricula vitae* of registered researchers, and found 41 researchers with doctoral degrees working in forensic entomology. There are several active research groups dedicated to forensic entomology in all five geographical regions of Brazil (south, southeast, northeast, central-western, and north) (Caneparo et al. 2012). According to Oliveira-Costa (2013), the following universities have researchers who are currently working in forensic entomology: in the south, Federal University of Paraná (State of Paraná) and Federal University of Pelotas (State of Rio Grande do Sul); in the southeast, Federal University of Rio de Janeiro, Oswaldo Cruz Foundation (State of Rio de Janeiro), Federal University of Uberlândia, Federal University of Juiz de Fora (State of Minas Gerais), University of Campinas, and São Paulo State University (State of São Paulo); in the central western region, University of Brasília (Federal District); in the northeast region, State University of Paraíba (State of Paraíba), Federal Rural University of Pernambuco, and Federal University of Pernambuco (State of Pernambuco); and in the north, National Institute of Research in the Amazon (State of Amazônia) and Federal University of Amapá (State of Amapá).

17.9 RECENT ACCOMPLISHMENTS IN FORENSIC ENTOMOLOGY IN BRAZIL

17.9.1 ABEF Foundation

In 2007, the Brazilian Association for Forensic Entomology (ABEF in Portuguese) was established. Based in Campinas (State of São Paulo), the organization's main goals are the following (extracted from

ABEF's website http://www.rc.unesp.br/ib/zoologia/abef/) : (i) bring together people interested in the development of studies and research in the area of forensic entomology; (ii) help to promote surveys of necrophagous arthropods at the national level; (iii) promote training for forensic entomologists and recognition of their techniques; (iv) support normalization of procedures used by forensic entomologists; (v) represent the community of Brazilian forensic entomologists at the national and international levels; (vi) promote meetings at the regional, national, and/or international levels; (vii) disseminate knowledge regarding forensic entomology; (viii) assist and advise authorities and private and civil society regarding the development of forensic entomology studies in its various sub-areas and specialties; and (ix) draw the attention of private and public bodies to the practice and implementation of forensic entomology. When it was founded, ABEF organized the First Brazilian Symposium of Forensic Entomology. In 2008, ABEF helped to promote the Second Brazilian Symposium of Forensic Entomology during the XXVII Brazilian Congress of Zoology, in Curitiba (State of Paraná). Unfortunately, ABEF has not been active since 2008.

17.9.2 Brazilian Books on Forensic Entomology

The book edited by Janyra Oliveira-Costa (2003), *Entomologia Forense: Quando os Insetos São Vestígios* (translation, *Forensic Entomology: When Insects Are Crime Evidence*), was the first book in Brazil to specifically address forensic entomology. The editor compiled the information available in the national and international literature. According to Pujol-Luz et al. (2008a), the most important contributions of this work were (1) dissemination of the theoretical basis of forensic entomology and (2) standardization of the language used by researchers and coroners. Oliveira-Costa released the second and third editions of *Entomologia Forense: Quando os Insetos São Vestígios* in 2008 and 2011, respectively. In 2013, Oliveira-Costa published another book: *Insetos Peritos: A Entomologia Forense no Brasil* (translation, *Insects as Criminal Experts: Forensic Entomology in Brazil*). This book analyzes advances in forensic entomology, emphasizing the regional differences across the country.

17.9.3 FAPESP Thematic Project: An Important Initiative in the State of São Paulo

Several research groups in São Paulo, the most productive Brazilian state with respect to forensic entomology research, have been advocating for the formation of a true forensic entomology database, with the aim of creating a solid biological and ecological foundation to aid in the definitive implementation of forensic entomology in the state (Moretti and Godoy 2013). Perhaps one of the most important steps in this implementation was the thematic project called "Forensic Entomology: The Utilization of Arthropods for Determining Time, Place, Cause, and Circumstances of Death," which was supported by FAPESP (São Paulo Research Foundation) between April 2005 and March 2010. This project involved researchers from various universities in the state of São Paulo, and its main goal was to expand the inventory of necrophagous insects in the state. Such local or regional surveys are fundamentally important to establishing baseline data for the implementation of forensic entomology in a given locality (Moretti and Godoy 2013).

Other objectives of the thematic project were to: (1) investigate the effects of toxic substances on the rate of development of necrophagous flies of forensic importance; (2) determine potential species that could function as forensic indicators for various types of habitats (urban, rural, and forest); (3) understand the dynamics, population variability, and interspecific associations in necrophagous flies; and (4) characterize and distinguish populations of flies of forensic importance from a variety of environments (urban, rural, and forest) and geographical regions in an attempt to create a method for determining whether there was displacement of a corpse after death.

17.10 MAIN CHALLENGES OF FORENSIC ENTOMOLOGY IN BRAZIL

17.10.1 Regional or Local Insect Inventories of Forensic Importance

The surveys of insects of forensic importance carried out in Brazil (e.g., Souza and Linhares 1997; Moretti et al. 2008a) suggest different decomposition and insect succession patterns compared with those in temperate regions (Anderson and VanLaerhoven 1996). These differences are due to the higher temperatures in tropical regions, which accelerate the development of insects and increase the attractiveness of carcasses, because they release more odors, which in turn increase the overall abundance of carrion-related insects. Consequently, the abundance of their natural enemies also rises, leading to more intense interspecific and trophic interactions. Tropical regions also have more diversified biomes, which allows for the existence of different species capable of developing in different substrates. This complex scenario is less likely to occur in temperate zones. Unfortunately, to our knowledge, no study has yet been designed to compare succession patterns over latitudinal gradients.

Because each habitat has its own carrion entomofauna and specific local features, regional entomofauna and their patterns of succession must be studied on carcasses before forensic entomology–related techniques can be put into practice (Pujol-Luz et al. 2008a; Aballay et al. 2012). Local and regional surveys are fundamental to establishing baseline data for the implementation of such techniques in a given locality. This step is particularly important in a country such as Brazil, with its enormous territory and wide variations in temperature, altitude, rainfall patterns, and population densities (Moretti and Godoy 2013). Discrepancies between the insects found on a corpse and the insect species at the site of discovery may indicate that the victim was transferred from his or her original location (Amendt et al. 2011).

Therefore, postmortem interval (PMI) estimates should be refined according to local or regional standards. Investigations in this area are still scarce in Brazil, preventing more widespread application of forensic entomology (Pujol-Luz et al. 2008a). Several surveys of carrion entomofauna focus on urban areas and overlook forest and rural environments. According to Moretti and Godoy (2013), it is crucial to study the necrophagous insects in the forest and rural areas because these types of environments are frequently used by criminals to dispose of cadavers.

Although the focus of inventories of forensically important insects so far has, understandably, been on the most traditional taxonomic groups because their forensic status is likely to remain globally unaltered, such surveys tend to be very conservative in Brazil (Moretti et al. 2013a). Usually these inventories almost exclusively comprise insects traditionally considered forensically important (mainly flies and beetles) and neglect taxa that are not commonly used in research with cadaveric fauna, even though biological and behavioral information on these ignored taxa is insufficient to definitively exclude them from the range of useful decomposing arthropods (Moretti et al. 2013a).

We believe that a broader approach is necessary, especially in tropical zones, where abundance and biodiversity patterns form a more complex scenario than in temperate regions (Moretti et al. 2013a). Apart from assessing the preferences of so-called insects of forensic importance with regard to a particular environment and type of bait, as well as analyzing their spatiotemporal variability, during the thematic project (see Section 17.9.3), we also considered arthropods that many people would consider of less forensic utility (Moretti et al. 2013a). Accordingly, we have published articles on social wasps (Moretti et al. 2008b, 2011), hover flies (Martins et al. 2010), and ants (Moretti et al. 2013b).

We are currently investigating the behavioral dynamics and relationships among hymenopterans and calliphorids and sarcophagids that regularly explore carcasses in tropical environments. In zones of high biodiversity, where trophic interactions (e.g., predation) are likely to substantially modify the occurrence of the most abundant species, less traditional taxa (such as bees and wasps) must be monitored.

17.10.2 Species Identification

Identification of flies of forensic importance in Brazil is hindered by two main problems: a lack of taxonomists and a lack of keys. These two problems may lead to occasional misidentification. The excellent key provided by Carvalho and Mello-Patiu (2008) helped to resolve this problem, at least for the most common families and species. Additional keys are necessary for the less common taxa, such as acalyptrates. The identification of forensically important Coleoptera also remains problematic.

17.10.3 Relationship between Universities and Police

According to Pujol-Luz et al. (2008a), the utility of forensic entomology in Brazil is limited by the lack of effective interaction between universities and professionals from the judiciary police service (coroners and medical examiners). The exceptions to this rule are the state of Rio de Janeiro and the Federal district, where the relationship between universities and the judiciary police has contributed to the incorporation of forensic entomology tools in the list of techniques routinely used for evaluating evidence in criminal investigations (Pujol-Luz et al. 2008a).

17.10.4 Reports of Real Cases

Only a very few reports of real-world cases involving the use of forensic entomology tools have been published in Brazil (Oliveira-Costa and Lopes 1999; Oliveira-Costa and Mello-Patiu 2004; Pujol-Luz et al. 2006, 2008b), which reflects the lack of close interaction between universities and police services. This is something that should be rectified if forensic entomology is to be used more frequently in future criminal investigations of South America.

ACKNOWLEDGMENTS

We thank Jeffery K. Tomberlin (Texas A&M University) and Mark E. Benbow (Michigan State University) for inviting us to prepare this chapter, José Roberto Pujol Luz (University of Brasília) for providing several rare publications, and FAPESP (São Paulo Research Foundation) for providing financial support. We dedicate this chapter to Ângelo Pires do Prado (University of Campinas) in view of his great contribution to Brazilian entomology and parasitology.

REFERENCES

Aballay, F.H., A.F. Murua, J.C. Acosta, and N.D. Centeno. 2012. Succession of carrion fauna in the arid region of San Juan province, Argentina and its forensic relevance. *Neotropical Entomology* 41: 27–31.

Almeida, L.M. and K.M. Mise. 2009. Diagnosis and key of the main families and species of South American Coleoptera of forensic importance. *Revista Brasileira de Entomologia* 53: 227–244.

Amendt, J., C.S. Richards, C.P. Campobasso, R. Zehner, and M.J.R. Hall. 2011. Forensic entomology: Applications and limitations. *Forensic Science, Medicine, and Pathology* 7: 379–392.

Anderson, G.S. and S.L. VanLaerhoven. 1996. Initial studies on insect succession on carrion in Southwestern British Columbia. *Journal of Forensic Sciences* 41: 617–625.

Battán-Horenstein, M., M.I. Arnaldos, B. Rosso, and M.D. García. 2005. Estudio preliminar de la comunidad sarcosaprófaga em Córdoba (Argentina): Aplicación a la entomología forense. *Anales de Biología* 27: 191–201.

Battán-Horenstein, M., A.X. Linhares, B.R. Ferradas, and D. García. 2010. Decomposition and dipteran suc-cession in pig carrion in central Argentina: Ecological aspects and their importance in forensic science. *Medical and Veterinary Entomology* 24: 16–25.

Belfort-Mattos, W. 1919. As Sarcophagas de S. Paulo. Doctoral thesis, Faculdade de Medicina e Cirurgia de São Paulo, Brazil.

Bornemissza, G.F. 1957. An analysis of arthropod succession in carrion and the effect of its decomposition on the soil fauna. *Australian Journal of Zoology* 5: 1–12.

Camacho, G. 2005. Sucesión de la entomofauna cadavérica y ciclo vital de *Calliphora vicina* (Diptera: Calliphoridae) como primera especie colonizadora, utilizando cerdo blanco (*Sus scrofa*) en Bogotá. *Revista Colombiana de Entomología* 31: 189–197.

Caneparo, M.F.C., R.C. Corrêa, K.M. Mise, and L.M. Almeida. 2012. Entomologia médico-criminal. *Estudos de Biologia, Ambiente e Diversidade* 34: 215–223.

Carvalho, C.J.B. and C.A. Mello-Patiu. 2008. Key to the adults of the most common forensic species of Diptera in South America. *Revista Brasileira de Entomologia* 52: 390–406.

Carvalho, L.M.L., P.J. Thyssen, A.X. Linhares, and F.B. Palhares. 2000. A checklist of arthropods associated with carrion and human corpses in southeastern Brazil. *Memórias do Instituto Oswaldo Cruz* 95: 135–138.

Centeno, N., M. Maldonado, and A. Oliva. 2002. Seasonal patterns of arthropods occurring on sheltered and unsheltered pig carcasses in Buenos Aires Province (Argentina). *Forensic Science International* 126: 63–70.

Foottit, R.G. and P.H. Adler. 2009. *Insect Biodiversity: Science and Society*. Chichester, United Kingdom: Blackwell Publishing.

Freire, O. 1914a. Algumas notas para o estudo da fauna cadaverica na Bahia. *Gazeta Medica da Bahia* 46: 110–125.

Freire, O. 1914b. Algumas notas para o estudo da fauna cadaverica na Bahia. *Gazeta Medica da Bahia* 46: 149–162.

Freire, O. 1923. Fauna cadaverica brasileira. *Revista de Medicina* 3–4: 15–40.

Gilliam, M., B.J. Lorenz, and G.V. Richardson. 1988. Digestive enzymes and micro-organisms in honey bees, *Apis mellifera*: Influence of streptomycin, age, season, and pollen. *Microbios* 55: 95–114.

Lord, W.D. and J.R. Stevenson. 1986. Directory of Forensic Entomologists. Defense pest management information analysis center. Washington, DC: Walter Reed Army Medical Center.

Luederwaldt, H. 1911. Os insectos necrophagos Paulistas. *Revista do Museu Paulista* 8: 414–433.

Luederwaldt, H. 1926. Observações biologicas sobre formigas brasileiras especialmente do estado de São Paulo. *Revista do Museu Paulista* 14: 185–303.

Mariluis, J.C. and J.A. Schnack. 1989. Ecology of the blow flies of an eusynanthropic habitat near Buenos Aires (Diptera, Calliphoridae). *Eos* 165: 93–101.

Mariluis, J.C. and J.A. Schnack. 1996. Elenco específico y aspectos ecológicos de Calliphoridae (Insecta, Diptera) de San Carlos de Bariloche, Argentina. *Boletín de la Real Sociedad Española de Historia Natural (Sección Biológica)* 92: 1–4.

Mariluis J.C., J.A. Schnack, I. Cervererizzo, and C. Quintana. 1994. *Cochliomyia hominivorax* (Coquerel, 1858) and *Phaenicia sericata* (Meigen, 1826) parasiting domestic animals in Buenos Aires and vicinities (Diptera, Calliphoridae). *Memórias do Instituto Oswaldo Cruz* 89: 139.

Martinez, E., P. Duque, and M. Wolff. 2007. Succession pattern of carrion-feeding insects in Páramo, Colombia. *Forensic Science International* 166: 182–189.

Martins, E., J.A. Neves, T.C. Moretti, W.A.C. Godoy, and P.J. Thyssen. 2010. Breeding of *Ornidia obesa* (Diptera: Syrphidae: Eristalinae) on pig carcasses in Brazil. *Journal of Medical Entomology* 47: 690–694.

Mégnin, P. 1894. *La Faune des Cadavres: application de l'entomologie a la* médecine légale. Encyclopédie Scientifique des Aide-Mémoire. Paris, France: G. Masson, Gauthier-Villars et Fils.

Monteiro-Filho, E.L.A. and J.L. Penereiro. 1987. A study on decomposition and succession on animal carcasses in an area of Sao Paulo, Brazil. *Brazilian Journal of Biology* 47: 289–295.

Moretti, T.C., E. Giannotti, P.J. Thyssen, D.R. Solis, and W.A.C. Godoy. 2011. Bait and habitat preferences, and temporal variability of social wasps (Hymenoptera: Vespidae) attracted to vertebrate carrion. *Journal of Medical Entomology* 48: 1069–1075.

Moretti, T.C. and W.A.C. Godoy. 2013. Spatio-temporal dynamics and preference for type of bait in necrophagous insects, particularly native and introduced blow flies (Diptera: Calliphoridae). *Journal of Medical Entomology* 50: 415–424.

Moretti, T.C., F.S. Nascimento, W.A.C. Godoy, and A.X. Linhares. 2013a. It is more than flies and beetles: Toward a less-conservative approach to insects of forensic importance in regions of high biodiversity. Presented at the 10th Meeting of the European Association for Forensic Entomology (EAFE), Coimbra, Portugal.

Moretti, T.C., O.B. Ribeiro, P.J. Thyssen, and D.R. Solis. 2008a. Insects on decomposing carcasses of small rodents in a secondary forest in Southeastern Brazil. *European Journal of Entomology* 105: 691–696.

Moretti, T.C., D.R. Solis, and W.A.C. Godoy. 2013b. Ants (Hymenoptera: Formicidae) collected with carrion-baited traps in Southeast Brazil. *The Open Forensic Science Journal* 6: 1–5.

Moretti, T.C., P.J. Thyssen, W.A.C. Godoy, and D.R. Solis. 2008b. Necrophagy by the social wasp *Agelaia pallipes* (Hymenoptera: Vespidae, Epiponini): Possible forensic implications. *Sociobiology* 51: 393–398.

Moura, M.O., C.J.B. Carvalho, and E.L.A. Monteiro. 1997. A preliminary analysis of insects of medico-legal importance in Curitiba, State of Parana. *Memórias do Instituto Oswaldo Cruz* 92: 269–274.

Oliva, A. 1997. Insectos de interés forense de Buenos Aires (Argentina). Primera lista ilustrada y datos bionómicos. *Revista del Museo Argentino de Ciencias Naturales 'Bernardino Rivadavia', Entomología* 7: 13–59.

Oliveira-Costa, J. 2003. *Entomologia forense: quando os insetos são vestígios*, 1st Ed. Campinas, São Paulo: Millennium Editora.

Oliveira-Costa, J. 2008. *Entomologia forense: quando os insetos são vestígios*, 2nd Ed. Campinas, São Paulo: Millennium Editora.

Oliveira-Costa, J. 2011. *Entomologia forense: quando os insetos são vestígios*, 3rd Ed. Campinas, São Paulo: Millennium Editora.

Oliveira-Costa, J. 2013. *Insetos peritos: a entomologia forense no Brasil*. Campinas, São Paulo: Millennium Editora.

Oliveira-Costa, J. and S. Lopes. 1999. A relevância da entomologia forense para a perícia criminal na elucidação de um caso de suicídio. *Entomologia y Vectores* 7: 203–209.

Oliveira-Costa, J. and C.A. Mello-Patiu. 2004. Estimation of PMI in homicide investigation by the Rio de Janeiro Police. *Aggrawal's Internet Journal of Forensic Medicine and Toxicology* 5: 40–44.

Pérez, S.P., P. Duque, and M. Wolff. 2005. Successional behavior and occurrence matrix of carrion associated arthropods in the urban area of Medellin, Colombia. *Journal of Forensic Sciences* 50: 448–454.

Pessôa, S. and F. Lane. 1941. Coleópteros necrófagos de interêsse médico-legal. Ensáio monográfico sôbre a família Scarabaeidae de S. Paulo e regiões vizinhas. *Arquivos de Zoologia do Estado de São Paulo* 2: 389–504.

Pujol-Luz, J.R., L.C. Arantes, and R. Constantino. 2008a. Cem anos da entomologia forense no Brasil (1908–2008). *Revista Brasileira de Entomologia* 52: 485–492.

Pujol-Luz, J.R., P.A.C. Francez, A. Ururahy-Rodrigues, and R. Constantino. 2008b. The black-soldier fly, *Hermetia illucens* (Diptera, Stratiomyidae), used to estimate the postmortem interval in a case in Amapá State, Brazil. *Journal of Forensic Sciences* 53: 476–478.

Pujol-Luz, J.R., H. Marques, A. Ururahy-Rodrigues, J.A. Rafael, F.H. Santana, L.C. Arantes, and R. Constantino. 2006. A forensic entomology case from the Amazon rain forest of Brazil. *Journal of Forensic Sciences* 51: 1151–1153.

Rafael, J.A., A.P. Aguiar, and D.S. Amorim. 2009. Knowledge of insect diversity in Brazil: Challenges and advances. *Neotropical Entomology* 38: 565–570.

Rafael, J.A., G.A. R. Melo, C.J.B. Carvalho, S.A. Casari, and R. Constantino. 2012. *Insetos do Brasil: diversidade e taxonomia*. Ribeirão Preto, São Paulo: Holos Editora.

Roquette-Pinto, E. 1908. Nota sobre a fauna cadaverica, no Rio de Janeiro. *A Tribuna Médica* 21: 413–417.

Salviano, R., R. Mello, R. Santos, L. Beck, and A. Ferreira. 1996. Calliphoridae (Diptera) associated with human corpses in Rio de Janeiro, Brazil. *Entomologia y Vectores* 3: 145–146.

Schnack, J.A., J.C. Mariluis, N. Centeno, and J. Muzón. 1995. Composición específica, ecología y sinantropía de Calliphoridae (Insecta: Diptera) en el Gran Buenos Aires. *Revista de la Sociedad Entomológica Argentina* 54: 161–171.

Smith, K.G.V. 1986. *A Manual of Forensic Entomology*. Ithaca, NY: Cornell University Press.

Souza, A.M. and A.X. Linhares. 1997. Diptera and Coleoptera of potential forensic importance in southeastern Brazil: Relative abundance and seasonality. *Medical and Veterinary Entomology* 11: 8–12.

Waiselfisz, J.J. 2011. *Mapa da violência 2012: os novos padrões da violência homicida no Brasil*. São Paulo, Brazil: Instituto Sangari.

Wolff, M., A. Uribe, A. Ortiz, and P. Duque. 2001. A preliminary study of forensic entomology in Medellin, Colombia. *Forensic Science International* 120: 53–59.

UNODC (United Nations Office on Drugs and Crime). 2011. Global Study on Homicide: Trends, Contexts, Data. Vienna, Austria.

CHAPTER 18

North America

John R. Wallace, Jason H. Byrd, Hélène N. LeBlanc, and Valerie J. Cervenka

CONTENTS

18.1 INTRODUCTION

Forensic entomology—the use of insects in criminal or medicolegal cases (related to death scene investigations) and civil investigations (urban and stored-product cases related to termite infestations and commercial food contamination)—has been portrayed to the general public in the mass media to be a relatively new field among the variety of subdisciplines within forensic science. However, it is perhaps one of the oldest in terms of applications in death scene investigations. Earliest records of the use of insects in medicolegal investigations date back to thirteenth-century China (McKnight 1981), to mid-nineteenth century Europe in two landmark publications by Bergeret (1855) and Mégnin (1894), and to the end of the nineteenth century in Canada (Johnston and Villeneuve 1897). Outside of published scientific research, the application of forensic entomology in the United States did not enter the courtroom until the 1970s and 1980s (Catts and Haskell 1990). The science began to grow significantly outside of empirical research since the 1980s when police became more aware of the use of this evidence via trainings, workshops, and conferences, and began to request entomologists to testify as expert witnesses (Solomon and Hackett 1996; Anderson 2005).

The purpose of this chapter is to provide a historical perspective of forensic entomology in North America, including the United States, Canada, and Mexico. This perspective includes a summary of important research contributions in these regions pre- and post-1950. Discussion in this chapter culminates with the current state of the field in North America with respect to forensic entomology organizations, growth of the profession, and concludes with hurdles that are present in each nation represented. Although this discussion is intended to be comprehensive, it is not possible to cover all aspects of forensic entomology in North America and so an attempt has been made to appropriately cover the highlights of the history and current status, while acknowledging that not everything will be covered in detail.

18.2 RESEARCH

18.2.1 Research Contributions in the United States, Pre-1950

Much of the early "forensic" entomological research was couched under the topic of decomposition-related research during the late nineteenth and early twentieth centuries and was restricted to ecological research describing the diversity and succession of animals associated with decomposing organic material, and a bibliography of this work was published in 1985 (Vincent et al. 1985). In many instances, the organic material decomposing was not vertebrate remains but plant material, such as a tree (Graham 1925), or the changes associated with ecological succession, such as a lake (Forbes 1887) or sand dune (Chapman et al. 1926).

Initial research in the area of decomposition ecology focused on individual arthropod species that are recognized today as forensically important. Generally speaking, taxonomic information on arthropod diversity was much more limited than it is today, so it would seem logical for many researchers to allocate their time toward describing and identifying larval and adult specimens (Dorsey 1940). This information was predominantly used to explain the natural history of a species (Davis 1915; Cole 1942) or to suppress their populations because of their impact on livestock production (Deonier 1940). In many instances, researchers used carrion as the resource to attract the species of interest (Deonier 1940; Dorsey 1940). However, a few studies were conducted specifically with the intent of applying information from such studies toward forensics.

Before 1950s in the United States, the discipline known today as medicolegal forensic entomology was emerging from a foundation of classic taxonomic work on two important dipteran families, Sarcophagidae and Calliphoridae (Aldrich 1916; Knipling 1936, 1939; Hall 1948), to an assortment of ecological, morphological, physiological, and other descriptive studies that focused on several insect orders such as Diptera, Coleoptera, and Collembola (Motter 1898; Folsom 1902; Davis 1915; Illingworth 1927; Steele 1927; Dorsey 1940). Early taxonomic and physiological studies produced during this period, both in the United States and in Europe, were essential to forensic entomologists, who would be called on as expert witnesses in the decades to come. By the middle of the twentieth century, the knowledge of insect development rates was paramount to understanding how to estimate a minimum postmortem interval (PMI_{min}). The first and one of the most widely used sources to determine a PMI_{min} estimate until the 1990s and early 2000s was Kamal's work on a wide collection of forensically important Diptera and their development rates at a constant temperature (Kamal 1958). Of special note around this period was perhaps the first graduate degree that focused on the succession of beetles on carrion in North Carolina completed by Howden (1950)—a harbinger of graduate projects to come (Reed 1958).

18.2.2 Research Contributions in the United States, 1950–Present

Before 1950, forensic entomological publications consisted of scattered peer-reviewed articles and notes. However, the last 20 years have provided a boom in research publications and forensic entomology textbooks, including two editions of the Catts and Haskell procedural guide (Catts and Haskell 1990; Haskell and Williams 2008) and two editions of Byrd and Castner's work (2001, 2010). Forensic entomology has entered the popular reading shelves with books such as Goff's *A Fly for the Prosecution* (2000) and Heinrich's stimulating *Life Everlasting* (2012). Today, most textbooks on forensic science include chapters on forensic entomology.

In the 1960s, Jerry Payne made a seminal contribution with his study on the stages of decomposition and insect succession during decomposition (Payne 1965, 1967). Payne's work also provided a broad entomological/ecological base, spanning work from aquatic systems to buried carcasses, and introduced the concept of using pigs as surrogates for humans in forensic entomological research. One of the more iconic products of Payne's work was a time-lapse video of pig decomposition he created in 1965. It is still widely referenced and watched today, has been archived in museums, and is available online (http://www.folkstreams.net/pub/FilmsByTitle.php). This early work would springboard the next four decades of empirical research (Payne and Crossley 1966; Payne et al. 1968a,b; Payne and King 1969a,b; Payne and Mason 1971; Payne and King 1972). Payne's PhD dissertation (Payne 1967) may be the first that dealt exclusively with decomposition ecology of vertebrate carrion with special reference to associated arthropods in the United States.

During the 1970s, like many of his peers interested in forensic entomology research at the time, C. Lamar Meek's work focused on medical entomology, specifically mosquito (Diptera: Culicidae) research and forensic entomology (Meek et al. 1983). Meek eventually published more than 20 articles about necrophilous arthropods and pioneered initial forensic entomology research related to wildlife forensics (Fuxa and Boethel 2000). He testified in more than 10 criminal trials, including the case on which the film *Dead Man Walking* was based. Meek was instrumental in organizing many of the early presentations and symposia on forensic entomology through the Entomological Society of America (ESA), for example, in 1981, Meek organized an ESA symposium, subsequently attracting some of the most prominent entomologists in forensic work to present their research. By this time, one such early presenter was E. Paul Catts, a faculty member of Washington State University. Catts developed a course entitled *Insects and People*, which is still taught today and is one of Catts' enduring legacies. Having consulted on several criminal cases, Catts had yet to delve extensively into the research aspects of forensic entomology. He later teamed up with Neal Haskell to produce the first forensic entomology text in the United States, entitled *Entomology & Death: A Procedural Guide* (Catts and Haskell 1990). This text, published in field guide format, was written for law enforcement and forensic investigators and became the cornerstone text for forensic entomology courses and workshops that would soon begin to spring up around the country.

During the 1970s and early 1980s, Bernard Greenberg at the University of Chicago was the only entomologist besides Meek with an active research program and ongoing forensic entomology casework. Greenberg contributed valuable taxonomic data on fly egg identification using scanning electron microscopy (Greenberg and Szyska 1984), and later a classic forensic entomology citation that provided valuable developmental data for several calliphorid taxa (Greenberg 1991). His textbook, *Entomology and the Law* (Greenberg and Kunich 2002), provides an excellent link between entomology and jurisprudence issues.

In 1981, the University of Tennessee opened the Anthropological Research Facility, which quickly became better known as the "Body Farm." Interestingly enough, forensic anthropologists, William Bass of the University of Tennessee and William Rodriguez, U.S. Armed Forces, were the first scientists in North America to publish information on the insect colonization of human remains

(Rodriguez and Bass 1985; Hall and Huntington 2010). During the 1980s, many other entomologists became involved in forensic science research and casework.

The forensic entomology research path was forged from the 1960s through the 1980s, but the last two decades (1990 to the present) have spawned an enormous library of forensic entomology publications. Because of the increase in criminal testimony by forensic entomologists, several aspects of the field began to be questioned in the court of law, such as the use of pigs as models (Schoenly et al. 1996a; Shahid et al. 2003), how well classic ecological succession models describe what really occurs on corpses (Schoenly and Reid 1987; Schoenly et al. 1996b; Schoenly et al 2005), the impact of drugs and toxins on insect development (Goff and Lord 2010), temperature effects on insect developmental rates (Higley and Haskell 2010), addressing statistical error rates and probability values when using developmental and succession data to estimate a PMI_{min} (LaMotte and Wells 2000; Tarone and Foran 2008), improved rearing and collecting procedures (Byrd et al. 2010; Byrd and Tomberlin 2010), weather and climate applications to forensic entomology (Shean et al 1993; Scala and Wallace 2010), the application of DNA in forensic entomology (Wells and Sperling 2001; Wells et al. 2001, 2007; Garrett 2008), and the role of aquatic insects in forensic investigations (Haskell et al. 1989; Hobischak and Anderson 2002; Wallace et al. 2008; Merritt and Wallace 2010). The increased research in the aquatic realm of forensic entomology modified the number of decomposition stages from six in Payne and King's article (1972) to five (Haefner et al. 2004). Although some of the earlier taxonomic works, for example, Hall (1948) and Hall and Townsend (1977) works remain essential to blow fly identification, a major contribution to enhance taxonomic accuracy has been Terry Whitworth's (2006) key to the genera and species of blow flies north of Mexico, which has subsequently been updated in the work of Byrd and Castner (2010). The application of entomological evidence in wildlife crimes (Anderson 1999; Watson and Carlton 2005; Tomberlin and Sanford 2012; Wallace and Ross 2012) has continued to evolve over the last 15 years.

With the exponential advancement of molecular tools and the use of DNA in practically all areas of forensic science, the use of genomics (determining the gene sequence of organisms) has been added to the forensic entomology toolbox, to elaborate on population biology and geographic variation in blow fly morphology (Tomberlin et al. 2011a), developmental rate variation in several species of calliphorids (Tarone et al. 2010), as well as the application of genomic studies outside of entomology to study microbial succession on forensically important insects such as black soldier flies, *Hermetia illucens* (L.), (Diptera: Stratiomyidae) (Zheng et al. 2013).

One topic in the last decade has generated considerable debate among forensic entomologists in the United States: time of death versus time since insect colonization. Tomberlin et al. (2006) initiated discussion on this concept and whether it was antiquated and warranted further analysis. A more recent publication by Tomberlin et al. (2011b) presented a paradigm, breaking down the decomposition process into phases within two intervals (pre- and postcolonization) allowing for specific questions to be asked with regard to this process. The goal was to develop terminology allowing for basic research in ecology to bridge with its application in forensics.

18.2.3 Research Contributions in Canada, Pre-1950

The French publication by Mégnin (1894), *La Faune des Cadavres*, inspired early research by Canadian medical doctors Wyatt Johnston and George Villeneuve. Mégnin, considered by many to have revolutionized forensic entomology, compiled 15 years of research and observations based on medicolegal work into a detailed publication listing eight successional waves. Johnston and Villeneuve (1897) embarked on numerous systematic entomological studies using human cadavers as well as making observations from various medicolegal cases (Benecke 2001). Having no other North American study to consult, they compared their findings to those observations made by Mégnin (1894). Johnston and Villeneuve (1897) noted that their findings closely followed the order of succession laid down by Mégnin; however, the duration of each wave, including the appearance of the associated species of insects, was significantly shorter than those described by Mégnin, by several weeks in some cases. Johnston and Villeneuve used insect development as a factor to indicate an approximate PMI (Johnston and Villeneuve 1897). When empty Diptera puparia were recovered, they concluded that the date of exposure must have been at least 1 month; the absence of puparia meant an exposure time of less than 1 month. Although this may not be entirely precise, they introduced the use of developmental stages to determine a more accurate PMI_{min}.

The ultimate purpose of the research undertaken by Johnston and Villeneuve was to gain information in their country and climate to make "safe deductions" in Canadian medicolegal investigations. Their fear was that Mégnin's "imitators" would use his observations to make conclusions in climates that may not have been supported by his findings. Finally, Johnston and Villeneuve recommended that research be conducted in different localities before applying any entomological data to legal medicine (Johnston and Villeneuve 1897), a recommendation that is still strongly upheld today and has spurred numerous forensic entomological studies all over the world (Anderson and VanLaerhoven 1996; Tantawi et al. 1996; Matuszewski et al. 2011; Bygarski and LeBlanc 2013).

Despite the principal work by Johnston and Villeneuve, very little else concerning medicolegal entomology was documented in Canada for the next century. Abbot (1927, 1937, 1938) contributed quite significantly to our understanding of olfaction and the general biology of many necrophilous Coleoptera. Chapman (1944) reported on cases of *Musca domestica* L. (Diptera: Muscidae) infestations in children's bedding. Although these were not medicolegal investigations, Chapman's findings identified a previously unidentified food source for *M. domestica*.

18.2.4 Research Contributions in Canada, 1950–Present

In the early 1990s, exploration of modern medicolegal or forensic entomology began with the first succession studies in Canada since 1887. In 1992, medical and veterinary entomologist Gail Anderson, Simon Frasier University, advised her undergraduate student Sherah VanLaerhoven, now at Windsor University, in insect succession studies that were conducted in the lower mainland of British Columbia in the Coastal Western Hemlock zone. It was immediately recognized that the findings differed from previously published work. In other temperate regions, such as Great Britain, Dermestidae (Coleoptera) and Piophilidae (Diptera) were reported to arrive 2–3 months after death (Smith 1986); Anderson and VanLaerhoven (1996) noted these insects near the end of the first month. Earlier colonization of Dermestidae and Piophilidae were reported in much more tropical regions such as Hawaii, (Goff and Flynn1991; Anderson 2001). These reported findings along with many others (Anderson and VanLaerhoven 1996) highlighted the need to gather data across Canada, during all seasons and in varying habitats, as the data cannot be transferred to other regions (Anderson 2001). The final conclusions of this study mirrored those of Johnston and Villeneuve (1897), stressing that there was a great need for research across all biogeoclimatic zones (Anderson and VanLaerhoven 1996).

The research conducted by Anderson and VanLaerhoven (1996) was not only the first of its kind in Canada in a century, but also the way in which the temperature inside the carcasses, particularly in larval (maggot) masses, was recorded. Previous research had recorded larval mass temperature daily, but only in a single instance each day. Anderson and VanLaerhoven (1996) monitored temperature using data-logger probes inserted inside the carcasses to continually record internal temperature. Using the data logger, it was possible to conclude that, as expected, internal carcass temperature increases considerably during active decay. It was also possible to establish that there was greater fluctuation exhibited in internal than ambient temperature, with daytime and nighttime differences of more than 35°C (Anderson and VanLaerhoven 1996). Continued research in this particular subject eventually led to the conclusion that the oldest larvae were unlikely to be affected by the larval masses as these older larvae left the carcass before the high temperatures were generated, especially in shaded areas or cooler seasons (Dillon and Anderson 1996a; Dillon 1997).

By the late 1990s, other applications of entomology in death scene investigations began to enter both the research and casework arenas: wildlife forensics and aquatic ecosystems. Although medicolegal entomology had previously been reserved for human cases, the area of wildlife forensics was beginning to emerge by the late 1990s. The killing of two bear cubs in Winnipeg, Manitoba, transformed the investigation of such cases with the help of *Phormia regina* (Meigen) (Diptera: Calliphoridae) eggs and forensic entomologist Gail Anderson (Anderson 1999). Entomological evidence was used to determine the time of death, linking two suspects to the scene where the cubs were recovered. This was the first time in Canada where a jail term was secured for the poachers, and in doing so, set a precedent (Anderson 1999).

After receiving several inquiries concerning criminal investigations involving corpses recovered from water, it was clear that research in an aquatic environment was needed. Thus aquatic forensic entomology began to blossom with studies both in freshwater streams and ponds (Hobischak 1997; Hobischak and Anderson 1999). Differences were noted between the two environments and it was clear that even greater differences would exist within a marine environment. With support from the Royal Canadian Mounted Police, in particular Corporal Bob Teather (retired), and the Canadian Coastguard, Vancouver Aquarium Marine Research Center, the Canadian Amphibious Research Team, as well as funding from the Canadian Police Research Center (CPRC), marine experiments were conducted in Howe Sound between Vancouver and Victoria, British Columbia, (Anderson and Hobischak 2004). The marine research has since dramatically expanded into other marine waters with the support and collaboration of the Victoria Experimental Network Under the Sea (VENUS), a cabled underwater laboratory (http://venus.uvic.ca/). The ongoing study began in the Saanich Inlet (Anderson 2008), northwest of Vancouver, British Columbia and is continuing in the Strait of Georgia (Anderson 2010).

Anderson (2001) formed the first laboratory in Canada dedicated to forensic entomology within the School of Criminology at Simon Fraser University in Burnaby, British Columbia. Through this lab Anderson directs research across Canada to develop databases of insect succession on carrion, encompassing the many biogeoclimatic zones (Anderson 2001). Since the initial research in British Columbia by Anderson and her team (Anderson and VanLaerhoven 1996; Dillon and Anderson 1996b; Hobischak and Anderson 1999; VanLaerhoven and Anderson 1999), additional succession studies have been conducted representing the more populated districts of this vast country. These include regions of Alberta (Hobischak and Anderson 2002), Saskatchewan (Sharanowski et al. 2008), Manitoba (Gill 2005), Ontario (Rosati and VanLaerhoven 2007), New Brunswick (Michaud and Moreau 2009), Nova Scotia (LeBlanc and Strongman 2002), and the Yukon (Bygarski and LeBlanc 2013). Their findings are incorporated into the national database. Research on this forensic entomology database was recently selected as one of top 10 projects funded by CPRC in last 10 years.

18.2.5 Research Contributions from Mexico

In the United States and Canada, forensic entomology is now readily accepted as a common part of many human or animal death scene investigations, whereas in Mexico and throughout Latin America the field is only recently becoming established. Forensic entomologists in Mexico are attempting to establish the field and educate law enforcement about the application of entomology in death or abuse investigations. Literature contributions in Mexico are scant. Except for a few early articles, most have appeared within the last dozen years and have characterized the entomological fauna involved in decomposition processes, akin to the way U.S. literature evolved in the 1960s (Morón and Terrón 1984; Narrarete-Heredia et al. 2002; Vasquez-Saucedo et al. 2007; Villamil-Ramirez et al. 2007; García-Espinoza et al. 2009; Pastrana-Ortíz et al. 2009). Interestingly, more recent publications have begun to spin-off research questions that include techniques pioneered in the United States such as employing molecular techniques to identify human DNA in larval blow fly tissue (Hernández-Cortez et al. 2007), succession effects on necrophagous insects (Martinez-Ruvalcaba et al. 2007), development of novel identification techniques to enhance taxonomy (Nuñez-Vasquez 2009), oviposition behaviors around human corpses (Vergara-Pineda et al. 2009a), and examination of mummified remains for insect fragments (Vergara-Pineda et al. 2009b). Most of these studies either used pigs or the heads of pigs to collect entomological data. However, just as paucity of data still exists regarding insect taxonomy, distribution, and variation in development rates as a function of geographic distribution in the United States and Canada, the same can be said for Mexico with published articles covering 4 or 5 of the 31 Mexican states.

For the United States, Canada, and Mexico, challenges exist for those involved in the research and practice of forensic entomology. Although the three regions appear to be disparate, communication and collaboration among researchers from these regions is strong. With continued diligence, forensic entomology in Mexico will reach the same level of appreciation found in the United States and Canada. For example, in 2008, Canadian forensic entomologist Gail Anderson and U.S. forensic entomologists Jeffery Tomberlin and John Wallace were invited to assist in an international forensic entomology conference based in Saltillo, Coahuila, Mexico. Such interactions between the United States, Canada, and Mexico are vital to advancing forensic entomology research and education so that the use of such evidence in their respective judicial systems is sound and consistent from a global perspective.

18.3 FROM RESEARCH TO PROFESSION

Clearly, empirical research on forensic entomology has increased exponentially over the past 20 years, and with the increased interest has been the birth of a profession that includes the development of a certifying board of forensic entomologists, a professional society, and a suite of universities with forensic entomology positions as well as graduate students focusing their research on this topic. The creation of the first full-time forensic entomologist position in a medical examiner's office in 2013 illustrates the continued advance of the significance of applied forensic entomology in medicolegal investigations.

In the 1980s, a small group of forensic entomologists began to gather for meals at the annual ESA meetings. Partly because of these meetings, the group was dubbed CAFE, or the Council of American Forensic Entomologists. Although not officially organized, the members all had something in common: they were … able to get up for breakfast and … look at pictures of decomposing bodies while eating (Goff 2000).

Paul Catts at Washington State University was instrumental in creating the CAFE, and although this informal organization was ultimately short-lived, it was the first attempt by practicing forensic entomologists to organize a professional group. Many of the forensic entomologists involved in the CAFE came together to create a Forensic Entomology Working Group.

The charismatic, if esoteric, group attracted others with strong stomachs and an interest in forensic entomology. Their informal meetings provided a forum for talking about casework experiences and research. In the early 1990s, the group now known as the Forensic Entomology Working Group began meeting more formally at the American Academy of Forensic Sciences (AAFS) annual meetings. The group included E. Paul Catts (Washington State University), Valerie J. Cervenka (University of Minnesota), M. Lee Goff (University of Hawaii, Manoa), Robert D. Hall (University of Missouri), Neal H. Haskell (Purdue University), K. C. Kim (Penn State University), Wayne Lord (Federal Bureau of Investigation), Kenneth Schoenly (International Rice Institute, Philippines), Theodore W. Suman (Anne Arundel Community College), and Jeffrey Wells (University of California, Berkeley).

Discussions at the 1994 AAFS meeting resulted in the decision to replace the working group with what would become the American Board of Forensic Entomology (ABFE). Those attending that meeting would become the organizing committee and most became the first board members: Gail S. Anderson, Val Cervenka, Lee Goff, Rob Hall, Neal Haskell, Wayne Lord, and Ted Suman. Over the next year, the group began to develop the organizational framework needed for formal recognition, incorporation, and function as a certifying board. The majority of the legwork was done by Lee Goff, who took on the role of "correspondent" with other certifying boards and with the Forensic Science Foundation. Without Goff's dogged persistence, patience, and dedication, it is doubtful there would have been an ABFE. On April 2, 1996, the corporate charter was signed by the Nevada Secretary of State, and the ABFE became official. The incorporation document bears the signatures of Gail Anderson, Paul Catts, Lee Goff, Rob Hall, and Wayne Lord.

18.4 BIRTH OF A SCIENTIFIC ORGANIZATION

In November 2002, while attending the ESA annual conference in Fort Lauderdale, Florida, John Wallace and Jeffery Tomberlin were at a café discussing a variety of topics that included the desire to bring the developing young discipline in entomology back to the science used to determine time since insect colonization, and inspired the organization of a stand-alone conference on forensic entomology. Initially written on a napkin, Wallace and Tomberlin presented the concept of a stand-alone forensic entomology conference at an ABFE meeting 4 months later in February 2003 at the AAFS conference in Chicago, Illinois, and received unanimous support from those in attendance.

With a shoestring budget and Wallace's credit card, Tomberlin and Wallace were able to organize the first North American Forensic Entomology Association (NAFEA) conference in Las Vegas, Nevada, in 2003. The site was chosen partly because of the wave on which forensic entomology was riding among the various forensic science-related television programs. This endeavor would not have been possible without Wallace's undergraduate student Lauren Way, in addition to Jason Byrd and Eric Benbow, who provided website development for the meeting and other logistical support. After several months of preparation, approximately 50 individuals representing 14 states and 3 countries were in attendance (Figure 18.1).

At the end of the conference, K. C. Kim, Pennsylvania State University, suggested that a professional organization be developed that would essentially take charge of this conference and make it an annual event. In 2005, at the NAFEA conference in Orlando, Florida, Wallace and Tomberlin organized and presented a set of bylaws that resulted in the formation of the NAFEA. The first officers were elected at the Orlando NAFEA conference. By 2011, with thanks in large part to Jason Byrd, NAFEA was a legally recognized 501(c)(3) nonprofit organization.

Figure 18.1   (See color insert.) Attendees of the First Annual North American Forensic Entomology Conference held in Las Vegas, Nevada, in August 2003.

18.5 FUTURE DIRECTIONS

Over the past half-century, forensic entomology has grown from an outstretched finger of general ecology and more specifically, insect ecology, to a blossoming discipline within the field of forensic science. With this emergence, a crop of academic professors and an entourage of students with specific forensic entomology research foci have developed. Historically, those scientists in the United States that claimed to be forensic entomologists were actually practicing forensic entomology as a research side interest, with their academic roots stemming from the area of general insect ecology or medical and veterinary entomology. Several entomologists now with research appointments at large institutions, such as Purdue University, Texas A&M University, Chaminade University in Hawaii, University of Florida, and Florida International University are academic offspring of the forensic entomology forefathers from the 1960s and 1970s. These faculty members are basing entire laboratory investigations and subsequent graduate student projects primarily addressing entomological issues in forensic science. The influx of forensic entomology graduate students has generated an exponential increase in Master's theses and PhD dissertations.

Before January 2013, there was no full-time forensic entomologist employed in medicolegal forensic entomology. The Harris County Institute of Forensic Sciences hired Michelle Sanford as the first full-time staff forensic entomologist in a medical examiner's office in the United States. This position provides unprecedented access to scene and autopsy death investigations, and the accessibility of entomology is increasing its visibility and the appreciation of the potential for insects to be useful in death investigations, expanding it beyond high-profile homicide investigations. Much similar to extension entomology, this position also provides educational opportunities as well as technology transfer opportunities to bring published academic methods into practice. The position also allows for strengthening the procedures and practices used in casework to move the profession toward accreditation as in other forensic disciplines.

Forensic entomology, like many specialized and relatively recently evolved areas of forensic science, is entering a new era in North America. This new era brings with it multidisciplinary issues with jurisprudence, social, and scientific origins that have created a potential judicial impasse potentially leading to proposed legislation designed to untangle this complex web of forensic issues.

In 2009, the National Research Council (NRC) released a much-anticipated, congressionally commissioned report on the current state of forensic science in the United States (Committee on Identifying the Needs of the Forensic Sciences Community, NRC Report 2009). The report was the culmination of multiple years of work by a panel of distinguished, independent forensics science experts, and it was anticipated that this report would focus on the advances in forensic science brought forth by molecular biology, digital forensics, and new standards for crime laboratory operations. However, when published, the report was an intense criticism of the nebulous and at times nonexistent foundations of many forensic subdisciplines.

If entomological evidence in the courtroom is allowed, despite the fact that it can be demonstrated to fail many of the minimum standards society expects from other scientific disciplines, how is the integrity of the forensic entomological evidence protected and improved in the courtroom? The NRC report responded to that question by stating that judges presiding over the presentation of this evidence need to better exercise their role as gatekeepers of physical evidence and expert witness testimony. In short, it is critical that judges require higher standards for forensic science testimony when it is used as legal evidence.

In the last decade, the media has created a sociological expectation with regards to conclusions drawn in the courtroom to "get it right," not just within the realm of forensic entomology but in all branches of forensic science (Jeff Tomberlin, pers. comm.). Legitimate phenomena such as the "CSI effect," a term coined from the plethora of television programs that attempt to encapsulate what real forensic investigators do to solve a crime in 1 hour have certainly facilitated this erroneous process. For example, in 2002, David Westerfield was convicted of kidnapping and killing 7-year-old Danielle Van Dam and was sentenced to death in San Diego, California. In this case, three entomologists testified for the prosecution and one for the defense with no consensus among them. This case not only generated considerable debate among the general public on how so many entomologists could disagree on their conclusions but also spawned significant discussion among entomologists in general. Certainly, every death investigation is different, and variation in conclusions among entomologists may reflect both the procedures used by each to arrive at a minimum period of insect activity and the natural variation of different geographical effects on insect populations. This aspect of variation, that is, the period of insect activity, has been recognized by forensic entomologists as an area of much needed empirical data and also demonstrates a need for standard operating procedures for all forensic entomologists. The lack of standard operating procedures was addressed in the National Academy of Science report and has been addressed by the European Association of Forensic Entomology (Amendt et al. 2007).

Pressure from the general public in the United States and Canada has resulted in a greater emphasis on analytical and hypothesis testing that emphasizes experimental design and complex statistical analyses while merging with other fields including genomics, molecular biology, and chemistry to name a few. Although the results being generated are strengthening the scientific foundation and credibility of forensic entomology as an applied science, the results also demonstrate that the complexity of nature and its application in forensics is far more tedious and challenging than previously appreciated (Jeff Tomberlin, pers. comm.).

18.6 CONCLUSIONS

Over the past century, the birth of a significant application and contribution of entomology specific to our judicial system and society in general has been witnessed. Like many areas of natural science, entomology has evolved from a taxonomically and ecologically focused discipline to an applied field as most recently exemplified by the theme for the 2013 ESA annual conference: Science Impacting a Connected World (http://www.entsoc.org/entomology2013). As the field of forensic entomology grows in the United States, Canada, and Mexico, so do the outreach and collaboration in Central and South America and elsewhere in the world. The numbers of forensic entomology textbooks have increased in addition to chapters on forensic entomology in other subdisciplines such as wildlife forensics and anthropology. The numbers of ABFE-certified entomologists have almost doubled since its inception in the late 1990s, undergraduate and graduate student research has greatly increased, and the recent inclusion of forensic entomologists in the medical examiner's setting, insures that the rigor of the science keeps pace as the application of the entomology in the court of law increases.

ACKNOWLEDGMENTS

We thank Carol Nuñez-Vasquez for her assistance in researching literature in Mexico and Rebecca McCabe of Millersville University for her assistance with U.S. literature searches.

REFERENCES

Abbot, C. E. 1927. Experimental data on the olfactory sense of Coleoptera, with special reference to the *Necrophori*. *Annals of the Entomological Society of America* 20: 207–216.
Abbot, C. E. 1937. The necrophilous habit of Coleoptera. *Bulletin of the Brooklyn Entomological Society* 32: 202–204.

Abbot, C. E. 1938. The development and general biology of *Creophilus villosus* Grav. *Journal of the New York Entomological Society* 46: 49–53.

Aldrich, J. M. 1916. *Sarcophaga and Allies in North America*, Vol. 1. La Fayette, IN: Thomas Say Foundation of Entomological Society of America.

Amendt, J., C. P. Campobasso, E. Gaudry, C. Reiter, H. LeBlanc, and M. J. R. Hall. 2007. Best practice in forensic entomology—Standards and guidelines. *International Journal of Legal Medicine* 121: 90–114.

Anderson, G. S. 1999. Wildlife forensic entomology: Determining time of death in two illegally killed black bear cubs, a case report. *Journal of Forensic Science* 44: 856–859.

Anderson, G. S. 2001. Forensic entomology in British Columbia: A brief history. *Journal of the Entomological Society of British Columbia* 98: 127–135.

Anderson, G. S. 2005. Forensic entomology. In: *Forensic Science: An Introduction to Scientific and Investigative Techniques*, eds. S. H. James and J. J. Nordby, pp. 135–164. Boca Raton, FL: Taylor & Francis.

Anderson, G. S. 2008. Investigation into the effects of oceanic submergence on carrion decomposition and faunal colonization using a baited camera. Part I, p. 126. Ottawa, Ontario, Canada: Canadian Police Research Centre.

Anderson, G. S. 2010. Decomposition and invertebrate colonization of cadavers in coastal marine environments. In: *Current Concepts in Forensic Entomology*, eds. J. Amendt, C. P. Campobasso, M. L. Goff, and M. Grassberger, pp. 223–272. New York, NY: Springer Science.

Anderson, G. S. and N. R. Hobischak. 2004. Decomposition of carrion in the marine environment in British Columbia, Canada. *International Journal of Legal Medicine* 118: 206–209.

Anderson, G. S. and S. L. VanLaerhoven. 1996. Initial studies on insect succession on carrion in southwestern British Columbia. *Journal of Forensic Sciences* 4: 617–625.

Benecke, M. 2001. A brief history of forensic entomology. *Forensic Science International*, 120: 2–14. Retrieved from http://www.ncbi.nlm.nih.gov/pubmed/11457602.

Bergeret, M. 1855. Infanticide, momification du cadavre. Découverte du cadavre d'un enfant nouveau – né dans une cheminée ou il sétait momifié. Determination de l'époque de la naissance par la présence de nymphes et des larves d'insectes dans le cadavre et par l'étude de leurs métamorphoses. *Annals of Hygiene and Legal Medicine* 4: 442–452.

Bygarski, K. and H. N. LeBlanc. 2013. Decomposition and arthropod succession in Whitehorse, Yukon Territory, Canada. *Journal of Forensic Sciences* 58: 413–418.

Byrd, J. H. and J. L. Castner. 2001. *Forensic Entomology: The Utility of Arthropods in Legal Investigations*, 1st Ed. Boca Raton, FL: CRC Press.

Byrd, J. H. and J. L. Castner. 2010. *Forensic Entomology: The Utility of Arthropods in Legal Investigations*, 2nd Ed. Boca Raton, FL: CRC Press.

Byrd, J. H., W. D. Lord, J. R. Wallace, J. K. Tomberlin, and N. Haskell. 2010. Collection of entomological evidence during legal investigations. In: *Forensic Entomology: The Utility of Arthropods in Legal Investigations*, eds. J. H. Byrd and J. L. Castner, pp. 127–176. Boca Raton, FL: CRC Press.

Byrd, J. H. and J. K. Tomberlin. 2010. Laboratory Rearing of Forensic Insects. In: *Forensic Entomology: The Utility of Arthropods in Legal Investigations*, eds. J. H. Byrd and J. L. Castner, pp. 177–200. Boca Raton, FL: CRC Press.

Catts, E. P. and N. H. Haskell (eds.). 1990. *Entomology and Death: A Procedural Guide*. Clemson, SC: Joyce's Print Shop.

Chapman, K. 1944. An interesting occurrence of *Musca domestica* L. larvae in infant bedding. *The Canadian Entomologist* 76: 230–232.

Chapman, R. N., C. E. Mickel, J. R. Parker, G. E. Miller, and E. G. Kelly. 1926. Studies in the ecology of sand dune insects. *Ecology* 7: 416–426.

Cole, A. C. 1942. Observations on three species of *Silpha* (Coleoptera: Silphidae). *American Midland Naturalist* 28: 161–163.

Committee on Identifying the Needs of the Forensic Sciences Community, National Research Council. 2009. *Strengthening Forensic Science in the United States: A Path Forward*. Washington, DC: National Academies Press.

Davis, W. T. 1915. *Silpha surinamensis* and *Creophilus villosus* as predacious insects. *Journal of New York Entomological Society* 23: 150–151.

Deonier, C. C. 1940. Carcass temperature and their relationship to winter blowfly populations and activity in the Southwest. *Journal of Economic Entomology* 33: 166–170.

Dillon, L. C. 1997. Insect Succession on Carrion in Three Biogeoclimatic Zones in British Columbia. MSc Thesis. Department of Biological Sciences, Simon Fraser University, Burnaby, British Columbia, Canada.

Dillon, L. C. and G. S. Anderson. 1996a. The use of insects to determine time of death of illegally killed wildlife. Technical Report. Toronto, Ontario, Canada.

Dillon, L. C. and G. S. Anderson. 1996b. Forensic entomology: A database for insect succession on carrion in northern and interior B. C. Technical Report TR-04-96.Ottawa, Ontario, Canada: Canadian Police Research Centre.

Dorsey, C. K. 1940. A comparative study of the larvae of six species of *Silpha* (Coleoptera: Silphidae). *Annals of the Entomological Society of America* 33: 120–139.

Folsom, J. 1902. Collembola of the grave. *Ohio Journal of Science* 49: 201–204.

Forbes, S. A. 1887. The lake as a microcosm. Reprinted in *Bulletin of the Illinois State Natural History Survey* (1925) 15: 537–50.

Fuxa, J. R. and D. J. Boethel. 2000. Chester Lamar Meek. *American Entomologist* 46: 271.

García-Espinoza, F., M. T. Valdés-Perezgasga, E. Pastrana-Ortiz, B. A. Cisneros-Flores. 2009. Abundancia estacional de géneros de sarcophagidae (Diptera) en el semi-desierto Coahuilense. En: *Entomología Mexicana* (E. G. Estrada-Venegas, A. Equihua-Martinez, M. P. Chaires-Grijalva, J. A. Acuña-Soto, J. R. Padilla-Ramirez, and A. Mendoza-Estrada, eds.) 8: 778–782.

Garrett, B. 2008. Judging innocence. *Columbia Law Review*, January, Retrieved from http://ssrn.com/abstract=999984

Gill, J. 2005. Decomposition and arthropod succession on above ground pig carrion in rural Manitoba.Ottawa, Ontario, Canada: Canadian Police Research Centre.

Goff, M. L. 2000. *A Fly for the Prosecution*. Cambridge, MA: Harvard University Press.

Goff, M. L. and M. M. Flynn. 1991. Determination of postmortem interval by arthropod succession: A case study from the Hawaiian islands. *Journal of Forensic Science* 36: 607–614.

Goff, M. L. and W. D. Lord. 2010. Entomotoxicology: Insects as indicators and the impact of drugs and toxins on insect development. In: *Forensic Entomology: The Utility of Arthropods in Legal Investigations*, eds. J. H. Byrd and J. L. Castner, pp. 427–436. Boca Raton, FL: CRC Press.

Graham, S. A. 1925. The felled tree trunk as an ecological unit. *Ecology* 6: 397–411.

Greenberg, B. 1991. Flies as forensic indicators. *Journal of Medical Entomology* 28: 565–77.

Greenberg, B. and J. C. Kunich. 2002. *Entomology and the Law: Flies as Forensic Indicators*. New York, NY: Cambridge University Press.

Greenberg, B. and M. L. Szyska. 1984. Immature stages and biology of fifteen species of Peruvian Calliphoridae (Diptera). *Annals of the Entomological Society of America* 77: 488–517.

Haefner, J. N, J. R. Wallace, and R. W. Merritt. 2004. Pig decomposition in lotic aquatic systems: The potential use of algal growth in establishing a postmortem submersion interval (PMSI)*. *Journal of Forensic Science* 49: 330–336.

Hall, D. G. 1948. *Blow Flies of North America*. Baltimore, MD: Thomas Say Foundation.

Hall, R. D. and T. E. Huntington. 2010. Introduction: Perceptions and status of forensic entomology. In: *Forensic Entomology: The Utility of Arthropods in Legal Investigations*, eds. J. H. Byrd and J. L. Castner, pp. 1–16. Boca Raton, FL: CRC Press.

Hall, R. D. and L. H. Townsend, Jr. 1977. The Blow Flies of Virginia. The Insects of Virginia, No. 11. *Virginia Polytechnic Institute Research Division Bulletin* No. 123: 48.

Haskell, N. H., D. G. McShaffrey, D. A. Hawley, R. E. Williams, and J. E. Pless. 1989. Use of aquatic insects in determining submersion interval. *Journal of Forensic Science* 34: 622–632.

Haskell, N. H. and R. E. Williams. 2008. *Entomology & Death: A Procedural Guide*. 2nd Ed. Clemson, SC: East Park Printing.

Heinrich, B. 2012. *Life Everlasting—The Animal Way of Death*. New York, NY: Houghton-Mifflin Harcourt.

Hernández-Cortez, H., J. R. Delgado, M. L. Chavez-Briones, P. Diaz-Torres, and M. Garzo y Garzo. Tipificación de humano de larvas de moscas en un cuerpo en descomposición en Nuevo León, México. En: *Entomología Mexicana* (E. G. Estrada-Venegas, A. Equihua-Martinez, C. Lund-Leon, and J. L. Rosas-Acevedo, eds.) 6: 851–855.

Higley, L. G. and N. H. Haskell. 2010. Insect development and forensic entomology. In: *Forensic Entomology: The Utility of Arthropods in Legal Investigations*, eds. J. H. Byrd and J. L. Castner, pp. 389–406. Boca Raton, FL: CRC Press.

Hobischak, N. R. 1997. Freshwater Invertebrate Succession and Decompositional Studies on Carrion in British Columbia. MPM Thesis. Department of Biological Sciences, Simon Fraser University, Burnaby, British Columbia, Canada.

Hobischak, N. R. and G. S. Anderson. 1999. Freshwater-related death investigations in British Columbia in 1995–1996. A review of coroner cases. *Canadian Society of Forensic Science* 32: 97–106.

Hobischak, N. R. and G. S. Anderson. 2002. Time of submergence using aquatic invertebrate succession and decompositional changes. *Journal of Forensic Sciences* 47: 142–151.

Howden, A. T. 1950. *The Succession of Beetles on Carrion.* MS Thesis. North Carolina State College, Raleigh, NC.

Illingworth, J. F. 1927. Insects attracted to carrion in southern California. *Proceedings of the Hawaiian Entomological Society* 6: 397–401.

Johnston, W. and G. Villeneuve. 1897. On the medico-legal application of entomology. *Montreal Medical Journal* 26: 81–90.

Kamal, A. S. 1958. Comparative study of thirteen species of sarcosaprophagous Calliphoridae and Sarcophagidae (Diptera). *Annals of the Entomological Society of America* 51: 261–271.

Knipling, E. F. 1936. Some specific taxonomic characters of common *Lucilia* larvae-Calliphorinae-Diptera. *Iowa State College Journal of Science* 10: 275–389.

Knipling, E. F. 1939. A key for blowfly larvae concerned in wound and cutaneous myiasis. *Annals of the Entomological Society of America* 32: 376–388.

LaMotte, L. R. and J. D. Wells. 2000. P-values for postmortem intervals from arthropod succession data. *Journal of Agricultural, Biological, and Environmental Statistics* 5: 58–68.

LeBlanc, H. and D. Strongman. 2002. Carrion insects associated with small pig carcasses during fall in Nova Scotia. *Canadian Society of Forensic Science Journal* 35: 145–152.

Martinez-Ruvalcaba, H., J. Escato-Rocha, and F. Tafoya. 2007. Sucesión de insectos necrófagos en *Sus scrofa*, durante el período estacional de primavera en la ciudad de Aguascalientes, México. En: *Entomología Mexicana*, (E. G. Estrada-Venegas, A. Equihua-Martinez, C. Lund-León, and J. L. Rosas-Acevedo, eds.) 6: 880–884.

Matuszewski, S., D. Bajerlein, S. Konwerski, and K. Szpila. 2011. Insect succession and carrion decomposition in selected forests of Central Europe. Part 3: Succession of carrion fauna. *Forensic Science International* 207: 150–163.

McKnight, B. E. 1981. *The Washing Away of Wrongs: Forensic Medicine in Thirteenth Century China.* Ann Arbor, MI: University of Michigan Press.

Meek, C. L., M. D. Andis, and C. S. Andrews. 1983. Role of the entomologist in forensic pathology, including a selected bibliography. *Bibliography Entomological Society of America* 1: 1–10.

Mégnin, P. 1894. La faune des cadavres: Application de l'entomologie à la médecine légale. In: *Encyclopedie scientifique des Aides-Memoire* (5-214), ed. G. Masson. Paris, Gauthier-Villards et Fils.

Merritt, R. W. and J. R. Wallace. 2010. The role of aquatic insects in forensic investigations. In: *Forensic Entomology: The Utility of Arthropods in Legal Investigations*, eds. J. H. Byrd and J. L. Castner, pp. 271–320. Boca Raton, FL: CRC Press.

Michaud, J. and G. Moreau. 2009. Predicting the visitation of carcasses by carrion-related insects under different rates of degree-day accumulation. *Forensic Science International* 185: 78–83.

Morón, R. M. A. and R. A. Terrón. 1984. Distribución altitudinal y estacional de los insectos necrófilos en la Sierra Norte de Hidalgo, México. *Acta Zoológica Mexicana* 3: 1–47.

Motter, M. G. 1898. A contribution to the study of the fauna of the grave. A study of one hundred and fifty disinterments, with some additional experimental observations. *Journal of the New York Entomological Society* 6: 201–233.

Navarrete-Heredia, J. L., A. F. Newton, M. K. Thayer, J. S. Ashe, D. S. Chandler. 2002. Guía ilustrada para los géneros de Staphylinidae (Coleoptera) de México, Universidad de Guadalajara y Conabio, México.

Nuñez-Vasquez, C. 2009. Tincion del esqueleto cefalofaringeo en larvas de Calliphoridae para diferenciacion taxonomica. En: *Entomología Mexicana* (E. G. Estrada- Venegas, A. Equihua-Martinez, M. P. Chaires-Grijalva, J. A. Acuña-Soto, J. R. Padilla-Ramirez, and A. Mendoza-Estrada, eds.) 8: 746–749.

Payne, J. A. 1965. A summer carrion study of the baby pig, *Sus scrofa* (L.). *Ecology* 46: 592–602.

Payne, J. A. 1967. A Comparative Ecological Study of Pig Carrion Decomposition and Animal Succession with Special Reference to the Insects. PhD Dissertation, Clemson University, SC.

Payne, J. A. and D. A. J. Crossley. 1966. Animal Species Associated with Pig Carrion. Oak Ridge, TN: Oak Ridge National Laboratory.

Payne, J. A. and E. W. King. 1969a. Lepidoptera associated with pig carrion. *Journal of Lepidoptera Society* 23:191–195.

Payne, J. A. and E. W. King. 1969b. Coleoptera associated with pig carrion. *Entomologist's Monthly Magazine* 105: 224–232.

Payne, J. A. and E. W. King. 1972. Insect succession and decomposition of pig carcasses in water. *Journal of the Georgia Entomological Society* 7: 153–162.

Payne, J. A., E. W. King, and G. Beinhart. 1968a. Arthropod succession and decomposition of buried pigs. *Nature* (London) 219: 1180–1181.

Payne, J. A. and W. R. M. Mason. 1971. Hymenoptera associated with pig decomposition. *Proceedings of the Entomological Society of Washington* 73: 132–141.

Payne, J. A., F. W. Mead, and E. W. King. 1968b. Hemiptera associated with pig carrion. *Proceedings of the Entomological Society of Washington* 61: 565–567.

Pastrana-Ortiz, F. M. T. Valdes-Perezgarsga, F. Garcia-Espinoza, F. J. Sanchez-Ramos. 2009. Abundancia estacional de especies de Calliphoridae en un area urbana abierta del semi-desierto de Coahuila. In: *Entomología Mexicana* (E. G. Estrada-Venegas, A. Equihua-Martinez, M. P. Chaires-Grijalva, J. A. Acuña-Soto, J. R. Padilla-Ramirez, and A. Mendoza-Estrada, eds.) 8: 783–787.

Reed, H. B. 1958. A study of dog carcass communities in Tennessee, with special reference to the insects. *American Midland Naturalist* 59: 213–245.

Rodriguez, W. C. and W. M. Bass. 1985. Decomposition of buried bodies and methods that may aid in their location. *Journal of Forensic Sciences* 30: 836–852.

Rosati, J. Y. and S. L. VanLaerhoven 2007. New record of *Chrysomya rufifacies* (Diptera: Calliphoridae) in Canada: Predicted range expansion and potential effects on native species. *Canadian Entomologist* 139: 670–677.

Scala, J. R. and J. R. Wallace. 2010. Forensic meteorology: The application of weather and climate. In: *Forensic Entomology: The Utility of Arthropods in Legal Investigations*, eds. J. H. Byrd and J. L. Castner, pp. 519–538. Boca Raton, FL: CRC Press.

Schoenly, K., M. L. Goff, J. D. Wells, and W. D. Lord. 1996b. Quantifying statistical uncertainty in succession-based entomological estimates of the postmortem interval in death scene investigations: A simulation study. *American Entomologist* 42: 106–112.

Schoenly, K., N. H. Haskell, and R. D. Hall. 1996a. Testing the reliability of an animal model for use in research and training programs in forensic entomology. Final Report No. 94-IJ-CX-0039. Washington, DC: U. S. National Institute of Justice.

Schoenly, K. and W. Reid. 1987. Dynamics of heterotrophic succession in carrion arthropod assemblages: Discrete series or continuum of change. *Oecologia (Berlin)* 73: 192–202.

Schoenly, K., S. A. Shahid, N. H. Haskell, and R. D. Hall. 2005. Does carcass enrichment alter community structure of predaceous and parasitic arthropods? A second test of the arthropod saturation hypothesis at the Anthropology Research Facility in Knoxville, Tennessee. *Journal of Forensic Sciences* 50: 134–141.

Shahid, S. A., K. G. Schoenly, N. H. Haskell, R. D. Hall, and W. Zhang. 2003. Carcass enrichment does not alter decay rates or arthropod community structure: A test of the arthropod saturation hypothesis at the Anthropology Research Facility in Knoxville, Tennessee. *Journal of Medical Entomology* 40: 559–569.

Sharanowski, B. J., E. G. Walker, and G. S. Anderson 2008. Insect succession and decomposition patterns on shaded and sunlit carrion in Saskatchewan in three different seasons. *Forensic Science International* 179: 219–40.

Shean, B. S., L. Messinger, and M. Papworth. 1993. Observations of differential decomposition on sun exposed v. shaded pig carrion in coastal Washington State. *Journal of Forensic Sciences* 38: 938–949.

Smith, K. G. V. 1986. *A Manual of Forensic Entomology*., London, United Kingdom: Trustees of the British Museum.

Solomon S. M. and E. J. Hackett. 1996. Setting boundaries between science and law: Lessons from Daubert v. Merrell. Dow Pharmaceuticals, Inc. *Science, Technology, and Human Values* 21: 131–156.

Steele, B. F. 1927. Notes on the feeding habits of carrion beetles. *Journal of the New York Entomological Society* 35: 77–81.

Tantawi, T. I., E. M. el-Kady, B. Greenberg, and H. A. el-Ghaffar. 1996. Arthropod succession on exposed rabbit carrion in Alexandria, Egypt. *Journal of Medical Entomology* 33: 566–580.

Tarone, A. M. and D. R. Foran. 2008. Generalized additive models and *Lucilia sericata* growth: Assessing confidence intervals and error rates in forensic entomology. *Journal of Forensic Science* 53: 942–948.

Tarone, A. M. and D. R. Foran. 2010. Gene expression during blow fly development: Improving the precision of age estimates in forensic entomology. *Journal of Forensic Sciences* 56: 112–122.

Tomberlin, J. K., M. E. Benbow, and A. M. Tarone. 2011a. Basic research in evolution and ecology enhances forensics. *Trends in Ecology and Evolution* 26: 53–55.

Tomberlin, J. K., R. Mohr, M. E. Benbow, A. M. Tarone, and S. VanLaerhoven. 2011b. A roadmap for bridging basic and applied research in forensic entomology. *Annual Review of Entomology* 56: 401–422.

Tomberlin, J. K. and M. R. Sanford. 2012. Forensic entomology and wildlife. In: *Wildlife Forensics: Methods and Applications*, eds. J. E. Huffman and J. R. Wallace, pp. 79–105. New York, NY: Wiley.

Tomberlin, J. K., J. Wallace, and J. H. Byrd. 2006. Forensic entomology: Myths busted! *Forensic Magazine* October/November: 10–14.

VanLaerhoven, S. L. and G. S. Anderson. 1999. Insect succession on buried carrion in two biogeoclimatic zones of British Columbia. *Journal of Forensic Sciences* 44: 31–34.

Vazquez-Saucedo, R., D. Stephano-Vera, C. H. Martín-Hernández, H. Quiroz-Martinez, A. Rodriquez-Castro, J. A. Flores-Mellado, and P. Diaz-Torres. 2007. Diptera necrófagos del estado de Nuevo León, México. En: *Entomología Mexicana* (E. G. Estrada-Venegas, A. Equihua-Martinez, C. Lund-León, and J. L. Rosas-Acevedo, eds.) 6: 885–888.

Vergara-Pineda, S., H. DeLeon-Muzquiz, O. Garcia-Martinez, M. Cantu-Sifuentes, B. H. Muhammad and J. K. Tomberlin. 2009a. Comportamiento de arribo de moscas necrófagas (Diptera: Calliphoridae) a un cadáver humano. En: *Entomología Mexicana* (E. G. Estrada- Venegas, A. Equihua-Martinez, M. P. Chaires-Grijalva, J. A. Acuña-Soto, J. R. Padilla-Ramirez, and A. Mendoza-Estrada, eds.) 8: 792–797.

Vergara-Pineda, S., J. M. Rojas-Chavez, J. Mansilla-Lory, and T. Campos de la Rosa. 2009b. Fragmentos de insectos asociados a momias prehispánicas. En: *Entomología Mexicana* (E. G. Estrada-Venegas, A. Equihua-Martinez, M. P. Chaires-Grijalva, J. A. Acuña-Soto, J. R. Padilla-Ramirez, and A. Mendoza-Estrada, eds.) 8: 798–800.

Victoria Experimental Network Under the Sea. Retrieved from http://venus.uvic.ca/.

Villamil-Ramirez, E. D., N. E. Galindo-Miranda, J. L. Navarrete-Heredia. 2007. Caracterización de la coleoptera fauna asociada a cadáveres de *Mus musculus* en la reserva ecológica del Pedregal de San Angel, México. En: *Entomología Mexicana* (E. G. Estrada Venegas, A. Equihua-Martinez, C. Lund-León, and J. L. Rosas-Acevedo, eds.) 6: 860–865.

Vincent, C., D. K. Kevan, M. Leclerq, and C. L. Meek. 1985. A bibliography of forensic entomology. *Journal of Medical Entomology* 22: 212–219.

Wallace, J. R., R. W. Merritt, R. Kimbarauskas, M. E. Benbow, M. McIntosh. 2008. Caddis fly cases assist homicide case: Determining a postmortem submersion interval (PMSI) using aquatic insects. *Journal of Forensic Sciences* 53: 1–3.

Wallace, J. R. and J. C. Ross. 2012. The application of wildlife evidence to forensic science. In *Wildlife Forensics: Techniques and Applications*, eds. J. E. Huffman and J. R. Wallace, pp. 35–50. London, United Kingdom: Wiley.

Watson, E. J. and C. E. Carlton. 2005. Insect succession and decomposition of wildlife carcasses during fall and winter in Louisiana. *Journal of Medical Entomology* 42: 193–203.

Wells, J. D., F. G. Introna, G. Di Vella, C. P. Campobasso, J. Haes, and F. A. H. Sperling. 2001. Human and insect mitochondrial DNA analysis from maggots. *Journal of Forensic Sciences* 46: 685–687.

Wells, J. D. and F. A. H. Sperling. 2001. DNA-based identification of forensically important Chysomyinae (Diptera: Calliphoridae). *Forensic Science International* 120: 110–115.

Wells, J. D., R. Wall, and J. R. Stevens. 2007. Phylogenetic analysis of forensically important *Lucilia* flies based on cytochrome oxidase 1: A cautionary tale for forensic species identification. *International Journal of Legal Medicine* 121: 1–8.

Whitworth, T. 2006. Keys to the genera and species of blow flies (Diptera: Calliphoridae) of America North of Mexico. *Proceedings of the Entomological Society of Washington* 108: 689–725.

Zheng, L., T. L. Crippen, B. Singh, M. Tarone, S. Dowd, T. K. Wood, and J. K. Tomberlin. 2013. A survey of bacterial diversity from successive life stages of black soldier fly (Diptera: Stratiomyidae) by using 16S rDNA pyrosequencing. *Journal of Medical Entomology* 50: 647–658.

PART V

Dimensions and Frontiers of Forensic Entomology

CHAPTER 19

Experimental Design, Inferential Statistics, and Computer Modeling

Gaétan Moreau, Jean-Philippe Michaud, and Kenneth G. Schoenly

CONTENTS

"Said the student to the professor: will you be my true confessor?
I have sinned a bit of late, and I have to know my fate.
Last week sans contemplation I did faulty replication.
All my data have been lumped - must my chi-squared test be dumped?
Is my study superficial? Was I blindly sacrificial?
What's the verdict - I can't wait - Did I pseudoreplicate?"

—S. H. Hurlbert (2009: 434)

19.1 INTRODUCTION

Within the last decade, the forensic sciences have undergone a groundswell of scrutiny and criticism of their methods, practices, and standards. Among those leading the charge were Saks and Koehler (2005), who insisted that "sound scientific foundations and justifiable protocols" replace "untested assumptions and semi-informed guesswork." Four years later, the long-anticipated, congressionally mandated report by the U.S. National Research Council (NRC 2009) concurred. The report found that forensic disciplines that had come from the biological or chemical sciences (e.g., nuclear DNA analysis, toxicology) had conducted more experimentation and validation of their methods than those derived from law enforcement (e.g., fingerprints, ballistics, toolmark analysis). In these forensic identification sciences, evidence is often presented to support conclusions of a match (i.e., to a particular person, firearm, etc.). Except for DNA analysis, however, these disciplines have rarely investigated the limits and uncertainties (i.e., error rates) of their methods or verified the assumptions that undergird their conclusions. Among the 13 recommendations in the NRC report was the need to "develop tools for advancing measurement, validation, reliability, information sharing, and proficiency testing in forensic science and to establish protocols for forensic examinations, methods, and practices" (NRC 2009, recommendation no. 6, p. 214).

Unlike the forensic identification sciences that "assume discernible uniqueness" (Saks and Koehler 2005), forensic entomology is an inferential science that depends on extrapolations from experimental data to estimate the postmortem interval (PMI) or to answer other insect-related questions (Michaud et al. 2012). Although forensic entomology escaped scrutiny of the NRC report, courtroom challenges of entomological testimony had already begun, motivated in part by a lack of statistical support in case study reports (i.e., error rates, confidence intervals, probabilities) and untested assumptions (Catts 1992; Archer 2003; Greenberg and Kunich 2005; VanLaerhoven 2008; Wells and LaMotte 2010). However, before the NRC report was published, researchers had anticipated the need to incorporate null models and randomization methods in hypothesis testing (Schoenly 1991, 1992; Schoenly et al. 1996), develop probability-based PMI estimation methods (Wells and LaMotte 1995; LaMotte and Wells 2000; Michaud and Moreau 2009), test the reliability of pig carcasses as model corpses (Schoenly et al. 2007), and conduct field validation tests of PMI estimation methods (VanLaerhoven 2008). After the NRC report was published, a few researchers beseeched their colleagues to design better field experiments (Michaud et al. 2012), improve their statistical models (Ieno et al. 2010; Michaud and Moreau 2011; Basqué and Amendt 2013), consistently report study variables (Tomberlin et al. 2012), and strengthen ties with ecologists and evolutionary biologists (Tomberlin et al. 2011a,b).

In a critique of experimental study designs in forensic entomology research, Michaud et al. (2012) found that the majority (78%) of field studies reviewed had design flaws and statistical errors that prevented valid inferences for future casework. In the life, behavioral, and physical sciences, adequate inference strength depends on clearly stated hypotheses and predictions, sound experimental design, and rigorous statistical analysis. However, for most of the twentieth century, forensic entomology did not adopt these concepts and methods and strayed from its original ecological roots (i.e., ecological succession theory) (Mégnin 1883, 1887, 1894). Instead, forensic entomologists emphasized insect taxonomy, typological concepts (e.g., decay stages), descriptive accounts, and case studies, perhaps due to underfunding by granting agencies or a misunderstanding of the consequences of inadequate design and statistics. With momentum growing to return forensic entomology to its former ecological (and empirical) foundations (Tomberlin et al. 2011a,b; Michaud et al. 2012), a critique of the discipline's experimental and statistical practices seems timely.

This chapter builds on an earlier critique of field and laboratory experiments in forensic entomology (Michaud et al. 2012) and follows a similar mock case study approach to identify design flaws (and other shortcomings) and solutions for increasing inference strength and informing future casework. Specific aims of this chapter are to (1) propose sound guidelines for achieving appropriate inference strength and high statistical power, (2) challenge core assumptions that undergird our conclusions, (3) identify common statistical errors and suggest ways to avoid them, and (4) review statistical and computer models developed for PMI estimation that rely on entomological data. It is acknowledged that field and laboratory experiments involve a complex mix of technical, logistical, and budgetary trade-offs; however, if a study is designed with future casework in mind, its design flaws and limitations need to be acknowledged.

19.2 EXPERIMENTAL DESIGN AND STATISTICAL INFERENCE

A hypothesis is defined as a proposed mechanism to explain a natural phenomenon. Assuming that a given hypothesis is true leads to prediction developments that can be tested through experimentation. Experimental design, in turn, is a term used to define the logical structure of an experiment. In other words, predictions can be tested, a treatment effect, if any, can be detected, variability can be controlled, and investigator error can be quantified. Ronald Fisher (1935), the father of modern statistics, revolutionized the sciences with his book *The Design of Experiments*, in which he introduced rules of hypothesis testing and procedures for designing experiments. These rules still hold true today, amidst enormous gains in statistical practices and computing power.

The goal of an experiment is to extract valid generalizations that are applicable to the wider world (i.e., inferences). Inference is a form of inductive reasoning in which conclusions drawn from available data (affected by random variation) are extrapolated to another time, place, and context. For example, a forensic entomologist called to assist in a murder investigation in British Columbia, Canada, could use Anderson and VanLaerhoven's (1996) successional data for the same region. Because the research study was conducted several kilometers from the crime scene, the biotic and abiotic conditions at both sites will not match up perfectly. Differences between sites could come from sampling practice (e.g., frequency, intensity), and random variation of biotic (e.g., regional species pool, vertebrate scavenging, predominant surrounding vegetation) and abiotic factors (e.g., daily temperatures, soil moisture, soil gradient). In applying Anderson and VanLaerhoven's (1996) data, the forensic entomologist must be certain that the study will allow proper extrapolations (i.e., inferences) to be drawn before findings are presented in court. To achieve adequate inference strength (i.e., to ensure that conclusions are valid and defendable in court), entomologists must comply with the following three rules: (1) adequate replication, (2) independence of experimental units, and (3) capturing a representative range of natural variability.

19.2.1 Replication and Pseudoreplication

In his landmark paper, Hurlbert (1984) coined the term "pseudoreplication" to describe errors of statistical practice or interpretation in ecological experiments in which "treatments are not replicated (though samples may be) or replicates are not statistically independent." High rates of pseudoreplication and confusion between observational and experimental units and their implications have been reported in other experimental disciplines, most notably animal behavior (Lombardi and Hurlbert 1996), economic entomology (Hurlbert and Meikle 2003), marine biology (Hurlbert and White 1993; Garcia-Berthou and Hurlbert 1999), fisheries (Millar and Anderson 2004), psychology (Hurlbert 2009, 2010), and, more recently, forensic entomology (Michaud et al. 2012). Other researchers have followed Hurlbert's lead in an effort to overcome this confusion (Searcy 1989; Heffner et al. 1996; Ramirez et al. 2000; Kroodsma et al. 2001; Kozlov 2003; Lazic 2010). Not everyone agrees, however, that pseudoreplication is a problem, as described in commentaries by Oksanen (2001, 2004); but see Hurlbert (2003), Coss (2009), and Schank and Koehnle (2009). Although these authors presented good arguments, we concur with Kroodsma et al. (2001) that the absence of pseudoreplication should be viewed as "a minimum requirement that should be met before the merits of an experiment are evaluated." For forensic entomology, this statement is particularly relevant given the impact a PMI estimate could have on a criminal death investigation.

The basis of any experiment is the experimental unit (n), which is defined as the smallest unit to which a treatment is applied (Hurlbert 1984). In turn, a treatment is a voluntary modification to an independent (explanatory) variable used to measure its effect on a dependent (response) variable. In manipulative studies, the investigator controls and assigns the treatments, which should be randomized among the different experimental units (e.g., the effect of different depth of burials on insect succession). In observational studies, the investigator does not control the treatment (e.g., the effect of different seasons on insect succession). It is important to note that a variable cannot be a treatment and an experimental unit at the same time (e.g., when comparing the insect fauna between a burnt and a control carcass, each carcass is the sole unit for a treatment). In some manipulative field studies, minimal replication ($n = 2$) may be sufficient to detect a treatment effect. In observational studies, however, sampling effort determines if the study has sufficient inferential strength.

To replicate a study, the number of experimental units has to be more than one. In simple pseudoreplication, a treatment is unreplicated (although samples may be replicated). In sacrificial pseudoreplication, the study is replicated but the data are pooled, thus ignoring variability between experimental units. In temporal pseudoreplication, the study may be replicated, but samples from experimental units are correlated in time. In these three cases, causal inference is unachievable and validation of treatment differences through the use of inferential statistics is impossible. In other words, the experimental error cannot be estimated, so differences between treatments cannot be compared. In Section 19.2.5, we use a mock case study approach, representative of the forensic entomology literature, to illustrate different design flaws in field and laboratory experiments.

19.2.2 Independence of Experimental Units

Independence of experimental units ensures that there is no unaccounted correlation (i.e., interdependence) between experimental units, a basic assumption of most statistical tests. Properly replicated studies can still suffer from interdependence if isolative segregation of treatments occurs, random effects (e.g., blocking variables) are unaccounted for, or repeated measurements from each experimental unit are erroneously treated as true replicates (i.e., temporal pseudoreplication).

In field studies on carrion-insect succession, independence of experimental units implies an adequate spacing distance between carcasses. Minimum distance for spatial independence has to be determined experimentally. Over the decades, a wide range of distances have been used, from a few meters (Tullis and Goff 1987; Leblanc and Strongman 2002; Shahid et al. 2003; Schoenly et al. 2005) to hundreds of meters (Braack 1981; Watson and Carlton 2003). Recent studies, however, have settled on a spacing distance of 50 m (Anderson and VanLaerhoven 1996; Bourel et al. 1999; Tabor et al. 2005; Martinez et al. 2007) because this distance minimizes cross-contamination by migrating fly larvae (Herms 1907; Cragg 1955; Norris 1959; Nuorteva 1977; Greenberg 1990; Tessmer and Meek 1996; Lewis and Benbow 2011). A similar outcome can be achieved by installing a ring of drift fencing (Chelgren et al. 2006) around each carcass. These two approaches, however, are inadequate for flying insects, especially blow flies (Diptera: Calliphoridae), given their ability to find decaying substrates several kilometers away (Gilmour et al. 1946; Lindquist et al. 1951; Schoof et al. 1952; Yates et al. 1952; Schoof and Mail 1953; Cragg and Hobert 1955; Norris 1959; Greenberg and Bornstein 1964). Unfortunately, this issue will remain unresolved until the effect of carcass interdependence is empirically tested. One way to establish such guidelines is to use replicate arrays of carcasses spaced at different distances (e.g., 5, 50, 500, and 5000 m between carcasses) and to evaluate changes in taxonomic composition, population abundances, or decomposition processes (see noncarrion studies by Heard [1998] and Horgan [2005]). Another method could expose carcasses 50 m apart inside sampling grids or along transects with population or community abundances analyzed using geostatistics (Rossi et al. 1992; Carroll and Pearson 1998; Crist 1998; Radeloff et al. 2000). A third method could involve the release of marked flies and beetles at different distances (e.g., 5, 50, 500, 5000 m) from carrion-baited traps. The distance corresponding to 95% of insect captures could represent the effective attraction radius for intercarcass spacing (for a noncarrion example, see Larsen and Forsyth [2005]). Until the effect of carcass interdependence is empirically tested for carrion insects in general, a minimum spacing distance of 50 m should be adopted for succession studies.

19.2.3 Capturing a Representative Range of Variability

The representativeness of experimental conditions is an important but often overlooked requirement. In manipulative studies, the goal is to spread natural variability evenly among treatments through randomization so that variability is not confounded or compounded with effect size. In observational studies, however, it is important to subject the experimental units to as much natural variability as possible to ensure valid inference at the level of the habitat, ecosystem, or geographic zone. For forensic entomology field studies, this implies that at least two study sites are needed to account for potential site effects and incorporate intersite variability. If only one study site is used, external validity is reduced and inferential power is limited to the study site alone.

The inherent problem with field studies is that the word "site" is difficult to define, partly because the effect of carcass interdependence has not been documented. Ideally, a minimum of two distant, unconnected study sites within the same geographic zone with similar habitat characteristics (e.g., predominant vegetation, topography, soil conditions) should be used. This will result in the experimental units being exposed to a larger range of natural variability. The same rationale applies to temporal replication in cases where interseasonal and interannual effects are examined. Indeed, it is well known that temperature patterns at the beginning of a summer are different from those at the end or in different summers in different years. In the context of forensic entomology field studies, tools invented by community and landscape ecologists, such as bias-corrected collection curves (Colwell 2006) and gradient-directed transects (Gillison and Brewer 1985), analyzed by boundary detection methods (Cornelius and Reynolds 1991; Zhang and Schoenly 1999), may prove useful.

19.2.4 Critique of Published Studies

Here, an enlarged sample and extended time series of a previous review of manipulative and observational field studies in forensic entomology (Michaud et al. 2012) is presented and includes papers published post-Hurlbert (1984) from 1985 through May 2013. Only papers in journals tracked

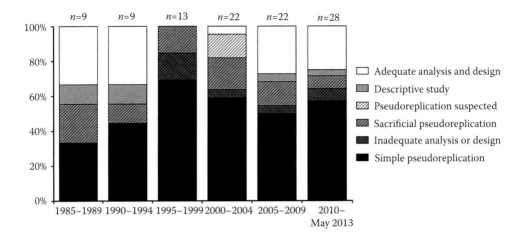

Figure 19.1 Frequency, in the forensic entomology literature, of studies on carrion-arthropod succession displaying (1) adequate analysis and design, (2) descriptive results only, (3) potential cases of simple pseudoreplication, (4) sacrificial pseudoreplication, (5) inadequate analysis or design, and (6) simple pseudoreplication. A total of 103 peer-reviewed reports were examined.

by Thomson Scientific (ISI) for at least 3 years (i.e., those with an impact factor) were considered. After evaluation, each article was placed into one of the six categories: (1) adequate design and analysis (i.e., an inferential study with valid design and analysis), (2) descriptive study (i.e., a study that did not draw inferences from the results), (3) pseudoreplication suspected (i.e., a study with ambiguous methodology that is suspected to be pseudoreplicated), (4) simple pseudoreplication (see previous definition), (5) sacrificial pseudoreplication (see previous definition), and (6) inadequate analysis or design (i.e., a study that included experimental or statistical errors). Finally, articles were sorted into 5-year blocks to analyze temporal trends.

In total, 103 studies were reviewed. Over the 30-year period, the number of published field studies increased, but the frequency of pseudoreplication remained unchanged (Figure 19.1). We found that 19% of the studies reviewed had adequate analysis and design, whereas 77% had flaws affecting inferential strength, including two of our own studies (Schoenly et al. 2007; Michaud and Moreau 2009). No inferences were made in 4% of the studies (see Figure 19.1). Simple pseudoreplication was due to the absence of treatment replication; thus, these studies could not be salvaged using linear mixed effects models (Chaves 2010; Behm et al. 2013). Except for sacrificial pseudoreplication, which can be reversed by analyzing unpooled replicates, the only remedy for simple pseudoreplication is to repeat the study.

19.2.5 Mock Cases

In this section, we use a mock case study approach to illustrate design flaws and suggest methodological remedies for increasing inference strength and informing future casework.

19.2.5.1 Case 1: Different Habitats

Forensic entomologists often compare different habitats within a given geographic zone to document differences in taxonomic composition, succession pattern, or decomposition process (Shean et al. 1993; Richards and Goff 1997; Davis and Goff 2000; Leblanc and Strongman 2002; Michaud et al. 2010). A common approach is to place a single carcass in each of two habitats (e.g., a forest and a meadow) and to examine differences in the variables. This practice qualifies as simple pseudoreplication. The "treatment" is the habitat and the experimental unit is a patch of ground within the habitat where the carcass lies. Thus, there is only one replicate per treatment. Daily insect samplings are subsamples of the experimental unit and are not true experimental units. Another common design is to simultaneously deploy multiple carcasses in each habitat (Figure 19.2a). Again, however, this practice qualifies as simple pseudoreplication. There are now multiple carcasses, but only one patch of habitat. As such, the number of experimental units is one. In this design, the multiple carcasses in each habitat are subsamples. If the goal of a study is to extrapolate results from carcasses placed in one habitat to a death investigation where remains were found in the same geographic area, at least one carcass is needed at multiple sites in two or more habitats, such as two forests and two meadows (Figure 19.2b). Statistically speaking, only one carcass is needed for each forest and meadow (i.e., four carcasses in total). However, this design makes the risky assumption that one carcass in each habitat will yield typical or representative results. This is impossible to confirm because intercarcass variation within each forest or meadow cannot be measured.

When habitats are selected, care should be taken to avoid isolative segregation of treatments (Figure 19.3a). For example, if buried carrion is compared to surface carrion (i.e., control), all experimental units should be randomized within the same field (Figure 19.3b). If, for example, a soil gradient or edge effect is present and is not the effect under study, isolated segregation of carcasses would confound the treatment effect (see Smith and Smith [2002] for general information on edge effects; see Fielder et al. [2008] for a casework example).

19.2.5.2 Case 2: Different Seasons

Forensic entomologists often study seasonal differences in taxonomic composition, succession pattern, and/or decomposition in one or more habitats within a specific geographic zone (Putnam 1983; Tantawi et al. 1996; Centeno et al. 2002; Grassberger and Frank 2004; Michaud and Moreau 2009). In such studies, a common design is to simultaneously expose two or more carcasses in each

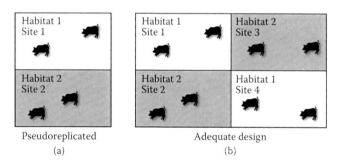

Figure 19.2 Schematic representations of field studies involving pseudoreplicated (a) and adequate (b) experimental designs in the study of habitats.

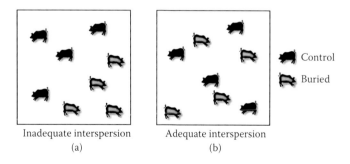

Figure 19.3 Schematic representations of field studies with inadequate (a) and adequate (b) interspersion of treatments.

season of the same year (Figure 19.4a). Again, this practice qualifies as simple pseudoreplication. In this case, the experimental unit is the season. Because only one of each season is sampled (i.e., one summer, one fall, etc.), the number of experimental units per treatment is one and carcasses qualify as subsamples, not experimental units.

Typically, the goal of a seasonal study is to extrapolate the results to human remains found in the same season but in a different year. To properly design such a study, it is necessary to expose carcasses in each of two summers, falls, winters, and springs of different years (see Figure 19.4b). Furthermore, at least two sites per season should be used. To account for season-related effects without introducing confounding site-related effects, the study sites should be the same for both years. It is also important to consider the importance of within-season variability, especially in temperate regions. Toward this end, it would be prudent to add more carcasses or expose carcasses sequentially (i.e., fortnightly, monthly) rather than simultaneously, throughout each season (Michaud et al. 2010).

19.2.5.3 Case 3: Different Crime Scene Circumstances

Forensic entomologists have also studied decomposition and insect succession involving different modes of death, such as hanging (Shalaby et al. 2000), burning (Avila and Goff 1998), or concealment (Payne and King 1968; VanLaerhoven and Anderson 1999; Voss et al. 2008; Anderson 2011). A common design would be to place a carcass inside and outside a shelter at ground level (Figure 19.5a). This practice, however, also qualifies as simple pseudoreplication. In this case, the treatment is the shelter and involves only one experimental unit per treatment (one carcass inside the shelter, one outside the shelter).

If the goal is to extrapolate the results to a case involving remains inside a shelter, the study requires at least two sheltered carcasses and two unsheltered carcasses (Figure 19.5b). Also, multiple carcasses should not be inside the same shelter; otherwise, pseudoreplication still prevails and the carcasses would qualify as subsamples. In other words, the treatment effect would be associated with

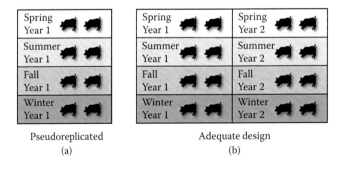

Figure 19.4 Schematic representations of field studies involving pseudoreplicated (a) and adequate (b) experimental designs in the study of seasonal effects.

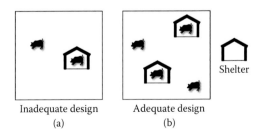

Figure 19.5 Schematic representations of studies examining the effects of a shelter with inadequate (a) and adequate (b) experimental designs.

the particular shelter used and not to shelters in general. As explained previously, temporal or spatial segregation of treatments should be avoided (i.e., sheltered carcasses at one time, unsheltered at another time; sheltered carcasses on one side, unsheltered on another). When mimicking casework, it would be preferable to expose carcasses at or near the same site and time of year as the case to reduce site- and season-related variability. This design, however, would not reduce interyear variation.

19.2.5.4 Case 4: Pooled Successional Data

Ever since Mégnin (1883, 1887, 1894), researchers have pooled successional data from replicate carcasses into "composite" figures or tables (Figure 19.6) without first testing insect species for temporal synchrony (Schoenly 1992; Nelder et al. 2009). This practice qualifies as sacrificial pseudoreplication because it "sacrifices the replicates." In other words, by pooling data into one figure or table, it prevents analysis of intercarcass variation. Such variation might be considerable, even for carcasses exposed simultaneously in the same site and season. Several measures for testing insect synchrony on carcasses include a taxon's residence time, its proportion of days present, or its probability of occurrence (Schoenly 1992; Michaud and Moreau 2009, 2011; Nelder et al. 2009). Adding means and error bars to each insect species arrival and departure times in succession diagrams or tables is one way to illustrate intercarcass variation (synchrony or asynchrony). Other practices for minimizing sacrificial pseudoreplication are discussed in Michaud et al. (2012).

19.2.5.5 Case 5: Development Time Studies

When dating human remains, forensic entomologists primarily use timetables from thermal development studies conducted in the laboratory to backtrack egg laying of adult flies. The rules of inference for laboratory studies differ slightly from field studies in that the former possess a high degree of internal validity, but little external validity, unless they are corroborated by the latter.

Rearing studies of a particular fly (Diptera) or beetle (Coleoptera) species are almost always conducted using a set of constant temperatures, consistent with the natural variability of the species geographic range (Anderson 2000; Byrd and Allen 2001a). The use of constant temperatures is invaluable in cases where bodies are found indoors, and perhaps even in tropical regions. A common design could involve a developmental study of a blow fly species raised in two growth chambers set at constant temperatures of 20°C and 25°C and each containing 200 individuals (Figure 19.7a). The experiment would then be "replicated" two more times using the same growth chambers set at the same temperatures. There are three problems with this practice. First, the experimental unit is the chamber (i.e., the temperature), but because there is only one chamber per temperature, there is only one experimental unit per treatment. Second, it is assumed that the growth chambers are identical and that any variation in growth rate is due to the treatment (i.e., the temperature). However, this may not be the case and it is impossible to know if the treatment effect is caused by differences

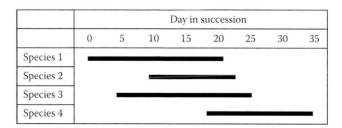

Figure 19.6 Schematic representation of a species occurrence matrix in a succession-based study that falls victim to sacrificial pseudoreplication (i.e., where replicate carcasses were pooled).

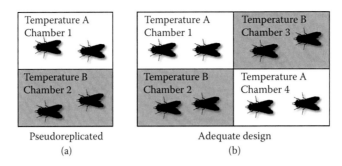

Pseudoreplicated   Adequate design
(a)   (b)

Figure 19.7 Schematic representations of laboratory studies involving pseudoreplicated (a) and adequate (b) experimental designs in studies using growth chambers.

in temperature or by differences in growth chambers. Third, if the same growth chamber is used for each treatment, then treatment differences could be caused by factors other than the treatment itself (i.e., nondemonic intrusion [Hurlbert 1984]). For example, chamber variation in lighting, humidity, vibration, or ventilation could cause the treatment effect instead of differences in temperature. Because all the samples for one temperature treatment are found in the same growth chamber (i.e., isolative segregation of treatment), they would all be affected by nondemonic intrusion without the investigator's knowledge. By randomizing the effect of nondemonic intrusion across growth chambers, a true effect size can still be deduced. An even better methodology could employ at least two more growth chambers (i.e., two growth chambers for each temperature). However, we recognize that this investment poses a financial challenge (Figure 19.7b). A less expensive alternative is to randomly assign temperatures to growth chambers in each trial. Each trial could then be treated as a block effect that could be subjected to a mixed-model analysis. For more advice on how to design experiments with growth chambers, see Lee and Rawlings (1982) and Hurlbert (1984).

Today, researchers use miniaturized data loggers to record subtle changes in microclimate inside rearing chambers instead of relying on (coarser) whole-chamber settings. In comparing these two approaches, Nabity et al. (2006) suggested that past practices (i.e., using whole-chamber settings) may have introduced more experimental error in final degree-day models than the more common problem of pseudoreplication.

19.2.6 Replication and Effect Size

There is no guarantee that treatment replication will be sufficient to detect significant effects even if an experiment is well designed and free of pseudoreplication. Insufficient replication, both in terms of experimental units (i.e., treatment replicates) and replicates within units (e.g., carcasses within experimental units), will limit the precision of estimates obtained from the study, as well as the capacity of statistical tests to detect significant effects (Lentner and Bishop 1986). The ability of a given test to detect significant treatment effects is referred to as power. The statistical power of a test will decrease more if the experiment involves an insufficient replication of experimental units than if it involves insufficient replication within units (Murray 1998). Considering the potential for high natural variability in field and laboratory measurements, forensic entomologists should always aim for a number of replicates that result in an adequate statistical power (see Section 19.4) to detect biologically significant effects.

When an experiment includes few replicates, it can be difficult to judge whether the lack of significance of a statistical test is due to the absence of an effect or to insufficient replication. However, if knowledge is available on the expected effect size and measurement variability (e.g., preliminary data, literature accounts), *a priori* power analysis (i.e., prospective power analysis) can be used to examine the adequacy of different replications (Lenth 2006–2009). Then again, if knowledge is not available, power analysis can still determine the number of replicates required to detect a predefined effect size with a given statistical test and power (Cohen 1992). By convention (Cohen 1988), a statistical power of 0.8 or greater is considered adequate to minimize the likelihood of a Type II error (1 − power = 0.2), which occurs when one fails to reject a false null hypothesis. This makes Type II error four times more likely than Type I error (i.e., when one rejects the null hypothesis when it is true), a reasonable reflection of the relative importance of both errors (Cohen 1992). Using published data from Kashyap and Pillay (1989), we showed that 16 cadavers were insufficient to detect a significant difference between different PMI estimation methods (power = 0.38) using a two-tailed, one sample t-test (Michaud et al. 2012). For such a test to be significant at $\alpha = 0.05$ with a power of 0.8, 43 cadavers would be required. Such an analysis is easy to conduct with a simple design (e.g., t-test, analysis of variance [ANOVA], correlation, linear regression) using freeware programs (Lenth 2006–2009). Conversely, *a priori* power analysis can be challenging with nested, multivariate, and mixed-model designs; in such cases, simulation or bootstrapping methods may be applicable (Thomas and Juanes 1996).

19.3 CHALLENGING CORE ASSUMPTIONS

19.3.1 Nonprimate Carcasses as Model Corpses

In the first half of the twentieth century, carrion researchers used carcasses from a variety of nonprimate animals to study insect colonization and succession patterns (e.g., Illingworth 1926; Fuller 1934; Reed 1958; Payne 1965). Shortly thereafter, promising forensic indicators were identified from successional data (Smith 1975; Nuorteva 1977; Rodriguez and Bass 1983; Tullis and Goff 1987). As human cadavers were infrequently available for field research, an issue still facing most researchers today, carcasses from different mammalian species were used, such as domestic pigs (Hewadikaram and Goff 1991; Komar and Beattie 1998; Leblanc and Strongman 2002; Shahid et al. 2003; Grassberger and Frank 2004), cats (Early and Goff 1986), rabbits (Tantawi et al. 1996), and rats (Tomberlin and Adler 1998; Kocarek 2003; Velasquez 2008). Without question, the domestic pig has become the universal surrogate (Catts and Goff 1992; Goff 1993; LaMotte and Wells 2000). In the literature, a domestic pig of 22–27 kg starting mass is cited as the best model because (1) its chest cavity approximates (in volume) the human torso where most decomposition occurs, (2) it has relatively hairless skin and is omnivorous suggesting it has a similar gut biota, (3) it is accessible and affordable, and (4) it rarely attracts negative public and media attention when used in field experiments (Catts and Goff 1992; Goff 1993; Anderson and VanLaerhoven 1996). More recently, Notter et al. (2009) found differences in decay rates and in fatty acid composition of pig and human adipose tissue after a 6-month immersion study in distilled water, but concluded that pigs are adequate mimics of humans in decomposition studies.

Given that forensic entomologists have embraced the domestic pig as a human analog, it is expected that outstanding empirical evidence exists to support this conclusion. However, due to obvious legal and security reasons, only one study has validated the pig-as-surrogate claim (Schoenly et al. 2007). Unfortunately, this comparative study sampled arthropods from only three study subjects (one human, two pigs). Although negligible preference was found for human over pig tissues by forensically important species, we caution that these results pertain only to pigs of 23–27 kg starting mass exposed in the summer under unconcealed conditions. Consequently, researchers who use pig carcasses that fall outside the 23–27 kg range and in circumstances different from those of Schoenly et al. (2007) cannot make the surrogacy claim. However, given the limited experimental data, plus anecdotal information from verified death cases (Smeeton et al. 1984; Erzinclioglu 1985; Lee 1989; Sukontason et al. 2001; Arnaldos et al. 2004), medium-sized pigs seem to be the best surrogate at present. Until more peer-reviewed studies are published that directly test the pig-as-surrogate claim, it appears that the use of medium-sized pig carcasses relies more on logic and practicality than experimental evidence.

The paucity of direct experimental comparisons between pig and human remains should also be a "talking point" for researchers who testify as expert witnesses. Indeed, such warnings appear repeatedly in courtroom proceedings (Catts and Goff 1992; Goff 1993). Moreover, studies have used pigs of different sizes, from as small as 0.75 kg (Leblanc and Strongman 2002) to as large as 160 kg (Komar and Beattie 1998). In his classic study, Payne (1965) reported that small mammal carcasses decompose at a faster rate than larger carcasses and that stages of decay are not recognizable in smaller carrion. Similarly, Komar and Beattie (1998) suggested that smaller pig carcasses (19–26 kg) decomposed more quickly than larger ones (36–162 kg), and that larger carcasses were better surrogates of human death cases when insect records were compared. Some evidence suggests that both carcass taxon and size alter insect species composition and abundance (Schoenly and Reid 1983; Kneidel 1984) (see Kuusela and Hanski [1982]), casting doubt on the use of surrogates other than pigs for forensic research. Obviously, more research is needed to verify the 23–27 kg weight range of pigs and test the extent to which PMI statistics vary between pig and human remains (Schoenly 2007). Until carcasses of rats, cats, and rabbits, which are smaller and furrier than pigs, have been tested alongside human cadavers, their use as human surrogates in medicolegal cases will remain in question (but see Goff and Flynn [1991]). In the meantime, we suggest that researchers obtain pig carcasses as close to the 23–27 kg standard as possible.

19.3.2 Decay Stages in Succession Studies

Using decay stages to summarize successional events in the carrion-arthropod community traces back to Mégnin's (1894) pioneering work on exhumed and exposed corpses. This concept was borrowed by carrion ecologists (e.g., Chapman and Sankey 1955; Bornemissza 1957; Reed 1958; Payne 1965) and carried over into forensic entomology (e.g., Anderson and VanLaerhoven 1996). The primary argument supporting decay stages was that stage boundaries marked abrupt changes in invertebrate taxonomic composition (e.g., Fuller 1934; Bornemissza 1957; Payne 1965; Wasti 1972; Coe 1978; Abell et al. 1982), a claim that was later rebuked (Schoenly and Reid 1987, 1989; Boulton and Lake 1988; Moura et al. 2005). In these studies, the existence and timing of authors' decay stages were verified in only a minority of cases.

Although many forensic entomologists acknowledge in their publications that decay stages misinform the more continuous successional and decompositional processes in carrion, they still find them useful heuristically, and perhaps pedagogically, even though their use evokes a stepwise (and necessarily abrupt) version of the successional process. Specific reasons for their continued use has been justified as providing "convenient descriptors" for summarizing postmortem changes (Mann et al. 1990; Anderson and VanLaerhoven 1996; Tibbett and Carter 2009) or as "reference points" for educating juries in the courtroom (Goff 1993). However, description and agreement varies on the number and duration of named decay stages, even in studies that share similar ecological, taxonomic, and sampling features (Schoenly and Reid 1987). Researchers have dedicated reams of journal space on stage descriptions and agreement/disagreement with other authors, a situation that evokes more scientific confusion than advancement. Recent research, however, has found new uses for decay stages in carrion studies. For example, Michaud and Moreau (2011) demonstrated that accumulated degree day (ADD) models can be used to predict decay stages. As such, it is our view that decay stages should be verified statistically or experimentally (Ordóñez et al. 2008; Lefebvre and Gaudry 2009; Michaud and Moreau 2011), not just inferred or carried over from related studies or disciplines.

19.3.3 Field Validation of Postmortem Interval Estimation Methods

To our knowledge, only VanLaerhoven (2008) and Núñez-Vásquez et al. (2013) have conducted validation studies of PMI estimation methods in the field. In three simulated cases, VanLaerhoven (2008) enlisted police and fire marshal personnel to collect and process life-cycle stages of the calliphorid *Phormia regina* (Meigen) from pig carcasses (with actual PMI's of 5, 7, and 8 days kept confidential from the author). PMI estimates were calculated using five degree-day models at three developmental thresholds and compared against laboratory rearing data from five published studies. Results showed that all five degree-day models were comparable, suggesting they might be interchangeable. PMI estimates also overlapped the actual PMI in at least three of the five rearing studies, with highest overlap occurring when the 6°C threshold was applied. In the Núñez-Vásquez et al. (2013) study, jars containing liver and vermiculate were inoculated with eggs of *Phormia regina*, with one jar randomly selected and destructively sampled every 12 hours for up to 8 days. PMI estimates were calculated by inverse prediction (Wells and LaMotte 1995) (see Section 19.4.2), using the authors' rearing data that tracked ages, weights, lengths, and instars of larvae at mean chamber temperatures of 14°C and 20.5°C. Results showed that confidence intervals for PMI estimates consistently overlapped the actual PMI when the 20.5°C rearing data and the oldest instar were used. Not surprisingly, larval length and weight, as continuous variables, were better predictors than larval instar. Although results from both validation studies are encouraging, mock cases were staged at one site (or in one cage!) and only one forensic entomologist was involved in determining PMI estimates. Inferences were also limited to carcasses or baits of young decompositional age (≤8 days). Consequently, we join both research groups in recommending more validation studies and suggest that future researchers incorporate a wider range of actual PMI's and species (and at multiple sites and seasons) to allow tests of both developmental and successional indicators.

19.4 STATISTICAL ANALYSIS AND COMPUTER MODELS

19.4.1 Common Statistical Errors and How to Avoid Them

Although more than two decades have passed since Gates (1991) published his "user's guide to misanalyzing planned experiments," his "cautionary tales" remain relevant and instructive today. In this section, we adapt and update Gates' common statistical errors to forensic entomology field experiments and offer solutions for avoiding them in the future.

19.4.1.1 Substituting Sampling Error for Experimental Error or Arbitrarily Combining These Two Errors

Studies often encompass subsamples as a means to reduce the error associated with statistical estimates without increasing the number of sampling units. For example, consider an experimental design that includes two subsamples for each experimental unit (Figure 19.2b) and its associated ANOVA table (Table 19.1). Erroneously treating subsamples as independent samples (see Table 19.2) will produce an analysis that is too liberal because the experimental error includes the sampling error. To avoid this error, we recommend that researchers verify the true experimental unit(s).

Table 19.1   Analysis of a Design with Subsampling (Figure 19.2b), Where *t* Is the Number of Treatments (i.e., 2), *r* Is the Number of Replicates Per Treatment (i.e., 2), and *s* Is the Number of Subsamples per Replication (i.e., 2)

Source of Variation	Theoretical Degrees of Freedom	Degrees of Freedom
Treatments	$t - 1$	1
Experimental error	$t(r - 1)$	2
Sampling error	$(t \times r)(s - 1)$	4
Total	$(t \times r \times s) - 1$	7

Table 19.2 Inappropriate Analysis of a Design with Subsampling (Figure 19.2b), Where t Is the Number of Treatments (i.e., 2), and r Is the Erroneous Number of Replicates per Treatment (i.e., 4)

Source of Variation	Theoretical Degrees of Freedom	Degrees of Freedom
Treatments	$t - 1$	1
Experimental error	$t(r - 1)$	6
Total	$(t \times r) - 1$	7

19.4.1.2 Overusing Multiple Comparison Tests

When more than one null hypothesis is tested, the use of multiple comparison procedures (e.g., Tukey–Kramer, Hochberg GT2 test) can adjust for this multiplicity and inflating alpha (α) can be avoided. However, when these procedures are misapplied or when many pairwise comparisons are needed, multiple comparison procedures can mislead or yield powerless tests of significance. Little (1978) provides an instructive paper on the overuse of multiple comparison tests. To avoid this error, we recommend the use of hypothesis-based contrasts or multivariate analyses instead of multiple comparisons.

19.4.1.3 Requiring Raw Data to Be Normally Distributed

This practice is often passed on from teachers (with a poor understanding of statistical models) to their students. In fact, except for t-tests and correlations, raw data analyzed by univariate tests do not have to be normally distributed. However, the residuals (e.g., error not explained by the statistical model) of parametric, univariate tests are required to be normally distributed. Fortunately, this condition is easier to meet than the former. Rather than trying to transform raw data to fit normal distributions, researchers should examine their data distributions (i.e., conduct exploratory data analysis, see Ellison [2001]) to find mistakes (e.g., aberrant data, typos, outliers) and verify whether generalized linear models are appropriate to model their data.

19.4.1.4 Failing to Recognize the Distinction between Nested and Crossed Factors

When an experimental design includes two or more factors, the classical approach is to use a crossed (factorial) design, in which each level of every factor is studied in every combination with other factors. However, if randomization or experimental constraints make it difficult or impossible to cross all of the factors of interest, nested designs may have to be used. Nested designs are also referred to as hierarchical designs because there is a hierarchy of factors in the experimental designs and in the associated ANOVA tables. For example, suppose that we want to study the effect of a shelter on a carcass using four pig and four rat carcasses. If shelters are unlimited, the crossed design (Figure 19.8a) and its associated ANOVA table (Table 19.3) represent the most parsimonious approach. However, if only two shelters are available, the nested design (Figure 19.8b) and its associated ANOVA table (Table 19.4) represent a viable alternative. This design is more difficult to analyze than the crossed design (Figure 19.8a) because the statistical model is not straightforward. The analysis will have little statistical power when the factor in the main plot is examined (i.e., shelter) but will be more powerful with the detection of subplot effects (i.e., species and the interaction Shelter × Species). Failing to recognize the distinction between nested and crossed factors can be avoided with enhanced training in experimental design.

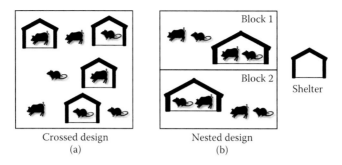

Crossed design
(a)

Block 1

Block 2

Shelter

Nested design
(b)

Figure 19.8 Schematic representations of field studies with crossed (a) and nested (b) experimental designs.

Table 19.3 **Analysis of a Crossed Design (Figure 19.8a), Where *r* Is the Replication (i.e., 2), *a* Is the Number of Levels of the Shelter Treatment (i.e., 2), and *b* Is the Number of Levels of the Species Treatment (i.e., 2)**

Source of Variation	Theoretical Degrees of Freedom	Degrees of Freedom
Shelter	$a - 1$	1
Species	$b - 1$	1
Shelter × Species	$(a - 1)(b - 1)$	1
Experimental error	$(a \times b)(r - 1)$	4
Total	$(r \times a \times b) - 1$	7

Table 19.4 **Analysis of a Nested Design (Figure 19.8b), Where *r* Is the Number of Blocks (i.e., 2), *a* Is the Number of Levels of the Shelter Treatment (i.e., 2), and *b* Is the Number of Levels of the Species Treatment (i.e., 2)**

Source of Variation	Theoretical Degrees of Freedom	Degrees of Freedom
Blocks	$r - 1$	1
Shelter (main plots)	$a - 1$	1
Main plot error	$(r - 1)(a - 1)$	1
Species (subplots)	$b - 1$	1
Shelter × Species	$(a - 1)(b - 1)$	1
Subplot error	$a(r - 1)(b - 1)$	2
Total	$(r \times a \times b) - 1$	7

19.4.1.5 Incorrectly Analyzing Split-Plot Designs

A split-plot design is a blocked experiment where the experimental units of a first factor include all the combinations of a second factor. It therefore represents a combination of a block design and a nested design. The nested design presented in the previous paragraph is also an example of a split-plot design (Figure 19.8b). Erroneously considering the split-plot design as a crossed design (Figure 19.8a) would result in a different ANOVA table (Table 19.5). Because the experimental error includes both a main plot error and a subplot error, the analysis of the main plot factor will be too liberal and the analysis of the subplot factors will be too conservative. Again, enhanced training in experimental design is the best way to avoid this error.

Table 19.5 Inappropriate Analysis of a Nested Design (Figure 19.8b), Where *r* Is the Number of Blocks (i.e., 2), *a* Is the Number of Levels of the Shelter Treatment (i.e., 2), and *b* Is the Number of Levels of the Species Treatment (i.e., 2)

Source of Variation	Theoretical Degrees of Freedom	Degrees of Freedom
Blocks	$r - 1$	1
Shelter	$a - 1$	1
Species	$b - 1$	1
Shelter × Species	$(a - 1)(b - 1)$	1
Experimental error	$(\{a \times b\} - 1)(r - 1)$	3
Total	$(r \times a \times b) - 1$	7

19.4.1.6 Incorrectly Analyzing Factorial Experiments That Have Both Qualitative and Quantitative Factors That Include the Zero Level

The analysis of this experimental design is complicated by the fact that treatments that include a zero level among the qualitative factors are all equivalent. For example, suppose that a study is conducted to determine the effects of drugs on larval development in decomposing bodies. In a well-replicated study, pig carcasses are each injected with one of four levels of three drugs (e.g., morphine, heroin, cocaine) before being euthanized. One of the injection levels is an untreated control containing only physiological saline (i.e., zero level of drug). Here, a problem arises because the injection of a zero amount of morphine is identical to the injection of a zero amount of heroin or cocaine. In other words, there is only one control treatment and not one for each drug. This design is referred to as an incomplete block design and must be acknowledged because it affects the analysis and interpretation of results. Addelman (1974) has proposed a mathematical solution to this problem. Briefly, it involves the computation of two different ANOVAs (with and without the control treatment) to derive sums of squares that are used to calculate F and *p* values.

19.4.1.7 Failing to Use Correct Error Terms in Regression Analyses

Regression analysis is a popular technique for analyzing the relationship between a dependent variable and one or more independent variables. This technique is closely related to ANOVA and construction of an ANOVA table derived from regression analysis is relatively straightforward, as is the reverse procedure. As such, regression models should be subjected to the same constraints as ANOVA tests and if random effects are present (i.e., a parameter that explains noise caused by hierarchical effects such as blocks and subject effects) the appropriate error term(s) should be used.

19.4.1.8 Failing to Analyze the Experiment as Designed

At the start of an analysis, an experiment should be analyzed as originally designed. Afterward, adjustments to the model may be necessary to improve its performance and eliminate negligible factors (Winer et al. 1991; Hines 1996). To avoid this error, a good understanding of different ANOVA designs (e.g., completely randomized designs, randomized block designs, Latin square designs, and incomplete block designs) and the distinction between fixed (treatments) and random effects is required.

19.4.2 Statistical and Computer Models for Postmortem Interval Estimation

Although descriptive models of carrion-arthropod succession and life-history tables for back-tracking blow fly development have been available to forensic entomologists since Mégnin (1894) and Kamal (1958), respectively, development of statistical and computer models with algorithms for PMI estimation first appeared in the 1980s. Among the first was a thermal model developed by Williams (1984) for aging fly larvae based on weight from growth rates of four species modeled with the logistic growth curve. Schoenly et al. (1992) wrote a computer algorithm that sorts baseline successional records to calculate PMI statistics from an inputted list of arthropod taxa recovered from a corpse. Using MATLAB® and Simulink® software, Byrd and Allen (2001b) invented a computer model for predicting thermal development of different blow fly species for each life-cycle stage from laboratory rearing data. Graphical outputs plot profiles of relative abundance of each life stage (egg to pupa) along a time axis. The advantage of this model is that it allows the user to vary starting conditions and visualize shifts in life-cycle development (i.e., PMI) for simulated crime scenes (e.g., victim deposited at dawn, average ambient temperatures of 20°C for entire developmental period).

In a PMI estimation method based on ADD, Michaud and Moreau (2009, 2011) developed composite ADD indices that were used as continuous variables in logistic regression to predict insect visitation patterns (Michaud and Moreau 2009) and decay stages (Michaud and Moreau 2011). The chief advantage of this approach is that it can generate statistical probabilities and confidence intervals for PMI estimates with just a few carcasses (i.e., as few as four). The model can also account for repeated measures on carcasses, reoccurring and nonreoccurring species on the carcass (Schoenly 1992), and within-season, between-season, and interannual variability. The model is also flexible enough to handle other decomposition-related phenomena, such as microbial succession, species abundance data, carcass temperature, and so on. However, one drawback of this model is that it requires greater statistical proficiency than the other methods.

Another PMI estimation method developed by Matuszewski (2011, 2012) and Matuszewski and Szafałowicz (2013) used local temperature records to predict the preappearance interval (PAI) (i.e., the period that precedes species appearance on a carcass or PAI) of carrion-frequenting beetles. The authors found significant inverse relationships between PAI and temperature in all but a few beetle species. This approach parallels the development of PMI estimation methods developed for fly taxa and may be most applicable in casework involving human remains discovered in more advanced periods of decomposition.

Despite the availability of methods for calculating the precision and uncertainty of PMI estimates, statistical support (i.e., error rates, confidence intervals, p values) rarely accompanies PMI estimates in courtrooms today (Wells and LaMotte 2010). Among the first was the inverse prediction method of Wells and LaMotte (1995), who constructed confidence intervals for larval age, based on weight. This approach requires rearing data that tracks ages and weights of larvae from egg hatch to postfeeding that depend on the assumptions that larval weight is normally distributed and that means and variances from those distributions can be linearly interpolated between sample ages. Although the authors provide a worked example using rearing data from the secondary screwworm fly, *Cochliomyia macellaria* Fabricius (Diptera: Calliphoridae), we are unaware of any uses in actual casework.

To derive succession-based PMI estimates, occurrence tables are required of forensically important species for different sites, seasons, and ecological circumstances (e.g., temperate forest, summer-exposed, unconcealed carcasses). In the most straightforward case, the lower and upper limits of the PMI correspond to the first and last days sampled taxa overlap in the succession (Schoenly 1992); however, these periods of overlap are not analogous to confidence intervals. To calculate the likelihood of a particular fauna's association on a human corpse requires construction of numerical probability tables from disaggregated successional data. LaMotte and Wells (2000) developed tables of minimum sample size, based on the likelihood-ratio statistic and Fisher's exact test, for deriving nominal p values from a few focal taxa (2–3). They determined that future field studies would need to deploy at least 25 carcasses to achieve a 10% level of significance for succession-based PMI estimates. This high level of replication will require sampling trade-offs (i.e., tracking only a few taxa) and close attention to intercarcass spacing to insure effective independence of experimental units. An empirical test of this approach using 53 pig carcasses exposed over three summers in north-central Indiana found that three insect taxa (i.e., larvae of two calliphorid species, one adult staphylinid species), when used singly or in pairs, showed forensic promise (Perez et al. 2014). The authors also found that two taxa yielded smaller (i.e., more precise) confidence intervals (that captured the true PMI) than one taxon (Perez et al. 2014). Although the forensic identification sciences were criticized for failing to test uncertainty and reliability of their methods (NRC 2009), forensic entomologists have begun to make significant gains in this area (see Chapter 29).

19.5 CONCLUSION

Scientific rigor in inferential sciences such as forensic entomology is only possible through appropriate experimental design, standardized sampling, and adequate statistical analysis. For authors, reviewers, and editors the absence of pseudoreplication should be viewed as a minimum requirement before the merits of an experiment are judged (Kroodsma et al. 2001). For inferences to be valuable in future casework and resistant to cross-examination, entomological methods demand validation, core assumptions require testing, and areas in need of urgent research need to be identified. Recent publications on forensic entomology include seven textbooks, six published in the English language (Smith 1986; Greenberg and Kunich 2005; Gennard 2007; Amendt et al. 2010; Byrd and Castner 2010; Rivers and Dahlem 2014) and one in French (Wyss and Cherix 2006), plus one handbook for law enforcement (Haskell and Williams 2008). However, none of these books devoted a single chapter to experimental design flaws and solutions for resolving them or included the terms "pseudoreplication" or "inference" in their glossaries or indices. Consequently, it is hoped that the coverage of these topics, first reviewed in Michaud et al. (2012), will help fill this vacuum. Beyond gaining familiarity with these topics, we join Ellison and Dennis (2010) in urging that researchers seek greater statistical fluency (not just statistical literacy), through coursework or self-study, and to become better ecological detectives (Hilborn and Mangel 1997).

ACKNOWLEDGMENT

We thank the editors for including a chapter on experimental design in their textbook and for inviting us to write it.

REFERENCES

Abell, D.H., S.S. Wasti, and G.C. Hartmann. 1982. Saprophagous arthropod fauna associated with turtle carrion. *Applied Entomology and Zoology* 17: 301–307.

Addelman, S. 1974. Computing the ANOVA table for experiments involving qualitative factors and zero amounts of quantitative factors. *American Statistician* 28: 21–22.

Amendt, J., M.L. Goff, C.P. Campobasso, and M. Grassberger. 2010. *Current Concepts in Forensic Entomology*. New York: Springer.

Anderson, G.S. 2000. Minimum and maximum developmental rates of some forensically important Calliphoridae (Diptera). *Journal of Forensic Sciences* 45: 824–832.

Anderson, G.S. 2011. Comparison of decomposition rates and faunal colonization of carrion in indoor and outdoor environments. *Journal of Forensic Sciences* 56: 136–142.

Anderson, G.S. and S.L. VanLaerhoven. 1996. Initial studies on insect succession on carrion in south western British Columbia. *Journal of Forensic Sciences* 41: 617–625.

Archer, M. 2003. Annual variation in arrival and departure times of carrion insects at carcasses: Implications for succession studies in forensic entomology. *Australian Journal of Zoology* 51: 569–576.

Arnaldos, M.I., E. Romera, J.J. Presa, A. Luna, and M.D. Garcia. 2004. Studies on seasonal arthropod succession on carrion in the south-eastern Iberian Peninsula. *International Journal of Legal Medicine* 118: 197–205.

Avila, F.W. and M.L. Goff. 1998. Arthropod succession patterns onto burnt carrion in two contrasting habitats in the Hawaiian islands. *Journal of Forensic Sciences* 43: 581–586.

Basqué, M. and J. Amendt. 2013. Strengthen forensic entomology in court: The need for data exploration and the validation of a generalized additive mixed model. *International Journal of Legal Medicine* 127: 213–223.

Behm, J.E., D.E. Edmonds, J.P. Harmon, and A.R. Ives. 2013. Multilevel statistical models and the analysis of experimental data. *Ecology* 94: 1479–1486.

Bornemissza, G.F. 1957. An analysis of arthropod succession in carrion and the effect of its decomposition on the soil fauna. *Australian Journal of Zoology* 5: 1–12.

Boulton, A.J. and P.S. Lake. 1988. Dynamics of heterotrophic succession in carrion arthropod assemblages: A comment on Schoenly and Reid (1987). *Oecologia* 76: 477–480.

Bourel, B., L. Martin-Bouyer, V. Hedouin, J.C. Cailliez, D. Derout, and D. Gosset. 1999. Necrophilous insect succession on rabbit carrion in sand dune habitats in Northern France. *Journal of Medical Entomology* 36: 420–442.

Braack, L. 1981. Visitation patterns of principal species of the insect complex at carcasses in the Kruger National Park. *Koedoe* 24: 33–49.

Byrd, J.H. and J.C. Allen. 2001a. The development of the black blow fly, *Phormia regina* (Meigen). *Forensic Science International* 120: 79–88.

Byrd, J.H. and J.C. Allen. 2001b. Computer modeling of insect growth and its application to forensic entomology. In *Forensic Entomology: The Utility of Arthropods in Legal Investigations*, eds. J.H. Byrd and J.L. Castner, pp. 303–330. Boca Raton, FL: CRC Press.

Byrd, J.H. and J.L. Castner. 2010. *Forensic Entomology: The Utility of Arthropods in Legal Investigations*. 2nd edition. Boca Raton, FL: CRC Press.

Carroll, S.S. and D.L. Pearson. 1998. The effects of scale and sample size on the accuracy of spatial predictions of tiger beetle (Cicindelidae) species richness. *Ecography* 21: 401–414.

Catts, E.P. 1992. Problems in estimating the postmortem interval in death investigations. *Journal of Agricultural Entomology* 9: 245–255.

Catts, E.P. and M.L. Goff. 1992. Forensic entomology in criminal investigations. *Annual Review of Entomology* 37: 253–273.

Centeno, N., M. Maldonado, and A. Oliva. 2002. Seasonal patterns of arthropods occurring on sheltered and unsheltered pig carcasses in Buenos Aires Province (Argentina). *Forensic Science International* 126: 63–70.

Chapman, R.F. and J.H.P. Sankey. 1955. The larger invertebrate fauna of three rabbit carcasses. *Journal of Animal Ecology* 24: 395–402.

Chaves, L.F. 2010. An entomologist guide to demystify pseudoreplication: Data analysis of field studies with design constraints. *Journal of Medical Entomology* 47: 291–298.

Chelgren, N.D., D.K. Rosenberg, S.S. Heppell, and A.I. Gitelman. 2006. Carryover aquatic effects on survival of metamorphic frogs during pond migration. *Ecological Applications* 16: 250–261.

Coe, M. 1978. The decomposition of elephant carcasses in the Tsavo (East) National Park, Kenya. *Journal of Arid Environments* 1: 71–86.

Cohen, J. 1988. *Statistical Power Analysis for the Behavioral Sciences*. 2nd edition. New Jersey: Routledge.

Cohen, J. 1992. Statistical power analysis. *Current Directions in Psychological Science* 1: 98–101.

Colwell, R.K. 2006. EstimateS: Statistical estimation of species richness and shared species from samples, version 8 (purl.oclc.org/estimates).

Cornelius, J.M. and J.F. Reynolds. 1991. On determining the statistical significance of discontinuities within ordered ecological data. *Ecology* 72: 2057–2070.

Coss, R.G. 2009. Pseudoreplication conventions are testable hypotheses. *Journal of Comparative Psychology* 123: 444–446.

Cragg, J.B. 1955. The natural history of sheep blowflies in Britain. *Annals of Applied Biology* 42: 197–207.

Cragg, J.B. and J. Hobart. 1955. A study of a field population of the blowflies *Lucilia caesar* (L.) and *L. sericata* (Mg.). *Annals of Applied Biology* 43: 645–663.

Crist, T.O. 1998. The spatial distribution of termites in shortgrass steppe: A geostatistical approach. *Oecologia* 114: 410–416.

Davis, J.B. and M.L. Goff. 2000. Decomposition patterns in terrestrial and intertidal habitats on Oahu Island and Coconut Island, Hawaii. *Journal of Forensic Sciences* 45: 836–842.

Early, M. and M.L. Goff. 1986. Arthropod succession patterns in exposed carrion on the island of O'ahu, Hawaiian islands, USA. *Journal of Medical Entomology* 23: 520–531.

Ellison, A.M. 2001. Exploratory data analysis and graphical display. In *Design and Analysis of Ecological Experiments,* 2nd edition, eds. S.M. Scheiner and J. Gurevitch, 37–62. Oxford, United Kingdom: Oxford University Press.

Ellison, A.M. and B. Dennis. 2010. Paths to statistical fluency for ecologists. *Frontiers in Ecology and the Environment* 8: 362–370.

Erzinclioglu, Y.Z. 1985. The entomological investigation of a concealed corpse. *Medicine Science and the Law* 25: 228–230.

Fielder, A., M. Halbach, B. Sinclair, and M. Benecke. 2008. What is the edge of a forest? A diversity analysis of adult Diptera found on decomposing piglets inside and on the edge of a Western German woodland inspired by a courtroom question. *Entomologie Heute* 20: 173–191.

Fisher, R.A. 1935. *The Design of Experiments*, 8th edition (1966). New York: Hafner Press.

Fuller, M.E. 1934. The insect inhabitants of carrion: A study of animal ecology. *Council for Scientific and Industrial Research (Australia)* 82: 1–62.

Garcia-Berthou, E. and S.H. Hurlbert. 1999. Pseudoreplication in hermit crab shell selection experiments: A reply to Wilber. *Bulletin of Marine Science* 65: 893–895.

Gates, C.E. 1991. A user's guide to misanalysing planned experiments. *HortScience* 26: 1261–1265.

Gennard, D.E. 2007. *Forensic Entomology: An Introduction*. West Sussex, United Kingdom: John Wiley.

Gillison, A.N. and K.R.W. Brewer. 1985. The use of gradient-directed transects or gradsects in natural resource surveys. *Journal of Environmental Management* 20: 103–127.

Gilmour, D., D.F. Waterhouse, and G.A. McIntyre. 1946. An account of experiments undertaken to determine the natural population density of the sheep blowfly, *Lucilia cuprina* Wied. *Council for Scientific and Industrial Research Bulletin* 195: 1–44.

Goff, M.L. 1993. Estimation of post mortem interval using arthropod development and successional patterns. *Forensic Science Review* 5: 81–94.

Goff, M.L. and M.M. Flynn. 1991. Determination of postmortem interval by arthropod succession: A case study from the Hawaiian Islands. *Journal of Forensic Sciences* 36: 607–614.

Grassberger, M. and C. Frank. 2004. Initial study of arthropod succession on pig carrion in a central European urban habitat. *Journal of Medical Entomology* 41: 511–523.

Greenberg, B. 1990. Behavior of postfeeding larvae of some Calliphoridae and a muscid (Diptera). *Annals of the Entomological Society of America* 83: 1210–1214.

Greenberg, B. and A.A. Bornstein. 1964. Fly dispersion from a rural Mexican slaughterhouse. *American Journal of Tropical Medicine and Hygiene* 13: 881–886.

Greenberg, B. and J.C. Kunich. 2005. *Entomology and the Law: Flies as Forensic Indicators*. New York: Cambridge University Press.

Haskell, N.H. and R.E. Williams. 2008. *Entomology and Death: A Procedural Guide*, 2nd edition. Clemson, SC: Forensic Entomology Partners.

Heard, S.B. 1998. Resource patch density and larval aggregation in mushroom-breeding flies. *Oikos* 81: 187–195.

Heffner, R.A., M.J. Butler IV, and C.K. Reilly. 1996. Pseudoreplication revisited. *Ecology* 77: 2558–2562.

Herms, W.B. 1907. An ecological and experimental study of Sarcophagidae with relation to lake beach debris. *Journal of Experimental Zoology* 4: 45–83.

Hewadikaram, K.A. and M.L. Goff. 1991. Effect of carcass size on rate of decomposition and arthropod succession patterns. *The American Journal of Forensic Medicine and Pathology* 12: 235–240.

Hilborn, R. and M. Mangel. 1997. *The Ecological Detective: Confronting Models with Data*. Monographs in Population Biology 28. Princeton, NJ: Princeton University Press.

Hines, W.G.S. 1996. Pragmatics of pooling in ANOVA tables. *American Statistician* 50: 127–139.

Horgan, F.G. 2005. Aggregated distribution of resources creates competition refuges for rainforest dung beetles. *Ecography* 28: 603–618.

Hurlbert, S.H. 1984. Pseudoreplication and the design of ecological field experiments. *Ecological Monographs* 54: 187–211.

Hurlbert, S.H. 2003. On misinterpretations of pseudoreplication and related matters: A reply to Oksanen. *Oikos* 104: 591–597.

Hurlbert, S.H. 2009. The ancient black art and transdisciplinary extent of pseudoreplication. *Journal of Comparative Psychology* 123: 434–443.

Hurlbert, S.H. 2010. Pseudoreplication capstone: Correction of 12 errors in Koehnle & Schank (2009). Department of Biology, San Diego State University, San Diego, CA. accessed October 17, 2014, http://www.bio.sdsu.edu/pub/stuart/stuart.html

Hurlbert, S.H. and W.G. Meikle. 2003. Pseudoreplication, fungi, and locusts. *Journal of Economic Entomology* 96: 533–535.

Hurlbert, S.H. and M.D. White. 1993. Experiments with freshwater invertebrate zooplanktivores: Quality of statistical analysis. *Bulletin of Marine Science* 53: 128–153.

Ieno, E.N., J. Amendt, H. Fremdt, A.A. Saveliev, and A.F. Zuur. 2010. Analysing forensic entomology data using additive mixed effects modelling. In *Current Concepts in Forensic Entomology*, eds. J. Amendt, M.L. Goff, C.P. Campobasso, and M. Grassberger, pp. 139–162. New York: Springer.

Illingworth, F.J. 1926. Insects attracted to carrion in southern California. *Proceedings of the Hawaiian Entomological Society* 6: 397–401.

Kamal, A.S. 1958. Comparative study of thirteen species of sarcosaprophagous Calliphoridae and Sarcophagidae (Diptera). 1. Bionomics. *Annals of the Entomological Society of America* 51: 261–271.

Kashyap, V.K. and V.V. Pillay, 1989. Efficacy of entomological method in estimation of postmortem interval: A comparative analysis. *Forensic Science International* 40: 245–250.

Kneidel, K.A. 1984. Influence of carcass taxon and size on species composition of carrion breeding Diptera. *American Midland Naturalist* 111: 57–63.

Kocarek, P. 2003. Decomposition and Coleoptera succession on exposed carrion of small mammal in Opava, the Czech Republic. *European Journal of Soil Biology* 39: 31–45.

Komar, D. and O. Beattie. 1998. Effect of carcass size on decay rates of shade and sun exposed carrion. *Canadian Society of Forensic Science Journal* 31: 343–385.

Kozlov, M.V. 2003. Pseudoreplication in Russian ecological research. *Bulletin of the Ecological Society of America* 84: 45–47.

Kroodsma, D.E., B.E. Byers, E. Goodale, S. Johnson, and W.-C. Liu. 2001. Pseudoreplication in playback experiments, revisited a decade later. *Animal Behavior* 61: 1029–1033.

Kuusela, S. and I. Hanski. 1982. The structure of carrion fly communities: The size and the type of carrion. *Holarctic Ecology* 5: 337–348.

LaMotte, L.R. and J.D. Wells. 2000. P-values for post mortem intervals from arthropod succession data. *Journal of Agricultural Biological Environmental Statistics* 5: 58–68.

Larsen, T.H. and A. Forsyth. 2005. Trap spacing and transect design for dung beetle biodiversity studies. *Biotropica* 37: 322–325.

Lazic, S.E. 2010. The problem of pseudoreplication in neuroscientific studies: Is it affecting your analysis? *BMC NeuroScience* 11: 5.

Leblanc, H.N. and D.B. Strongman. 2002. Carrion insects associated with small pig carcasses during fall in Nova Scotia. *Canadian Society of Forensic Science Journal* 35: 145–152.

Lee, C.-S. and J.O. Rawlings. 1982. Design of experiments in growth chambers—Uniformity trials in the North Carolina State University Phytotron. *Crop Science* 22: 551–558.

Lee, H.L. 1989. Recovery of forensically important entomological specimens from human cadavers in Malaysia—An update. *Malaysian Journal of Pathology* 11: 33–36.

Lefebvre, F. and E. Gaudry. 2009. Forensic entomology: A new hypothesis for the chronological succession pattern of necrophagous insect on human corpses. *Annales de la Société Entomologique de France* 45: 377–392.

Lenth, R.V. 2006–2009. Java Applets for Power and Sample Size. accessed October 17, 2014, http://www.stat.uiowa.edu/~rlenth/Power.

Lentner, M. and T. Bishop. 1986. *Experimental Design and Analysis*. Blacksburg, VA: Valley Book.

Lewis, A.J. and M.E. Benbow. 2011. When entomological evidence crawls away: *Phormia regina* en masse larval dispersal. *Journal of Medical Entomology* 48: 1112–1119.

Lindquist, A.W., W.W. Yates, R.A. Hoffmann, and J.S. Butts. 1951. Studies on the flight habits of three species of flies tagged with radioactive phosphorus. *Journal of Economic Entomology* 44: 397–400.

Little, T. 1978. If Galileo published in HortScience. *HortScience* 13: 504–506.

Lombardi, C.M. and S.H. Hurlbert. 1996. Sunfish cognition and pseudoreplication. *Animal Behavior* 52: 419–422.

Mann, R.W., W.M. Bass, and L. Meadows. 1990. Time since death and decomposition of the human body: Variables and observations in case and experimental field studies. *Journal of Forensic Sciences* 35: 103–111.

Martinez, E., P. Duque, and M. Wolff. 2007. Succession pattern of carrion-feeding insects in Paramo, Colombia. *Forensic Science International* 166: 182–189.

Matuszewski, S. 2011. Estimating the pre-appearance interval from temperature in *Necrodes littoralis* L. (Coleoptera: Silphidae). *Forensic Science International* 212:180–88.

Matuszewski, S. 2012. Estimating the preappearance interval from temperature in *Creophilus maxillosus* L. (Coleoptera: Staphylinidae). *Journal of Forensic Sciences* 57: 136–145.

Matuszewski, S. and M. Szafałowicz. 2013. Temperature-dependent appearance of forensically useful beetles on carcasses. *Forensic Science International* 229: 92–99.

Mégnin, J.P. 1883. L'application de l'entomologie à la médecine légale. [The application of entomology to forensic medicine.] *Gazette Hebdomadaire de Médecine et de Chirurgie* 29: 480–82.

Mégnin, P. 1887. La faune des tombeaux. [Fauna of the tombs.] *Séances de l'Académie des Sciences* 105: 948–951.

Mégnin, P. 1894. *La Faune des Cadavres.* [Fauna of Cadavers.] *Encyclopédie Scientifique des Aides-Mémoires.* Paris: G. Masson, Gauthier-Villars et Fils.

Michaud, J.-P., C.G. Majka, J.-P. Privé, and G. Moreau. 2010. Natural and anthropogenic changes in the insect fauna associated with carcasses in the North American Maritime Lowlands. *Forensic Science International* 202: 64–70.

Michaud, J.-P. and G. Moreau. 2009. Predicting the visitation of carcasses by carrion-related insects under different rates of degree-day accumulation. *Forensic Science International* 185: 78–83.

Michaud, J.-P. and G. Moreau. 2011. A statistical approach based on accumulated degree-days to predict decomposition-related processes in forensic studies. *Journal of Forensic Sciences* 56: 229–232.

Michaud, J.-P., K.G. Schoenly, and G. Moreau. 2012. Sampling flies or sampling flaws? Experimental design and inference strength in forensic entomology. *Journal of Medical Entomology* 49: 1–10.

Millar, R.B. and M.J. Anderson. 2004. Remedies for pseudoreplication. *Fisheries Research* 70: 397–407.

Moura, M.O., E.L.A. Monteiro-Filho, and C.J.B. Carvalho. 2005. Heterotrophic succession in carrion arthropod assemblages. *Brazilian Archives of Biology and Technology* 48: 477–86.

Murray, D.M. 1998. *Design and Analysis of Group Randomized Trials.* New York: Oxford University Press.

Nabity, P.D., L.G. Higley, and T.M. Heng-Moss. 2006. Effects of temperature on development of *Phormia regina* (Diptera: Calliphoridae) and use of developmental data in determining time intervals in forensic entomology. *Journal of Medical Entomology* 43: 1276–1286.

National Research Council. 2009. *Strengthening Forensic Science in the United States: A Path Forward.* Washington DC: The National Academies Press.

Nelder, M.P., J.W. McCreadie, and C.S. Major. 2009. Blow flies visiting decaying alligators: Is succession synchronous or asynchronous? *Psyche* 2009: 575362.

Norris, K.R. 1959. The ecology of sheep blowflies in Australia. In *Biogeography and Ecology in Australia*, ed. A. Keast, R.L. Crocker, and C.S. Christian, 514–544. Monographia Biologicae 8. The Hague, The Netherlands: Junk Publishers.

Notter, S.J., B.H. Stuart, R. Rowe, and N. Langlois. 2009. The initial changes of fat deposit during the decomposition of human and pig remains. *Journal of Forensic Sciences* 54: 195–201.

Núñez-Vázquez, C., J.K. Tomberlin, M. Cantú-Sifuentes, and O. García-Martínez. 2013. Laboratory development and field validation of *Phormia regina* (Diptera: Calliphoridae). *Journal of Medical Entomology* 50: 252–260.

Nuorteva, P. 1977. Sarcosaprophagous insects as forensic indicators In *Forensic Medicine: A Study in Trauma and Environmental Hazards*, eds. C.G. Tedeschi, W.G. Eckert, and L.G. Tedeschi, 1072–1095. New York: Saunders.

Oksanen, L. 2001. Logic of experiments in ecology: Is pseudoreplication a pseudoissue? *Oikos* 94: 27–38.

Oksanen, L. 2004. The devil lies in details: A reply to Stuart Hurlbert. *Oikos* 104: 598–605.

Ordóñez, A., M.D. Garcia, and G. Fagua. 2008. Evaluation of efficiency of Schoenly trap for collecting adult sarcosaprophagous dipterans. *Journal of Medical Entomology* 45: 522–532.

Payne, J.A. 1965. A summer carrion study of the baby pig *Sus scrofa* Linnaeus. *Ecology* 46: 592–602.

Payne, J.A. and E.W. King. 1968. Arthropod succession and decomposition of buried pigs. *Nature* 219: 1180–1181.

Perez, A., N.H. Haskell, and J.D. Wells. 2014. Evaluating the utility of hexapod species for calculating a confidence interval about a succession based postmortem interval estimate. *Forensic Science International* 241: 91–95.

Putnam, R.J. 1983. *Carrion and Dung: The Decomposition of Animal Wastes.* London, United Kingdom: Edward Arnold.

Radeloff, V.C., T.F. Miller, H.S. He, and D.J. Mladenoff. 2000. Periodicity in spatial data and geostatistical models: Autocorrelation between patches. *Ecography* 23: 81–91.

Ramírez, C.C., D. Fuentes-Contreras, L.C. Rodríguez, and H.M. Niemeyer. 2000. Pseudoreplication and its frequency in olfactometric laboratory studies. *Journal of Chemical Ecology* 26: 1423–1431.

Reed, H.B. 1958. Study of dog carcass communities in Tennessee, with special reference to the insects. *American Midland Naturalist* 59: 213–245.

Richards, E.N. and M.L. Goff. 1997. Arthropod succession on exposed carrion in three contrasting tropical habitats on Hawaii Island, Hawaii. *Journal of Medical Entomology* 34: 328–339.

Rivers, D.B. and G.A. Dahlem. 2014. *The Science of Forensic Entomology.* New York: Wiley-Blackwell.

Rodriguez, W.C. and W.M. Bass. 1983. Insect activity and its relationship to decay rates of human cadavers in East Tennessee. *Journal of Forensic Science* 28: 423–432.

Rossi, R.E., D.J. Mulla, A.G. Journel, and E.H. Franz. 1992. Geostatistical tools for modeling and interpreting ecological spatial dependence. *Ecological Monographs* 62: 277–314.

Saks, M.J. and J.J. Koehler. 2005. The coming paradigm shift in forensic identification science. *Science* 309: 892–895.

Schank, J.C. and T.J. Koehnle. 2009. Pseudoreplication is a pseudoproblem. *Journal of Comparative Psychology* 123: 421–433.

Schoenly, K. 1991. Food web structure in dung and carrion arthropod assemblages, null models and Monte Carlo simulation: Application to medical/veterinary entomology. *Journal of Agricultural Entomology* 8: 227–249.

Schoenly, K. 1992. A statistical analysis of successional patterns in carrion-arthropod assemblages: Implications for forensic entomology and determination of the postmortem interval. *Journal of Forensic Sciences* 37: 1489–1513.

Schoenly, K. 2007. Quantifying between-subject repeatability of succession- and development-based PMI statistics: A computer demonstration. Paper presented at the annual meeting of the North American Forensic Entomology Association, Burnaby, British Columbia.

Schoenly, K., M.L. Goff, and M. Early. 1992. A BASIC algorithm for calculating the postmortem interval from arthropod successional data. *Journal of Forensic Sciences* 37: 808–823.

Schoenly, K., M.L. Goff, J.D. Wells, and W.D. Lord. 1996. Quantifying statistical uncertainty in succession-based entomological estimates of the post mortem interval in death scene investigations: A simulation study. *American Entomologist* 42: 106–112.

Schoenly, K., N.H. Haskell, R.D. Hall, and R.J. Gbur. 2007. Comparative performance and complementarity of four sampling methods and arthropod preference tests from human and porcine remains at the Forensic Anthropology Center in Knoxville, Tennessee. *Journal of Medical Entomology* 44: 881–894.

Schoenly, K. and W. Reid.1983. Community structure of carrion arthropods in the Chihuahuan Desert. *Journal of Arid Environments* 6: 253–263.

Schoenly, K. and W. Reid. 1987. Dynamics of heterotrophic succession in carrion arthropod assemblages: Discrete seres or a continuum of change? *Oecologia* 73: 192–202.

Schoenly, K. and W. Reid. 1989. Dynamics of heterotrophic succession in carrion revisited: A reply to Boulton and Lake (1988). *Oecologia* 79: 140–142.

Schoenly, K., S.A. Shahid, N.H. Haskell, and R.D. Hall. 2005. Does carcass enrichment alter community structure of predaceous and parasitic arthropods? A second test of the arthropod saturation hypothesis at the Anthropology Research Facility in Knoxville, Tennessee. *Journal of Forensic Sciences* 50: 134–142.

Schoof, H.F. and G.A. Mail. 1953. Dispersal habits of *Phormia regina* in Charleston, West Virginia. *Journal of Economic Entomology* 46: 258–262.

Schoof, H.F., R.E. Siverly, and J.A. Jensen. 1952. House fly dispersion studies in metropolitan areas. *Journal of Economic Entomology* 45: 675–683.

Searcy, W.A. 1989. Pseudoreplication, external validity and the design of playback experiments. *Animal Behavior* 38: 715–717.

Shahid, S.A., K. Schoenly, N.H. Haskell, R.D. Hall, and W. Zhang. 2003. Carcass enrichment does not alter decay rates or arthropod community structure: A test of the arthropod saturation hypothesis at the Anthropology Research Facility in Knoxville, Tennessee. *Journal of Medical Entomology* 40: 559–569.

Shalaby, O.A., L.M.L. de Carvalho, and M.L. Goff. 2000. Comparison of patterns of decomposition in an hanging carcass and a carcass in contact with soil in a xerophytic habitat on the island of Oahu, Hawaii. *Journal of Forensic Sciences* 45: 1267–1273.

Shean, B.S., B.A. Messinger, and M. Papworth. 1993. Observations of differential decomposition on sun exposed v. shaded pig carrion in coastal Washington State. *Journal of Forensic Sciences* 38: 938–949.

Smeeton, W.M.I., T.D. Koelmeyer, B.A. Holloway, and P. Singh. 1984. Insects associated with exposed human corpses in Auckland, New Zealand. *Medicine Science and the Law* 24: 167–174.

Smith, K.G.V. 1975. The faunal succession of insects and other invertebrates on a dead fox. *Entomologist's Gazette* 26: 277–287.

Smith, K.G.V. 1986. *A Manual of Forensic Entomology*. London, United Kingdom: British Museum and Cornell University Press.

Smith, R.L. and T.M. Smith. 2002. *Ecology and Field Biology*. New York: Benjamin Cumming.

Sukontason, K., K. Sukontason, K. Vichairat, S. Piangjai, S. Lertthamnongtham, R.C. Vogtsberger, and J.K. Olson. 2001. The first documented forensic entomology case in Thailand. *Journal of Medical Entomology* 38: 746–748.

Tabor, K., R. Fell, and C. Brewster. 2005. Insect fauna visiting carrion in southwest Virginia. *Forensic Science International* 150: 73–80.

Tantawi, T.I., E.M. El-Kady, B. Greenberg, and H.A. El-Ghaffar. 1996. Arthropod succession on exposed rabbit carrion in Alexandria, Egypt. *Journal of Medical Entomology* 33: 566–580.

Tessmer, J.W. and C.L. Meek. 1996. Dispersal and distribution of Calliphoridae (Diptera) immatures from animal carcasses in southern Louisiana. *Journal of Medical Entomology* 33: 665–669.

Thomas, L. and F. Juanes. 1996. The importance of statistical power analysis: An example from Animal Behaviour. *Animal Behavior* 52: 856–859.

Tibbett, M. and D.O. Carter. 2009. Research in forensic taphonomy: A soil-based perspective. In *Criminal and Environmental Soil Forensics*, eds. K. Ritz, L. Dawson, and D. Miller, 317–331. New York: Springer Science.

Tomberlin, J.K. and P.H. Adler. 1998. Seasonal colonization and decomposition of rat carrion in water and land in an open field in South Carolina. *Journal of Medical Entomology* 35: 704–709.

Tomberlin, J.K., M.E. Benbow, A.M. Tarone, and R.M. Mohr. 2011a. Basic research in evolution and ecology enhances forensics. *Trends in Ecology and Evolution* 26: 53–55.

Tomberlin, J.K., J.H. Byrd, J.R. Wallace, and M.E. Benbow. 2012. Assessment of decomposition studies indicates need for standardized and repeatable research methods in forensic entomology. *Journal of Forensic Research* 3: 147. doi: 4172/2157-7145. 1000137.

Tomberlin, J.K., R. Mohr, M.E. Benbow, A.M. Tarone, and S. VanLaerhoven. 2011b. A roadmap for bridging basic and applied research in forensic entomology. *Annual Review of Entomology* 56: 401–421.

Tullis, K. and M.L. Goff. 1987. Arthropod succession in exposed carrion in a tropical rainforest on Oahu Island, Hawaii. *Journal of Medical Entomology* 24: 332–339.

VanLaerhoven, S.L. 2008. Blind validation of postmortem interval estimates using developmental rates of blow flies. *Forensic Science International* 180: 76–80.

VanLaerhoven, S.L. and G.S. Anderson. 1999. Insect succession on buried carrion in two biogeoclimatic zones of British Columbia. *Journal of Forensic Sciences* 44: 32–43.

Velasquez, Y. 2008. A checklist of arthropods associated with rat carrion in a montane locality of Northern Venezuela. *Forensic Science International* 174: 67–69.

Voss, S.C., S.L. Forbes, and I.R. Dadour. 2008. Decomposition and insect succession on cadavers inside a vehicle environment. *Forensic Science Medicine and Pathology* 4: 22–32.

Wasti, W.S. 1972. A study of the carrion of the common fowl, *Gallus domesticus*, in relation to arthropod succession. *Journal of the Georgia Entomological Society* 7: 221–229.

Watson, E.J. and C.E. Carlton. 2003. Spring succession of necrophilous insects on wildlife carcasses in Louisiana. *Journal of Medical Entomology* 40: 338–347.

Wells, J.D. and L.R. LaMotte. 1995. Estimating maggot age from weight using inverse prediction. *Journal of Forensic Sciences* 40: 585–590.

Wells, J.D. and L.R. LaMotte. 2010. Estimating the postmortem interval. In *Forensic Entomology: The Utility of Arthropods in Legal Investigations*, 2nd edition, eds. J.H. Byrd and J.L. Castner, 367–388. Boca Raton, FL: CRC Press.

Williams, H. 1984. A model for the aging of fly larvae in forensic entomology. *Forensic Science International* 25: 191–199.

Winer, B.J., D.R. Brown, and K.M. Michels. 1991. *Statistical Procedures in Experimental Design*. New York: McGraw-Hill.

Wyss, C. and D. Cherix. 2006. *Traité d'entomologie Forensique. Les Insectes sur la Scène de Crime* [Forensic Entomology Treatise. The Insects on a Crime Scene]. Lausanne, Suisse: Presses polytechniques et universitaires romandes.

Yates, W.W., A.W. Lindquist, and J.S. Butts. 1952. Further studies on dispersion of flies tagged with radioactive phosphoric acid. *Journal of Economic Entomology* 45: 547–548.

Zhang, W.J. and K.G. Schoenly. 1999. *Biodiversity software series III. BOUNDARY: a program for detecting boundaries in ecological landscapes*. IRRI Technical Bulletin No. 3. Manila, Philippines: International Rice Research Institute.

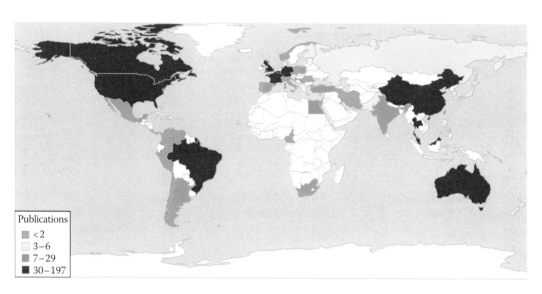

Figure 1 Presentation of publications on forensic entomology throughout the world from 1974 to 2012. (Figure courtesy of Jennifer Pechal)

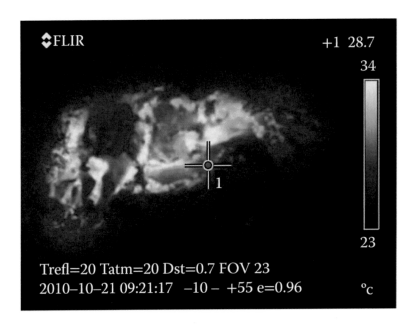

Figure 6.2 Infrared camera image showing heating (temperatures indicated on right) in a 10-kg pig carcass infested with larvae of *Chrysomya rufifacies* and located in a 23°C constant-temperature room.

Figure 7.2 Belgian artist Jan Fabre (a beetle, right) and Natural History Museum, London, entomologist Martin Hall (a fly, left) appearing in *A Consilience*. (From Fabre, J., *A Consilience*, Film Installation by Jan Fabre at The Natural History Museum, London, U.K. 13/01/2000-29/02/2000. Commissioned by the Arts Catalyst, United Kingdom, 2000 [© The Trustees of the Natural History Museum and Angelos bvba/Jan Fabre.]).

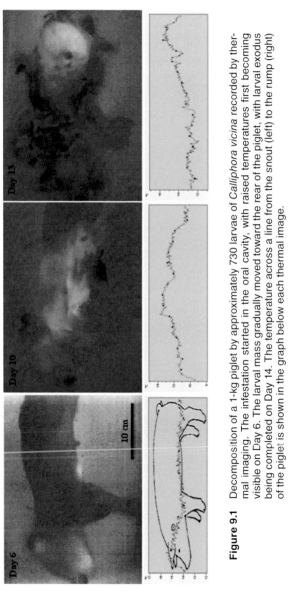

Figure 9.1 Decomposition of a 1-kg piglet by approximately 730 larvae of *Calliphora vicina* recorded by thermal imaging. The infestation started in the oral cavity, with raised temperatures first becoming visible on Day 6. The larval mass gradually moved toward the rear of the piglet, with larval exodus being completed on Day 14. The temperature across a line from the snout (left) to the rump (right) of the piglet is shown in the graph below each thermal image.

Figure 18.1 Attendees of the First Annual North American Forensic Entomology Conference held in Las Vegas, Nevada, in August 2003.

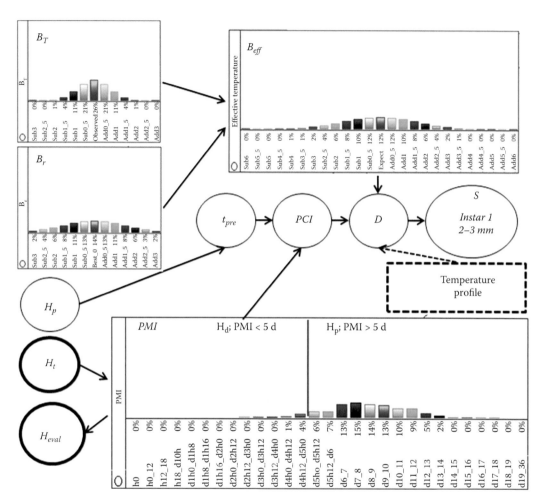

Figure 20.4 Details from the BBN applied to case 1. The bar charts in nodes B_T and B_r shows the probability distribution for the possible deviation between expected and true body temperature and development rate, respectively, assuming $SD_T = 0.75$ and $SD_r = 1.5$. Simulation was performed with 25 alternative temperature profiles ranging from 6°C below the expected to 6°C above. The expected temperature profile in this case is 2°C below the curve in Figure 20.3. In node S, the oldest sampled stages are fixed as instar 1 at the length of 1–2 mm. The bar chart in node postmortem interval (PMI) indicates that the posterior probability of PMI <5 days is approximately 5% (≈1% in interval 4 days–4 days 12 h; ≈4% in interval 4 days 12 h–5 days). The posterior probability of PMI less than 5 days is approximately 95%.

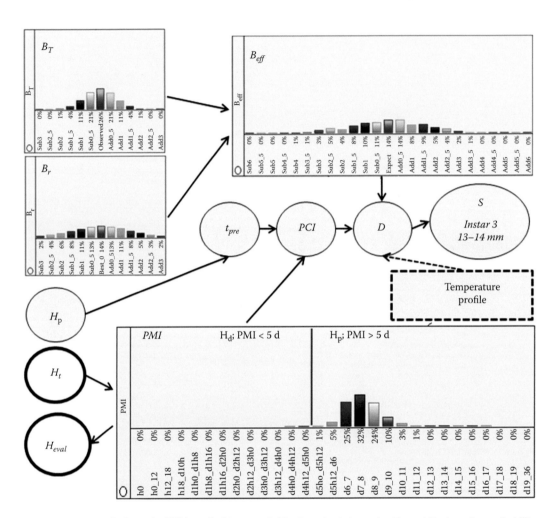

Figure 20.5 Details from the BBN applied to case 2. The bar charts in nodes B_T and B_r show the probability distribution for the possible deviation between expected and true body temperature and development rate, respectively, assuming $SD_T = 0.75$ and $SD_r = 1.5$. Simulation was performed as in case 1. The expected temperature profile in this case is 8°C above the curve in Figure 20.3. In node S, the oldest sampled stages are fixed as instar 3 at the length of 13–14 mm. The bar chart in node PMI indicates that the posterior probability of PMI less than 5 days is <1%. A closer look at the values in node H_{eval} shows that the posterior probability for PMI more than 5 days is 99.96%.

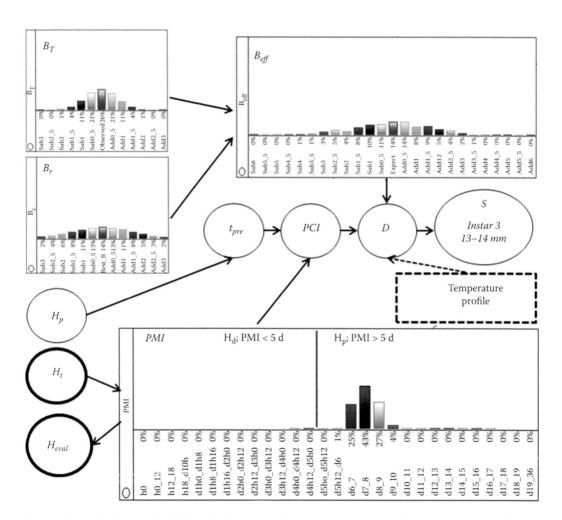

Figure 20.6 Details from the BBN applied to case 3. Simulation was performed as in case 3 but the distribution for the possible deviation from expected development rate (B_r) rate is more narrow ($SD_r = 0.5$) reflecting the investigator's higher confidence in the estimate. As in case 2, the expected temperature profile in this case is 8°C above the curve in Figure 20.3. In node S, the oldest sampled stages are fixed as instar 3 at the length of 13–14 mm. The bar chart in node PMI indicates that the posterior probability of PMI less than 5 days is <1%. A closer look at the values in node H_{eval} shows that the posterior probability for PMI more than 5 days is 99.9998%.

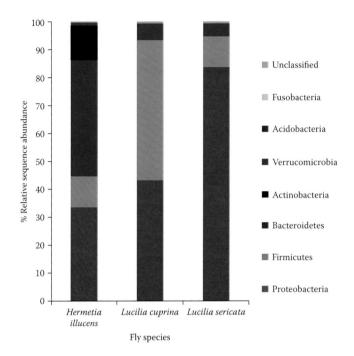

Figure 21.1 Graph showing the diversity of bacterial phyla associated with three species of Diptera: *Hermetia illucens*, *Lucilia sericata*, and *Lucilia cuprina*, identified by 454-pyrosequencing. (Modified from Zheng, L. et al., *J. Med. Entomol.*, 50, 647–658, 2013b and Singh et al., *Appl. Microbiol. Biotechnol.*, 20: 1–15, 2014.)

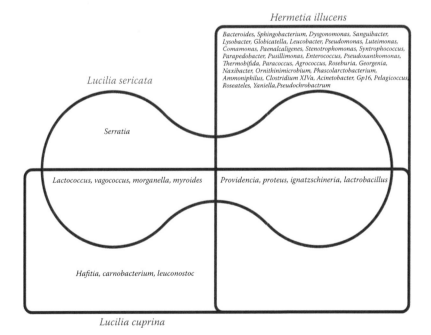

Figure 21.2 Venn diagram showing differences and similarities in bacterial genera associated with combined successive life stages (egg, larvae pupae, adult) of the Diptera: *Hermetia illucens* (red rectangle), *Lucilia sericata* (pink Cassinian ovals), and *Lucilia cuprina* (blue rectangle), identified by 454-pyrosequencing. (Modified from Zheng, L. et al., *J. Med. Entomol.*, 50, 647–658, 2013b and Singh et al., *Appl. Microbiol. Biotechnol.*, 20: 1–15, 2014.)

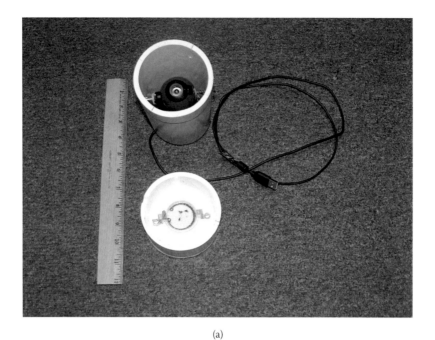

(a)

(b)

Figure 25.1 (a) Wasp Hound assembled showing fan on top for pulling sample air into the instrument. (b) Wasp Hound chemical detector showing the top removed to expose the camera and wasp cartridge, where odor enters into the instrument. Air enters through the bottom. USB connector plugs into computer running real-time behavioral analysis program.

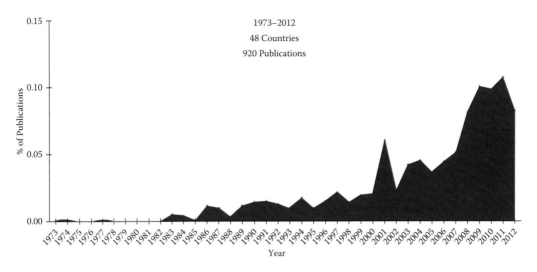

Figure 31.1 Publications from 1973 through 2012 that reference "forensic entomology."

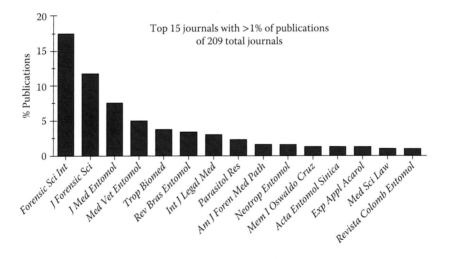

Figure 31.2 Journals that published manuscripts from 1973 through 2012 referencing "forensic entomology."

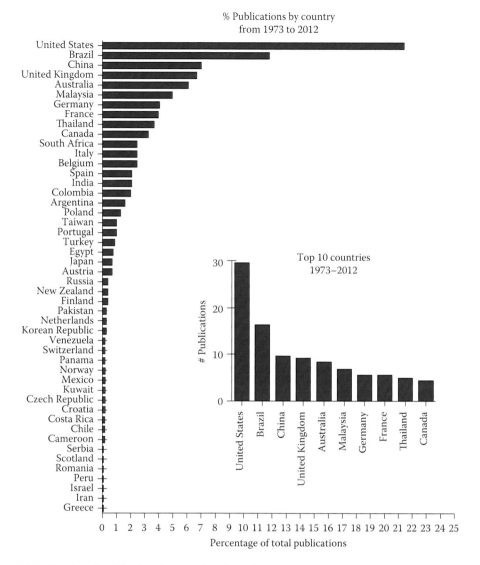

Figure 31.3 Number of publications from each nation referencing "forensic entomology."

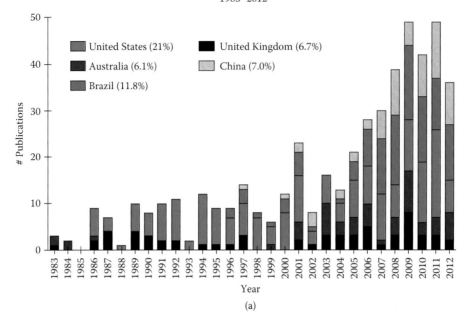

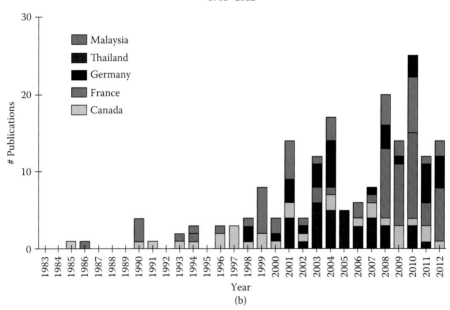

Figure 31.4 Articles pertaining to forensic entomology related topics that were published from 1983 to 2012: (a) the top five nations and (b) the following five nations.

CHAPTER 20

Bayesian Statistics and Predictive Modeling

Anders Lindström and Gunnar Andersson

CONTENTS

20.1 INTRODUCTION

An important issue in the investigation of suspect crime is having appropriate standards for interpreting and weighting forensic evidence (Koblentz 2010). In recent years, the state of the art in forensic interpretation has been to evaluate forensic evidence using likelihood ratios in the framework of Bayesian hypothesis testing. In this framework, it is used to evaluate to what extent results from forensic investigation speak in favor of the prosecutors or defendants hypotheses

(Taroni et al. 2006; Aitken 2010). The Bayesian approach has been applied to a wide range of forensic problems including evidence based on DNA analysis (Aitken 2010); mass spectroscopy (Jarman et al. 2008); transfer of glass, fibers, and paint (Aitken 2010); and microbial counts (Keats et al. 2006). In this work, the application of Bayesian statistics is presented and discussed for forensic entomology.

The following scenario is entirely hypothetical but represents a typical case in which an entomologist could be asked by the police to help with establishment of a timeline. A woman is reported missing by her boyfriend. He claims that he saw her the last time on the evening of 10, August, and he thinks that she has left him for someone else. A witness claims to have seen the woman 2 days after she disappeared (12 August). On 18, August (8 days after she went missing) a man walking his dog alerts the police to what he suspects are human remains. The remains are found close to a road in a desolate forested area 30 km from the home of the girl and her boyfriend. The police question the boyfriend who denies any knowledge of the crime. There are no signs of struggle or blood in the apartment they shared. A forensic entomologist is called to the scene and collects blow fly (Diptera: Calliphoridae) larvae from the corpse. The oldest larvae turn out to be *Calliphora vicina* Robineau-Desvoidy (Diptera: Calliphoridae), and their mean length is 12.9 mm. It is crucial for the investigation and for the boyfriend's alibi to establish if the woman was alive when the witness claims to have seen her. So what are the odds given the temperatures and the age of the larvae that the woman was alive at the time of the witness sighting? Traditional forensic entomology will not allow us to draw this conclusion as it will only provide a minimum age of the larvae. There are also several uncertainties in the calculation that are difficult to quantify using traditional modeling of larval age such as the variation in the temperature at the scene, larval mass temperature, and variation in growth rate of the larvae compared to the data extracted from literature.

20.2 INTRODUCTION TO BAYESIAN NETWORKS IN FORENSICS

20.2.1 Bayesian Statistics in Forensic Science

Misunderstandings related to statistical information and probabilities have contributed toward serious miscarriages of justice, and it has been deemed necessary to develop guidelines for statistical evaluation of evidence (Aitken 2010). Forensic statistics is typically based on Bayesian hypothesis testing (Aitken 2010), which may be performed using *Bayesian Belief Networks* (BBNs). In this framework, information about the expected probability of obtaining a particular result is used to calculate the *posterior probability* of a hypothesis.

The extent to which the piece of evidence should modify your belief is expressed as the *value of evidence* (*V*). *V* is typically expressed as a likelihood ratio but may be transformed into a qualitative statement using verbal scales (Aitken et al. 2010; Nordgaard 2012). The relationship between the value of evidence and the posterior probability and the significance for correct interpretation of results is illustrated in the following example. First we make a simplistic calculation to give an intuitive understanding of the results and then we show how the calculations are performed in the Bayesian framework.

A hypothetical blood test used by the police that will give a positive result with 99% probability if a person has used drug X. However, it will also give a positive result with 1% probability if the person is innocent. Thus, the true positive (TP) rate is 0.99 and the false positive (FP) rate is 0.01 and it is 990 times more likely to get a positive test result from a person that has taken drug X than from an innocent person. But, as illustrated in the example, this does not imply that the odds that a person who tested positive has used drug X is 990 to 1.

In the first case, the test is applied on a group of suspect users where 75% of the tested persons have taken the drug (*n* = 10,000).

The expected number of true positives (n_{tp}) and false positives (n_{fp}) is given by the following formulas:

$$n_{tp} = n \times \text{proportion of users} \times \text{TP} = 10,000 \times 0.75 \times 0.99 = 7,425$$

$$n_{fp} = n \times (1 - \text{proportion of users}) \times \text{FP} = 10,000 \times 0.25 \times 0.01 = 25$$

Thus, in this case, the probability that a positive test result is a false positive, $P(\text{fp})$ would be

$$P(\text{fp}) = \frac{25}{25 + 7,425} = 0.0034$$

In the second case, the same test is applied on randomly selected persons from a population where 0.01% of the tested persons have taken the drug ($n = 100,000$). In this case

$$n_{tp} = n \times \text{proportion of users} \times \text{TP} = 100,000 \times 0.0001 \times 0.99 \approx 10$$

$$n_{fp} = n \times (1 - \text{proportion of users}) \times \text{FP} = 100,000 \times 0.9999 \times 0.001 \approx 10$$

Thus, in this case, the probability that a positive test result is a false positive $P(\text{fp})$ would be

$$P(\text{fp}) = \frac{99}{10 + 99} \approx 0.92$$

From these examples it should be clear that knowledge about the test results and the performance of the test is not sufficient to draw conclusions about the guilt of a suspect, in this case about whether or not they have taken drug X. Evidently, it is necessary to evaluate the test results in light of other knowledge. In this case, whether the test was applied to a randomly selected person or a suspect user.

In forensic statistics, the evaluation of evidence is based on *Bayes' theorem* that was proved in the eighteenth century by Reverend Thomas Bayes (Taroni et al. 2006). A common form of expressing this theorem is the *odds form*:

$$\frac{P(H_p/E)}{P(H_d/E)} = \frac{P(E/H_p)}{P(E/H_d)} \times \frac{P(H_p)}{P(H_d)}$$

where

H_p is the hypothesis of interest, commonly referred to as the *prosecutor's hypothesis*. In the example, H_p is *the person has used drug X*.

H_d is the relevant alternative to H_p, commonly referred to as the *defendant's hypothesis*. In the example, H_d is *the person has **not** used the drug*.

$P(H_p)$ is the *prior probability* that H_p is true. In other words, $P(H_p)$ is the expected probability of H_p being true without knowing the test results. In the cases above, $P(H_p)$ is 0.75 and 0.0001, respectively.

$P(H_d)$ is the *prior probability* that H_d is true. In other words, the expected probability of H_d being true without knowing the test results. In the cases above, $P(H_d)$ is 0.25 and 0.9999, respectively.

E is the observed result(s). In the example, E is a positive result with the specific test.

$P(E/H_p)$ is the probability to obtain the particular piece of evidence given that H_p is true. In the examples, $P(E/H_p) = TP = 0.75$.

$P(E/H_d)$ is the probability to obtain the particular piece of evidence given that H_d is true. In the examples, $P(E/H_d) = FP = 0.25$.

$P(H_d/E)$ is the probability that H_p is true given the evidence (a positive test result). From the calculations above, we know that in case 1 $P(H_d/E) = (1 - P(fp)) = (1 - 0.0034) = 0.9966$ and in case 2, $P(H_d/E) = (1 - P_{FP}) = (1 - 0.92) = 0.08$.

$P(H_d/E)$ is the probability that H_d is true given the evidence (a positive test result). From the calculations above, we know that in case 1 $P(H_d/E) = 0.034$ and in case 2 $P(H_d/E) = 0.092$.

To evaluate the cases above using Bayes' theorem, we start by defining the *prior odds*.
For the first case, the prior odds are

$$\frac{P(H_p)}{P(H_d)} = \frac{0.75}{0.25} = \frac{3}{1} = 3$$

For the second case, the prior odds are

$$\frac{P(H_p)}{P(H_d)} = \frac{0.0001}{0.9999} = \frac{1}{9999} \approx 0.0001$$

The next step is to calculate the value of evidence (V). In the example, V is determined by the characteristics of the test and is equal for the two cases.

$$V = \frac{P(E/H_p)}{P(E/H_d)} = \frac{TP}{FP} = \frac{0.99}{0.001} = 990$$

Finally, we calculate the posterior odds by multiplying the prior odds and the value of evidence. For the first case:

$$\frac{P(H_p/E)}{P(H_d/E)} = \frac{P(E/H_p)}{P(E/H_d)} \times \frac{P(H_p)}{P(H_d)} = 990 \times 3 = 2970$$

For the second case:

$$\frac{P(H_p/E)}{P(H_d/E)} = \frac{P(E/H_p)}{P(E/H_d)} \times \frac{P(H_p)}{P(H_d)} = 990 \times 0.0001 = 0.099$$

Setting the prior odds and performing the final calculation is the duty of the court or possibly of the investigator or prosecutor. Thus, in the example the forensic laboratory would report the test result and explain the value of evidence (V) for the analysis, which in both cases is 990.

In practice, the value of evidence is typically not reported as a number, and the evaluation of evidence by the court is rarely done by multiplying numbers. However, even if the calculation is not performed explicitly the logic still applies. Typically, the value of evidence is expressed in words using expressions like: "the results provide limited/moderate/strong support for the prosecutor's hypothesis over the defendant's hypothesis." From the example above, it should be clear that a forensic scientist should not write a statement like: "the results show that there is a high probability that the H_p is true (that the person used the drug)" as this would involve an assumption about the

Table 20.1   Example of a Scale of Conclusions Proposed by Evett

Interval			Verbal expression
1	< V <	10	Limited evidence to support
10	< V <	100	Moderate evidence to support
100	< V <	1000	Moderate strong evidence to support
1000	< V <	10000	Strong evidence to support
10000	< V		Very strong evidence to support

prior odds. On the contrary expressions like: "the results supports H_p over H_d" or "the results speak in favor of H_p over H_d" would be appropriate and analogous to reporting the value of evidence.

To promote consistent use of qualitative expressions between experts, it has been proposed (Aitken et al. 2010; Nordgaard 2012) to use scales of conclusions where fixed verbal expressions correspond to specified intervals for V. An example of a scale proposed by Evett and cited in the work of Aitken et al. (2010) is shown in Table 20.1.

In the ideal case, the likelihood for observing a particular piece of evidence would be calculated using data from an explicit reference database. However, the logical approach to evaluate the evidence provided by forensic statistics is applicable also in the majority of forensic cases where an explicit reference dataset is not available (Nordgaard 2012). The opinion rendered by the expert witness is often based on an estimate of the likelihood of an observation under alternative propositions (hypothetical scenarios) based on experience, training, and scientific literature. A prerequisite for this is of course that there is sufficient knowledge available to support an estimate of the likelihood, or at least provide a range or magnitude.

20.2.2 Introduction to Bayesian Networks

In the abovementioned example, the calculations are simple and can be performed by hand. However, in many cases there are complex relationships between the propositions and the observations involving several uncertain parameters. In such situations, BBNs have proven to be useful tools for visualizing complex relations between parameters and assessing the value of evidence accounting for the combined effect of several uncertainties (Taroni et al. 2006; Jarman et al. 2008). The visual representation also serves to promote transparency and logic (Sjerps and Berger 2012).

A BBN is a set of nodes representing uncertain state variables (Taroni et al. 2006) that may be binary (yes/no), nominal (nonnumerical categories), numerical, or continuous (e.g., time, temperature, concentration). In the latter case, the number of possible states is infinite, but can be reduced by dividing the scale into intervals. The nodes in the network are linked by arrows representing causal or evidential relationships (Figure 20.1). If a node A receives arrows from other nodes (B_1, B_2, B_3...B_n) the former is referred to as a *child* and the latter as *parents*. In this case, the probability of a particular state or value at the child node is conditional on the states of the parental nodes. In practice, each node A has a probability table indicating the probability of each state for each combination of states in the parental nodes (B_1, B_2, B_3...B_n).

20.3 APPLICABILITY TO FORENSIC ENTOMOLOGY

At first sight it is not obvious how this framework would apply to forensic entomology where a goal is to estimate the postmortem interval (PMI). The following scenario explains the connection between estimation of PMI and reporting the value of evidence.

If there is a question regarding the time of death like in the scenario outlined in the beginning of the chapter (What are the odds that the deceased was alive or dead on a given date?), the forensic

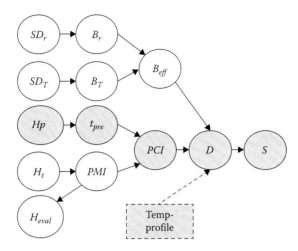

Figure 20.1 Outline of the Bayesian Belief Network (BBN). Gray nodes: colonization, development, sampling. White nodes: other information relevant for process. Described in Table 20.2.

entomologists have the opportunity to help the police with a more qualified estimation based on data from the case and from the larvae that were sampled. It also allows forensic entomologists to report their findings in graphic and easily understandable ways that express the uncertainties in the evidence and in the calculations and assumptions. This makes it easier for the police to conduct their investigations and for the court to evaluate the evidence.

In this example, based on the findings, the forensic entomologist estimated PMI to be 8 days and may choose to report this to the court. However, a major issue in the case is a witness claiming to have seen the victim alive 6 days before the body was found in which case the defendant would have an alibi. Thus, to determine whether or not the entomological evidence is sufficient to falsify the witness statement, it is necessary to know the precision of the PMI estimate. In other words, the court needs to know how strongly the entomological evidence speaks against the defendant's hypothesis (H_d). In the following sections, it is shown how the Bayesian framework and BBN can be applied to such scenarios.

In the Bayesian framework, probabilities are considered as measures of belief and the approach enables the combination of *objective probabilities* based on data and *subjective probabilities* based on knowledge and experience (Taroni et al. 2006). In forensic entomology, it is possible to calculate objective probabilities for many, but not all, variables. In comparison to experts in many other forensic disciplines (Taroni et al. 2006), forensic entomologists often have substantial scientific data to substantiate estimates of the subjective probabilities. The likelihood of observing an insect larva at a particular stage given a proposition on the time of death depends on the time for colonization, the temperature experienced by the insects on the remains, and the temperature-dependent growth rate of the species in question (Amendt et al. 2007). Exact numbers or empirical distributions for these parameters are not readily available from databases.

However, with the help of meteorological data and observations on the crime scene it is usually possible to estimate a temperature interval that covers the true temperature of the carcass. Similarly, when data on the growth rate of a particular species at the relevant temperature are not available the combination of information on growth rates of related species, the ecology of the species, and sporadic case data can often inform an expert that the growth rate is likely to be within a certain range (Fremdt et al. 2012).

20.4 CONSTRUCTING THE BAYESIAN NETWORK

The Bayesian network described below was constructed in GeNie 2.0 (Decisions Systems Laboratory, University of Pittsburgh, Pennsylvania). This software is designed for discrete data. That is to say, a measurement can only take one of a limited number of predefined values. This is solved by dividing the scales for time, temperature, and length into intervals as in the work by Taroni et al. (2006).

20.4.1 Sources of Data

Data for development rates of *C. vicina* are available in the literature. Numerous authors have published developmental data for *C. vicina* at different temperatures. However, there have been no published studies on how to combine the various data into one analysis. There are difficulties as different authors report on development of different life stages and some report just on the total development from egg to adult. Because a meta-analysis cannot be performed on these data sets, a simulation of the growth of larvae using the data presented in the various papers was conducted. The temperature range covered stretches from 4°C to 35°C and is likely to cover most situations in which *C. vicina* will be encountered.

20.4.2 Data for Development Rate of *Calliphora vicina*

We used nine different literature sources (Kamal 1958; Nuorteva 1977; Reiter 1984; Greenberg 1991; Davies and Ratcliffe 1994; Anderson 2000; Greenberg and Kunich 2002; Ames and Turner 2003; Niederegger et al. 2010) for duration of development of *C. vicina* from egg to adult. To estimate development rates at different temperatures we used only data from articles in which the whole duration of development from egg to adult was reported. Most authors report on development to different instars and other life history stages rather than body length. An attempt was made to transform the data to lengths because that provides higher precision in the estimates of larval age.

20.4.3 Data on Length in Different Intervals

It is surprisingly difficult to find good data on body length variation of different instars of *C. vicina*, no doubt, due to the overlap between instars in body length. It seems that either developmental stage or length is reported. However, Erzinçlioğlu (1996) reports on typical lengths of *C. vicina*.

20.4.4 Data to Support Assumptions on Precolonization Time

As pointed out by Tomberlin et al. (2011), there are limited useful data on the actual time it takes for the insects to colonize a corpse, or the precolonization interval. On the other hand, we know about many of the factors affecting the colonization dynamics (Campobasso et al. 2001) and can often, based on experience, give a good estimate of the precolonization time. Factors affecting the colonization can be weather, temperature, if the corpse is indoors or outdoors, and so on. This can be weighed in to the estimate of precolonization and hence PMI.

Linear regression of the development rate (1/total development time) as a function of temperature was performed using the function lm() in R 2.15.2 (R Core Team 2012). The results indicated that a linear model described the relationship (Figure 20.2). The deviation of observations from the regression line is relatively constant for all temperatures supporting a model with constant standard deviation. The deviation of observations from the fitted line is likely to represent variability within the species, and to obtain a conservative estimate for the interval within which the development rate for a new isolate is likely to be found prediction intervals were generated using function predict(). On the basis of the

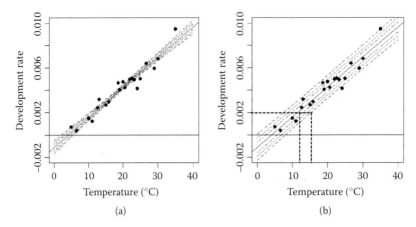

Figure 20.2   Result from linear regression of development rate of *Calliphora vicina* as function of temperature. (a) Regression line with 50%, 75%, and 90% confidence interval for the average rate. (b) 50%, 75%, and 90% prediction interval for expected new observations.

90% prediction interval, it was estimated that the temperature at which a particular development rate is observed (e.g., 0.002) would be within approximately ±3°C from the expected (Figure 20.2b).

20.5 BAYESIAN NETWORK FOR FORENSIC ENTOMOLOGY

20.5.1 Model Structure

The knowledge is structured in a BBN as outlined in Figure 20.1. Each node can take two or more states, which may be nominal (categories without order), discrete, or interval. For decision nodes (H_t, H_p, SD_T, SD_t), the state is decided by the user. For chance nodes the conditional probability of state x given each combination of states in the parental nodes is set in the probability table. The states of each node and functions used to set probabilities are summarized in Table 20.2.

The central node in the BBN is the PMI, which is the parameter of interest. The prior probabilities for PMI will depend on the prior knowledge of the case and after entering the evidence posterior probabilities for each time interval will define a Bayesian confidence interval for the PMI. The length of the PMI is the sum of the precolonization time (t_{pre}) and the *postcolonization interval* (PCI). If the circumstances in the case can justify the assumptions about the precolonization time, this may be implemented by selecting alternative (b) in node H_p and assign prior probabilities to the time intervals of node t_{pre} according to some formula. The formula in the example is based on the assumptions that (1) the probability that a fly finds the carcass at a given time point is constant and (2) the probability that the body is colonized within 24 hours is approximately 80%. If the circumstances do not justify such an assumption, the state of the node H_p is set to 0 hour and the reported time interval will be identical to the PCI and should be interpreted accordingly.

The node E_{eval} facilitates the calculation of the value of evidence in relation to the prosecutors hypothesis (H_p) that "the person was dead 5 days ago" against the defendants hypothesis (H_d) that "the person was alive 5 days ago."

The node *oldest developmental stage* (D) defines the probability that the insect has reached a particular stage of development at the time of sampling. The development stage of the insect on detection depends on its age and on the temperature of the carcass that is estimated by simulation along the observed temperature profile (Figure 20.3) for each combination of temperature interval and effective temperature bias (B_{eff}).

Table 20.2   Nodes in the Bayesian Network Model. The Values of Decision Nodes Are Set by the User whereas Chance Nodes Has a Probability for Each State

Parameter	Description	Type	States/function
H_t	Model for prior on PMI	Decision-Nominal	a. p = k*Δt (proportional to interval length) b. p = k*Δt, 1 d < PMI < 18 d
PMI	Post Mortem Interval	Chance-Interval	0 h 0 h–12 h 12 h–18 h … 19 days–36 days (Probability function according to Ht)
Heval	General hypothesis	Chance-Binary	a. Procecutors hypothesis, H_p, PMI <5 d b. Defendants hypothesis, H_d, PMI >5 d
H_p	Precolonization hypothesis	Decision-Nominal	(a) t_{pre} = 0 (b) $t_{pre} \sim e^{-k(t\text{-}t_{min})}$ h; k = 0.17, t_{min} 6 h
t_{pre}	Pre-colonization time	Chance-Interval	Intervals as PMI. (Probability function according to H_p)
PCI	Post Colonization Interval	Chance-Interval	Intervals as PMI ($PCI = PMI - t_{pre}$)
D	Oldest developmental stage	Discrete	States from table 2 (Probability table from simulation, as table 3)
S	Oldest specimen sampled	Discrete	States from table 2 (Probabilities from user defined table)
S_{DT}	Standard deviation of observed temperature profile	Decision-Discrete	0.5°C 0.75°C 1°C 1.25°C 1.5°C
B_T	Bias in observed temperature	Chance-Discrete	Observed profile Observed +0.5; +1; +1.5 … Observed −0.5; −1; −1.5 … ($BT \sim$ Normal(0,SDT) (Truncated))
SD_r	Standard deviation of temperature that gives specific growth rate	Decision-Discrete	0.5°C 0.75°C 1°C 1.25°C 1.5°C
B_r	Bias in estimate of temperature that gives a specific growth rate	Discrete	Observed profile Observed +0.5; +1; +1.5 … Observed −0.5; −1; −1.5 … ($Br \sim$ Normal(0,SD_r) (Truncated))
B_{eff}	Combined effect of uncertainty from H_T and H_r	Discrete	Observed profile Observed +0.5; +1; +1.5 … Observed −0.5; −1; −1.5 … ($T_r \sim B_T + B_r$)

The node *oldest specimen sampled* (S) represents the oldest developmental stage sampled at time of sample collection. The probability of not finding the oldest specimen present depends on many factors including the experience of the person performing sampling, and the *subjective probabilities* will reflect the expert opinion on the particular case. Accounting for nonzero probability of missing the oldest specimen is important for not overestimating confidence in the upper limit of the PMI. The probability table of this node will be set by the user accounting for experience and case-specific circumstances. In this example, the table from the work of Andersson et al. (2013) was used.

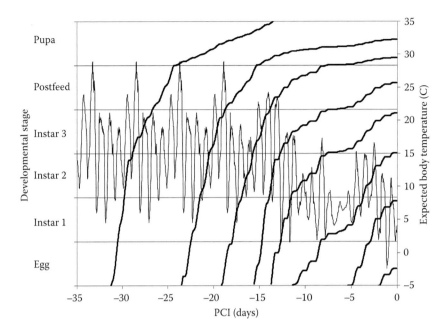

Figure 20.3 Right axis: Temperature profile from meteorological observation used as approximation of the temperature in the corpse. Left axis: Predicted developmental stage at the time of sampling ($t = 0$) based on simulation. The graphs show predicted development of *C. vicina* following colonization at approximately 2, 6, 12, 14, 16, 19, 23, and 32 days before sampling. For example, if colonization took place approximately 16 days before sampling ($t = -16$), the larvae will be halfway through the postfeeding stage at $t = 0$.

The node B_T represents the uncertainty about the body temperature. Each state represents a difference between the expected and the true temperature. In the current implementation, it takes values from 3.0°C below ambient temperature to 3.0°C above ambient temperature around in steps of 0.5°C. The prior probability that for the different values is approximated from the normal distribution with standard deviation (SD) set by the user in node SD_T.

The prediction interval from the linear regression indicates that for a new observation the temperature needed to reach a specific development rate may deviate at least ±3.0°C from the expected, corresponding to SD_r approximately 1.5°C. The node B_{eff} represents the combined error from about temperature and development rate and is the sum of B_T is B_r.

20.5.2 Calculating Probabilities

The probability table in node D was generated using simulation using a Perl 5.16.1 (Active state) script modified from the script described in the work by Andersson et al. (2013). The input values were specified in the following text files:

1. Configuration script specifying the number of simulations and the temperature curves to use.
2. Table of the duration of each stage at temperatures from −5°C to +40°C (selection of rows shown in Table 20.3).
3. File defining substages and indicating the percentage of the total stage duration after which the development has proceeded from one substage to the next (Table 20.3).
4. File with meteorological data showing the air temperature at the nearest station during the period of interest (Figure 20.2).

Table 20.3 Duration of Stages (h) at Selected Temperatures. The Figures Are Point Estimates Based on Linear Regression. For the Stages Instar 1 – Instar 3 the Value under Proportion Indicates the Percentage of the Total Stage Duration after Which the Development Has Proceeded from One Substage to the Next, e.g., after 50% of the Duration of Stage Instar 1 the Estimated Size Interval for the Largest Larvae Will Go from 2–3 mm to 3–4 mm

Stage	Substage Size interval	Proportion	0°C	5°C	10°C	15°C	20°C	25°C	30°C	35°C	40°C
Egg	–	100%	∞	472	71	38	26	20	16	14	12
Instar 1	1–2 mm	25%	∞	441	67	36	25	19	15	13	11
	2–3 mm	50%									
	3–4 mm	75%									
	4 mm	100%									
Instar 2	4–5 mm	25%	∞	596	90	49	33	25	20	17	15
	5–6 mm	50%									
	6–7 mm	75%									
	7–8 mm	100%									
Instar 3	8–9 mm	12.5%	∞	1080	163	88	60	46	37	31	27
	9–10 mm	25 %									
	10–11 mm	37.5%									
	11–12 mm	50%									
	12–13 mm	61.5%									
	13–14 mm	75%									
	14–15 mm	87.5%									
	>15 mm	100%									
Postfeed	–	100%	∞	1595	241	130	89	68	55	46	39
Pupa	–	100%	∞	5719	862	466	320	243	196	164	141

As the software for Bayesian networks is based on discrete data the time scale was divided into intervals, and during the simulation exact values for colonization time were randomly sampled from the time interval. During the simulation the temperature at each 30-min interval was compared with the development rate at a specific temperature according to Table 20.3 and the predicted stage of development was recorded (Figure 20.3).

The results from a large number of simulations were summarized in a table indicating the probability that development has reached a particular stage at the time of sampling given that colonization occurred in a particular time interval and given the assumptions on the temperature profile (Table 20.4). To complete a BBN for the particular case the table with simulation results for all temperature profiles in node T_{eff} and all time intervals in node PCI was copied into the probability table of D.

The value of evidence V for the main hypothesis H_p against the alternative hypothesis H_d is calculated in the node from values of prior and posterior probabilities of H_p and H_d in the node H_{eval}.

$$V = B = \frac{(P(H_p/E))/(P(H_d)/E))}{(P(H_p)/P(H_d))}$$

Calculating the value of evidence was facilitated by the introduction of the evaluation node H_{eval}. The probability table is a set of 1 and 0, and the node takes the value 1 if the PMI is in an interval fulfilling the criteria for H_p (e.g., PMI < 12 days) else 0.

20.5.3 Applying the BBN to a Case

To investigate the impact of the different uncertainties on the conclusions from the forensic entomology investigation, the constructed BBN was applied to some examples of cases based on the

Table 20.4 Detail of Probability Table Obtained by Simulation Using Temperature Profile in Figure 20.3. The Values Indicated Are the Probabilitity That the Development Has Reached a Particular (sub-) Stage at the Time of Sampling If the Corps Was Colonized in the Given Time Interval. For Example, If the Corps Was Colonized in the Interval 9 to 10 Days Prior to Sampling the Probability Is 0.78 That the Oldest Specimen Would Be Instar 2 4–5 mm and 0.22 That the Oldest Specimen Would Be Instar 2 at the Size If 5–6 mm

	Time interval						
Stage	51/2–6 d	6–7 d	7–8 d	8–9 d	9–10 d	10–11 d	11–12 d
Egg	0	0	0	0	0	0	0
Instar 1 1–2 mm	0	0	0	0	0	0	0
Instar 1 2–3 mm	0.38	0	0	0	0	0	0
Instar 1 3–4 mm	0.62	1	0.65	0	0	0	0
Instar 1 4 mm	0	0	0.35	0.5	0	0	0
Instar 2 4–5 mm	0	0	0	0.5	0.78	0	0
Instar 2 5–6 mm	0	0	0	0	0.22	1	0.05
Instar 2 6–7 mm	0	0	0	0	0	0	0.76
Instar 2 7–8 mm	0	0	0	0	0	0	0.19
Instar 3 8–9 mm	0	0	0	0	0	0	0

case in the introduction but with differences in the details that affect the uncertainty in PMI estimation and the value of evidence in relation to the hypothesis.

20.5.3.1 Case 1

In this case, the temperature in the area has been relatively low, corresponding to a temperature 2.0°C below the curve in Figure 20.3. On the body, the investigator finds instar 1 larvae with a length of 1–2 mm. A comparison between the temperature profiles from several nearby meteorological stations and the temperature in the body at the time of discovery indicated that the difference between the temperature profile and the true temperature in the body has most likely been in the interval ±2.0°C, corresponding to $SD_T = 0.75$. When this information is entered, the model *posterior probability distribution* for PMI, represented by the bar chart, is centered around 8–9 days but a 99% interval spans from approximately 4.0 to 14.0 days (Figure 20.4).

Calculations were performed under the assumption that the person had been dead for 1–18 days and that any time interval was equally probable. Under these assumptions the prior odds (odds before considering evidence) for H_p:H_d is 76:26 = 2.9 (H_p = PMI < 5 days; H_d = PMI < 5 days). The posterior odds (odds after considering the entomological evidence) for H_p:H_d is 94.5:5.5 = 17.2. Thus, using Bayes formula the value of evidence for the entomological results in relation to the question can be calculated as follows:

$$V = \frac{P(H_p/E)/P(H_d/E)}{P(H_p)/P(H_d)} = \frac{94.5/5.5}{76/26} = 5.9$$

According to the scale in Table 20.1, a value of V between 1 and 10 is expressed as "limited evidence to support" H_p over H_d. Therefore, although a point estimate of PMI is 8 days, it is not sufficient to exclude the possibility that the person was alive 5 days ago. The large uncertainty in this case is due to uncertainty about development rate that is high at low temperatures (Figure 20.2) and also that that a few degrees uncertainty about temperature has a large impact near threshold for development.

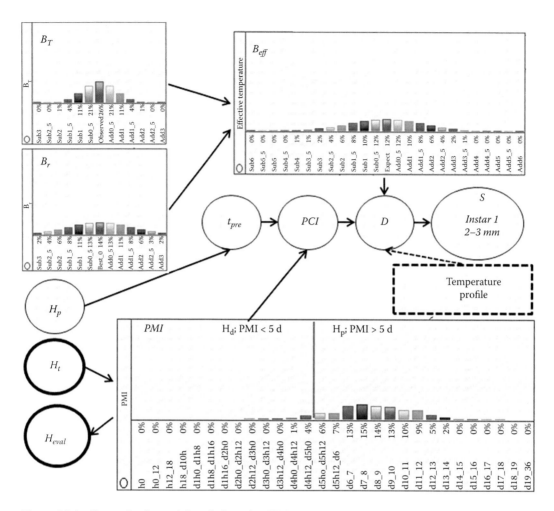

Figure 20.4 (**See color insert.**) Details from the BBN applied to case 1. The bar charts in nodes B_T and B_r show the probability distribution for the possible deviation between expected and true body temperature and development rate, respectively, assuming $SD_T = 0.75$ and $SD_r = 1.5$. Simulation was performed with 25 alternative temperature profiles ranging from 6°C below the expected to 6°C above. The expected temperature profile in this case is 2°C below the curve in Figure 20.3. In node S, the sampled stages are fixed as instar 1 at the length of 1–2 mm. The bar chart in node postmortem interval (PMI) indicates that the posterior probability of PMI <5 days is approximately 5% (≈1% in interval 4 days–4 days 12 h; ≈4% in interval 4 days 12 h–5 days). The posterior probability of PMI less than 5 days is approximately 95%.

20.5.3.2 *Case 2*

In this case, the temperature in the area of the body has been relatively high, corresponding to a temperature 8.0°C above the curve in Figure 20.3. On the body, the investigator finds instar 3 larvae ranging 13–14 mm. With the same assumptions as in case 1 the posterior probability distribution for PMI is again centered around 8–9 days but the 99% interval is shorter, 5–11 days (Figure 20.5). In this case, the posterior odds for H_p against H_d is 99957/43. The value of evidence is calculated as

$$V = \frac{P(H_p/E)/P(H_d/E)}{P(H_p)/P(H_d)} = \frac{99957/43}{76/26} = 795$$

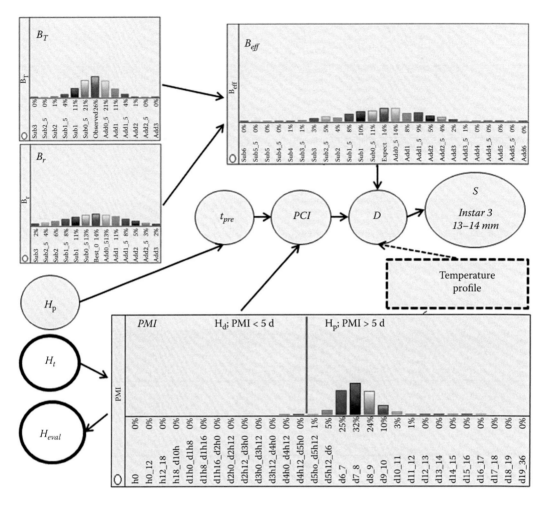

Figure 20.5   **(See color insert.)** Details from the BBN applied to case 2. The bar charts in nodes B_T and B_r show the probability distribution for the possible deviation between expected and true body temperature and development rate, respectively, assuming $SD_T = 0.75$ and $SD_r = 1.5$. Simulation was performed as in case 1. The expected temperature profile in this case is 8°C above the curve in Figure 20.3. In node S, the oldest sampled stages are fixed as instar 3 at the length of 13–14 mm. The bar chart in node PMI indicates that the posterior probability of PMI less than 5 days is <1%. A closer look at the values in node H_{eval} shows that the posterior probability for PMI more than 5 days is 99.96%.

According to the scale of conclusions, this may be expressed as "moderate strong evidence to support" the hypothesis H_p, that PMI > 5 days.

20.5.3.3 *Case 3*

In this case, the observations (temperature and entomological) are the same as in case 2 but the investigator has reasons to believe that the development rate of the *C. vicina* in the area is known with a precision of ±1.0° ($SD_r = 0.5$°C). Again the interval for PMI is centered around 8-9 days but with an even narrower 99% interval, approximately 6-10 days (Figure 20.6). In this case, the posterior odds for H_p against H_d is 999998/2. The value of evidence is calculated as

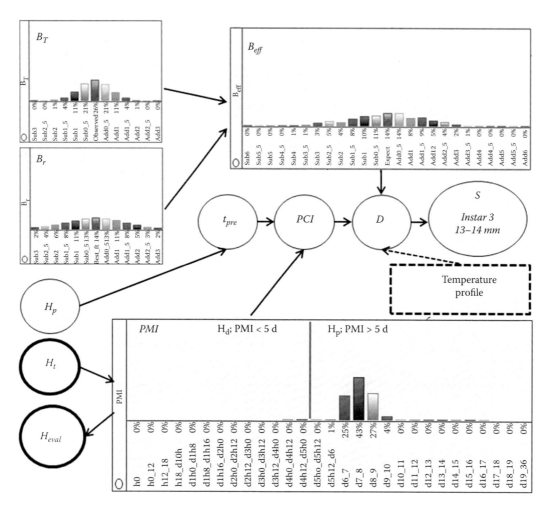

Figure 20.6 **(See color insert.)** Details from the BBN applied to case 3. Simulation was performed as in case 3 but the distribution for the possible deviation from expected development rate (B$_r$) rate is more narrow (SD$_r$ = 0.5) reflecting the investigator's higher confidence in the estimate. As in case 2, the expected temperature profile in this case is 8°C above the curve in Figure 20.3. In node S, the oldest sampled stages are fixed as instar 3 at the length of 13–14 mm. The bar chart in node PMI indicates that the posterior probability of PMI less than 5 days is <1%. A closer look at the values in node H$_{eval}$ shows that the posterior probability for PMI more than 5 days is 99.9998%.

$$V = \frac{P(\mathrm{H_p}/\mathrm{E})/P(\mathrm{H_d}/\mathrm{E})}{P(\mathrm{H_p})/P(\mathrm{H_d})} = \frac{999998/2}{76/26} = 171052$$

According to the scale of conclusions, this is "very strong evidence to support" the hypothesis that PMI > 5 days.

In the analysis of these cases, the prior probability for the precolonization time t_{pre} was assumed to follow the formula:

$$t_{\mathrm{pre}} : \mathrm{e}^{-k(t-t_{\mathrm{min}})}\mathrm{h}; \quad k = 0.17, t_{\mathrm{min}} = 6\,\mathrm{h}$$

With these settings there is 64% probability of colonization within 12 h, 81% within 24 h, and 99% within 48 h. If instead the precolonization time had been set to 0, the bar chart in node PMI would have been identical to the probability distribution for the PCI. Because the PCI is shorter the evidence for a PCI less than 5 days would be weaker. For example in case 1 the posterior odds for PCI > 5 days against PCI < 5 days is approximately 903:97 = 9.3 and the value of evidence (V) calculated as above would be 3.2.

20.6 DISCUSSION

As illustrated by the examples, the uncertainty in a PMI estimate may vary depending on the circumstances, and failure to account for this may result in misinterpretation. The example shows how the BBN framework can be applied to forensic entomology for combining different sources of information and includes information on uncertainty. Depending on the question, the results may be represented as a posterior probability distribution for PMI or PCI or as the value of evidence in relation to a pair of hypotheses. Naturally, there will be cases where the uncertainty about some parameter is too large to apply this approach. Whether the available information and data are of sufficient quality to justify the presentation value of evidence in court as a likelihood ratio or on corresponding verbal scale (Nordgaard 2012) will have to be assessed on a case-by-case basis.

At first glance it may seem convenient to present entomological evidence as confidence intervals for PMI or PCI. However, the traditional confidence intervals on the form $x \pm y$ are clearly inappropriate because the probability density distribution for PMI or PCI (represented as bar charts in nodes PMI and PCI) are intrinsically asymmetric, especially in the tails. Most importantly the probability density function for PMI may have a well-defined lower bound determined by the maximum development rate, whereas the upper bound may be extremely uncertain because the precolonization time is unknown, although the experienced forensic entomologist often can give a good approximation. In addition, uncertainty about whether the oldest life stage of insect has been found and systematic variation in temperature over the period of interest may further skew the probability distribution. Besides this, presenting uncertainty about evidence as confidence intervals has been deemed inappropriate in court proceedings because it disregards evidence falling outside an arbitrarily chosen confidence (e.g., 95%) (Aitken et al. 2010). Importantly, in a legal context a statement that the hypothesis PMI (e.g., 5.0 days) would fall outside a 95% confidence interval would only be limited or moderate evidence.

The BBN presented here represents only one of several possible designs. A forensic problem can generally be addressed at different levels of detail resulting in different model structures (Taroni et al. 2006). In this work, the information about growth rate of *C. vicina* at different temperatures was provided as a table. Because the development rate of all stages show the same temperature dependence, it would have been possible to build a BBN where development is modeled via accumulated degree days (ADD) over a temperature threshold as in the work by Damos and Savopoulu-Soultani (2012). However, the ADD approach only works when the temperature dependence, including the critical temperature is the same for all stages something which may not be true for all species. In an earlier work, the analysis of data from the work by Grassberger and Reiter (2001) on development rate of *Lucilia sericata* (Meigen) (Diptera: Calliphoridae) indicated that the temperature threshold was higher at the later development stages (Andersson et al. 2013), and the present approach was chosen as being more general.

The assumptions made about the uncertainties of the different variables will have a major impact on the design of the BBN. In this work, the analysis of development data for *C. vicina* indicated that major uncertainty was associated with the development rate whereas the relative duration of stages was of lesser importance. Further, the regression analysis indicated that it was reasonable to assume that the uncertainty about development rate is constant at all temperatures. In combination, these assumptions made it possible to design a simple model where uncertainty about development rate is modeled as a possible uncertainty about temperature. In our previous work (Andersson et al. 2013), a comparison between data on *L. sericata* development from the work by Anderson (2000) and Grassberger and Reiter (2001) indicated that the uncertainty increased with the development rate resulting in a different network structure.

20.7 LIMITS AND POSSIBILITIES

The work shown is aimed primarily as a proof of principle, and the routine application of the proposed framework to real cases would require the development of validated scripts and BBNs as well as a database of curated data on development rate, PCI, and so on. Significant data exist on the development of insects of forensic interest as reviewed in the work by Amendt et al. (2007), but the data on insect development is typically not complete and data may be missing for a portion of the relevant temperature interval. A special problem for the statistical analysis is that different publications use different experimental designs and different means of measuring development and expressing uncertainty. Also different studies have been performed using different population of the insect species that may be adapted to different climate. Thus, to design generally applicable software, it will be necessary to process the available data into a standardized format. Achieving this will require some collaborative effort among forensic entomologists and we hope that this work will stimulate a discussion about the suitable model assumptions and data formats. Importantly, the issues about data format and model assumptions should be addressed together to ensure that models and data are compatible.

20.8 FUTURE PROSPECTS

BBNs can be used in many cases and data from different scientific fields can be weighed together to give a more precise estimate of the PMI. For example, we could use forensic taphonomy describing the decompositional changes of human remains as a complement to the results from the entomological investigations. Forensic palynology is another field that could lend itself to PMI estimations and could be combined with BBN approaches.

ACKNOWLEDGMENTS

This research was supported by/executed in the framework of the EU-project AniBioThreat (Grant Agreement: Home/2009/ISEC/AG/191) with the financial support from the Prevention of and Fight against Crime program of the European Union, European Commission—Directorate General Home Affairs. This publication reflects the views only of the author, and the European Commission cannot be held responsible for any use, which may be made of the information contained therein.

REFERENCES

Aitken, C., P. Roberts, and G. Jackson. 2010. *Practitioner Guide No 1. Fundamentals of Probability and Statistical Evidence in Criminal Proceedings. Guidance for Judges, Lawyers, Forensic Scientists and Expert Witnesses*. Royal Statistical Society's Working Group on Statistics and the Law. Royal Statistical Society, United Kingdom.

Amendt, J., C.P. Campobasso, E. Gaudry, C. Reiter, H. LeBlanc, and M. Hall. 2007. Best practice in forensic entomology—Standards and guidelines. *International Journal of Legal Medicine* 121: 90–104.

Ames, C. and B. Turner. 2003. Low temperature episodes in development of blowflies: Implications for postmortem interval estimation. *Medical and Veterinary Entomology* 17: 178–186.

Anderson, G.S. 2000. Minimum and maximum development rates of some forensically important Calliphoridae (Diptera). *Journal of Forensic Sciences* 45(4): 824–832.

Andersson, M.G., A. Sundström, and A. Lindström. 2013. Bayesian networks for evaluation of evidence from forensic entomology. *Biosecurity and Bioterrorism: Biodefense Strategy, Practice, and Science* 11(Suppl. 1): S64–S77.

Campobasso C.P., G. Di Vella, and F. Introna. 2001. Factors affecting decomposition and Diptera colonization. *Forensic Science International* 120: 18–27.

Damos, P. and M. Savopoulu-Soultani. 2012. Temperature-driven models for insect development and vital thermal requirements. *Psyche* 2012: 13, Article ID 123405.

Davies, L. and G. Ratcliffe. 1994. Development rates of some pre-adult stages in blowflies with reference to low temperatures. *Medical and Veterinary Entomology* 8: 245–254.

Fremdt, H., K. Szpila, J. Huijbregts, A. Lindström, R. Zehner, and J. Amendt. 2012. *Lucilia silvarum* Meigen, 1826 (Diptera: Calliphoridae)—A new species of interest for forensic entomology in Europe. *Forensic Science International* 222: 335–339.

Grassberger, M. and C. Reiter. 2001. Effect of temperature on *Lucilia sericata* (Diptera: Calliphoridae) development with special reference to the isomegalen- and isomorphen-diagram. *Forensic Science International* 120(1–2): 32–36.

Greenberg, B. 1991. Flies as forensic indicators. *Journal of Medical Entomology* 28: 565–577.

Greenberg, B. and J. Kunich. 2002. *Entomology and the Law: Flies as Forensic Indicators*. Cambridge University Press.

Jarman, K.H., H.W. Kreuzer-Martin, D.S. Wunschel, N.B. Valentine, J.B. Cliff, C.E. Petersen, H.A. Colburn, and K.L. Wahl. 2008. Bayesian-integrated microbial forensics. *Applied and Environmental Microbiology* 74(11): 3573–3582.

Keats, A., F.S. Lien, and E. Yee. 2006. Source determination in built-up environments through Bayesian inference with validation using the MUST array and Joint Urban 2003 Tracer experiments. *Proceedings of 14th Annual Conference of the Computational Fluid Dynamics Society of Canada*, July 16–18, Kingston, Canada.

Koblentz, G.D. and J.B. Tucker 2010. Tracing an attack: The promise and pitfalls of microbial forensics. *Survival* 52(1): 159–186.

Niederegger, S., J. Pastuschek, and G. Mall. 2010. Preliminary studies of the influence of fluctuating temperatures on the development of various forensically relevant flies. *Forensic Science International* 199: 72–78.

Nordgaard, A., R. Ansell, W. Drotz, and L. Jaeger. 2012. Scale of conclusions for the value of evidence. *Law, Probability & Risk* 11: 1–24.

Nuorteva, P. 1977. Sarcosaprophagous insects as forensic indicators. In *Forensic Medicine: A Study in Trauma and Environmental Hazards*, Vol. 2., eds. G.C. Tedeshi, W.G. Eckert, and L.G. Tedeschi, pp. 1072–1095. Philadelphia, PA: Saunders.

R Core Team. 2012. *R: A language and environment for statistical computing*. R Foundation for Statistical Computing, Vienna, Austria. ISBN 3-900051-07-0. http://www.R-project.org/.

Reiter, C. 1984. Zum Wachstumsverhalten der Maden der blauen Schmeißfliege *Calliphora vicina*. *Zeitung für Rechtsmedizin* 91: 295–308.

Sjerps, M.J. and C.H.E Berger. 2012. How clear is transparent? Reporting expert reasoning in legal cases. *Law, Probability & Risk* 11: 317–329.

Taroni, F., C. Aitken, P. Garbolino, and A. Biedermann. 2006. *Bayesian Networks and Probabilistic Inference in Forensic Science*. Chichester, United Kingdom: Wiley.

Tomberlin, J.K, M.E. Benbow, A.M. Tarone, and R. Mohr. 2011. Basic research in evolution and ecology enhances forensics. *Trends in Ecology and Evolution* 26: 53–55.

CHAPTER 21

Forensic and Decomposition Microbiology

Tawni L. Crippen and Baneshwar Singh

CONTENTS

21.1 INTRODUCTION

Prokaryotes are vital to Earth's biota. They are ubiquitous in terrestrial, atmospheric, and aquatic environments and even share space on and within other organisms. Estimates of the number of organisms existing in habitats on Earth are 12×10^{28} in aquatic, 355×10^{28} in oceanic subsurface, 26×10^{28} in terrestrial, and 250×10^{28} in terrestrial subsurface habitats, totaling a remarkable 6.4×10^{30} organisms (Whitman et al. 1998). They represent a large portion of Earth's genetic diversity; they swap genes and evolve at a rate not possible by organisms of the Animalia and Plantae Kingdoms. They metabolize matter and obtain energy by feeding on other organic substances, including decaying resources; many members manufacture their own organic compounds and some live in symbiotic relationships with other organisms. An entire community of bacteria, representing thousands of species, can exist in a mere gram of soil, most of which are yet uncharacterized by science. In the past, our ability to probe this microscopic universe has been limited, but technology now exists to investigate their vibrant community. This chapter and the next (Chapter 22) will explore the ecology of microbes involved in the decomposition of ephemeral resources and their interkingdom effects with insects along with the application of analytical microbiological techniques to forensic entomology.

21.2 CLASSICAL FORENSIC MICROBIOLOGY

21.2.1 Biosecurity Threats

21.2.1.1 Diagnosing Microbes of Bioterrorism

Classically, microbial forensics has dealt with characterizing physical microbiological evidence from bioterrorism crime scenes, after the use of pathogenic microorganisms or their toxins as weapons of targeted or mass destruction (Breeze et al. 2005; Budowle et al. 2005). In general, the Department of Defense, the Intelligence Community, the Department of Homeland Security, and the Federal Bureau of Investigation (FBI) are the primary authorities having jurisdiction over criminal activity involving biological agents. Forensic investigation of microbiological crimes involves standard operating procedure protocols of evidence collection, chain of custody handling and preservation, analysis and interpretation of results, and court presentation. In addition, they also have an epidemiological component to determine the etiology of the illness leading backward to the causative agent and eventually the source where dissemination of the agent originated.

Biological warfare is not a new tactical weapon. A multitude of substances have been used as bioweapons, such as plant toxins (ricin), bacteria (anthrax), and viruses causing bubonic plague, smallpox, and measles (Fenn 2000; FBI 2008, 2011). Toxins, such as ricin, anthrax, and botulinum neurotoxin, are cheap to produce, relatively easy to distribute, and difficult to protect against. Disease-causing microorganisms have been used from early in the history of humankind in an attempt to gain advantage during military engagements. A memoir by the Italian Gabriele de' Mussi claims that the Mongol army hurled corpses infected with black plague, caused by the Gram-negative bacterium *Yersinia pestis*, into the besieged Crimean city of Caffa (Feodosija, Ukraine) (Wheelis 2002). In a narrative by Gabriele De' Mussi, presumed to have been written circ. 1348, he stated "… they ordered corpses to be placed in catapults and lobbed into the city in the hope that the intolerable stench would kill everyone inside." From this it is clear that purposeful pathogen transmission may not have been envisioned, but the result of their actions was the bioterroristic dissemination of the plague to the inhabitants of the city (Wheelis 2002).

Although much of microbial forensics encompasses the diagnosis of particular bacteria, viruses, or fungi that might be the source of bioterrorism for military applications, it also includes inadvertent release, such as food-poisoning cases, and naturally occurring infections that may cause human, livestock, or crop pandemics and epidemics. Genomes of bacterial pathogens generally contain a few million base pairs up to the largest prokaryotic genome described thus far, a member of the class Ktedonobacteria. *Ktedonobacter racemifer* has a 13,661,586 bp genome and is an aerobic, filamentous, nonmotile, spore-forming Gram-positive heterotroph isolated from soil (Chang et al. 2011). Eighty-five to ninety percent of the prokaryotic genome is composed of nonrepetitive DNA, therefore the genome primarily consists of coding DNA, unlike a eukaryotic genome in which repetitive DNA is common (Koonin and Wolf 2010). Different bacterial genomic regions from both highly conserved regions, such as the small subunit ribosomal RNA genes (16S rRNA), to regions that evolve very rapidly, offer potential sites for analysis. The 16S rRNA gene encodes for an RNA molecule of approximately 1500 nucleotides. The gene is present in all bacteria and contains conserved regions that allow identification to the bacterial phylum by using a universal primer for polymerase chain reaction. The 16S rRNA gene also has variable regions that allow identification to the genus or species taxa level by comparison to a large public sequence database that covers most known bacterial species. There are limits to the database as not all existing bacteria are represented, therefore some comparisons only infer phylogenetic relations to bacterial sequences within the database and do not lead to an actual identification. However, a drawback of the structural consistency of the 16S rRNA region is that using this sequence site usually does not allow for distinguishing closely related pathogen strains, such as serotypes, thus making pinpointing source or identification of a recently emerged pathogen difficult. The use of other more rapidly evolving and thus more discriminating regions are required for a more precise identification. However, even by analyzing a variable region, a recently emerged bacterial strain can present a challenge as it generally has less genomic variation available to differentiate it from existing strains (Zhou et al. 2013).

21.2.1.2 Using Microbial Phylogeny for Determination of Source of Infection

Genomic and bioinformatics methodologies for use in microbial forensics are expanding rapidly. Recent advances in technologies have allowed greater resolution in microbial sequencing and this directly affects forensic efficacy. Amerithrax was the FBI code name for the bioterrorist case occurring in September and October 2001 just post 9/11, in which letters laced with quantities of *Bacillus anthracis* were mailed to U.S. Senators Patrick Leahy and Thomas Daschle in the District of Columbia, as well as to media organizations located in New York City, New York and Boca Raton, Florida. At least 22 victims contracted anthrax from the bacteria on these envelopes and another 31 people tested positive for exposure to anthrax spores. Eleven were diagnosed with inhalational anthrax and 11 suffered cutaneous exposure. Five of the inhalational victims, from postal workers to individuals at the recipient locations and hospital employees, died from their infections. Forensics use of DNA sequencing analyses determined that a particular strain of anthrax (RMR-1029) used in the attacks was identical to a strain used by the U.S. Army Medical Research Institute for Infectious Diseases (USAMRIID) at Fort Detrick, Maryland (Read et al. 2002). In August 2002, the Task Force assembled to investigate the crime suspected a researcher working at USAMRIID from 1997 to 1999 (DOJ 2010). The suspect had virtually unrestricted access to the bacterium and the required knowledge and skills necessary for successful dissemination. Ultimately however, genetic analysis of the organism used in the attacks excluded him conclusively as a suspect (DOJ 2010). Early on, they assumed that any individual with access to the USAMRIID biocontainment laboratories also had access to the strain of anthrax used in the attack. However, advancements in genetic analyses later determined that the suspect only had access to the parent anthrax and never had access to the spore-batch strain used in the mailings. This new information allowed the Task Force to refocus their investigation on researchers at the USAMRIID laboratory with access and the opportunity to process and mail this strain of bacteria. Evidence obtained from this and other investigative leads narrowed down the list of suspects to the individual who had actually committed the crime.

The remainder of this chapter discusses corpse/carrion decomposition microbiology, which examines the ecology of microbes involved in the decay of ephemeral resources. It discusses the use of current metagenomic technology, which is key to a more comprehensive identification of fluctuations in the microbial community structure and includes investigations into interkingdom effects associated with the decomposition process, such as bacterial–insect interactions. These new considerations about the influence of microbes during the process of decomposition may transform forensic entomology by improving the precision of analytical measurements, such as the estimation of minimum postmortem interval (PMI_{min}).

21.3 DECOMPOSITION MICROBIOLOGY

21.3.1 Basic Corpse/Carrion Decomposition Microbiology

21.3.1.1 Bacterial Metabolism and Decomposition

The distribution of indigenous bacteria on the human body is primarily within the oral cavity, upper respiratory tract, and the vaginal and gastrointestinal tract (Wilson 2005). The skin is dominated by Gram-positive species, but sparsely populated considering its large surface area. Microbes associated with the body number approximately 10^{14} cells (Walker et al. 2005; Wilson 2005). This is ten times the number of mammalian cells comprising the entire body structures. It is estimated that 3.2×10^{11} prokaryotes live in the human colon and approximately 10^3–10^4 cells/cm^2 on the skin, except in the groin and axilla, where it is 10^6 cells/cm^2, totaling about 3.3×10^8 prokaryotes per human adult (Whitman et al. 1998). The complexity of the microbial community is difficult to characterize as variation between individuals is extensive, dependent on factors such as age, clothing, diet, genotype, physiology, health status, gender, lifestyle, and the surrounding environment (Human Microbiome Project 2012; Maurice et al. 2013). In addition, the characterization of the associated diversity of microbial communities poses considerable technical difficulties. Although reliable culturing and biochemical techniques exist to identify some species of bacteria, these techniques result in the characterization of only a small fraction of the species occupying the human microbiome (Staley and Konopka 1985; Amann 2000). However, the development of metagenomic sequencing has enhanced our ability to characterize the vast number of microbial species, which have gone undefined by previous analytical techniques (MacDougall 2012; Morgan et al. 2013; Schloissnig et al. 2013).

Upon death, the decomposition of organic tissues begins with cellular autolysis by hydrolytic enzymes resulting in the release of carbohydrates, proteins, minerals, and fat from body structures. Autolysis of tissue is closely followed by putrefaction, which is the utilization of these products through enzymatic action and metabolism by bacteria present in and around the body at the time of death (Dent et al. 2004; Garg et al. 2005). Many microbes metabolize carbohydrates, as they are a readily available source of nutrition. The common families of bacteria that exploit carbohydrates are Clostridiaceae, Enterobacteriaceae, Staphylococcaceae, Streptococcaceae, Propionibacteriaceae, Lactobacillaceae, Bifidobacteriaceae, Pseudomonaceae, Neisseriaceae, and Corynebacteriaceae (Gottschalk 1986; Boumba et al. 2007; Paczkowski and Schutz 2011). Common proteolytic bacteria, which use proteins, are of the genera *Pseudomonas, Bacillus, Nitrosomonas, Nitrobacter*, and *Micrococcus* (Dent et al. 2004). Common lipolytic bacteria, using fats, are of the genera *Clostridia, Pseudomonas, Acinetobacter*, and *Bacillus* (Gottschalk 1986; Boumba et al. 2007). The time frame of the putrefaction process is dependent on many factors both internal, such as gastrointestinal and respiratory microbes, and external, such as scavengers, soil microbes, and environmental conditions. During the process, aerobic microorganisms first use the available oxygen within the tissues. The oxygen depleted tissues set a favorable environment for anaerobic microorganisms, such as *Clostridium* and *Bacteroides*, to continue the decomposition of this ephemeral resource (Dent et al. 2004; Carter et al. 2007). In addition, the metabolism of sulfur-containing amino acids, cysteine and methionine, can occur in several bacterial genera on carrion, such as lactic acid bacteria and *Staphylococcus* (Schultz and Dickschat 2007).

Organic acids, alcohols, carbon dioxide, water, and other gases, such as methane, cadaverine, putrescine, hydrogen sulfide, and ammonia, are the primary end products of this decompositional bacterial metabolism (Dent et al. 2004; Schultz and Dickschat 2007; Dekeirsschieter et al. 2009). However, depending on the bacteria present, a vast range of organic acids and other bacterial metabolites can result from these processes. The production of gases from anaerobic bacteria within the gastrointestinal tract adds to the inflation of the abdomen (bloat). The process continues through liquefaction and disintegration of soft tissues, which purges fluids and microbial biomass through orifices to the outside of the body. Gas pressure and necrophagous insect activity eventually ruptures the skin allowing exposure of internal surfaces to oxygen, which again supports expansion of aerobic bacterial activity. This is followed by drying and skeletonization, that leaves only bone, teeth, hair, nails, and cartilage. This drier environment does not support the same rich concentration of microbes, and the community abundance decreases. Organic collagen can be removed by the work of bacterial collagenases, such as those produced by *Clostridia* spp. (Shi et al. 2010). Keratin-rich hair and nails are broken down by keratinases, which are produced by several bacteria, many of which are from the genera *Streptomyces* and *Bacillus*, found in soils (Carter et al. 1988; Brandelli 2008).

21.3.2 Advanced Corpse/Carrion Decomposition Microbiology

21.3.2.1 *Bacterial Community Structure and Function during Decomposition*

Decomposition microbiology integrates longitudinal analyses of the microbial communities exploiting carrion and corpses and their interactions with the community structure, physiology, and behavior of other organisms, such as insects. Before a discussion of the specifics of corpse/carrion resources, the microbial diversity as a whole requires consideration. Microbes first drew the attention of scientists because they function as cyclic transformers of the major chemical elements required by living organisms on Earth and because they are responsible for infectious diseases in plants and animals. Prokaryotic cells created the global oxygenated environment providing the foundation of the biosphere and the organisms living with it. Prokaryotes engage in a versatile array of metabolic processes in aerobic, as well as anaerobic conditions and can survive in extreme environments (Pikuta et al. 2007). They also carry out biochemical reactions as endosymbionts that are essential to the survival of higher organisms, such as insects (Kikuchi 2009; Clark et al. 2010).

Carrion are nutrient rich resources for both microbes and insects. Microbes, once considered only nutrient recyclers are now known to participate in many ecological processes in nature (Lindeman 1942). During decomposition, insects interact directly with bacteria, and it is hypothesized that some bacteria may even manipulate such ecological interactions for their benefit (Janzen 1977). Unfortunately, the scientific data supporting this hypothesis were lacking because of the absence of technology to identify the multitude of species within a microbial community. However, with the advent of next-generation sequencing, specific microbial species or communities can now be uniquely linked to a geographical region, to a decomposing carcass, or to a specific insect species, gender, or life stage (Figures 21.1 and 21.2). This technology is advancing rapidly and presently allows for the confident, simultaneous identification of a community of bacteria at the genera level.

Janzen (1977) suspected microbes compete with other consumers of carrion by producing compounds affecting the appeal of the resource to reduce competition. For example, the burying beetle *Nicrophorus vespilloides* (Coleoptera: Silphidae) has behavioral counterstrategies to suppress microbial communities and reduce competition for small carrion resources on which to breed; suppression of these behaviors compromises larval development and adult reproductive success (Rozen et al. 2008). Some blow fly (Diptera: Calliphoridae) species (including *Lucilia sericata*, *Calliphora vicina*, and *Phormia regina*) reduce competition on carcasses with microbes by the production of antimicrobial substances to bacteria such as, *Staphylococcus*, *Streptococcus*, and *Pseudomonas* (van der Plas et al. 2007; Jones et al. 2008). This ability has led to the blow fly larvae being used in debridement therapy treatment for bacterial skin infections (Sherman et al. 2000; Mumcuoglu et al. 2001). Similarly, house fly, *Musca domestica*, L. (Diptera: Muscidae) larvae release enzymes that reduce the growth of competing fungi in targeted decomposing remains (Lam et al. 2007). Eventually scavengers evolved behavioral and physiological counter measures to compete for the decomposing remains and the competition continues (Lam et al. 2007; Haine et al. 2008; Rozen et al. 2008). But because microbes must also compete for this rich nutritional source, could they be

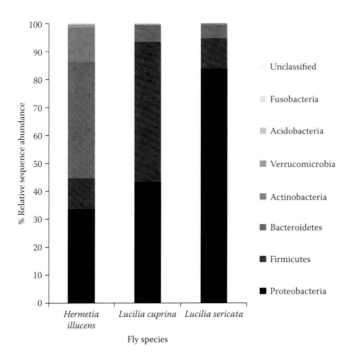

Figure 21.1　(See color insert.) Graph showing the diversity of bacterial phyla associated with three species of Diptera: *Hermetia illucens*, *Lucilia sericata*, and *Lucilia cuprina*, identified by 454-pyrosequencing. (Modified from Zheng, L. et al., *J. Med. Entomol.*, 50, 647–658, 2013b and Singh et al., *Appl. Microbiol. Biotechnol.*, 20: 1–15, 2014.)

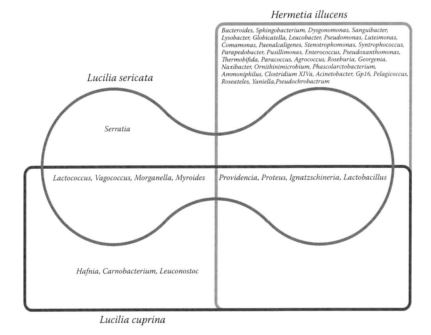

Figure 21.2　(See color insert.) Venn diagram showing differences and similarities in bacterial genera associated with combined successive life stages (egg, larvae pupae, adult) of the Diptera: *Hermetia illucens* (red rectangle), *Lucilia sericata* (pink Cassinian ovals), and *Lucilia cuprina* (blue rectangle), identified by 454-pyrosequencing. (Modified from Zheng, L. et al., *J. Med. Entomol.*, 50, 647–658, 2013b and Singh et al., *Appl. Microbiol. Biotechnol.*, 20: 1–15, 2014.)

manipulating the behavior of the insects to attract those that benefit their survival and repel those that impede their growth (Zheng et al. 2013a)?

21.3.2.2 Microbial Volatile Organic Compounds and Insects

Microbes produce microbial volatile organic compounds (MVOCs) during their normal metabolic processes. These result in the release of a diverse variety of volatiles produced from an extensive phylogenetic array of microbes using a decomposing resource. Arthropods are known to respond to these MVOCs (Davis et al. 2013). For example, MVOCs emitted by microbes on decomposing remains can be used by blow flies to locate ephemeral resources (Ashworth and Wall 1994; Wooldridge et al. 2007). The production of these metabolites also has a temporal aspect. Hide beetles, *Dermestes maculatus* (De Geer) (Coleoptera: Dermestidae), are attracted to MVOCs emitted from pig carcasses, but the attraction varies with the stage of decay. The beetles were most attracted to the carcasses during the postbloat stage corresponding with a significant production of benzyl butyrate that could be produced by the bacterium *Clostridium butyricum*. This bacterium performs anaerobic fermentation of glucose into acetic acid, hydrogen, carbon dioxide, and butyric acid in association with the decomposition of carrion. Burying beetles, *Nicrophorus vespillo* (Linnaeus) and *N. vespilloides* (Herbst) (Coleoptera: Silphidae), respond to sulfur-containing MVOCs commonly emitted by fresh carcasses (Kalinová et al. 2009). The metabolism of sulfur-containing amino acids can occur in several bacterial genera found inhabiting the carcasses, for example, *Brevibacterium*, *Corynebacterium*, *Micrococcus*, *Staphylococcus*, *Arthrobacter*, and lactic acid bacteria (Schultz and Dickschat 2007). So arthropods appear to recognize unique time-based MVOC profiles and use them to determine if a resource is viable for their needs (Chaudhury et al. 2010; Tomberlin et al. 2012; Zheng et al. 2013a). The identification of arthropods colonizing a corpse/carrion and characterization of their life stage is an important factor to estimate time of death, location, and other parameters for forensic investigations (Tomberlin et al. 2012; Benbow et al. 2013). If MVOCs produced by decomposition bacteria are being used by insects to evaluate resource quality, then this temporal aspect of microbial utilization needs to be considered when assessing insect species composition in forensic studies (Ma et al. 2012; Tomberlin et al. 2012). In addition, an insect locating and using a resource by olfactory perception is influenced by other variables including the nutritional status and age of the insect before encountering the stimulus (Tomberlin et al. 2012). Adding to this complexity, is the environment in which the carrion is found, which has a large influence on both the insects and microbial species participating in the decomposition process (Beasley et al. 2012; Benbow et al. 2013; Pechal et al. 2013). Thus, the succession of carrion ecology is a complex process, wherein decomposing remains potentially affect the physiology of associated microbial communities, whose resulting MVOC production is exploited by scent monitoring arthropods to evaluate the quality of the resource. It is therefore important to investigate bacterial metabolic activity along with species concentrations, because a species' importance to the process may not be directly related to its numerical dominance. The unraveling and documentation of this system of interkingdom signaling should greatly clarify arthropod succession on carrion during the pre- and postcolonization intervals and thus, the physiological timing of decomposition stages of vertebrate remains, consequently advancing the field of forensic entomology (Tomberlin et al. 2011).

21.3.2.3 Linking the Necrobiome to Carrion Decomposition Ecology

Pechal et al. (2014) have begun to characterize fluctuating epinecrotic communities during carrion decomposition using a metagenomic approach. The epinecrotic community is defined as those organisms residing on the surface of decomposing remains, such as the skin or the mucous membrane of a cavity, and is primarily represented by prokaryotes, protists, and fungi.

It represents part of the necrobiome, defined as the community of prokaryotic and eukaryotic species associated with decomposing heterotrophic biomass, including animal carrion and human corpses (Benbow et al. 2013). In their study, Pechal et al. (2014) reported that epinecrotic bacterial communities had distinct profile shifts within the first day of decomposition. The authors proposed a framework to conduct replicate studies in other biomes and during different seasons to add to a larger database in a range of environmental conditions, with the potential to use bacterial communities in estimating the PMI_{min} of carcasses, including those of humans. Bacterial community analyses may be a forensically important tool and could potentially identify key measurable community interactions during specific periods within the overall sequence of carrion decomposition ecology. Certainly, microbial community analysis will be a useful analytical measurement when insects have yet to colonize the remains.

21.3.3 Microbiome Diversity

21.3.3.1 Microbial Diversity between Humans

If arthropods participate in the uptake and dispersal of microbes from remains and may be used as a measurement tool of the PMI_{min} and other parameters important for forensic investigations, then it is important to understand the diversity and consistency of microbes among individuals and populations. Scientific development is moving toward a day when information gleaned by microbial DNA fingerprinting may be forensically relevant. Such technology could greatly assist forensic investigations, if microbial DNA could be retrieved from areas where someone physically touched an object and be used as a unique signature to identify that individual. If bacterial species/strains collected at crime scenes have geographically restricted distributions, then they can supply information about a suspect's place of origin or recent movements. The pool of missing persons could be narrowed significantly, if microbial DNA could supplement genetic markers to provide information about the ancestry of an unidentified individual and assist in the individual's identification.

21.3.3.2 Microbial Remnants Deposited on Physical Objects

Forensic scientists are now exploring the possibility of identification of individuals based on the residual bacteria left by an individual after touching a surface (Fierer et al. 2005). Humans harbor a large ecosystem of bacteria on their skin, living in distinct niches and exercising a range of physiological and metabolic activities (Grice et al. 2008; Segre 2006). This bacterial community is vital to the innate immune system and serves as an initial barrier against infectious agents and dehydration (Segre 2006; Gallo and Nakatsuji 2011). Despite this barrier, humans can acquire pathogenic agents and infectious disease. Studies have demonstrated that transfer of skin-associated pathogenic bacteria from an individual to an inanimate surface is not uncommon (Brooke et al. 2009; Jeon et al. 2013); but can these bacteria persist, can they be effectively collected from the skin, and is the bacterial community distinctive enough to identify individuals?

So, how long can bacteria persist on environmental surfaces? In studies of nosocomial pathogens, most Gram-positive and many Gram-negative bacteria, as well as fungal pathogens, for example, *Candida albicans*, could survive for months on dry surfaces (Kramer et al. 2006; Otter and French 2009). Spore-forming bacteria in particular are built to survive long term in the environment, whereas bacteria, such as *Bordetella pertussis*, *Vibrio cholerae*, *Haemophilus influenzae*, and *Proteus vulgaris* can survive only days (Kramer et al. 2006). From a forensic perspective, "days" may be long enough to assist in an investigation and in addition the lack of certain common species may give clues to determining time frames of demise.

Can bacteria be effectively collected from the skin of individuals? Several bacterial collection techniques have been devised. Comparison of bacterial 16S rRNA sequences of DNA extracted from the skin of individuals by swab, scraping, and punch biopsy, showed no significant differences in composition (Grice et al. 2008). All techniques resulted in collection of sufficient material to perform sequence analyses.

Is the bacterial community distinctive enough to identify individuals? The comparison of bacterial 16S rRNA sequences of DNA extracted from arm samples determined that diversity exists between topographical regions on the skin surface of individuals (Grice et al. 2008, 2009). Local skin anatomy, lipid and moisture content of sebum secretions, and pH correlate with distinct microbiota (Grice et al. 2009). Not surprisingly due to the difference in environment (anaerobic versus aerobic), the skin microbiome is different from the gut microbiome (Grice et al. 2008). Microbiome refers to the entire community of microbes, including bacteria, archaea, fungi, and viruses (Kong 2011). Analyzing the bacterial community structure of the palms of humans, Fierer et al. (2008) determined that although there appears to be a core set of bacterial phylotypes common to all individuals, there is a large amount of heterogeneity in community structure between the individuals. Out of 51 individuals, any two individuals only shared 13% of the bacterial phylotypes. Most of the noncore taxa on the palms were rare taxa that could be transient or colonizers present in low abundance. In addition, handedness (left or right) had a significant influence on the bacterial composition of opposite hands of the same individual, as did the gender of the individual, and the time since their last hand washing. Gao et al. (2007) determined that samples from the arms of individuals were more similar to each other than to other subjects. The individuals' skin bacteria differed among the six subjects sampled. In addition, temporal sampling (8–10 months) demonstrated a higher relatedness between multiple samples (forearm skin) taken from an individual at the same time point than samples obtained at different time points, showing that the bacterial biota present on an individual is dynamic over time. The studies indicated a stable core of a few predominant genera of bacteria, whose relative abundance on any individual might differ, cohabitating with a high frequency of more rare transient bacteria.

To add to the variability, bacteria can exist in two states: as planktonic, free-floating, single cells; or as sessile biofilm communities of multiple species encased in an extracellular matrix with a sophisticated communication system (Sutherland 2001). The skin is colonized by bacteria that exist in biofilms, and these biofilms can alter the metabolic activity of their residents, as well as create an environment that supports fastidious anaerobes (Vlassova et al. 2011). A study was done to link bacterial communities recovered from computer keyboards to the three individuals who touched the keys by comparison of high-throughput pyrosequencing analyses of the bacteria swabbed from the keys and from the individuals' fingertips (Fierer et al. 2010). The relatedness of the specific individual to the keys they touched was far more similar than to the keys and fingertips of other individuals. Although a small study, it shows promise as a possible forensic tool. However, in another small study comparing microbiota from the inner elbow of five subjects, Grice et al. (2008) did not find significant differences in the community structure, suggesting the possibility of a common core microbiome. The complexity of each individual's skin microbial ecosystem indicates that specific sampling strategies would have to be universally standardized for comparison and use of these samples in any forensic investigation. It must be determined if a core community precludes the use of the microbiome for individual identification or if there are regions on the body where the microbiome is individually unique (Li et al. 2013).

21.3.3.3 Distinctive Geo-Signatures from Microbiomes

The basis for microbial geo-signatures would be to determine the microbial DNA of specific marker species or community profiles that would show differential genetic diversity in the form of genotypes that clustered in correlation with geographic origin of the host. A study sampling the fingertip microflora of five Dutch European and three Bangladeshi adults was conducted in an attempt to prove the feasibility of such a technique (Tims et al. 2010). The study had limited success and concluded that the transient and resident microflora on human fingertips was too dynamic to be clear markers. However, a second study used gut microbial communities instead, and examined 531 individuals (children and adults) from Venezuela, Malawi, and the United States (Yatsunenko et al. 2012). In this study, differences in the gut microbial ecology from individuals corresponding to age and geography/cultural traditions were determined.

A large database of microbial constituents from a variety of individuals and body and geographic regions would be required for using either the skin or gut microbes for forensic identification (Aagaard et al. 2013). Although a core gut microbial community would be useful to determine geo-signatures, a unique community structure on the skin, or other body regions, is required for determination of individual signatures. Many confounding factors would have to be explored to determine the stability of the microbial fingerprint if it withstands perturbations allowing consistent identification. For example, what are the effects of antibiotics and external disinfection on the fingerprint and does the microbial fingerprint change over time with natural evolution of microbial species, with changes in lifestyle, or with relocation to other geographical areas?

21.3.3.4 Source Tracking

Insects have always coexisted with human populations, and the spread of many of the disease-causing microbes can be augmented by insects (Blazar et al. 2011). In these situations, forensic source tracking to determine the primary locations of bacterial contamination causing disease or pollution is important to public health (Stapleton et al. 2007; Foley et al. 2009). Bacterial DNA, both pathogenic and nonpathogenic, can be isolated from insects (Sant'Anna et al. 2012; Zheng et al. 2013b). Bacteria will adapt its genetic material to cohabitate with its specific host, which results in a unique genetic variation that can be used to identify the source. For example, the DNA from bacteria found in excrement, such as *Escherichia coli*, was used to track the source of environmental pollution of waterways (Stoeckel et al. 2004). Such methods require the development of databases of known isolates from a variety of species: human, domestic animals, and wildlife, to classify bacterial isolates collected from the environment (Scott et al. 2002; Simpson et al. 2002). If specific bacteria can be linked to a geospatial and temporal event, then the bacterial contents of flies or other insects using resources of forensic interest, may provide useful clues or evidence.

Using the DNA extracted from the gut of insects to identify their past location, diet, and interaction is a recent and still developing technique. It is estimated that the hindgut of a single termite contains 2.7×10^6 prokaryotes (Whitman et al. 1998). The technique relies on DNA barcoding that uses short sequences of DNA to differentiate and assign taxonomies to the DNA extracted from specimens. DNA barcode libraries have been used to reconstruct the herbivorous insect diet and interactions with plants (Jurado-Rivera et al. 2009; Garcia-Robledo et al. 2013) to measure biodiversity for ecological conservation monitoring (Ji et al. 2013) and to solve wildlife forensics crimes by correctly identifying unknown samples of arthropods (Rolo et al. 2013). Given that the extraction techniques to harvest microbial DNA from insect gastrointestinal contents is well developed and has been used extensively in other disciplines (Ohkuma et al. 1999; Brennan et al. 2004), the merging of DNA barcoding techniques with the characterization of microbial contents from arthropods using corpse/carrion to answer forensic geospatial and identity questions is likely forthcoming.

21.4 CONCLUSION

Exploration and understanding of the microscopic world of microbial ecology is in its infancy. It is clear that microbes participate with other organisms during decomposition on a more sophisticated level than first thought. As technology advances, it will become easier to understand the intricacies of microbial interactions with organisms from other kingdoms, such as arthropods.

Ultimately, understanding Earth's microbial community and the extent of its interdependency and exchange of information with arthropods will lead to the development of analytical tools to help forensic investigators answer questions about the progression of decomposition, the source of a disease-causing microbe, and possibly about the identity of a victim or the perpetrator of a crime.

REFERENCES

Aagaard, K., J. Petrosino, W. Keitel, M. Watson, J. Katancik, N. Garcia, S. Patel et al. 2013. The human microbiome project strategy for comprehensive sampling of the human microbiome and why it matters. *FASEB Journal* 27: 1012–1022.

Amann, R. 2000. Who is out there? Microbial aspects of diversity. *Systematic and Applied Microbiology* 23: 1–8.

Ashworth, J.R. and R. Wall. 1994. Response of the sheep blowflies *Lucilia sericata* and *L. cuprina* to odour and the development of semiochemical baits. *Medical Veterinary Entomology* 8: 303–309.

Beasley, J., T. Olson, and T. DeVault. 2012. Carrion cycling in food webs: Comparisons among terrestrial and marine ecosystems. *Oikos* 121: 1021–1026.

Benbow, M.E., A.J. Lewis, J.K. Tomberlin, and J.L. Pechal. 2013. Seasonal necrophagous insect community assembly during vertebrate carrion decomposition. *Journal of Medical Entomology* 50: 440–450.

Blazar, J., E. Lienau, and M. Allard. 2011. Insects as vectors of foodborne pathogenic bacteria. *Terrestrial Arthropod Reviews* 4: 5–19.

Boumba, V., K. Ziavrou, and T. Vougiouklakis. 2007. Biochemical pathways generating post-mortem volatile compounds co-detected during forensic ethanol analyses. *Forensic Science International* 174: 133–151.

Brandelli, A. 2008. Bacterial keratinases: Useful enzymes for bioprocessing agroindustrial wastes and beyond. *Food Bioprocessing Technology* 1: 105–116.

Breeze, R.G., B. Budowle, and S.E. Schutzer. 2005. *Microbial Forensics*, 425 pp. San Diego, CA: Elsevier Academic Press.

Brennan, Y., W.N. Callen, L. Christoffersen, P. Dupree, F. Goubet, S. Healey, M. Hernandez et al. 2004. Unusual microbial xylanases from insect guts. *Applied and Environmental Microbiology* 70: 3609–3617.

Brooke, J., J. Annand, A. Hammer, K. Dembkowski, and S. Shulman. 2009. Investigation of bacterial pathogens on 70 frequently used environmental surfaces in a large urban U.S. university. *Journal of Environmental Health* 71: 17–22.

Budowle, B., R. Murch, and R. Chakraborty. 2005. Microbial forensics: The next forensic challenge. *International Journal of Legal Medicine* 119: 317–320.

Carter, D., D. Yellowlees, and M. Tibbett. 2007. Cadaver decomposition in terrestrial ecosystems. *Naturwissenschaften* 94: 12–23.

Carter, T., D. Best, and K. Seal. 1988. Studies on the colonization and degradation of human hair by *Streptomyces fradiae*. In *Biodeterioration* 7, eds D. Houghton, R. Smith, H. Eggins, pp. 171–79. The Netherlands: Elsevier Science Publishers.

Chang, Y., M. Land, L. Hauser, O. Chertkov, D.R. Tg, M. Nolan, A. Copeland et al. 2011. Non-contiguous finished genome sequence and contextual data of the filamentous soil bacterium *Ktedonobacter racemifer* type strain (SOSP1-21). *Standards in Genomic Sciences* 5: 97–111.

Chaudhury, M., S. Skoda, A. Sagel, and J. Welch. 2010. Volatiles emitted from eight wound-isolated bacteria differentially attracted gravid screwworms (Diptera: Calliphoridae) to oviposit. *Journal of Medical Entomology* 47: 349–354.

Clark, E., A. Karley, and S. Hubbard. 2010. Insect endosymbionts: Manipulators of insect herbivore trophic interactions? *Protoplasma* 244: 25–51.

Davis, T.S., T.L. Crippen, R.W. Hofstetter, and J.K. Tomberlin. 2013. Microbial volatile emissions as insect semiochemicals. *Journal of Chemical Ecology* 39: 840–859.

Dekeirsschieter, J., F.J. Verheggen, M. Gohy, F. Hubrecht, L. Bourguignon, G. Lognay, and E. Haubruge. 2009. Cadaveric volatile organic compounds released by decaying pig carcasses (*Sus domesticus* L.) in different biotopes. *Forensic Science International* 189: 46–53.

Dent, B., S. Forbes, and B. Stuart. 2004. Review of human decomposition processes in soil. *Environmental Geology* 45: 576–585.

DOJ. 2010. Amerithrax Investigative Summary.

FBI. 2008. Cases example: Ricin case. In *Weapons of Mass Destruction*. Washington, DC: Federal Bureau of Investigation.

FBI. 2011. Amerithrax or Anthrax investigation. In *Famous Cases and Criminals*. Washington, DC: Federal Bureau of Investigation.

Fenn, E.A. 2000. Biological warfare in eighteenth-century North America: Beyond Jeffery Amherst. *The Journal of American History* 86: 1552–1580.

Fierer, N., M. Hamady, C.L. Lauber, and R. Knight. 2008. The influence of sex, handedness, and washing on the diversity of hand surface bacteria. *Proceedings of the National Academy of Sciences of the United States of America* 105: 17994–17999.

Fierer, N., J.A. Jackson, R. Vilgalys, and R.B. Jackson. 2005. Assessment of soil microbial community structure by use of taxon-specific quantitative PCR assays. *Applied and Environmental Microbiology* 71: 4117–4120.

Fierer, N., C.L. Lauber, N. Zhou, D. McDonald, E.K. Costello, and R. Knight. 2010. Forensic identification using skin bacterial communities. *Proceedings of the National Academy of Sciences of the United States of America* 107: 6477–6481.

Foley, S.L., A.M. Lynne, and R. Nayak. 2009. Molecular typing methodologies for microbial source tracking and epidemiological investigations of Gram-negative bacterial foodborne pathogens. *Infection Genetics and Evolution* 9: 430–440.

Gallo, R.L. and T. Nakatsuji. 2011. Microbial symbiosis with the innate immune defense system of the skin. *Journal of Investigative Dermatology* 131: 1974–1980.

Gao, Z., C.H. Tseng, Z. Pei, and M.J. Blaser. 2007. Molecular analysis of human forearm superficial skin bacterial biota. *Proceedings of the National Academy of Sciences of the United States of America* 104: 2927–2932.

Garcia-Robledo, C., D.L. Erickson, C.L. Staines, T.L. Erwin, and W.J. Kress. 2013. Tropical plant-herbivore networks: Reconstructing species interactions using DNA barcodes. *PLOS One* 8: e52967.

Garg, S., A. Arora, and B. Dubey. 2005. A study of serum enzymal changes after death and its correlation with time since death. *Journal of the Indian Academy of Forensic Medicine* 27: 16–18.

Gottschalk, G. 1986. Bacterial fermentations. In: *Bacterial Metabolism*, pp. 208–282. New York: Springer-Verlag.

Grice, E.A., H.H. Kong, S. Conlan, C.B. Deming, J. Davis, A.C. Young, G.G. Bouffard et al. 2009. Topographical and temporal diversity of the human skin microbiome. *Science* 324: 1190–1192.

Grice, E.A., H.H. Kong, G. Renaud, A.C. Young, G.G. Bouffard, R.W. Blakesley, T.G. Wolfsberg, M.L. Turner, and J.A. Segre. 2008. A diversity profile of the human skin microbiota. *Genome Research* 18: 1043–1050.

Haine, E.R., Y. Moret, M.T. Siva-Jothy, and J. Rolff. 2008. Antimicrobial defense and persistent infection in insects. *Science* 322: 1257–1259.

Human Microbiome Project. 2012. Structure, function and diversity of the healthy human microbiome. *Nature* 486: 207–214.

Janzen, D.H. 1977. Why fruits rot, seeds mold, and meat spoils. *American Naturalist* 111: 691–713.

Jeon, Y.S., J. Chun, and B.S. Kim. 2013. Identification of household bacterial community and analysis of species shared with human microbiome. *Current Microbiology* 67: 557–563.

Ji, Y., L. Ashton, S.M. Pedley, D.P. Edwards, Y. Tang, A. Nakamura, R. Kitching et al. 2013. Reliable, verifiable and efficient monitoring of biodiversity via metabarcoding. *Ecology Letters* 16: 1245–1257.

Jones, R.T., K.F. McCormick, and A.P. Martin. 2008. Bacterial communities of Bartonella-positive fleas: Diversity and community assembly patterns. *Applied and Environmental Microbiology* 74: 1667–1670.

Jurado-Rivera, J.A., A.P. Vogler, C.A. Reid, E. Petitpierre, and J. Gomez-Zurita. 2009. DNA barcoding insect-host plant associations. *Proceedings of the Royal Society B: Biological Sciences* 276: 639–648.

Kikuchi, Y. 2009. Endosymbiotic bacteria in insects: Their diversity and culturability. *Microbes and Environments* 24: 109–204.

Kong, H. 2011. Skin microbiome: Genomics-based insights into the diversity and role of skin microbes. *Trends in Molecular Medicine* 17: 302–328.

Koonin, E.V. and Y.I. Wolf. 2010. Constraints and plasticity in genome and molecular-phenome evolution. *Nature Reviews Genetics* 11: 487–498.

Kramer, A., I. Schwebke, and G. Kampf. 2006. How long do nosocomial pathogens persist on inanimate surfaces? A systematic review. *BMC Infectious Diseases* 6: 130.

Lam, K., D. Babor, B. Duthie, E. Babor, M. Moore, and G. Gries. 2007. Proliferating bacterial symbionts on house fly eggs affect oviposition behaviour of adult flies. *Animal Behaviour* 74: 81–92.

Li, K., M. Bihan, and B.A. Methe. 2013. Analyses of the stability and core taxonomic memberships of the human microbiome. *PLOS One* 8: e63139.

Lindeman, R.L. 1942. The trophic-dynamic aspect of ecology. *Ecology* 23: 399–418.

Ma, Q., A. Fonseca, W. Liu, A.T. Fields, M.L. Pimsler, A.F. Spindola, A.M. Tarone, T.L. Crippen, J.K. Tomberlin, and T.K. Wood. 2012. *Proteus mirabilis* interkingdom swarming signals attract blow flies. *ISME Journal* 6: 1356–1366.

MacDougall, R. 2012. NIH Human Microbiome Project defines normal bacterial makeup of the body. National Institutes of Health News: National Human Genome Research Institute, National Institutes of Health. http://www.nih.gov/news/health/jun2012/nhgri-13.htm (accessed May 08, 2014).

Maurice, C.F., H.J. Haiser, and P.J. Turnbaugh. 2013. Xenobiotics shape the physiology and gene expression of the active human gut microbiome. *Cell* 152: 39–50.

Morgan, X.C., N. Segata, and C. Huttenhower. 2013. Biodiversity and functional genomics in the human microbiome. *Trends in Genetics* 29: 51–58.

Mumcuoglu, K.Y., J. Miller, M. Mumcuoglu, M. Friger, and M. Tarshis. 2001. Destruction of bacteria in the digestive tract of the maggot of *Lucilia sericata* (Diptera: Calliphoridae). *Journal of Medical Entomology* 38: 161–166.

Ohkuma, M., T. Iida, and T. Kudo. 1999. Phylogenetic relationships of symbiotic spirochetes in the gut of diverse termites. *FEMS Microbiology Letters* 181: 123–129.

Otter, J.A. and G.L. French. 2009. Survival of nosocomial bacteria and spores on surfaces and inactivation by hydrogen peroxide vapor. *Journal of Clinical Microbiology* 47: 205–207.

Paczkowski, S. and S. Schutz. 2011. Post-mortem volatiles of vertebrate tissue. *Applied Microbiology and Biotechnology* 91: 917–935.

Pechal, J.L., T.L. Crippen, M.E. Benbow, A.M. Tarone, S. Dowd, and J.K. Tomberlin. 2014. The potential use of bacterial community succession in forensics as described by high throuput metagenomic sequencing. *International Journal of Legal Medicine* 128: 193–205.

Pikuta, E., R. Hoover, and J. Tang. 2007. Microbial extremophiles at the limits of life. *Critical Reviews in Microbiology* 33: 183–209.

Read, T.D., S.L. Salzberg, M. Pop, M. Shumway, L. Umayam, L. Jiang, E. Holtzapple et al. 2002. Comparative genome sequencing for discovery of novel polymorphisms in *Bacillus anthracis*. *Science* 296: 2028–2033.

Rolo, E.A., A.R. Oliveira, C.G. Dourado, A. Farinha, M.T. Rebelo, and D. Dias. 2013. Identification of sarcosaprophagous Diptera species through DNA barcoding in wildlife forensics. *Forensic Science International* 228: 160–164.

Rozen, D.E., D.J.P. Engelmoer, and P.T. Smiseth. 2008. Antimicrobial strategies in burying beetles breeding on carrion. *Proceedings of the National Academy of Sciences of the United States of America* 105: 17890–17895.

Sant'Anna, M.R., A.C. Darby, R.P. Brazil, J. Montoya-Lerma, V.M. Dillon, P.A. Bates, and R.J. Dillon. 2012. Investigation of the bacterial communities associated with females of Lutzomyia sand fly species from South America. *PLOS One* 7: e42531.

Schloissnig, S., M. Arumugam, S. Sunagawa, M. Mitreva, J. Tap, A. Zhu, A. Waller et al. 2013. Genomic variation landscape of the human gut microbiome. *Nature* 493: 45–50.

Schultz, S. and J. Dickschat. 2007. Bacterial volatiles: The smell of small organisms. *Natural Product Reports* 24: 814–842.

Scott, T.M., J.B. Rose, T.M. Jenkins, S.R. Farrah, and J. Lukasik. 2002. Microbial source tracking: Current methodology and future directions. *Applied and Environmental Microbiology* 68: 5796–5803.

Segre, J.A. 2006. Epidermal barrier formation and recovery in skin disorders. *Journal of Clinical Investigation* 116: 1150–1158.

Sherman, R.A., M.J. Hall, and S. Thomas. 2000. Medicinal maggots: An ancient remedy for some contemporary afflictions. *Annual Review of Entomology* 45: 55–81.

Shi, L., R. Ermis, A. Garcia, D. Telgenhoff, and D. Aust. 2010. Degradation of human collagen isoforms by Clostridium collagenase and the effects of degradation products on cell migration. *International Wound Journal* 7: 87–95.

Simpson, J.M., J.W. Santo Domingo, and D.J. Reasoner. 2002. Microbial source tracking: state of the science. *Environmental Science and Technology* 36: 5279–5288.

Singh, B., T.L. Crippen, L. Zheng, A.T. Fields, Z. Yu, Q. Ma, T.K. Wood, S.E. Dowd, M. Flores, J.K. Tomberlin, and A.M. Tarone. 2014. A metagenomic assessment of the bacteria associated with Lucilia sericata and Lucilia cuprina (Diptera: Calliphoridae). *Applied Microbiology and Biotechnology* 20: 1–15. 10.1007/s00253-014-6115-7.

Staley, J.T. and A. Konopka. 1985. Measurement of in situ activities of nonphotosynthetic microorganisms in aquatic and terrestrial habitats. *Annual Review of Microbiology* 39: 321–346.

Stapleton, C.M., M.D. Wyer, D. Kay, J. Crowther, A.T. Mcdonald, M. Walters, A. Gawler, and T. Hindle. 2007. Microbial source tracking: A forensic technique for microbial source identification? *Journal Environmental Monitoring* 9: 427–439.

Stoeckel, D., M. Mathes, K. Hyer, C. Hagedorn, H. Kator, J. Lukasik, T. O'brien, T. Fenger, M. Samadpour, K. Strickler, and B. Wiggins. 2004. Comparison of seven protocols to identify fecal contamination sources using *Escherichia coli*. *Environmental Science and Technology* 38: 6109–6117.

Sutherland, I.W. 2001. The biofilm matrix: An immobilized but dynamic microbial environment. *Trends in Microbiology* 9: 222–227.

Tims, S., W. van Wamel, H.P. Endtz, A. van Belkum, and M. Kayser. 2010. Microbial DNA fingerprinting of human fingerprints: Dynamic colonization of fingertip microflora challenges human host inferences for forensic purposes. *International Journal of Legal Medicine* 124: 477–481.

Tomberlin, J.K., T.L. Crippen, A.M. Tarone, B. Singh, K. Adams, Y.H. Rezenom, M.E. Benbow et al. 2012. Interkingdom responses of flies to bacteria mediated by fly physiology and bacterial quorum sensing. *Animal Behaviour* 84: 1449–1456.

Tomberlin, J.K., R. Mohr, M.E. Benbow, A.M. Tarone, and S. VanLaerhoven. 2011. A roadmap for bridging basic and applied research in forensic entomology. *Annual Review of Entomology* 56: 401–421.

van der Plas, M.J.A., A.M. Van Der Does, M. Baldry, H.C.M. Dogterom-Ballering, C. Van Gulpen, J.T. Van Dissel, P.H. Nibbering, and G.N. Jukema. 2007. Maggot excretions/secretions inhibit multiple neutrophil pro-inflammatory responses. *Microbes and Infection* 9: 507–514.

Vlassova, N., A. Han, J.M. Zenilman, G. James, and G.S. Lazarus. 2011. New horizons for cutaneous microbiology: The role of biofilms in dermatological disease. *British Journal of Dermatology* 165: 751–759.

Walker, B.G., P.D. Boersma, and J.C. Wingfield. 2005. Field endocrinology and conservation biology. *Integrative and Comparative Biology* 45: 12–18.

Wheelis, M. 2002. Biological warfare at the 1346 siege of Caffa. *Emerging Infectious Diseases* 8: 971–975.

Whitman, W.B., D.C. Coleman, and W.J. Wiebe. 1998. Prokaryotes: The unseen majority. *Proceedings of the National Academy of Sciences of the United States of America* 95: 6578–6583.

Wilson, M. 2005. *Microbial Inhabitants of Humans. Their Ecology and Role in Health and Disease*, 455 pp. Cambridge, United Kingdom: Cambridge University Press..

Wooldridge, J., L. Scrase, R. Wall. 2007. Flight activity of the blowflies, *Calliphora vomitoria* and *Lucilia sericata*, in the dark. *Forensic Science International* 172: 94–97.

Yatsunenko, T., F.E. Rey, M.J. Manary, I. Trehan, M.G. Dominguez-Bello, M. Contreras, M. Magris et al. 2012. Human gut microbiome viewed across age and geography. *Nature* 486: 222–227.

Zheng, L., T.L. Crippen, L. Holmes, B. Singh, M.L. Pimsler, M.E. Benbow, A.M. Tarone et al. 2013a. Bacteria mediate oviposition by the black soldier fly, *Hermetia illucens* (L.), (Diptera: Stratiomyidae). *Scientific Reports* 3: 2563.

Zheng, L., T.L. Crippen, B. Singh, A.M. Tarone, S. Dowd, Z. Yu, T.K. Wood, and J.K. Tomberlin. 2013b. A survey of bacterial diversity from successive life stages of black soldier fly (Diptera: Stratiomyidae) by using 16S rDNA pyrosequencing. *Journal of Medical Entomology* 50: 647–658.

Zhou, Z., A. Mccann, E. Litrup, R. Murphy, M. Cormican, S. Fanning, D. Brown, D.S. Guttman, S. Brisse, and M. Achtman. 2013. Neutral genomic microevolution of a recently emerged pathogen, *Salmonella enterica* serovar Agona. *PLoS Genetics* 9: e1003471.

CHAPTER 22

Methodologies in Forensic and Decomposition Microbiology

Baneshwar Singh and Tawni L. Crippen

CONTENTS

22.1 INTRODUCTION

Culturable microorganisms represent only 0.1%–1% of the total microbial diversity of the biosphere (Whitman et al. 1998). This has severely restricted the ability of scientists to study the microbial biodiversity associated with the decomposition of ephemeral resources in the past (Vass 2001). Innovations in technology are bringing in a new analytical depth to the study of microbial ecology. Currently, next-generation sequencing tools allow scientists to sequence and characterize a majority of microbes at the community level directly from the natural environment, thus bypassing the limitations of traditional microbiology and its culture-based techniques. The microbial utilization and interface with insects that occurs during the decay of vertebrate remains is an understudied area of decomposition ecology. Recent metagenomic studies have revealed enormous microbial diversity on carrion and associated insects, including a significant number of microbes not previously described (Pechal et al. 2014; Zheng et al. 2013). This chapter and Chapter 21 explore the ecology of microbes involved in the decomposition of ephemeral resources and their relationships with insects, along with the application of analytical microbiological techniques to forensic entomology.

22.2 SAMPLE PROCESSING

22.2.1 Sample Collection

Microbial sample collection is the first and most important aspect in any metagenomic analysis. A microbial sample should be collected aseptically and in a way that is sure to collect as many of the representative of the community as possible. Persons collecting samples are required to wear personal protective gears, such as booties, gloves, aprons, sleeves, safety glasses, surgical masks, and hair covers, to prevent contamination of the sample with extraneous microbes. All metadata (such as who, where, when, and under what conditions) related to a sample should be recorded. Sample collection procedures vary according to the environment or host. For example, skin microbes from healthy humans can be collected by swabbing, scraping, or punch biopsy, with each method yielding nearly identical microbial profiles (Grice et al. 2008). In a majority of forensic cases, it is better not to disturb evidence, and hence noninvasive methods such as swabbing with a sterile cotton swab or Catch-All™ sample collection swabs (Epicentre Biotechnologies, Madison, Wisconsin) on specific body parts are preferred. Microbes are primarily present in gastrointestinal tract, oral cavity, vagina, and skin in decreasing order (Wilson 2009); therefore, samples collected from these areas should provide optimal microbial information for carrion decomposition studies and forensic investigations. Because the amount of time spent sampling could affect the microbial species compositions, all samples should be standardized in terms of the time spent swabbing (e.g., 30 seconds or 1 minute). Furthermore, if possible, samples should be collected in duplicate or triplicate to ensure that sufficient DNA is collected and is therefore useful in the detection of polymerase chain reaction (PCR) and sequencing errors (see 22.6.2).

Apart from microbes associated with actual carrion, insects that feed on carrion and soil under the carrion also contribute significantly to the microbial decomposition of carcasses (Damann 2010; Zheng et al. 2013). Microbial samples from the different life stages of insects of forensic importance can be collected by using the method described by Zheng et al. (2013). Soil below the cadaver is generally rich in nutrients leached from cadavers and constitutes microbes from both the soil community and the cadaver (Parkinson et al. 2009; Carter et al. 2010). However, when designing experiments for the collection and analysis of soil microbes, replicated carrion should be placed at appropriate distances and inclines to avoid contamination resulting from leaching. Because of this, all confounding factors should be documented when collecting evidence at crime scenes.

Because diversity and density of soil microbes vary greatly with soil depth and quantity of soil sampled (Ranjard et al. 2003; Kakirde et al. 2010), it is important to sample appropriate quantity of soil from an appropriate soil layer, so that it is reproducible and representative. Although soil sample size has less effect on the characterization of the bacterial community, it can significantly affect archaeal and fungal community analyses (Nicol et al. 2003; Ranjard et al. 2003). Also, microbial cell density is generally higher in surface soil than in subsurface soil (Ranjard et al. 2003; Parkinson et al. 2009; Kakirde et al. 2010; Wallenius 2011). In general, 5–10 g of soil is collected from 0 to 5 cm of the soil surface layer using a sterile tool, such as a 4–6 cm tube/pipe having a diameter of 2–4 cm. In research scenarios, soil samples should be collected below the cadaver/carrion by carefully tilting the body and using an implement (e.g., sterile tube, shovel, spatula, and soil corer) to recover the desired quantity of soil (Parkinson et al. 2009). To prevent contamination, it is important to use a separate collection device for each sample or to thoroughly treat each device with a sterilizing agent, such as alcohol, bleach, or Lysol™, between samples (Kakirde et al. 2010). In repeated sampling designs (e.g., sampling at regular time intervals), it is important to avoid resampling from previously sampled sites to prevent collection of microbes from different soil layers in different samples (Parkinson et al. 2009). If the cost of sequencing is a concern and natural spatial variation of soil microbes within each replication is not important, then a composite sample (two to three individual samples pooled together) is preferred over individual samples to obtain a mean microbial diversity from the site (Kakirde et al. 2010; Wallenius 2011).

22.2.2 Sample Storage

There is no universally accepted or validated method to process or store samples at this time, and conflicting studies of various procedures are reported on the efficacy. The preferred method is to process samples within 2 hours of collection (Rochelle et al. 1994); however, in cases where this is impractical the most prevalent storage method used includes some level of refrigeration. In community-level physiological profiling, when immediate sample processing is not possible, storage at 4°C for up to 10 days is recommended (Insam and Goberna 2004). For RNA-based studies, such as in metatranscriptomic analyses, samples can be stored in RNA*later*™ (Ambion, Texas) solution for about a week at room temperature, or at −80°C for long term. For DNA-based studies, such as in metagenomic analyses, samples can be stored at −20°C for up to 2 weeks (Lauber et al. 2010) or at −80°C for up to 6 months (Kakirde et al. 2010; Carroll et al. 2012) without a significant change in the bacterial community structure. Although the majority of published work recommends freezing (Lauber et al. 2010; Wu et al. 2010; Carroll et al. 2012), there are some studies that suggest changes in microbial community structure occur as a result of freezing (Roesch et al. 2009; Cardona et al. 2012). Kwambana et al. (2011) observed significant effects of freezing on samples collected from human females but little or no effects on those collected from males. Some species of microbes may to be more sensitive to damage at freezing temperatures. Once a sample is frozen, it is important to keep it frozen until DNA extraction can be performed. Repeat freeze-thaw cycles can cause DNA fragmentation (Cardona et al. 2012) and can significantly affect the characterization of microbial diversity analysis especially at genus and species levels (Männistö et al. 2009; Cardona et al. 2012).

22.3 DNA EXTRACTION

Selection of the appropriate DNA extraction method that yields high-quality DNA from a majority of the microbes within a sample is vital for all downstream metagenomic analyses (Inceoglu et al. 2010). Extraction from environmental samples is especially challenging because each environment has unique physiological, biological, and chemical properties (Kakirde et al. 2010). There are two main types of bacteria based on the characteristics of their cell walls, gram-positive and gram-negative bacteria. Both cell wall types contain a peptidoglycan layer, a multilayered, rigid structure of glycan chains cross-linked with flexible peptide bridges. This structure maintains the cell shape but also has elasticity and flexibility to counter intracellular turgor pressure and prevent osmotic lysis. However, the quantity, thickness, length distribution, and degree of cross-linking are more extensive in gram-positive cells (Vollmer and Holtje 2004; Cabeen and Jacobs-Wagner 2005). Thus, chosen lysis methods should not favor the gram-negative bacteria (easier to lyse) (Mahalanabis et al. 2009).

DNA can be extracted from environmental samples by either direct (Ogram et al. 1987) or indirect (cell extraction) methods (Holben et al. 1988; Jacobsen and Rasmussen 1992). During indirect extractions prokaryotic cells are initially separated from the sample using a density gradient followed by lysis and purification, whereas direct extraction methods lyse the entire sample. Although these methods differ and have several advantages and disadvantages, they provide almost similar DNA diversities (Delmont et al. 2011). The indirect method only isolates prokaryotic cells (e.g., bacteria and archaea); hence, it is not suitable if target groups include eukaryotic organisms (e.g., Fungi) (Delmont et al. 2011). Among these two methods, the direct method has been used more frequently because it is fast, requires less sample input (<1 g of soil), and provides higher DNA yields (up to 100 times). The indirect method is laborious, requires greater sample input (>60 g of soil), and results in smaller quantities of DNA (Roh et al. 2006; Delmont et al. 2011). The major disadvantages of direct methods are the possible co-extraction of many PCR-inhibitory substances (e.g., humic acid, polyphenols, and polysaccharides), particularly from soil samples, and the substantial shearing of the extracted DNA (approximately 12-kb fragments) (Miller et al. 1999; Roh et al. 2006; Kakirde et al. 2010). For next-generation sequencing of 16S rDNA, sheared DNA is not a major concern because target fragment size is only a few hundred base pairs (bp). But if the goal is to construct large-insert clone libraries (e.g., fosmids and cosmids clones), then indirect methods are preferred (Kakirde et al. 2010).

To extract DNA from environmental samples, the first step is to break down the microbial cell walls and then isolate and purify the nucleic acids from the mixture containing cell debris, humic acids, polyphenols, polysaccharides, and other contaminants (Kakirde et al. 2010). Microbial cells can be lysed by physical, chemical, and/or enzymatic methods, each of which can be utilized separately or in combination (Kuczynski et al. 2012). Based on several studies, DNA yields are generally higher from soil and swab samples utilizing combined cell lysis methods (Kuczynski et al. 2012). Physical disruption techniques such as freeze-thawing (Erb and Wagner-Dobler 1993), bead-mill homogenization (Ogram et al. 1987), sonication (Picard et al. 1992; Frostegard et al. 1999), and grinding (Volossiouk et al. 1995; Frostegard et al. 1999) disrupt soil structure and expose the entire bacterial community, even those which may be hidden in soil microaggregates, for chemical and enzymatic breakdown (Robe et al. 2003). Among the aforementioned physical disruption techniques, freeze-thawing and bead-mill homogenization methods are the most common (Miller et al. 1999).

Although chemical and enzymatic reagents vary greatly in published literature, generally they consist of a buffer (Tris or phosphate) at pH = 7.0 to 8.0, a chelating agent (Chelex 100 or ethylenediaminetetraacetic acid), a detergent (sodium dodecyl sulfate or Sarkosyl), sodium chloride (NaCl), and enzymes (e.g., RNase A, proteinase K, and lysozyme) (Ausubel et al. 1988; Miller et al. 1999). A buffer maintains the pH of the solution, whereas chelating agents bind to divalent metal ions (Mg^{++}, Ca^{++}: cofactors of DNase), and hence protect DNA from DNase activity. Detergents dissolve hydrophobic constituents of the cell membrane and augment the breakdown of the cell. Lysozyme hydrolyzes glycosidic bonds, proteinase K digests cellular proteins, and RNaseA digests RNA. NaCl provides Na^+ ions that neutralize the negative charges of the phosphates in DNA, which in turn decrease electrophobic interactions, and helps stabilize the DNA.

In addition to chemical and enzymatic reagents, some lysis solutions, particularly those used for plant cells, may also contain cetyltrimethyl-ammonium bromide (CTAB) or polyvinylpolypyrrolidone (PVPP), which helps in the partial removal of PCR inhibitors, such as humic compounds, for downstream procedures (Frostegard et al. 1999; Robe et al. 2003). CTAB is preferred because PVPP can cause DNA loss (Zhou et al. 1996). CTAB forms an insoluble complex with residual proteins and polysaccharides and helps in the removal of these molecules during purifications steps (Ausubel et al. 1988). But, to avoid DNA loss it is important to maintain a 0.5 M or greater concentration of NaCl. Homogenized samples and lysis solutions are generally incubated from 2 to 24 hours at 65°C; 10–12 hours of incubation is optimal for efficient enzymatic activity.

Following lysis, the next step is to isolate the DNA from cellular debris. DNA is generally purified by sequential extraction in an equal volume of phenol (pH: 7.8) followed by a phenol:chloroform:isoamyl alcohol mixture, and finally in a chloroform:isoamyl alcohol mixture. Phenol efficiently denatures protein; chloroform also denatures proteins and, in addition, stabilizes the boundary between the resulting aqueous and organic phases. The phenol:chloroform mixture reduces the amount of aqueous phase retained in organic phase (more efficiently than just pure phenol) and hence increases DNA yield. The isoamyl alcohol reduces foaming due to mixing of the solution and thus improves the definition between the aqueous and organic phases (Ausubel et al. 1988). The resulting extracted DNA can be precipitated in either ethanol or isopropyl alcohol (Ausubel et al. 1988).

The quality and quantity of the DNA produced by any extraction method have to be appropriate for the eventual downstream analyses. For example, although CTAB and PVPP help to remove majority of PCR inhibitors during the extraction phase, the purging is not complete and residual inhibitors may result in interference with the production of PCR amplicons for sequencing (Braid et al. 2003). Several additional methods, such as hydroxyapatite columns, cesium chloride density centrifugations, chromatographic separation, chemical flocculation, and extraction of excised DNA bands from agarose gel electrophoresis methods, have been proposed to facilitate the elimination of inhibitors. However, each method has its disadvantages in terms of time and DNA recovery (Braid et al. 2003). Several commercial soil DNA extraction and purification kits (e.g., MoBio PowerSoil® DNA isolation kit, MP FastDNA® SPIN Kit, and Epicentre SoilMaster™ DNA Extraction) are available that yield high-quality DNA, but low quantity of DNA, from environmental samples. However, there are instances where the combination of an organic extraction method and a commercial genomic DNA cleanup kit results in higher quantity and quality of DNA from environmental samples (Pechal et al. 2014).

22.4 POLYMERASE CHAIN REACTION AMPLIFICATION

PCR has revolutionized microbial ecology because of its ability to amplify a target gene from a mixture of environmental nucleic acids. Extracted community DNA can be used either in the amplification of marker genes (e.g., *16S rRNA/18S rRNA/23S rRNA* genes) and/or for whole genome sequencing (WGS/metagenomics) (Tringe and Rubin 2005). The marker gene approach provides better taxonomic coverage and hence aids in elucidating the microbial community structure, whereas the WGS approach can infer the microbial community function (Kuczynski et al. 2012). The *16S rRNA* gene is considered a gold standard for bacterial and archaeal community diversity studies, whereas the *18S rRNA* gene and internal transcribed spacer (ITS) are the most commonly used markers for fungal community diversity studies. The *16S rRNA* gene (1541 bp long in *Escherichia coli* [Brosius et al. 1978]) is present in all prokaryotic organisms and generally in multiple copies. It can be amplified from partially degraded samples and includes conserved regions, good for universal primer design, and nine variable regions, good for discrimination of closely related taxa (Figure 22.1a) (Baker et al. 2003; Schloss et al. 2011; Kuczynski et al. 2012; Vos et al. 2012; Vetrovsky and Baldrian 2013). In addition, several reference databases, such as Greengenes (DeSantis et al. 2006), SILVA (Bauer et al. 2006), and the Ribosomal Database Project (RDP) (Cole et al. 2009), are available for *16S rRNA*–based taxonomical identification of bacteria and archaea.

The *18S rRNA* gene (1798 bp long in *Saccharomyces cerevisiae* [Mankin et al. 1986]) is similar to 16S in that it is present in multiple copies, can be amplified from partially degraded samples (Figure 22.1b) (Mankin et al. 1986), and contains conserved and variable regions. However, variable regions of the *18S rRNA* gene are comparatively less variable than the ITS region, advantageous for alignment and phylogenetic analyses purposes. Unfortunately, analysis of *18S rRNA* is generally inadequate for discernment of lower taxonomic levels (genus/species) compared to the ITS region (Rousk et al. 2010). Therefore, most prefer the ITS region when the aim is to identify fungal genus/species from an environmental sample (Buee et al. 2009; LaTuga et al. 2011; Bokulich and Mills 2013; Schmidt et al. 2013). Multiple copies of marker genes, common in a majority of microbial genomes, can also create problems. Multiple gene copies can result in an overestimation of microbial diversity because relative species abundance may not always equal gene sequence abundance (Pei et al. 2010). *16S rRNA* gene copies (1–15 copies per bacterial genome) differ between different bacterial species and can even differ within a bacterial species

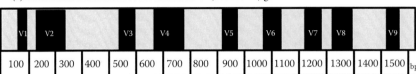

(a) Bacterial and archaeal 16S ribosomal RNA (16S rRNA) gene

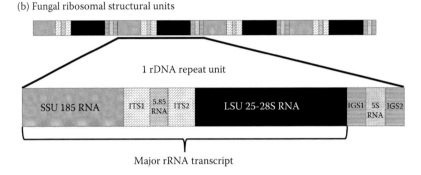

(b) Fungal ribosomal structural units

Figure 22.1   Ribosomal RNA gene structure showing different conserved and variable regions of (a) bacteria and archaea and (b) fungi. Note: the position and length of these regions are not to exact scale.

(Pei et al. 2010; Vos et al. 2012; Vetrovsky and Baldrian 2013). One way to avoid this problem is to use multiple single copy genes for the quantification of relative species abundance, as implemented in MetaPhlAn (Segata et al. 2012).

Several factors should be considered for unbiased microbial community analyses. Most importantly, care should be taken to avoid contamination of PCR reagents from extraneous DNA. Thus, it is very important to follow standard laboratory practices, such as autoclaving of solutions, tips, and tubes; use of DNase/RNase/pyrogen-free plasticware; separation of workspaces and equipment for pre- and post-PCR work; and use of aerosol-resistant pipette tips. In addition, a separate PCR preparatory area that limits air disturbance should be employed, along with frequent changes of gloves and aliquoting reagents, quick centrifuging of tubes to avoid contamination due to liquid dispersing during opening of the tube cap, and the inclusion of both positive and negative control with each PCR run.

Both the quality and quantity of DNA plays an important role during the PCR amplification process. In general, 0.1–10 ng of DNA template yields sufficient PCR product. DNA extracted from environmental samples (especially soil samples) using direct methods are sheared, and the sample often contains PCR inhibitors, such as humic acids, proteins, polyphenols, and polysaccharides. Sheared, short DNA fragments or an incomplete PCR product can serve as nonspecific primers during PCR amplification, resulting in chimera formation (Devereux and Wilkinson 2004; Haas et al. 2011; Schloss et al. 2011). The rate of chimera formation ranges from 5% to 45%, and it is of major concern in microbial diversity studies (Haas et al. 2011; Schloss et al. 2011). Although chimera formation during PCR cannot be controlled, several software packages are available to help identify and remove chimeric sequences (see Section 22.6.2). Many PCR inhibitors can be removed using commercially available DNA cleanup kits, but with some types of samples it is difficult to remove all the inhibitors. There are several PCR enhancers (bovine serum albumin [BSA], dimethyl-sulfoxide [DMSO], formamide, and glycerol) that can be added to the PCR reaction to improve amplification of these problematic samples; however, these are time-consuming steps (Devereux and Wilkinson 2004). BSA (0.01% to 0.1% [w/v]) binds to inhibitors and hence helps amplification by making

DNA polymerase available (Devereux and Wilkinson 2004). DMSO (1%–10% [v/v]), formamide (1.25%–10% [v/v]), and glycerol (5%–20% [v/v]) alter primer-annealing temperatures for enhancement of specificity of primer-template hybridization (Devereux and Wilkinson 2004).

The selection of primers is the most important factor to consider during the amplification of environmental samples. Normally, 5–20 picomoles of each primer are used in a 25-μL PCR reaction. A good primer should have 40%–60% GC content (or guanine–cytosine content), a noncomplementary 3' end, a length of 18–24 bases, no internal secondary structure, and an annealing temperature between 55°C and 65°C (Devereux and Wilkinson 2004). Several universal primers are available for the amplification of different target regions of bacteria, archaea, and fungi (Table 22.1). However, some perform poorly in the amplification of difficult environmental samples (e.g., soil, biofilm associated with stones in running waters).

Most primers were designed based on existing sequence databases, such as RDP (Cole et al. 2009) and Greengene databases (DeSantis et al. 2006), which are incomplete, so even those that perform well with difficult samples have problems because they may not amplify all microorganisms from an environment (Liu et al. 2007). The efficacy of selected universal primer pairs at enriching for all microbial groups may vary with the type of environmental sample; hence, knowledge of the sample target community is very important during selection of universal primers (Liu et al. 2008). For example, for skin samples the widely used universal primer pair F27-R338 performs better than another widely used primer pair F515-R806, which performs poorly for amplification of a common skin bacteria *Propionibacterium*. Alternatively, the primer pair F27-R338 performs poorly for a common soil bacteria such as *Verrucomicrobia* (Bergmann et al. 2011; Kuczynski et al. 2012). Multiple sequence alignment of a reference database that constitutes target communities can be used as an input for designing a barcoded PCR primer (using the software package PrimerProspector) (Walters et al. 2011). However, it is always recommended to evaluate newly designed primers for their intended specificity before utilizing them in an actual study.

PCR reagents (e.g., PCR buffer, dNTP's deoxyribonucleotide triphosphates, and Taq polymerase enzyme) and PCR conditions (e.g., annealing temperature and extension time) require optimization of conditions for ideal amplification. In a standard PCR reaction, 1X PCR buffer with 1.5 mM $MgCl_2$, 0.2 mM of each dNTP's, and 0.5–2.5 units of Taq polymerase enzyme are used. The concentration of $MgCl_2$ is very important during amplification. $MgCl_2$ supplies Mg^{++}, which is a cofactor for DNA polymerase enzyme, and plays a crucial role during primer annealing and extension steps. $MgCl_2$ concentration ranges from 0.5 to 4.0 mM and is always greater than the dNTP's concentration in a PCR reaction. A final concentration of 1.5 mM of $MgCl_2$ works very well for most PCR reactions, but a too high $MgCl_2$ concentration may cause unspecific amplification, whereas a too low $MgCl_2$ concentration may cause weak or no amplification. Among all PCR conditions, annealing temperature is very crucial and depends on the primer.

22.5 DNA SEQUENCING

An amplified PCR product can be sequenced by using several existing sequencing platforms (Table 22.2). There are several factors to consider before making a selection of a particular sequencing platform for microbial diversity studies, such as the goal of the sequencing, sequencing cost, sequencing time, read length, coverage, and the error rate. Generally, the platform that produces more accurate sequence reads (minimum error rate), and yields relatively longer reads, is preferred for amplicon sequencing–based microbial diversity studies. Obtaining sequence reads with maximum accuracy is very important in amplicon sequencing, because in microbial diversity studies each unique sequence is considered as a novel microorganism (no sequence assembly is performed for obtaining a consensus sequence), and hence sequencing error can lead to increased microbial diversity and a higher percentage of unclassified sequences (Kunin et al. 2010; Schloss et al. 2011).

Table 22.1   List of Important Universal Primers for Amplification of Bacterial, Archaeal, and Fungal Ribosomal Genes

Target Organism	Marker	Primer Name	Primer Sequence (5′-3′)	References
Bacteria[a]	*16S rRNA (V1-V3)*	27F	A GAG TTT GAT CMT GGC TCA G	Youssef et al. (2009)
		28F	GAG TTT GAT CNT GGC TCA G	Handl et al. (2011)
		334F	CCA GAC TCC TAC GGG AGG CAG C	Rudi et al. (1997), Baker et al. (2003)
		338F	ACT CCT ACG GGA GGC AGC AG	Huse et al. (2008)
		341F	CCT ACG GGA GGC AGC AG	Muyzer et al. (1995)
		338R	TGC TGC CTC CCG TAG GAG T	Hunt et al. (2011)
		355R	ACT CCT ACG GGA GGC AGC	Youssef et al. (2009)
		519R	GTN TTA CNG CGG CKG CTG	Handl et al. (2011)
		533R	TTA CCG CGG CTG CTG GCA C	Huse et al. (2008)
	16S rRNA (V4-V6)	515F	GTG CCA GCM GCC GCG GTA A	Caporaso et al. (2012)
		519F	GTG CCA GCT GCC GCG GTA ATA C	Kumar et al. (2011)
		530F	ACG CTT GCA CCC TCC GTA TT	Youssef et al. (2009)
		805F	GAC TAC CAG GGT ATC TAA TCC	Youssef et al. (2009)
		917F	GAA TTG ACG GGG RCC C	Liu et al. (2007)
		967F	CAA CGC GAA GAA CCT TACC	Youssef et al. (2009)
		806R	GGA CTA CHV GGG TWT CTA AT	Caporaso et al. (2012)
		907R	CCG TCA ATT CCT TTR AGT TT	Muyzer et al. (1995)
		939R	CTT GTG CGG GCC CCC GTC AAT TC	Rudi et al. (1997), Baker et al. (2003)
		1046R	AGG TGN TGC ATG GRT GTC G	Huse et al. (2008)
		1065R	AGG TGC TGC ATG GCT GT	Youssef et al. (2009)
		1100R	AGG GTT GCG CTC GTT G	Turner et al. (1999)
		1114R	GGG TTG CGC TCG TTG C	Kumar et al. (2011)
	16S rRNA (V7-V9)	1046F	ACA GCC ATG CAG CAC CT	Youssef et al. (2009)
		1099F	GYA ACG AGC GCA ACC C	Liu et al. (2007)
		1238R	GTA GCR CGT GTG TMG CCC	Youssef et al. (2009)
		1391R	GAC GGG CGG TGT GTR CA	Turner et al. (1999)
		1406R	GAC GGG CGG TGW GTR CA	Youssef et al. (2009)
		1492R	GGT TAC CTT GTT ACG ACT T	Turner et al. (1999)
Archaea[a]	*16S rRNA gene*	A1F	TCY GKT TGA TCC YGS CRG AG	Embley et al. (1992)
		A21F	TTC CGG TTG ATC CYG CCG GA	DeLong (1992)
		A109F	ACK GCT CAG TAA CAC GT	Whitehead and Cotta (1999)
		A340F	CCC TAY GGG GYG CAS CAG	Gantner et al. (2011)
		A571F	GCY TAA AGS RIC CGT AGC	Baker et al. (2003)
		A751F	CCG ACG GTG AGR GRY GAA	Baker et al. (2003)

Table 22.1 **List of Important Universal Primers for Amplification of Bacterial, Archaeal, and Fungal Ribosomal Genes (*Continued*)**

Target Organism	Marker	Primer Name	Primer Sequence (5′-3′)	References
		A1098F	GGC AAC GAG CGM GAC CC	Reysenbach and Pace (1995) (as cited in Baker et al. [2003])
		A348R	CCC CGT AGG GCC YGG	Barns et al. (1994)
		U529R	ACC GCG GCK GCT GGC	DasSarma and Fleischmann (1995) (as cited in Baker et al. [2003])
		A958R	YCC GGC GTT GAM TCC AAT T	Delong (1992)
		A1000R	GGC CAT GCA CYW CYT CTC	Gantner et al. (2011)
		A1100R	TGG GTC TCG CTC GTT G	Embley et al. (1992)
		UA1204R	TTM GGG GCA TRC IKA CCT	Baker et al. (2003)
		UA1406R	ACG GGC GGT GWG TRC AA	Baker et al. (2003)
Fungi[b]	18S rRNA gene	F-SSU-20F (NS1)	GTA GTC ATA TGC TTG TCT C	White et al. (1990)
		F-SSU-573R (NS2)	GGC TGC TGG CAC CAG ACT TGC	White et al. (1990)
		F-SSU-553F (NS3)	GCA AGT CTG GTG CCA GCA GCC	White et al. (1990)
		F-SSU-1150R (NS4)	CTT CCG TCA ATT CCT TTA AG	White et al. (1990)
		F-SSU-1129F (NS5)	AAC TTA AAG GAA TTG ACG GAA G	White et al. (1990)
		F-SSU-1436R (NS6)	GCA TCA CAG ACC TGT TAT TGC CTC	White et al. (1990)
		F-SSU-1413F (NS7)	GAG GCA ATA ACA GGT CTG TGA TGC	White et al. (1990)
		F-SSU-1788R (NS8)	TCC GCA GGT TCA CCT ACG GA	White et al. (1990)
		F-SSU-0817F	TTA GCA TGG AAT AAT RRA ATA GGA	Borneman and Hartin (2000), Rousk et al. (2010)
		F-SSU-1196R	TCT GGA CCT GGT GAG TTT CC	Borneman and Hartin (2000), Rousk et al. (2010)
		F-SSU-1536R	ATT GCA ATG CYC TAT CCC CA	Borneman and Hartin (2000)
		F-SSU-75F (NSSU97)	TAT ACG GTG AAA CTG CGA ATG GC	Kauff and Lutzoni (2002)
		F-SSU-107F (NSSU131)	CAG TTA TCG TTT ATT TGA TAG TAC C	Kauff and Lutzoni (2002)
		F-SSU-705R (NSSU634)	CCC CAG AAG GAA AGI CCC GIC C	Kauff and Lutzoni (2002)
		F-SSU-897F (NSSU897R)	AGA GGT GAA ATT CTT GGA	Kauff and Lutzoni (2002)
		F-SSU-1088R (NSSU1088)	TGA TTT CTC GTA AGG TGC CG	Kauff and Lutzoni (2002)
	ITS	F-SSU-1778F (ITS1F12)	GAA CCW GCG GAR GGA TCA	Schmidt et al. (2013)

(Continued)

Table 22.1   List of Important Universal Primers for Amplification of Bacterial, Archaeal, and Fungal Ribosomal Genes (*Continued*)

Target Organism	Marker	Primer Name	Primer Sequence (5'-3')	References
		F-SSU-1731F (ITS1F)	CTT GGT CAT TTA GAG GAA GTA A	Gardes and Burn (1993), Buee et al. (2009)
		F-SSU-1769F (ITS1)	TCC GTA GGT GAA CCT GCG G	White et al. (1990)
		F-SSU-1745F (ITS5)	GGA AGT AAA AGT CGT AAC AAG G	White et al. (1990)
		F-5.8SR (ITS2)	GCT GCG TTC TTC ATC GAT GC	White et al. (1990), Buee et al. (2009), Schmidt et al. (2013)
		F-SSUF (BITS)	ACC TGC GGA RGG ATC A	Bokulich and Mills (2013)
		F-SSUR (B58S3)	GAG ATC CRT TGY TRA AAG TT	Bokulich and Mills (2013)
		F-5.8SF (ITS3)	GCA TCG ATG AAG AAC GCA GC	White et al. (1990), LaTuga et al. (2011)
		F-LSU-60R (ITS4)	TCC TCC GCT TAT TGA TAT GC	White et al. (1990), LaTuga et al. (2011)

[a] Numbers associated with bacterial and archaeal primers correspond to the *16S rRNA* gene of Escherichia coli (Brosius et al. 1978).
[b] Numbers associated with fungal primers correspond to ribosomal gene positions of *Saccharomyces cerevisiae* (Mankin et al. 1986).

Table 22.2   Advantages and Disadvantages of Different Sequencing Platforms

Sequencing Platforms	Method	Approximate Read Length	*16S rDNA* Gene Coverage	Comments
Sanger sequencing	Di-deoxy terminator sequencing	750 bp	Two to three reads cover full length	Low throughput, with longer reads and minimal error rate (0.001%)
Roche-454	Pyrosequencing	400 bp	Each read can cover up to two to three variable regions	High throughput, with longer reads, but prone to error because of homopolymers
Illumina (HiSeq/ MiSeq)	Sequencing by synthesis	150 bp	Paired-end sequencing can cover up to two variable regions	High throughput, but with shorter reads and minimal error
Ion-torrent	Semiconductor sequencing	400 bp	Each read can cover up to two to three variable regions	High throughput, low cost, and rapid, but prone to error because of homopolymers
PacBio	SMRT sequencing	1000–3000 bp	One read can cover full length	High throughput, longest reads, but highest error rate (up to 15%)

Although traditional Sanger sequencing is very accurate and produces longer reads, it is very time consuming (requiring an additional cloning step), yields relatively less sequence reads per sample, and is very costly. Hence, it is not a preferred sequencing platform for decomposition microbiology studies. Until recently the 454 pyrosequencing platform (454 Life Sciences, Branford, Connecticut) was the most commonly used sequencing platform for microbial diversity studies, mainly because of longer reads and high throughput sequencing. The main disadvantage with the 454 pyrosequencing method is the occurrence of a comparatively higher error rate, because of its inability to correctly sequence a long homopolymer stretch (reported error rate is 0.01–0.02) (Margulies et al. 2005; Schloss et al. 2011). The Ion-torrent™ semiconductor sequencing platform (Life Technologies, Grand Island, New York), a sequencer similar to the 454 pyrosequencing platform, is also prone to error because of homopolymers, although it is cheaper and faster than the 454 pyrosequencing output. Recently, Illumina® HiSeq/MiSeq sequencing platforms (Illumina® Inc., San Diego, California) have gained more popularity for microbial diversity studies, mainly because of the lowest cost per mega base pairs data generated, the lowest error rate among available next-generation sequencing platforms, high coverage, and availability of user-friendly analytical pipelines. Also, paired end sequencing by Illumina® HiSeq/MiSeq can yield comparatively longer reads (approximately 300 bp; 150 bp from each direction), which results in good coverage of at least two variable regions of the *16S rRNA* gene. PacBio® sequencing platform (Pacific Biosystems® Menlo Park, California) is another sequencing platform that shows promise (a single read can cover the complete *16S rRNA* gene), but it is rarely used in decomposition microbiology research mainly because of the high (up to 15%) error rate (Kuczynski et al. 2012). As the majority of next-generation sequencing platforms are constantly changing, in the future most sequencing platforms will be able to overcome the previously mentioned limitations and will generate longer and more accurate sequence reads.

22.6 METAGENOMIC ANALYSES

22.6.1 Metagenomic Sequence Analysis Pipelines

Several pipelines are available for the analysis of amplicon sequencing data obtained from bacteria, archaea, and fungi using various next-generation sequencing platforms, but the most commonly used pipelines are Mothur (Schloss et al. 2009), Quantitative Insights into Microbial Ecology (QIIME) (Caporaso et al. 2010b), and the RDP's RDPipeline (Cole et al. 2009). Each pipeline has some advantages and disadvantages, but in general all are capable of processing sequences from start to end. Selection of a pipeline depends on size of the dataset, user bioinformatics skills, and user preference. RDPipeline is very user friendly as many tools in the RDPipeline are web based, but it is very time consuming and relatively less efficient for large datasets. Both QIIME (Caporaso et al. 2010b) and Mothur (Schloss et al. 2009) are capable of handling large datasets; however, Mothur is more streamlined. This is mainly because QIIME uses "original implementation" of several independent bioinformatics tools, whereas Mothur integrates edited versions of the source codes from several independent bioinformatics tools (Schloss et al. 2009; Caporaso et al. 2010b). Also, Mothur can be used either in a command line-based or in a graphical user interface-based environment, whereas QIIME is available in only a command line environment.

22.6.2 Minimizing Polymerase Chain Reaction and Sequencing Errors

As mentioned previously (see Sections 22.4 and 22.5), both PCR and sequencing steps add errors in a sequence; hence, the first step after obtaining a sequence from a sequencing platform is to minimize these errors. There are several ways to reduce sequencing errors, and generally a combination of these approaches produces good-quality sequences. The first step is to remove all sequences with ambiguous base calls and primer mismatches and sequences of unusually long and short reads (Huse et al. 2007; Schloss et al. 2011). The second step is to remove regions of the sequences that have low-quality scores. The determination of quality score cutoff (Q score) value is very subjective and varies with user, but in general a Q score ≥ 27 yields quality sequences

(Chou and Holmes 2001; Kunin et al. 2010). The third step is to use different denoising programs such as PyroNoise (Quince et al. 2009) or Denoiser (Reeder and Knight 2010) for the correction of base calls resulting from the modeling of original flow grams.

To remove chimeric sequence (see Section 22.3 for information on chimera formation), several software packages such as ChimeraSlayer (Haas et al. 2011), Perseus (Quince et al. 2011), Uchime (Edgar et al. 2011), and Decipher (Wright et al. 2012), are available. ChimeraSlayer uses the ends of the query sequences (30% from both sides) in a search against a chimera-free reference database for the detection of chimeric sequences. This works very well if the chimeric sequence originated from up to two parents, but this performs poorly for cases where the chimera has greater than two parents (Schloss et al. 2011). Perseus detects chimera *de novo* (does not require any reference datasets) and is based on the assumption that parent sequences will be more common than chimeric sequences if they originated from a single PCR reaction. Decipher also uses a search-based approach against a chimera-free reference database, but it is computationally very demanding and thus not recommended for large datasets. Uchime can detect chimera either *de novo* or by using reference databases and is even capable of detecting those chimeric sequences that have more than two parents. Although Uchime's sensitivity is similar to that of Perseus, it is much faster (>100 times) (Edgar et al. 2011). Even among reference data–based chimera check tools, Uchime performs better and faster than others (Edgar et al. 2011). Because no reference dataset is complete and 100% chimera-free, it is difficult to be certain that any particular chimera check program detected all the chimeric sequences present.

22.6.3 Metagenomic Sequence Clustering

Following quality trimming and chimera removal, the remaining sequences are grouped or binned by taxonomy, operational taxonomic unit (OTU), and phylogeny or sometimes by a combination of more than one method for diversity estimations. In the taxonomy-based method, sequences are binned into taxonomic groups based on their similarity to those sequences whose taxonomy is known in reference databases such as the RDP database (http://rdp.cme.msu.edu), Greengene database (DeSantis et al. 2006) (http://greengenes.lbl.gov/cgi-bin/nph-index.cgi), SILVA database (Quast et al. 2013) (http://www.arb-silva.de/), and National Center for Biotechnology Information (NCBI) taxonomy database (http://www.ncbi.nlm.nih.gov/taxonomy) for the *16S rRNA* gene of bacteria and archaea and in the UNITE database (Tedersoo et al. 2011) (http://unite.ut.ee/) for the ITS region of fungi.

Hierarchical classification of sequences can be performed by RDP classifier (Wang et al. 2007) either as a stand-alone package or via pipelines, such as QIIME or Mothur. RDP classifier uses a naive Bayesian classifier, which can be trained based on chosen external reference databases. Alternatively, RTAX (Soergel et al. 2012) can also be used for taxonomic assignment of short sequences. The success of taxonomy-based methods depends on how much coverage a particular database has for a particular environment. Because microbes associated with decomposing remains are comparatively less abundant in reference databases, samples collected from these environments may yield a higher percentage of unclassified sequences, especially at the genus level. The label "unclassified" does not necessarily imply a bad sequence but instead an, as yet, undiscovered and previously unsequenced organism.

Taxonomy-based binning is generally preferred if the purpose of the study is to compare the taxonomic information obtained from different studies or from different regions of the *16S rRNA* gene. Alternatively, in the OTU-based clustering method sequences are clustered *de novo* at different sequence similarity thresholds (OTUs) such as at ≥97%, at ≥95%, at ≥90%, and at ≥85%, which hypothetically represent species, genus, family, and order, respectively (Cressman et al. 2010). Because the OTU-based method does not provide exact taxonomic information, it is not recommended to compare different studies or different regions of the *16S rRNA* gene (Kuczynski et al. 2012). However, for diversity estimations OTU-based clustering is preferred over taxonomy-based clustering because it utilizes all available sequence information irrespective of whether sequences are classified or not. In general, taxonomy- and OTU-based binning methods complement each other, and thus the combination of both methods provides more information than any single method alone.

Phylogeny-based approaches cluster sequences based on the phylogenetic relationship between DNA sequences. Phylogenetic relationships can be inferred either *de novo* using different alignments and tree building tools or based on reference databases, such as the Greengene or SILVA. For *de novo* phylogeny construction, DNA sequences are first aligned either using a *de novo* alignment tool, such as Muscle (Edgar 2004) or MAFFT (Katoh and Standley 2013), or using a reference-based alignment tool, such as Infernal (Nawrocki et al. 2009), PyNAST (Caporaso et al. 2010a), and Greengene core reference alignment (DeSantis et al. 2006). *De novo* tools are not recommended for alignment of very large DNA sequence datasets, mainly because these tools are computationally demanding and time consuming.

Once sequences are aligned, it can be used as an input file either in maximum likelihood–based tree building tools, such as FastTree (Price et al. 2010) or RAxML (Stamatakis 2006), or in relaxed neighbor joining-based tree building tools, such as Clearcut (Sheneman et al. 2006). FastTree and Clearcut are much faster than traditional comprehensive phylogeny packages, such as PAUP (Swofford 2003) and PHYLIP (Felsenstein 1989), and hence are better choices for very large datasets. If aligned datasets are smaller (<15,000 sequences), then the comprehensive phylogenetic packages such as PAUP (Swofford 2003) and PHYLIP (Felsenstein 1989) can also perform well for construction of phylogenetic tree. A phylogenetic tree can also help in the determination of close relatives of those sequences that were not assigned to any taxonomic group during taxonomy-based clustering.

22.6.4 Microbial Diversity Estimation

Sequences clustered based on any binning method mentioned previously (see Section 22.6.3) can be utilized for various α and β diversity estimations (Achtman and Wagner 2008), but for unbiased diversity estimations it is very important to normalize the datasets so that all samples have equal numbers of sequences (i.e., rarefaction) because next-generation sequencing library sizes vary greatly from sample to sample (Gihring et al. 2012). An alternative approach to visualize β diversity between different samples is to draw an ordination plot, such as principal coordinate analyses (PCoAs) or nonmetric multidimensional scaling (nMDS). These ordination plots can be drawn using either popular metagenomics pipelines, such as QIIME (Caporaso et al. 2010b) or Mothur (Schloss et al. 2009), or the Vegan package (Oksanen et al. 2012) in R (R Foundation for Statistical Computing 2011). PCoA uses an eigenvector-based method that tries to maximize linear correlation between distance measures and distance in the ordination space to represent multidimensional data (i.e., from multiple communities) into the smallest possible number of dimensions, whereas nMDS tries to maximize rank order correlation between distance measures and distance in the ordination space to represent data from multiple dimensions (i.e., from multiple communities) to a user-defined limited number of dimensions (generally two to three dimensions specified in advance). The quality of any PCoA plot is determined by the R^2 value between distance measures and distance in the ordination space (higher is better) and by the proportion of total variance represented by the first two or three axes of the plot. The quality of an nMDS plot is determined by the R^2 value (higher is better) and stress, which is a measure of mismatch between distance measures and distance in the ordination space (lower is better). Generally, stress values increase with increasing quantity of data and decrease with increasing number of dimensions. Hence, there is no cutoff value of good or bad stress, but generally a stress value below 0.20 is considered a reasonable representation of data. Sometimes a stress value can be reduced by data transformations (e.g., logarithmic, square root, arcsine, reciprocal, and squared), but before using any transformation method it is important to ensure that the particular transformation method is appropriate for the dataset in question. Ordination plots are just a visualization method; but if the purpose is to test if the spatial separation between different samples is statistically significant, then statistical tools, such as Analysis of Molecular Variance and Multiresponse Permutation Procedures, may be performed. Popular metagenomics pipelines, such as QIIME (Caporaso et al. 2010b) or Mothur (Schloss et al. 2009), or Vegan (Oksanen et al. 2012) in R (R Foundation for Statistical Computing 2011) can be used to perform the tests of significance.

22.7 CONCLUSION AND FUTURE PROSPECTS

The study of multilevel bacterial community interactions with insects utilizing vertebrate carrion is in its infancy. Currently, what we know about decomposition microbiology is primarily based on *16S rDNA*-based amplicon sequencing of bacteria from decomposing remains, insects, and the associated environment. However, bacteria are not the only microbes associated with decaying remains. To include all microbes, their gene expressions and relative abundances in appropriate detail, it is important to have sequencing platforms and software that are capable of generating and analyzing enormous datasets in an efficient manner. In the last few years, sequencing methods and associated analysis software have developed very rapidly, and presently we have technology to begin generating and analyzing whole metagenomic and metatranscriptomic data for the study of microbial structure and its correlation to microbial functions (Shi et al. 2009; Martinez et al. 2013) associated with decomposing remains. Meanwhile, the current information obtained from these technologies is beginning to answer fundamental questions about decomposition ecology, forensic entomology, and food web dynamics. For example, now it is well known that bacterial succession on carrion, in part, is determined by insects associated with carrion; several bacteria associated with insects and carrion release interkingdom signaling molecules to communicate with insects; and this, in turn, determines insect arrival and oviposition patterns (Ma et al. 2012; Tomberlin et al. 2012). This information offers promise for more precise postmortem interval estimation methods, using insects and/or microbes.

REFERENCES

Achtman, M. and M. Wagner. 2008. Microbial diversity and the genetic nature of microbial species. *Nature Reviews Microbiology* 6 (6): 431–440.

Ausubel, F.M., R. Brent, R.E. Kingston, D.D. Moore, J.G. Seidman, and K. Struhl. 1988. *Current Protocols in Molecular Biology*. Hoboken, NJ: John Wiley & Sons.

Baker, G.C., J.J. Smith, and D.A. Cowan. 2003. Review and re-analysis of domain-specific 16S primers. *Journal of Microbiological Methods* 55 (3): 541–555.

Barns, S.M., R.E. Fundyga, M.W. Jeffries, and N.R. Pace. 1994. Remarkable Archaeal Diversity Detected in a Yellowstone-National-Park Hot-Spring Environment. *Proceedings of the National Academy of Sciences of the United States of America* 91 (5): 1609–1613. doi: 10.1073/pnas.91.5.1609.

Bauer, M., M. Kube, H. Teeling, M. Richter, T. Lombardot, E. Allers, C.A. Würdemann et al. 2006. Whole genome analysis of the marine Bacteroidetes'*Gramella forsetii*' reveals adaptations to degradation of polymeric organic matter. *Environmental Microbiology* 8 (12): 2201–2213.

Bergmann, G.T., S.T. Bates, K.G. Eilers, C.L. Lauber, J.G. Caporaso, W.A. Walters, R. Knight, and N. Fierer. 2011. The under-recognized dominance of Verrucomicrobia in soil bacterial communities. *Soil Biology & Biochemistry* 43 (7): 1450–1455.

Bokulich, N.A. and D.A. Mills. 2013. Improved selection of internal transcribed spacer-specific primers enables quantitative, ultra-high-throughput profiling of fungal communities. *Applied and Environmental Microbiology* 79 (8): 2519–2526.

Borneman, J. and R.J. Hartin. 2000. PCR primers that amplify fungal rRNA genes from environmental samples. *Applied and Environmental Microbiology* 66 (10): 4356–4360. doi: 10.1128/Aem.66.10.4356–4360.2000.

Braid, M.D., L.M. Daniels, and C.L. Kitts. 2003. Removal of PCR inhibitors from soil DNA by chemical flocculation. *Journal of Microbiological Methods* 52 (3): 389–393.

Brosius, J., M.L. Palmer, P.J. Kennedy, and H.F. Noller. 1978. Complete nucleotide sequence of a 16S ribosomal RNA gene from *Escherichia coli*. *Proceedings of the National Academy of Sciences of the United States of America* 75 (10): 4801–4805.

Buee, M., M. Reich, C. Murat, E. Morin, R.H. Nilsson, S. Uroz, and F. Martin. 2009. 454 Pyrosequencing analyses of forest soils reveal an unexpectedly high fungal diversity. *The New Phytologist* 184 (2): 449–456.

Cabeen, M.T. and C. Jacobs-Wagner. 2005. Bacterial cell shape. *Nature Reviews Microbiology* 3 (8): 601–610.

Caporaso, J.G., K. Bittinger, F.D. Bushman, T.Z. DeSantis, G.L. Andersen, and R. Knight. 2010a. PyNAST: A flexible tool for aligning sequences to a template alignment. *Bioinformatics* 26 (2): 266–267.

Caporaso, J.G., J. Kuczynski, J. Stombaugh, K. Bittinger, F.D. Bushman, E.K. Costello, N. Fierer et al. 2010b. QIIME allows analysis of high-throughput community sequencing data. *Nature Methods* 7 (5): 335–336.

Caporaso, J.G., C.L. Lauber, W.A. Walters, D. Berg-Lyons, J. Huntley, N. Fierer, S. M. Owens et al. 2012. Ultra-high-throughput microbial community analysis on the Illumina HiSeq and MiSeq platforms. *ISME J* 6 (8): 1621–1624. doi: 10.1038/ismej.2012.8.

Cardona, S., A. Eck, M. Cassellas, M. Gallart, C. Alastrue, J. Dore, F. Azpiroz, J. Roca, F. Guarner, and C. Manichanh. 2012. Storage conditions of intestinal microbiota matter in metagenomic analysis. *BMC Microbiology* 12: 158.

Carroll, I.M., T. Ringel-Kulka, J.P. Siddle, T.R. Klaenhammer, and Y. Ringel. 2012. Characterization of the fecal microbiota using high-throughput sequencing reveals a stable microbial community during storage. *Plos One* 7 (10): e46953.

Carter, D.O., D. Yellowlees, and M. Tibbett. 2010. Moisture can be the dominant environmental parameter governing cadaver decomposition in soil. *Forensic Science International* 200 (1–3): 60–66.

Chou, H.H. and M.H. Holmes. 2001. DNA sequence quality trimming and vector removal. *Bioinformatics* 17 (12): 1093–1104.

Cole, J.R., Q. Wang, E. Cardenas, J. Fish, B. Chai, R.J. Farris, A.S. Kulam-Syed-Mohideen et al. 2009. The Ribosomal Database Project: Improved alignments and new tools for rRNA analysis. *Nucleic Acids Research* 37 (Database issue): D141–145.

Cressman, M.D., Z. Yu, M.C. Nelson, S.J. Moeller, M.S. Lilburn, and H.N. Zerby. 2010. Interrelations between the microbiotas in the litter and in the intestines of commercial broiler chickens. *Applied and Environmental Microbiology* 76 (19): 6572–6582.

Damann, F.E. 2010. Human Decomposition Ecology at the University of Tennessee Anthropology Research Facility. Dissertation, University of Tennessee, Knoxville.

DasSarma, S. and E.F. Fleischmann. 1995. *Archaea: A Laboratory Manual—Halophiles*, 269–272. New York: Cold Spring Harbour Laboratory Press.

Delmont, T.O., P. Robe, I. Clark, P. Simonet, and T.M. Vogel. 2011. Metagenomic comparison of direct and indirect soil DNA extraction approaches. *Journal of Microbiological Methods* 86 (3): 397–400.

DeLong, E.F. 1992. Archaea in coastal marine environments. *Proceedings of the National Academy of Sciences of the United States of America* 89 (12): 5685–5689.

DeSantis, T.Z., P. Hugenholtz, N. Larsen, M. Rojas, E.L. Brodie, K. Keller, T. Huber, D. Dalevi, P. Hu, and G.L. Andersen. 2006. Greengenes, a chimera-checked 16S rRNA gene database and workbench compatible with ARB. *Applied and Environmental Microbiology* 72 (7): 5069–5072.

Devereux, R. and S.S. Wilkinson. 2004. Amplification of ribosomal RNA sequences. In *Molecular Microbial Ecology Manual*. Dordrecht, The Netherlands: Kluwer Academic Publishers.

Edgar, R.C. 2004. MUSCLE: Multiple sequence alignment with high accuracy and high throughput. *Nucleic Acids Research* 32 (5): 1792–1797.

Edgar, R.C., B.J. Haas, J.C. Clemente, C. Quince, and R. Knight. 2011. UCHIME improves sensitivity and speed of chimera detection. *Bioinformatics* 27 (16): 2194–2200.

Embley, T.M., B.J. Finlay, R.H. Thomas, and P.L. Dyal. 1992. The use of rRNA sequences and fluorescent probes to investigate the phylogenetic positions of the anaerobic ciliate Metopus palaeformis and its archaeobacterial endosymbiont. *Journal of General Microbiology* 138 (7): 1479–1487. doi: 10.1099/00221287-138-7-1479.

Erb, R.W. and I. Wagner-Dobler. 1993. Detection of polychlorinated biphenyl degradation genes in polluted sediments by direct DNA extraction and polymerase chain reaction. *Applied and Environmental Microbiology* 59 (12): 4065–4073.

Felsenstein, J. 1989. PHYLIP—phylogeny inference package (Version 3.2). *Cladistics* 5: 164–166.

Frostegard, A., S. Courtois, V. Ramisse, S. Clerc, D. Bernillon, F. Le Gall, P. Jeannin, X. Nesme, and P. Simonet. 1999. Quantification of bias related to the extraction of DNA directly from soils. *Applied and Environmental Microbiology* 65 (12): 5409–5420.

Gantner, S., A.F. Andersson, L. Alonso-Saez, and S. Bertilsson. 2011. Novel primers for 16S rRNA-based archaeal community analyses in environmental samples. *Journal of Microbiological Methods* 84 (1): 12–8. doi: 10.1016/j.mimet.2010.10.001.

Gardes, M. and T.D. Bruns. 1993. ITS primers with enhanced specificity for basidiomycetes—application to the identification of mycorrhizae and rusts. *Molecular Ecology* 2 (2): 113–8.

Gihring, T.M., S.J. Green, and C.W. Schadt. 2012. Massively parallel rRNA gene sequencing exacerbates the potential for biased community diversity comparisons due to variable library sizes. *Bauer, Margarete* 14 (2): 285–290.

Grice, E.A., H.H. Kong, G. Renaud, A.C. Young, G.G. Bouffard, R.W. Blakesley, T.G. Wolfsberg, M.L. Turner, and J.A. Segre. 2008. A diversity profile of the human skin microbiota. *Genome Research* 18 (7): 1043–1050.

Haas, B.J., D. Gevers, A.M. Earl, M. Feldgarden, D.V. Ward, G. Giannoukos, D. Ciulla et al. 2011. Chimeric 16S rRNA sequence formation and detection in Sanger and 454-pyrosequenced PCR amplicons. *Genome Research* 21 (3): 494–504.

Handl, S., S.E. Dowd, J.F. Garcia-Mazcorro, J. M. Steiner, and J.S. Suchodolski. 2011. Massive parallel 16S rRNA gene pyrosequencing reveals highly diverse fecal bacterial and fungal communities in healthy dogs and cats. *FEMS Microbiology Ecology* 76 (2): 301–10. doi: 10.1111/j.1574-6941.2011.01058.x.

Holben, W.E., J.K. Jansson, B.K. Chelm, and J.M. Tiedje. 1988. DNA probe method for the detection of specific microorganisms in the soil bacterial community. *Applied and Environmental Microbiology* 54 (3): 703–711.

Hunt, K.M., J.A. Foster, L.J. Forney, U.M. Schutte, D.L. Beck, Z. Abdo, L.K. Fox, J.E. Williams, M.K. McGuire, and M.A. McGuire. 2011. Characterization of the diversity and temporal stability of bacterial communities in human milk. *PLoS One* 6 (6): e21313. doi: 10.1371/journal.pone.0021313.

Huse, S.M., L. Dethlefsen, J.A. Huber, D. Mark Welch, D.A. Relman, and M.L. Sogin. 2008. Exploring microbial diversity and taxonomy using SSU rRNA hypervariable tag sequencing. *PLoS Genetics* 4 (11): e1000255. doi: 10.1371/journal.pgen.1000255.

Huse, S.M., J.A. Huber, H.G. Morrison, M.L. Sogin, and D.M. Welch. 2007. Accuracy and quality of massively parallel DNA pyrosequencing. *Genome Biology* 8 (7): R143.

Inceoglu, O., E.F. Hoogwout, P. Hill, and J.D. van Elsas. 2010. Effect of DNA extraction method on the apparent microbial diversity of soil. *Applied and Environmental Microbiology* 76 (10): 3378–82.

Insam, H. and M. Goberna. 2004. *Use of Biolog® for the Community Level Physiological Profiling (CLPP) of Environmental Samples*, 853–860. Dordrecht, The Netherlands: Kluwer Academic Publishers.

Jacobsen, C.S. and O.F. Rasmussen. 1992. Development and application of a new method to extract bacterial DNA from soil based on separation of bacteria from soil with cation-exchange resin. *Applied and Environmental Microbiology* 58 (8): 2458–2462.

Kakirde, K.S., L.C. Parsley, and M.R. Liles. 2010. Size does matter: Application-driven approaches for soil metagenomics. *Soil Biology and Biochemistry* 42 (11): 1911–1923.

Katoh, K. and D.M. Standley. 2013. MAFFT multiple sequence alignment software version 7: Improvements in performance and usability. *Molecular Biology and Evolution* 30 (4): 772–780.

Kauff, F. and F. Lutzoni. 2002. Phylogeny of the Gyalectales and Ostropales (Ascomycota, Fungi): among and within order relationships based on nuclear ribosomal RNA small and large subunits. *Molecular Phylogenetics and Evolution* 25 (1): 138–156. doi: Pii S1055-7903(02)00214-2. doi 10.1016/S1055-7903(02)00214-2.

Kuczynski, J., C.L. Lauber, W.A. Walters, L.W. Parfrey, J.C. Clemente, D. Gevers, and R. Knight. 2012. Experimental and analytical tools for studying the human microbiome. *Nature Reviews Genetics* 13 (1): 47–58.

Kumar, P.S., M.R. Brooker, S.E. Dowd, and T. Camerlengo. 2011. Target region selection is a critical determinant of community fingerprints generated by 16S pyrosequencing. *Plos One* 6 (6): e20956. doi: ARTN e20956. doi 10.1371/journal.pone.0020956.

Kunin, V., A. Engelbrektson, H. Ochman, and P. Hugenholtz. 2010. Wrinkles in the rare biosphere: Pyrosequencing errors can lead to artificial inflation of diversity estimates. *Environmental Microbiology* 12 (1): 118–123.

Kwambana, B.A., N.I. Mohammed, D. Jeffries, M. Barer, R.A. Adegbola, and M. Antonio. 2011. Differential effects of frozen storage on the molecular detection of bacterial taxa that inhabit the nasopharynx. *BMC Clinical Pathology* 11: 2.

LaTuga, M.S., J.C. Ellis, C.M. Cotton, R.N. Goldberg, J.L. Wynn, R.B. Jackson, and P.C. Seed. 2011. Beyond bacteria: A study of the enteric microbial consortium in extremely low birth weight infants. *PLoS One* 6 (12): e27858.

Lauber, C.L., N. Zhou, J.I. Gordon, R. Knight, and N. Fierer. 2010. Effect of storage conditions on the assessment of bacterial community structure in soil and human-associated samples. *FEMS Microbiology Letters* 307 (1): 80–86.

Liu, Z., T.Z. DeSantis, G.L. Andersen, and R. Knight. 2008. Accurate taxonomy assignments from 16S rRNA sequences produced by highly parallel pyrosequencers. *Nucleic Acids Research* 36 (18): e120.

Liu, Z., C. Lozupone, M. Hamady, F.D. Bushman, and R. Knight. 2007. Short pyrosequencing reads suffice for accurate microbial community analysis. *Nucleic Acids Research* 35 (18): e120. doi: 10.1093/nar/gkm541.

Ma, Q., A. Fonseca, W. Liu, A.T. Fields, M.L. Pimsler, A.F. Spindola, A.M. Tarone, T.L. Crippen, J.K. Tomberlin, and T.K. Wood. 2012. *Proteus mirabilis* interkingdom swarming signals attract blow flies. *The ISME Journal* 6 (7): 1356–1366.

Mahalanabis, M., H. Al-Muayad, M.D. Kulinski, D. Altman, and C.M. Klapperich. 2009. Cell lysis and DNA extraction of gram-positive and gram-negative bacteria from whole blood in a disposable microfluidic chip. *Lab Chip* 9 (19): 2811–2817.

Mankin, A.S., K.G. Skryabin, and P.M. Rubtsov. 1986. Identification of ten additional nucleotides in the primary structure of yeast 18S rRNA. *Gene* 44 (1): 143–145.

Männistö, M.K., M. Tiirola, and M.M. Häggblom. 2009. Effect of freeze-thaw cycles on bacterial communities of arctic tundra soil. *Microbial Ecology* 58 (3): 621–631.

Margulies, M., M. Egholm, W.E. Altman, S. Attiya, J.S. Bader, L.A. Bemben, J. Berka et al. 2005. Genome sequencing in microfabricated high-density picolitre reactors. *Nature* 437 (7057): 376–380.

Martinez, A., L.A. Ventouras, S.T. Wilson, D.M. Karl, and E.F. DeLong. 2013. Metatranscriptomic and functional metagenomic analysis of methylphosphonate utilization by marine bacteria. *Frontiers in Microbiology* 4: 340.

Miller, D.N., J.E. Bryant, E.L. Madsen, and W.C. Ghiorse. 1999. Evaluation and optimization of DNA extraction and purification procedures for soil and sediment samples. *Applied Environmental Microbiology* 65 (11): 4715–4724.

Muyzer, G., A. Teske, C.O. Wirsen, and H.W. Jannasch. 1995. Phylogenetic relationships of Thiomicrospira species and their identification in deep-sea hydrothermal vent samples by denaturing gradient gel electrophoresis of 16S rDNA fragments. *Archives of Microbiology* 164 (3): 165–72.

Nawrocki, E.P., D.L. Kolbe, and S.R. Eddy. 2009. Infernal 1.0: inference of RNA alignments. *Bioinformatics* 25 (10):1335–1337.

Nicol, G.W., L.A. Glover, and J.I. Prosser. 2003. Spatial analysis of archaeal community structure in grassland soil. *Applied and Environmental Microbiology* 69 (12): 7420–7429.

Ogram, A., G.S. Sayler, and T. Barkay. 1987. The extraction and purification of microbial DNA from sediments. *Journal of Microbiological Methods* 7 (2–3): 57–66.

Oksanen, J., G.F. Blanchet, R. Kindt, P. Legendre, P.R. Minchin, R.B. O'Hara, G.L. Simpson, P.M. Solymos, H.H. Stevens, and H. Wagner. 2012. Vegan: Community Ecology Package. R package version 2.0–4.

Parkinson, R.A, K.R. Dias, J. Horswell, P. Greenwood, N. Banning, M. Tibbett, and A. Vass. 2009. Microbial community analysis of human decomposition on soil. In *Criminal and Environmental Soil Forensics*, edited by K. Ritz, L. Dawson, and D. Miller, 379–394. Dordrecht, The Netherlands: Springer.

Pechal, J.L., T.L. Crippen, M.E. Benbow, A.M. Tarone, S. Dowd, and J.K. Tomberlin. 2014. The potential use of bacterial community succession in forensics as described by high throughput metagenomic sequencing. *International Journal of Legal Medicine* 128: 193–205.

Pei, A. Y., W.E. Oberdorf, C.W. Nossa, A. Agarwal, P. Chokshi, E.A. Gerz, Z.D. Jin et al. 2010. Diversity of 16S rRNA genes within individual prokaryotic genomes. *Applied and Environmental Microbiology* 76 (15): 5333–5333.

Picard, C., C. Ponsonnet, E. Paget, X. Nesme, and P. Simonet. 1992. Detection and enumeration of bacteria in soil by direct DNA extraction and polymerase chain reaction. *Applied and Environmental Microbiology* 58 (9): 2717–22.

Price, M.N., P.S. Dehal, and A.P. Arkin. 2010. FastTree 2—Approximately maximum-likelihood trees for large alignments. *PLoS One* 5 (3): e9490. doi:10.1371/journal.pone.0009490.

Quast, C., E. Pruesse, P. Yilmaz, J. Gerken, T. Schweer, P. Yarza, J. Peplies, and F.O. Glockner. 2013. The SILVA ribosomal RNA gene database project: Improved data processing and web-based tools. *Nucleic Acids Research* 41 (D1): D590–D596.

Quince, C., A. Lanzen, T.P. Curtis, R.J. Davenport, N. Hall, I.M. Head, L.F. Read, and W.T. Sloan. 2009. Accurate determination of microbial diversity from 454 pyrosequencing data. *Nature Methods* 6 (9):639–641.

Quince, C., A. Lanzen, R.J. Davenport, and P.J. Turnbaugh. 2011. Removing noise from pyrosequenced amplicons. *BMC Bioinformatics* 12: 38.

R Foundation for Statistical Computing. 2011. R: A language and environment for statistical computing. R Foundation for Statistical Computing,, Vienna, Austria. Available at http://www.R-project.org

Ranjard, L., D.P. Lejon, C. Mougel, L. Schehrer, D. Merdinoglu, and R. Chaussod. 2003. Sampling strategy in molecular microbial ecology: Influence of soil sample size on DNA fingerprinting analysis of fungal and bacterial communities. *Environmental Microbiology* 5 (11): 1111–1120.

Reeder, J. and R. Knight. 2010. Rapidly denoising pyrosequencing amplicon reads by exploiting rank-abundance distributions. *Nature Methods* 7 (9): 668–669.

Reysenbach, A.L. and N.R. Pace. 1995. *Archaea: A Laboratory Manual—Thermophiles*, edited by F.T. Robb and A.R. Place. New York: Cold Spring Harbour Laboratory Press.

Robe, P., R. Nalin, C. Capellano, T.M. Vogel, and P. Simonet. 2003. Extraction of DNA from soil. *European Journal of Soil Biology* 39 (4): 183–190.

Rochelle, P.A., B.A. Cragg, J.C. Fry, R.J. Parkes, and A.J. Weightman. 1994. Effect of sample handling on estimation of bacterial diversity in marine-sediments by 16S ribosomal-RNA gene sequence-analysis. *FEMS Microbiology Ecology* 15 (1, 2): 215–225.

Roesch, L.F., G. Casella, O. Simell, J. Krischer, C.H. Wasserfall, D. Schatz, M.A. Atkinson, J. Neu, and E.W. Triplett. 2009. Influence of fecal sample storage on bacterial community diversity. *The Open Microbiology Journal* 3: 40–6.

Roh, C., F. Villatte, B.G. Kim, and R.D. Schmid. 2006. Comparative study of methods for extraction and purification of environmental DNA from soil and sludge samples. *Applied Biochemistry and Biotechnology* 134 (2): 97–112.

Rousk, J., E. Baath, P.C. Brookes, C.L. Lauber, C. Lozupone, J.G. Caporaso, R. Knight, and N. Fierer. 2010. Soil bacterial and fungal communities across a pH gradient in an arable soil. *The ISME Journal* 4 (10): 1340–1351.

Rudi, K., O.M. Skulberg, F. Larsen, and K.S. Jakobsen. 1997. Strain characterization and classification of oxyphotobacteria in clone cultures on the basis of 16S rRNA sequences from the variable regions V6, V7, and V8. *Applied and Environmental Microbiology* 63 (7): 2593–2599.

Schloss, P.D., D. Gevers, and S.L. Westcott. 2011. Reducing the effects of PCR amplification and sequencing artifacts on 16S rRNA-based studies. *PLoS One* 6 (12): e27310.

Schloss, P.D., S.L. Westcott, T. Ryabin, J.R. Hall, M. Hartmann, E.B. Hollister, R.A. Lesniewski et al. 2009. Introducing mothur: Open-source, platform-independent, community-supported software for describing and comparing microbial communities. *Applied and Environmental Microbiology* 75 (23): 7537–41.

Schmidt, P.A., M. Bálint, B. Greshake, C. Bandow, J. Römbke, and I. Schmitt. 2013. Illumina metabarcoding of a soil fungal community. *Soil Biology and Biochemistry* 65 (0): 128–132.

Segata, N., L. Waldron, A. Ballarini, V. Narasimhan, O. Jousson, and C. Huttenhower. 2012. Metagenomic microbial community profiling using unique clade-specific marker genes. *Nature Methods* 9 (8): 811–814.

Sheneman, L., J. Evans, and J.A. Foster. 2006. Clearcut: A fast implementation of relaxed neighbor joining. *Bioinformatics* 22 (22): 2823, 2824.

Shi, Y., G.W. Tyson, and E.F. DeLong. 2009. Metatranscriptomics reveals unique microbial small RNAs in the ocean's water column. *Nature* 459 (7244): 266–269.

Soergel, D.A., N. Dey, R. Knight, and S.E. Brenner. 2012. Selection of primers for optimal taxonomic classification of environmental 16S rRNA gene sequences. *The ISME Journal* 6 (7): 1440–1444.

Stamatakis, A. 2006. RAxML-VI-HPC: Maximum likelihood-based phylogenetic analyses with thousands of taxa and mixed models. *Bioinformatics* 22 (21): 2688–2690.

Swofford, D.L. 2003. PAUP*. Phylogenetic Analysis Using Parsimony (*and Other Methods). Version 4: Sinauer Associates.

Tedersoo, L., K. Abarenkov, R.H. Nilsson, A. Schussler, G.A. Grelet, P. Kohout, J. Oja et al. 2011. Tidying up international nucleotide sequence databases: Ecological, geographical and sequence quality annotation of its sequences of mycorrhizal fungi. *PLoS One* 6 (9): e24940.

Tomberlin, J.K., T.L. Crippen, A.M. Tarone, B. Singh, K. Adams, Y.H. Rezenom, M.E. Benbow et al. 2012. Interkingdom responses of flies to bacteria mediated by fly physiology and bacterial quorum sensing. *Animal Behaviour* 84 (6): 1449–1456.

Tringe, S.G. and E.M. Rubin. 2005. Metagenomics: DNA sequencing of environmental samples. *Nature Reviews Genetics* 6 (11): 805–14.

Turner, S., K.M. Pryer, V.P.W. Miao, and J.D. Palmer. 1999. Investigating deep phylogenetic relationships among cyanobacteria and plastids by small submit rRNA sequence analysis. *Journal of Eukaryotic Microbiology* 46 (4): 327–338. doi: doi 10.1111/j.1550-7408.1999.tb04612.x.

Vass, A. 2001. Beyond the grave: Understanding human decomposition. *Microbiology Today* 28 (1): 190–192.

Vetrovsky, T. and P. Baldrian. 2013. The variability of the 16S rRNA gene in bacterial genomes and its consequences for bacterial community analyses. *PLoS One* 8 (2): e57923.

Vollmer, W. and J.V. Holtje. 2004. The architecture of the murein (peptidoglycan) in gram-negative bacteria: Vertical scaffold or horizontal layer(s)? *Journal of Bacteriology* 186 (18): 5978–5987.

Volossiouk, T., E.J. Robb, and R.N. Nazar. 1995. Direct DNA extraction for PCR-mediated assays of soil organisms. *Applied and Environmental Microbiology* 61 (11): 3972–3976.

Vos, M., C. Quince, A.S. Pijl, M. de Hollander, and G.A. Kowalchuk. 2012. A comparison of rpoB and 16S rRNA as markers in pyrosequencing studies of bacterial diversity. *Plos One* 7 (2):e30600.

Wallenius, K. 2011. Microbiological Characterisation of Soils: Evaluation of Some Critical Steps in Data Collection and Experimental Design. PhD Dissertation, Department of Food and Environmental Sciences, University of Helsinki, Helsinki.

Walters, W.A., J.G. Caporaso, C.L. Lauber, D. Berg-Lyons, N. Fierer, and R. Knight. 2011. PrimerProspector: de novo design and taxonomic analysis of barcoded polymerase chain reaction primers. *Bioinformatics* 27 (8):1159–61.

Wang, Q., G.M. Garrity, J.M. Tiedje, and J.R. Cole. 2007. Naive Bayesian classifier for rapid assignment of rRNA sequences into the new bacterial taxonomy. *Applied and Environmental Microbiology* 73 (16): 5261–5267.

White, T., T. Bruns, S. Lee, and J. Taylor. 1990. Amplification and direct sequencing of fungal ribosomal RNA genes for phylogenetics. In *PCR Protocols: A Guide to Methods and Applications*, edited by M.A. Innis, D. Gelfand, J. Shinsky, and T. White, 315–322. San Diego, CA: Academic Press.

Whitehead, T.R. and M.A. Cotta. 1999. Phylogenetic diversity of methanogenic archaea in swine waste storage pits. *FEMS Microbiology Letters* 179 (2): 223–226. doi: doi 10.1111/j.1574-6968.1999.tb08731.x.

Whitman, W.B., D.C. Coleman, and W.J. Wiebe. 1998. Prokaryotes: The unseen majority. *Proceedings of the National Academy of Sciences* 95 (12): 6578–6583.

Wilson, M. 2009. *Bacteriology of Humans: An Ecological Perspective*. Malden, MA: Wiley/Blackwell Publishing.

Wright, E.S., L.S. Yilmaz, and D.R. Noguera. 2012. DECIPHER: A search-based approach to chimera identification for 16S rRNA sequences. *Applied and Environmental Microbiology* (3): 717–725.

Wu, G.D., J.D. Lewis, C. Hoffmann, Y.Y. Chen, R. Knight, K. Bittinger, J. Hwang et al. 2010. Sampling and pyrosequencing methods for characterizing bacterial communities in the human gut using 16S sequence tags. *BMC Microbiology* 10: 206.

Youssef, N., C.S. Sheik, L.R. Krumholz, F.Z. Najar, B.A. Roe, and M.S. Elshahed. 2009. Comparison of species richness estimates obtained using nearly complete fragments and simulated pyrosequencing-generated fragments in 16s rRNA gene-based environmental surveys. *Applied and Environmental Microbiology* 75 (16): 5227–5236. doi: doi 10.1128/Aem.00592-09.

Zheng, L., T.L. Crippen, B. Singh, A.M. Tarone, S. Dowd, Z. Yu, T.K. Wood, and J.K. Tomberlin. 2013. A survey of bacterial diversity from successive life stages of black soldier fly (Diptera: Stratiomyidae) by using 16S rDNA pyrosequencing. *Journal of Medical Entomology* 50 (3): 647–658.

Zhou, J., M.A. Bruns, and J.M. Tiedje. 1996. DNA recovery from soils of diverse composition. *Applied and Environmental Microbiology* 62 (2): 316–22.

CHAPTER 23

Applications of Soil Chemistry in Forensic Entomology

Jacqueline A. Aitkenhead-Peterson, Michael B. Alexander, Joan A. Bytheway, David O. Carter, and Daniel J. Wescott

CONTENTS

23.1 INTRODUCTION

The current best method to estimate postmortem interval (PMI) of human remains or their surrogates is forensic entomology (Reibe et al. 2010; Brown et al. 2012; Magni et al. 2012; Richards et al. 2012). However, the earliest recorded method for estimating early PMI of human remains was a rate-of-change method based on the most easily observed postmortem changes. The cooling of the body after death (algor mortis), the gradual stiffening of the body (rigor mortis), and fixed pooling of the blood resulting in discoloration of the lower portions of the body (livor mortis) could all be easily assessed. Decomposition of a cadaver is a complex process that begins immediately following death and proceeds beyond the time of skeletonization. Decomposition can be broken down into two major stages; the first stage, soft tissue decomposition, is caused by autolysis and putrefaction. Autolysis is the digestion of tissue by cellular enzymes and digestive processes normally present in the organism. Putrefaction is the digestion of whole tissue and this process is caused by fungal and bacterial enzyme activity either as part of the cadaver microbiome or surrounding environment. Early postmortem changes in soft tissue together with temperature and humidity on the day the corpse is discovered can be used to provide an estimate of the PMI until skeletonization is reached (Vass 2011). The rate of soft tissue decomposition can be dramatically affected, however, by factors that impact the body such as cause of death, animal scavenging, environmental conditions (temperature, rainfall, humidity, soil type), presence or absence of clothing, body mass, mummification, and adipocere formation (Rodriguez and Bass 1985; Micozzi 1986; Mant 1987; Vass et al. 1992; Komar 1998; Campobasso et al. 2001). As a result, these easily assessed postmortem changes have yet to serve as reliable estimators of PMI.

Forensic entomology is a well-established science often used for determining time since death in short PMIs soon after death (Myskowiak et al. 1999; Kapil and Reject 2013). There are two approaches for estimating the PMI using insect evidence (Schoenly and Reid 1987). The first method involves the analysis of the pattern of colonization of the cadaver by the succession of insect and other arthropod taxa, and the second method examines the development of immature flies (Diptera) that are deposited on the cadaver shortly after death (Kapil and Reject 2013). Both approaches depend on climate, location of the corpse, and physical condition of the corpse such as presence or absence of clothing, burial or surface dumping, animal scavenging, and cause of death. There is, however, the potential for issues that might arise from the use of arthropod evidence in the determination of the PMI. For example, growth rates of blow flies (Diptera: Calliphoridae) differ when feeding on different body tissues (Clark et al. 2006) and a drug overdose or certain medication taken prior to death may influence their development rate (Goff et al. 1991; O'Brien and Turner 2004). Also, insects might be delayed or prevented from colonizing a cadaver because of environmental parameters such as ambient temperature or burial. All of these scenarios could lead to an under- or over-estimation of the PMI. Once the dry remains stage is reached, the precision of entomological methods for determining the PMI can be compromised. Insects can be used to estimate the season of death, but not necessarily the year of death. Insects are attracted to a cadaver by a change in cadaver chemistry, oftentimes noted as a release of volatile organic compounds (VOCs) (Statheropolous et al. 2007; Dekeirsschieter et al. 2009).

23.2 CHEMISTRY AND ENTOMOLOGY

One strategy to enhance our ability to estimate PMI begins by recognizing the importance of decomposition chemistry. The chemistry of the decomposition environment is crucial to entomology because a cadaver represents a nutrient-rich resource that can be used by many organisms to meet their metabolic needs. Cadaver-derived chemicals serve as sources of energy and nutrients, signals to attract insects, and modulators of ecological succession (Carter et al. 2007). Initially a carcass represents a nutrient-rich habitat patch that is colonized rapidly by insects. These insects are attracted to a cadaver by the numerous chemicals released after death, many of which are VOCs (Vass et al. 2004; Statheropolous et al. 2007; Dekeirsschieter et al. 2009). This attraction facilitates the decomposition of cadavers and formation of gravesoil because insects are effective decomposers and are quite successful at outcompeting scavengers (Devault et al. 2004). However, we still understand relatively little about the dynamics of these volatile chemicals, as they have only recently been explored in detail. These chemical profiles are being characterized to aid in the search of clandestine graves, mass disaster victims, and the estimation of PMI (Statheropolous et al. 2007; Dekeirsschieter et al. 2009; Vass 2012).

Another poorly understood aspect of insect colonization is the role of microorganisms in the signaling of insects to a cadaver. It is well known that many of the putrefactive bacteria can release these volatile compounds, particularly the obligate fermenter *Clostridium* spp. But it has only recently been observed that bacteria can directly mediate oviposition of carrion insects (Ma et al. 2012; Zheng et al. 2013). Thus, one way chemistry plays a crucial role in forensic entomology is because many chemicals attract insects to cadavers and some of these chemicals are the products of microbial metabolism.

Chemistry also modulates entomological resources because it selects for a specialized insect community with the ability to use a cadaver as a resource. During the early PMI, a Cadaver Decomposition Island (CDI) is produced (Carter et al. 2007). The CDI is a patch with high temperature, alkalinity, and reduction potential (Carter et al. 2007). So it is not surprising that the chemistry of this environment selects for a community with the ability to compete in this highly local and ephemeral resource. Decomposition facilitates the activity of many arthropods and inhibits the activity of others. These effects, however, are not static. The organisms that are attracted and repelled do not remain constant throughout the course of cadaver breakdown. The structure of insect and microbial communities changes as the chemical composition of a cadaver changes. It is this process that drives insect succession and is a primary factor when using insect succession to estimate PMI.

Many of these relationships between chemistry and entomology are clearly related to the consumption of a cadaver by insects. Decomposer communities will change as the cadaver is consumed. In addition, significant amounts of cadaver material are also released into soil beneath or surrounding the cadaver. This process is important to recognize because it shows that the decomposition environment comprises two primary habitats, the cadaver itself and the soil into which cadaveric materials are released. These changes in soil chemistry alter the structure of soil insect communities, which are typically associated with a decrease in richness over time (Bornemissza 1957; Anderson and VanLaerhoven 1996). Decomposition has been recently observed to have a similar effect on the microbial diversity of mice in lab settings (Metcalf et al. 2013) and swine, *Sus domesticus* and *Sus scrofa*, in the natural environment (Pechal et al. 2013). These changes in soil chemistry during decomposition (discussed in Section 23.6) represent an opportunity to bridge entomology and soil chemistry as estimators of PMI.

23.3 SOIL CHEMISTRY IN THE CADAVER DECOMPOSITION ISLAND

The CDI is defined as a highly concentrated island of fertility (Carter et al. 2007). It is that region in the soil below and around a decomposing cadaver (Carter et al. 2007; Aitkenhead-Peterson et al. 2012). Soil chemistry under decomposing or dry remains of swine and humans has previously been used to estimate PMIs (Vass et al. 1992; Wilson et al. 2007; Pringle et al. 2010). Unfortunately, every grave site is different in terms of its soil type, which is based on local environmental factors such as vegetation and its associated soil microbiology, climate, geology, topography or aspect, and time since soil formation (Jenny 1945). Thus the effect of the soil chemistry itself on the decomposition products may have an effect on PMI predictions using CDI soil models. Very little research has been conducted on water extractable soil chemistry under human remains in different soil orders. Currently, the Vass et al. (1992) study that examined seven unautopsied, unembalmed cadavers at a research facility at the University of Tennessee, Knoxville, Tennessee is the only one to illustrate the use of soil chemistry in the CDI for determining human PMI. Soils in the Vass et al. (1992) study were a fine, mixed, thermic Typic Paleudalf (Alfisol) sampled to a depth of 3–5 cm, and soil samples were taken from beneath the remains for a period of approximately 2 years. Vass et al. (1992) suggested that soil solution volatile fatty acids (VFA), anions, and cations could be used to estimate PMI. Other research has concentrated on volatile organic carbon (Vass 2008, 2012; Hoffman et al. 2009) where the major classes of compounds detected were alcohols and aldehydes (Hoffman et al. 2009), and cyclic and halogenated compounds in early decomposition and aldehydes and alkanes as decomposition progressed (Vass 2012). Several other studies have examined the soil chemistry and microbiology of the CDI of a range of animals as surrogates for human decomposition (Melis et al. 2007; Howard et al. 2010; Pringle et al. 2010). However, according to Vass (2012), there are markers for VOC release that are present only in human decomposition, which suggests that decomposition chemistry of humans and animals may differ in several respects.

More recent work examined the lateral extent of the CDI under two human subjects with PMIs of 248 and 288 days and accumulated degree days (ADD) of 5469 and 5799 in a forest environment (Aitkenhead-Peterson et al. 2012). After the purge stage, the soil below the cadaver became anaerobic. The huge increase in microbial activity as a response to new substrate uses all available oxygen in the CDI, which means that nitrate and sulfate are used as an electron acceptor or an oxygen source. This reduces the nitrate to nitrous oxide or dinitrogen and sulfate to hydrogen sulfide, which are released as gases (Vass 2012). It can be expected, because of these anaerobic conditions, in undisturbed soil under decomposing intact human remains that the dominant nitrogen species will be organic-N. In anaerobic soil the organic-N cannot be mineralized (converted to ammonium-N) and most of the nitrate-N will have been reduced and released as gas. Furthermore, the process of nitrification (conversion of ammonium-N to nitrate-N) can only be carried out in aerobic soils. In soil with a good supply of oxygen (aerobic), such as CDI soil that has been continually disturbed through the removal of soil cores for analysis or animal burrowing; organic-N is mineralized to ammonium-N through extracellular enzyme cleavage of the amino groups. In these aerobic conditions, ammonium-N is then either transformed to (1) nitrate-N or (2) converted back to organic-N through microbial immobilization or plant uptake.

Recent studies have suggested that ninhydrin reactive nitrogen (NRN) concentrations in soil solutions beneath rats, *Rattus rattus*, and swine may be useful in estimating the PMI (Carter et al. 2008; Van Belle et al. 2009; Spicka et al. 2011). NRN is in the form of proteins, peptides, amino acids, amines, and ammonium, which are all forms that can be released from swine cadaver during decomposition (Van Belle et al. 2009). Van Belle et al. (2009) examined the CDI of surface-placed swine cadavers and sampled soil daily (days 1–10 PMI), twice weekly (days 11–16 PMI), and weekly (days 17–97 PMI) and found that NRN was significantly higher in the CDI relative to control soil after the first 3 days postmortem until the end of their 97-day study. For the first 2 days postmortem, NRN was not significantly different from control soils. NRN peaked at 14 days PMI (92 mg·kg^{-1} soil) and leveled off for 60 days (approximately 80 mg·kg^{-1} soil) before starting to decline to 48 mg·kg^{-1} soil at the end of their study. Examination of the vapor phase of NRN was conducted with success where a higher gaseous NRN release was reported from buried rats compared to exposed rats (Lovestead and Bruno 2011).

23.4 ESTIMATING POSTMORTEM INTERVAL USING SOIL CHEMISTRY

We examined cold water extractable soil carbon, nitrogen, and phosphorus in the CDI below 14 human cadavers at two sites in Texas for this chapter. Reference soils were taken from these sites as soil nutrients do vary seasonally. It was hoped to compare the variation in concentration of each nutrient, in comparison to those in CDI soils. Soil from beneath six cadavers was retrieved from the Southeast Texas Applied Forensic Science (STAFS) Facility, Huntsville, Texas, and from beneath eight cadavers at the Forensic Anthropology Center at Texas State (FACTS), San Marcos, Texas. Each facility is located on soils quite unique from one another. The STAFS facility is located within a 1 km^2 compound at the Center for Biological Field Studies at Sam Houston State University, Walker County, Texas, which is within the Sam Houston National Forest Wildlife Management Area. Vegetation is primarily Loblolly and short leaf pine and soil taken at the facility is classified as loamy, silicious, semiactive, thermic arenic plinthic paleudalfs of the Depcor series (Alfisols). Annual precipitation is 1100 mm, and mean annual temperature is 20.5°C.

The FACTS facility is located on Freeman Ranch, San Marcos, Texas. Vegetation is perennial grassland, which has been invaded by ashe juniper, and soil is a shallow stony clay of the Rumple-Comfort association, a clayey-skeletal, mixed, thermic Udic Argiustolls (Mollisol), which has weathered from dolomitic limestone. Annual precipitation is 857 mm, and mean annual temperature is 19.4°C.

Soils were collected from beneath the torso of cadavers at various stages of decomposition, which ranged from 18 to 580 days postmortem at STAFS and 43 to 407 days postmortem at FACTS (Table 23.1). One sample was taken from a new and previously undisturbed CDI to reduce the aeration effect of multiple sampling. Each sample (reference and CDI soil) was cold-water extracted at a 1:10 soil:water ratio and analyzed. To assess the concentrations of human decomposition in the CDI, we deducted reference soil chemistry from the CDI soil chemistry collected on the same date. Values for average or ambient soil are of an average reference soil concentrations that are set at zero (Figure 23.1) to show magnitudes of increase or decrease in relation to ambient soil concentration.

Table 23.1 Details of Cadavers Sampled at Southeast Texas Applied Forensic Science (STAFS) and Forensic Anthropology Center at Texas State (FACTS) Facilities. If Sampled Date Is after the Removed Date, Then the Cadaver Decomposition Island with No Cadaver Was Sampled. If Sampled Date Is before the Removed Date, Then the CDI Was Sampled Beneath the Torso of the Cadaver

Site	Weight (kg)	Sex	PMI (d)	ADD	Placed Date	Removed Date	Sampled Date
FACTS	122.02	M	348	24,825	February 17, 2011	September 13, 2011	January 30, 2012
FACTS	98.88	M	196	13,507	July 19, 2011	January 2, 2012	January 30, 2012
FACTS	92.97	F	176	11,741	August 8, 2011	Still out	January 30, 2012
FACTS	90.72	M	407	27,593	December 20, 2010	Still out	January 30, 2012
FACTS	47.63	M	90	4,943	November 2, 2011	Still out	January 30, 2012
FACTS	61.24	F	90	4,943	November 2, 2011	Still out	January 30, 2012
FACTS	86.18	M	82	5,175	December 19, 2011	Still out	March 9, 2012
FACTS	65.77	F	43	2,250	December 19, 2011	Still out	January 30, 2012
STAFS	56.7	F	459	31,128	August 17, 2009	December 15, 2009	November 18, 2010
STAFS	131.54	M	580	36,902	August 9, 2009	February 22, 2011	March 11, 2011
STAFS	72.57	F	337	22,904	December 7, 2009	September 20, 2010	November 18, 2010
STAFS	181.44	M	297	21,030	January 26, 2010	Still out	November 18, 2010
STAFS	72.57	F	18	1,199	March 8, 2011	Still out	March 25, 2011
STAFS	72.57	M	18	1,199	March 8, 2011	Still out	March 25, 2011

Note: Still out indicates that at the time of sampling the cadaver was in place.
ADD, accumulated degree days; PMI, postmortem interval.

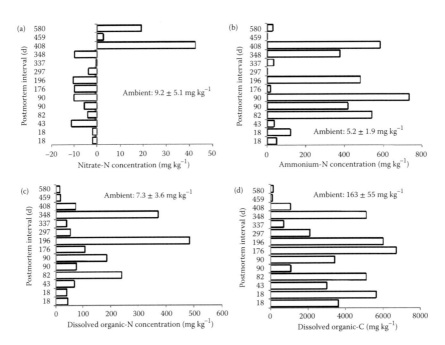

Figure 23.1 Water extractable human decomposition products in the Cadaver Decomposition Island (CDI) of 14 individual cadavers with differing postmortem interval (PMI). Values represent CDI soil extract minus reference soil extract with reference concentrations set at zero for (a) nitrate-N, (b) ammonium-N, (c) dissolved organic nitrogen, and (d) dissolved organic carbon (mg·kg⁻¹ soil). Ambient concentrations are mean concentrations for reference soils ± standard deviation across the two sites for period of the study.

23.4.1 Soil Nitrogen

Concentrations of water extractable soil nitrate-N remained below ambient (zero on Figure 23.1a) soil concentrations for approximately 1 year (365 days postmortem) and thereafter increased above ambient soil concentrations (Figure 23.1a). A similar observation was reported by Anderson et al. (2013) in the CDI of decomposing, 20-kg swine carcasses. Because the purge initiates anaerobic conditions in the CDI, nitrate-N is reduced ($NO_{3\ (l)}^{-} \rightarrow NO_{2\ (l)}^{-} \rightarrow NO_{(g)} + N_2O_{(g)} \rightarrow N_{2(g)}$). The increase in nitrate-N concentration in the CDI after a year was likely due to burrowing by macroinvertebrates and plant root growth inducing oxygenated conditions. Ammonium-N cannot be nitrified in soil containing purge fluid as the reaction of nitrification is aerobic. While it is expected that ammonium-N in the CDI will remain high, it is also the form of nitrogen most used by soil bacteria and fungi (Sagara 1976; Hopkins et al. 2000; Tibbett and Carter 2003). Therefore, as ammonium-N increases as the cadaver decomposes it can also be expected to decrease due to microbial uptake (Figure 23.1b). Of the nitrogen species, dissolved organic nitrogen (DON) tended to increase in the CDI up to approximately 196 days PMI and then decrease (Figure 23.1c). Only certain bacteria can mineralize DON to ammonium-N under the anaerobic conditions in the CDI, and our data suggest that both mineralization (DON → NH_4–N) and immobilization (NH_4–N → DON) may occur after 176 days PMI (Figure 23.1b and c). The decline in ammonium-N and DON with a subsequent increase in nitrate-N after approximately 1 year suggests that normal aerobic soil conditions are in place.

23.4.2 Soil Carbon

Carbon is the second most abundant element in the human body. For every kilogram of body weight, there is approximately 230 g carbon. In the CDI, dissolved organic carbon (DOC) tends to remain high for about year (Figure 23.1d), although extremely high concentrations of DOC ($120-150$ mg·kg^{-1}) have been reported at the bottom of graves in a 27-year-old burial site in Germany (Fiedler et al. 2004). DOC is a substrate for soil microorganisms along with NH_4–N and expectations are that it would be eventually mineralized to CO_2 but not until aerobic conditions are restored in the CDI. The slow breakdown of DOC and potential loss as methane gas under anaerobic conditions makes it a good option for predicting PMI.

A relatively good predictive model was achieved by deducting reference soil DOC from CDI soil extracts and then dividing the remaining DOC by the initial mass of the cadaver (Figure 23.2a and b). However, a better predictive model was achieved when using the decomposition products DOC:DON ratio divided by the initial mass of the cadaver (Figure 23.2c and d). Observed versus predicted PMI values varied. The best model used the decomposition products DOC:DON ratio divided by the initial mass of the cadaver (Figure 23.2c); here 50% of the predicted values were within 60 days PMI,14% within 100 days PMI, 29% within 150 days PMI, and one sample was underpredicted by 165 days PMI. While the use of the DOC:DON ratio does not achieve perfect predictions of PMI, it does recognize the typical biogeochemical cycling that occurs within soils. One of the major issues of using DOC and DON concentrations for PMI predictions is that both constituents show a hyperbolic curve over time and so concentrations could be similar at early and extended decomposition periods. Nevertheless, based on the examination of soil chemistry of the 14 cadavers at two sites in Texas, clearly fluctuations in important microbial substrates and nutrients occur. Another approach may be a method using soil chemistry whereby several chemistries or functional groups can be assessed.

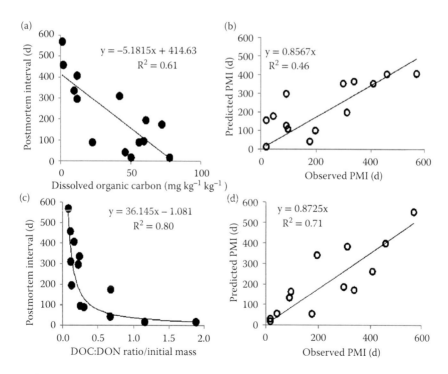

Figure 23.2 Predictive models for PMI using, (a) decomposition product dissolved organic carbon (DOC) (CDI soil DOC minus ambient [reference soil] DOC) divided by initial cadaver mass (kg), and (c) decomposition product DOC:DON ratio divided by initial cadaver mass. The relative success of predictive models is shown for (b) DOC and (d) DOC:DON ratio, where PMI was estimated using the model equations.

23.5 ESTIMATING POSTMORTEM INTERVAL USING DIFFUSE REFLECTANCE NEAR-INFRARED SPECTROSCOPY

Recognition that DOC and DON may be valuable indicators of grave soil and useful for determining PMI, we turned to diffuse reflectance near-infrared (DR-NIR) spectroscopy as an alternative, nondestructive method of capturing the whole CDI soil spectra under decomposing cadavers. Over the last decade, DR-NIR spectroscopy has proven a useful tool for quantifying several soil characteristics. Recent research has shown that diffuse reflectance is correlated with variable soil properties (Ben-Dor and Banin 1995; Reeves et al. 2000; Zornoza et al. 2008). A benefit of using this method of analysis is that it is nondestructive (i.e., soil can be used for other analyses) and relatively small sample sizes (approximately 3–10 g) can be used. There has been considerable success in predicting soil nitrogen ($r^2 = 0.95$), calcium, and magnesium ($r^2 = 0.95$–0.91), phosphatase ($r^2 = 0.93$), and total phospholipids fatty acid (PLFA) ($r^2 = 0.91$) (Zornoza et al. 2008). It has also been suggested that near-infrared spectroscopy is often more precise than standard laboratory assays because operator error is removed (Foley et al. 1998).

Instead of using DR-NIR spectroscopy as a tool to predict chemical components in soil, our preliminary research examined the whole spectral signature of soil under human remains. High correlations ($R = 0.70$ to 0.80 and −0.70 to −0.80) between individual peaks or troughs in the spectral signature and known PMI suggest that this could be an invaluable method to estimate PMI using single wavelengths in empirical models or several wavelengths in partial least squares (PLS) or backward linear multiple (BLM) regression analysis to predict PMI.

For this exercise, DR-NIR spectroscopy was used (Labspec 500, ASD Inc, Boulder, CO.) to scan repeated collections of soil from under the torso of a single cadaver at the STAFS facility over a period of 20 months ($n = 16$). The cadaver used in this study was a male of approximately 122.5 kg at death. He was laid unclothed on the surface of the soil on August 18, 2009 and the skeletal remains removed on February 22, 2011. Soil sampling of the CDI commenced on November 3, 2011 at 580 days PMI and finished on December 13, 2012 at 1269 PMI. Significant differences were apparent in the soil spectra between the first and last PMI soil scans (Figure 23.3) further suggesting that DR-NIR spectroscopy might be a useful tool for the prediction of PMI.

The 16 soils were scanned three times each, rotating the soil 90° between each scan. Raw spectral data were transformed to the first derivative and each 1-nm wavelength in the spectra was

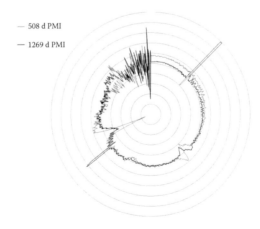

Figure 23.3 Diffuse reflectance soil spectral images in the near-infrared range (800–2400 nm) for soils collected at the Southeast Texas Applied Forensic Science facility from the same CDI at 580 and 1269 days postmortem interval.

averaged. The resultant wavelength data for (1) full spectra (350–2500 nm), (2) UV–Visible spectra (350–700 nm), and (3) near-infrared spectra (700–2500 nm) were used with PMI values as the dependent variable in PLS regression analysis with a full cross-validation (Figure 23.4). Standard error for the full spectra ranged from 106 to 117 days PMI (Figure 23.4a), and for the UV–Visible spectra, the standard error for PMI ranged from 162 to 175 days PMI (Figure 23.4b). The near-infrared spectra provided the best results (Figure 23.4c) with a standard error of between 13 and 16 days PMI and the observed PMI fell within the window of predicted PMI for all dates.

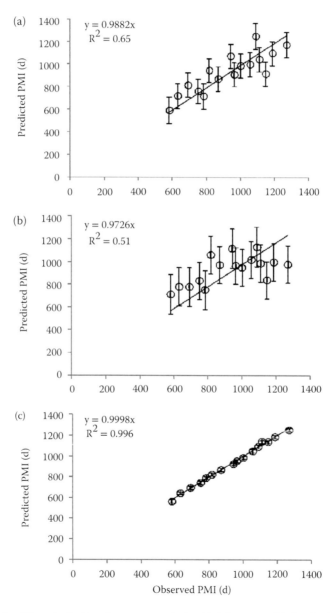

Figure 23.4 Results of the principal least squares analysis with full cross-validation for the estimation of PMI using diffuse reflectance spectroscopy for (a) wavelengths 350–2500 nm and (b) wavelengths 350–700 nm. Backward multiple linear regressions analysis was used to identify wave lengths for predicting PMI for (c) wavelengths 700–2500 nm. Error bars are standard error. The trend lines intercept at zero to assess under- and overpredictions.

Data in the near-infrared portion of the spectra were then used in backward stepwise multiple regression analysis with known PMIs as the dependent variable to identify which wavelengths had the most influence on predicting PMI. The backward stepwise analysis presented 15 individual 1-nm wavelengths that could be used in a predictive model for PMI (Table 23.2).

The use of DR-NIR spectroscopy to predict PMI in soils beneath human remains appears to be successful, but more research needs to be done with this method, which includes different soil types and climates. The study here spanned 580–1269 days PMI and using the near-infrared portion of the spectra was able to predict PMI to within 13–16 days PMI, which could present a useful bridge in PMI predictions for forensic entomologists when presented with human remains over a year old.

Table 23.2 Coefficients and Wavelengths Used for Predicting Postmortem Interval (PMI) from Diffuse Reflectance Spectroscopy of a Single Cadaver Decomposition Island over a Sampling Period of 20 Months. Wavelengths in the Near-Infrared (700–2500 nm) Were Used in Backward Stepwise Regression Analysis to Isolate Important Wavelengths for PMI Prediction. This Model Was Used to Predict PMI in Figure 23.4c

Wavelength (nm)	Unstandardized Coefficients	Standard Error	Standardized Coefficients Beta	Vibrational Modes	Functional Groups
(Constant)	1277.8	0			
970	624222.0	0	0.31	Third overtone CH bands	CH, R–OH
984	−1750993.8	0	−0.91	Third overtone CH bands	CH, R–OH
998	85849.1	0	0.44	Third overtone CH bands	CH, R–OH
1000	195359.6	0	1.02	Second overtone NH bands	R–NHR
1002	−11433.6	0	−0.06	Second overtone NH bands	R–NHR
1828	−429521.5	0	−1.39	First overtone CH bands	CH_2
1829	−771838.8	0	−2.51	First overtone CH bands	CH_2
1830	−584766.4	0	−1.84	First overtone CH bands	CH_2
1831	1717470.4	0	5.20	First overtone CH bands	CH_2
1832	−270541.3	0	−0.77	First overtone CH bands	CH_2
1833	−147665.5	0	−0.38	First overtone CH bands	CH_2
1862	−356163.5	0	−0.25	Sixth overtone C–Cl bands	C–Cl
2094	1361614.7	0	0.81	First overtone of CO and O–H combination bands	R–OH, R–CONH,
2294	−450968.6	0	−0.63	C–H + C–H combination bands	CH, CH_2, CH_3
2353	−480887.3	0	−0.69	C–H + CC combination bands	CH, CH_2,

23.6 OPPORTUNITIES TO BRIDGE ENTOMOLOGY AND SOIL-CHEMISTRY-ESTIMATED POSTMORTEM INTERVAL IN FORENSIC RESEARCH

There appears to be great potential for bridging the forensic application of entomology and soil chemistry, particularly during the extended PMI after insect larvae have migrated. Ultimately, there are two approaches to bridging these two applications. First, soil chemistry can be used to complement entomological estimates of PMI. This approach has value during early and extended PMIs. During the early PMI, soil chemistry can be used to corroborate the estimates based on the development of blow fly larvae. Changes in soil chemistry can occur during the initial 72 hours postmortem while insects are actively feeding and developing (Spicka et al. 2011). In this scenario, blow fly larval data would probably serve as the primary estimator of PMI while the soil chemical profile would be used to demonstrate that the decomposer environment was consistent with a maggot mass of said PMI.

It is envisioned that the use of soil chemistry and entomology would be slightly different during the extended PMI, after blow fly larvae have migrated. In this scenario, soil chemistry could be used as the primary estimator of PMI. For example, using the data presented in Figure 23.1, an elevated concentration of nitrate-N would be consistent with a PMI of approximately 408 days. These data could then be compared to blow fly puparia and the insect communities inhabiting the corpse to determine if they corroborate insect development and succession data, respectively. Anderson et al. (2013) briefly described this approach when using soil NRN to estimate PMI up to 3 years postmortem.

The other primary strategy in which soil chemistry can complement entomological estimates of PMI is when insect data are not available due to season, burial, or lack of expertise. In this scenario, soil chemistry using wet chemistry and DR-NIR would be used as described in the extended PMI scenario mentioned previously, but there would be no entomology data with which to corroborate. Gravesoil chemistry would need to be compared to the physical condition of the cadaver to determine if chemical concentrations are consistent with the degree of breakdown. For DR-NIR spectra, statistical comparisons of known gravesoil spectra would be important, such as those described in this chapter. This strategy will require a more detailed understanding of cadaver breakdown in the absence of insects, probably to include the relationships between soil chemistry and scavenging. Yet the most likely investigative scenario in which this would be used is the processing of a clandestine burial. Because of this, we are in dire need of more detailed investigations into the relationships between the arthropods and soil chemistry of burials.

23.7 CASE STUDIES

While there are some published case studies on the use of entomology to determine PMI, there are few case studies on the use of soil or water-extractable soil solution from the CDI for the determination of PMI. The pattern of insect succession together with autopsy findings were important in determining PMI in a case of two missing boys found in a dry cistern in Italy (Introna et al. 2011). In this particular case, the potential of soil or sediment at the bottom of the well to further confirm PMI may not have been applicable because of the effect of a very cool temperature on decomposition rate. A second case study reported insect evidence from a cold case some 9 years after it had been collected but not considered (Magni et al. 2012). Here, on examining the desiccated insect evidence, the authors called for the appropriate preservation of evidence collected.

Case studies where gravesoil chemistry has been used as physical evidence to understand taphonomy are few. Vass et al. (1992) described a number of investigations in support of their method to estimate PMI. More recently, Carter et al. (2009) used gravesoil NRN to identify the decomposition site of a suicide in southeastern Nebraska. This work was done in conjunction with forensic entomologists that used blow fly puparia to estimate the season of death. The results obtained were consistent with one another, which allowed the investigators to test their investigative hypothesis that the death occurred within close proximity of the location of discovery and the remains had been undiscovered for several months.

Because chemistry and entomology are strongly related, we predict that the number of investigations using soil chemistry will increase greatly over the coming decades. This will require collaborative work among researchers examining human decomposition using insects, soil chemistry, and volatiles.

23.8 CONCLUSION AND FUTURE RESEARCH NEEDS

In future studies, multiple replicates of each soil sample should be extracted so that a mean and standard deviation can be calculated, which will enable a window of PMI to be estimated. This of course is time extensive and analysis can be expensive.

The preliminary examination of the use of DR-NIR spectroscopy on soil retrieved from the CDI suggests that this may be an alternative method for predicting PMI under remains that have reached the skeletal stage. Much more research needs to be done using DR-NIR spectroscopy for recent and extended PMI in conjunction with forensic entomology in buried and surface human decomposition scenarios and across different soil types and climates.

ACKNOWLEDGMENTS

We thank the families who have donated their loved ones. Without them, none of the research described in this chapter could have been completed. We thank students and staff at the STAFS and FACTS facilities for access to CDIs for the collection of soil samples. We thank Nina Stanley for the wet chemical analysis of soil samples collected and Texas A&M undergraduate Austin Moore for help in field soil collections. Collection and analysis of soils was unfunded.

REFERENCES

Aitkenhead-Peterson, J.A., C.G. Owings, M.B. Alexander, N. Larison, and J. A. Bytheway. 2012. Mapping the lateral extent of cadaver decomposition islands with soil chemistry. *Forensic Science International* 216: 127–134.

Anderson, B., J. Meyer, and D.O. Carter. 2013. Dynamics of ninhydrin-reactive nitrogen and pH in gravesoil during the extended postmortem interval. *Journal of Forensic Sciences* 58: 1348–1352.

Anderson, G.S. and S.L. VanLaerhoven. 1996. Initial studies on insect succession on carrion in Southwestern British Columbia. *Journal of Forensic Sciences* 41: 617–625.

Ben-Dor, E. and A. Banin 1995. Near infrared analysis as a rapid method to simultaneously evaluate several soil properties. *Soil Science Society of America Journal* 59: 364–372.

Bornemissza, G.F. 1957. An analysis of arthropod succession in carrion and the effect of its decomposition on the soil fauna. *Australian Journal of Zoology* 5: 1–12.

Brown, K., A. Thorne, and M. Harvey. 2012. Preservation of *Calliphora vicina* (Diptera: Calliphoridae) pupae for use in post-mortem interval estimation. *Forensic Science International* 223: 176–183.

Campobasso, C.P., G. Di Vella, and F. Introna. 2001. Factors affecting decomposition and Diptera colonization. *Forensic Science International* 120: 18–27.

Carter, D.O., J. Filippi, T.E. Huntington, M.I. Okoye, M. Scriven, and J. Bliemeister. 2009. Using ninhydrin to reconstruct a disturbed outdoor death scene. *Journal of Forensic Identification* 59: 190–195.

Carter, D.O., D. Yellowlees, and M. Tibbett. 2007. Cadaver decomposition in terrestrial ecosystems. *Naturewissenschaften* 4: 12–24.

Carter, D.O., D. Yellowlees, and M. Tibbett. 2008. Using ninhydrin to detect gravesoil. *Journal of Forensic Science* 53: 397–400.

Clark, K., L. Evans, and R. Wall. 2006. Growth rates of the blowfly, *Lucilia sericata*, on different body tissues. *Forensic Science International* 156: 145–149.

Dekeirsschieter, J., F.J. Verheggen, M. Gohy, F. Hubrecht, L. Bourguignon, G. Lognay, and E. Haubruge. 2009. Cadaveric volatile compounds released by decaying pig carcasses (*Sus domesticus* L.) in different biotopes. *Forensic Science International* 189: 46–53.

Devault, T.L., I.L. Brisbin Jr., and O.E. Rhodes. 2004. Factors influencing the acquisition of rodent carrion by vertebrate scavengers and decomposers. *Canadian Journal of Zoology* 82: 502–509.

Fiedler, S., K. Schneckenberger, and M. Graw. 2004. Characterization of soils containing adipocere. *Archives of Environmental Contaminant Toxicology* 47: 561–568.

Foley, W.J., A. McIlwee, and I. Lawler, L. Aragones, A.P. Woolnough, and N. Berding. 1998. Ecological applications of near infrared reflectance spectroscopy a tool for rapid, cost-effective prediction of the composition of plant and animal tissues and aspects of animal performance. *Oecologia* 116: 293–305.

Goff, M.L., W.A. Brown, K.A. Hewadikaram, and A.I. Omori. 1991. Effect of heroin in decomposing tissues on the development rate of *Boettcherisca peregrina* (Diptera, Sarcophagidae) and implications of this effect on estimation of postmortem intervals using arthropod development patterns. *Journal of Forensic Science* 36: 537–542.

Hoffman, E.M., A.M. Curran, N. Dulgerian, R.A. Stockam, and B.A. Eckenrode. 2009. Characterization of the volatile organic compounds present in the headspace of decomposing human remains. *Forensic Science International* 186: 6–13.

Hopkins, D.W., P.E. J. Wiltshire, and B.D. Turner. 2000. Microbial characteristics of soils from graves: an investigation at the interface of soil microbiology and forensic science *Applied Soil Ecology* 14: 283–288.

Howard, G.T., B. Duos, and E.J. Watson-Horzelski. 2010. Characterization of the soil microbial community associated with the decomposition of a swine carcass. *International Biodeterioration and Biodegradation* 64: 300–304.

Introna, F., A. De Donno, V. Santoro, S. Corrado, V. Romano, F. Porcelli, and C.P. Campobasso. 2011. The bodies of two missing children in an enclosed underground environment. *Forensic Science International* 207: E40–E47.

Jenny, H. 1945. Arrangement of soil series and types according to functions of soil forming factors. *Soil Science* 61: 375–391.

Kapil, V. and P. Reject. 2013. Assessment of post mortem interval, (PMI) from forensic entomotoxicological studies of larvae and flies. *Entomology, Ornithology and Herpetology* 2: 104.Komar, D.A. 1998. Decay rates in a cold climate region: A review of cases involving advanced decomposition from the Medical Examiner's Office in Edmonton, Alberta. *Journal of Forensic Science* 43: 57–61.

Lovestead, T.M. and T.J. Bruno. 2011. Detecting gravesoil with headspace analysis with adsorption on short porous layer open tubular (PLOT) columns. *Forensic Science International* 204: 156–161.

Ma, Q., A. Fonseca, W. Liu, A.T. Fields, M.L. Pimsler, A.F. Spindola, A.M. Tarone, T.L. Crippen, J.K. Tomberlin, and T.K. Wood. 2012. *Proteus mirabilis* interkingdom swarming signals attract blow flies. *The ISME Journal* 6: 1356–1366.

Magni, P.A., M.L. Harvey, L. Saravo, and I.R. Dadour. 2012. Entomological evidence: Lessons to be learnt from a cold case review. *Forensic Science International* 223: E31–E34.

Mant, A.1987. Knowledge acquired from post-War exhumations, p. 249. In A. Boddington, A. Garland (ed.), *Death, Decay, and Reconstruction*. Manchester University Press, Manchester, England.

Melis, C., N. Selva, I. Teurlings, C. Skarpe, J.D.C. Linnell, and R. Andersen. 2007. Soil and vegetation nutrient response to bison carcasses in Bialowieza primeval forest, Poland, *Ecological Research* 22: 807–813.

Metcalf, J., L. Wegener-Parfrey, A. Gonzalez, C.L. Lauber, D. Knights, G. Ackermann, G. Humphrey et al. 2013. A microbial clock provides an accurate estimate of the postmortem interval in a mouse model system. *eLIFE* 2: e1104.

Micozzi, M.S.1986. Experimental study of postmortem change under field conditions: Effects of freezing, thawing, and mechanical injury. *Journal of Forensic Science* 31:953–61.

Myskowiak, J.B., B. Chauvet, T. Pasquerault, C. Rocheteau, and J.M. Vian. 1999. A summary of six years of forensic entomology field work. The role of necrophagous insects in forensic science. *Annales de la Societe Entomologique de France* 35: 569–572.

O'Brien, C. and B. Turner. 2004. Impact of paracetamol on *Calliphora vicina* larval development. *International Journal of Legal Medicine* 118: 188–189.

Pechal, J.L., T.L. Crippen, M.E. Benbow, A.M. Tarone, S. Dowd, and J.K. Tomberlin. 2013. The potential use of bacterial community succession in forensics as described by high throughput metagenomic sequencing. *International Journal of Legal Medicine*. 1–13.

Pringle, J.K., J.P. Cassella, and J.R. Jervis. 2010. Preliminary soilwater conductivity analysis to date clandestine burials of homicide victims, *Forensic Science International* 198: 126–133.

Reeves, J.B., G.W. McCarty, and J.J. Meisinger. 2000. Near infrared reflectance spectroscopy for the determination of biological activity in agricultural soils. *Journal of Near Infrared Spectroscopy* 8:161–170.

Reibe, S., P. von Doetinchem, and B. Madea. 2010. A new simulation-based model for calculating post-mortem intervals using development; data for *Lucilia sericata* (Dipt: Calliphoridae. *Parasitology Research* 107: 9–16.

Richards. C.S., T.J. Simonsen, R.L. Abel, M.J. R. Hall, D.A. Schwyn, and M. Wicklein. 2012. Virtual forensic entomology: Improving estimates of minimum post-mortem interval with 3D micro-computed tomography. *Forensic Science International* 220: 251–264.

Rodriguez, W.C., 3rd and W.M. Bass. 1985. Decomposition of buried bodies and methods that may aid in their location. *Journal of Forensic Science* 30: 836–52.

Sagara, N. 1976. Presence of a buried mammalian carcass indicated by fungal fruiting bodies. *Nature* 262: 816.

Schoenly, K. and W. Reid. 1987. Dynamics of heterotrophic succession in carrion arthropod assemblages: Discrete series or a continuum of change? *Oecologia* 73: 192–202.

Spicka, A., R. Johnson, J. Bushing, L.G. Higley, and D.O. Carter. 2011. Carcass mass can influence rate of decomposition and release of ninhydrin-reactive nitrogen into gravesoil. *Forensic Science International* 209: 80–85.

Statheropoulos, M., A. Agapiou, C. Spiliopoulou, G.C. Pallis, and E. Sianos. 2007. Environmental aspects of VOCs evolved in the early stages of human decomposition. *Science of the Total Environment* 385: 221–227.

Tibbett, M. and D. O. Carter. 2003 Mushrooms and taphonomy: The fungi that mark woodland graves. *Mycologist* 17(1): 20–24.

Van Belle, L.E., D.O. Carter, and S.L. Forbes. 2009. Measurement of ninhydrin reactive nitrogen influx into gravesoil during aboveground and belowground carcass (*Sus domesticus*) decomposition. *Forensic Science International* 193: 37–41.

Vass, A.A. 2008. Odor analysis of decomposing buried human remains. *Journal of Forensic Science* 53: 384–391.

Vass A.A. 2011. The elusive universal post-mortem interval formula. *Forensic Science International* 204: 34–40.

Vass, A.A. 2012. Odor Mortis. *Forensic Science International* 222: 234–241.

Vass, A.A., W.M. Bass, J.D. Wolt, J.E. Foss, and J.T. Ammons. 1992. Time since death determinations of human cadavers using soil solution. *Journal of Forensic Science* 37: 1236–1253.

Vass, A.A., R.R. Smith, C.V. Thompson, M.N. Burnett, D.A. Wolf, J.A. Synstelien, N. Dulgerian, and B.A. Eckenrode. 2004. Decompositional odor analysis database. *Journal of Forensic Science* 49: 760–769.

Wilson, A.S., R.C. Janaway, A.D. Holland, H.I. Dodson, E. Baran, A.M. Pollard, and D.J. Tobin. 2007. Modelling the buried human body environment in upland climes using three contrasting field sites, *Forensic Science International* 169: 6–18.

Zheng, L., T.L. Crippen, L. Holmes, B. Singh, M.L. Pimsler, M.E. Benbow, A.M. Tarone et al. 2013. Bacteria mediate oviposition by the black soldier fly, *Hermetia illucens* (L.), (Diptera: Stratiomyidae). *Scientific Reports* 3: Article Number 2563.

Zornoza, R., C. Guerro, J. Mataix-Solera, K.M. Skow, V. Arcenegui, and J. Mataix-Beneyto. 2008. Near infrared spectroscopy for determination of various physical, chemical and biochemical properties in Mediterranean soils. *Soil Biology and Biochemistry* 40: 1923–1930.

CHAPTER 24

Molecular Biology in Forensic Entomology

Aaron M. Tarone, Baneshwar Singh, and Christine J. Picard

CONTENTS

24.1 INTRODUCTION

In 2005, a convergence of legal and scientific pressures on the forensic sciences was cited as the beginning of the end of business as usual for their implementation (Saks and Koehler 2005). One major reason for this perception of change in forensic sciences research was the rise of DNA as the "gold standard." In part, this success of DNA analysis in the courtroom was due to the robust state of the population genetics community already in existence when forensic DNA research came of age. They provided a clear path for investigators to follow to "get it right" with their interpretations of DNA evidence (National Research Council [U.S.]. Committee on DNA Forensic Science: An Update and National Research Council [U.S.]. Commission on DNA Forensic Science: An Update 1996). The genetics community example highlighted weaknesses in the nature of inferences derived from other fields, culminating in a National Research Council (NRC) report on forensic sciences in 2009 (National Research Council [U.S.] Committee on Identifying the Needs of the Forensic Science Community et al. 2009), which called for an increase in basic research in forensic sciences.

In the context of the NRC report, forensic entomology fares better than some forensic disciplines, as the research community has produced decades of published scientific research testing identifiable, repeatable hypotheses. However, there is room for improvement, which has been highlighted by a number of researchers in the field (Tomberlin et al. 2011a,b; Michaud et al. 2012; Barton et al. 2013; Fiene et al. 2014). These papers identified basic areas of research specifically targeting genetics, evolution, ecology, and statistics as areas that can move the field forward. These fields converge with molecular biology research on some level; thus, this chapter addresses the molecular biology aspects of each of these areas, specifically with relation to molecular work with nucleic acids (DNA and RNA). The first section of this chapter will highlight the state of research in DNA-based species identifications and provide a framework for the applications of DNA sequence analysis to the identification of evidentiary samples. Subsequent sections deal with newer applications of genetic and genomic analyses to forensic entomology research, including analyses of RNA.

24.2 SYSTEMATICS AND PHYLOGENETICS

24.2.1 Evolutionary Relationships of Insects

The first order of business in a forensic analysis of insect evidence is to determine the species in evidence. This is a critical step in evaluation, which can adversely affect results if it is done incorrectly. Accordingly, there is a considerable amount of effort devoted toward understanding the best ways to identify insect species. Although several insect orders are forensically relevant, members of Coleoptera and Diptera are used most frequently in forensic investigations (Byrd and Castner 2010). Thus, the evolution of the species within these orders is of critical importance to their relevance in the courtroom. Both orders are very diverse and together constitute more than half of all known insects (Lawrence et al. 2011; Wiegmann et al. 2011). Traditional methods of species identification employ the use of morphological characters. In some cases though, evidentiary samples may be damaged, making morphological analyses difficult or impossible. Furthermore, even with the best-quality evidence, few morphological keys exist for immature insects of forensic importance and often only taxonomic specialists are capable of correct identifications. In such an instance, if the insect evidence is collected and a forensic entomologist is contacted immediately the immature insects can be reared to adulthood for identification. However, this is very time consuming or not possible in many insects (i.e., some *Lucilia* spp. [Diptera: Calliphoridae] do not rear well in the laboratory, and morphological keys exist only for male adult flies in the Sarcophagidae [Pape 1987; Stamper et al. 2013]). Alternatively, DNA-based species identifications are useful as they can be applied for identification regardless of life stage, sex, or the quality of the sample. To use DNA-based species identification methods, it is important that markers are informative of intraspecific genetic variation, not interspecific variation, and to understand the evolutionary relationships among species in these insect orders.

24.2.1.1 Coleoptera

With 360,000 known species (approximately 25% of all known organisms on the Earth), beetles constitute the largest clade in the Tree of Life and have a worldwide presence with the exception of Antarctica (Beutel and Haas 2000). Coleopterans have diverse feeding habits that include, but are not limited to, phytophagy, saprophagy, and parasitism, making them likely associates of human remains. Coleoptera is divided into 4 suborders, 17 superfamilies, and 168 families

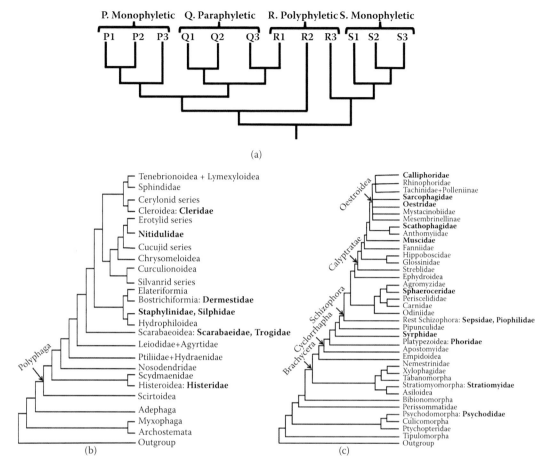

Figure 24.1 Phylogenetic trees showing relationships between insects of forensic importance. (a) Phylogeny of hypothetical species P (P1, P2, and P3), Q (Q1, Q2, and Q3), R (R1, R2, and R3) and S (S1, S2, and S3). The nature of species relationships can help investigators understand if the information they are using is capable of reliably identifying species of interest. Reliable identifications require monophyletic relationships in a phylogenetic tree. (b) Phylogenetic relationships among the beetles, adapted from Hunt et al. Families that include forensically important beetle species are indicated in bold fonts. (c) Phylogenetic relationships among the true flies. Groups of forensic importance are indicated in bold fonts. (b: Adapted from Hunt, T. et al. *Science* 318, 1913–1916, 2007); (c: Adapted from Wiegmann, B.M. et al. *Proceedings of the National Academy of Sciences of the United States of America* 108, 5690–5695, 2001, and Singh, B. and J.D. Wells, *Journal of Medical Entomology*, 50, 15–23, 2013.)

(Hunt et al. 2007). Although the monophyly (see Figure 24.1 and Section 24.2.2) of Coleoptera is well supported by both morphology (Beutel and Haas 2000) and molecular data (Hunt et al. 2007; Song et al. 2010), the relationships between suborders differ both within and between molecular and morphological analyses (Beutel and Haas 2000; Hunt et al. 2007; Song et al. 2010; Lawrence et al. 2011). Morphologically, the monophyly of all suborders is well supported, but in molecular systematic analyses only the monophyly of Adephaga is well supported (Hunt et al. 2007; Song et al. 2010). Extensive molecular systematic analysis by Hunt et al. (2007) supports the monophyly of Polyphaga, but it is contradicted in recent molecular systematic analyses by Song et al. (2010). Of the 168 known families, 8 families are of forensic importance, and all belong to the suborder Polyphaga (see Figure 24.1b, with names in bold font).

24.2.1.2 Diptera

With more than 150,000 species (>10% of all known organisms on Earth) from more than 150 families, Diptera (true flies/two-winged flies) are one of the most diverse groups of organisms on Earth (Wiegmann et al. 2011). They are distributed in all continents, including Antarctica (Baust and Edwards 1979) and constitute a well-supported monophyletic group (Figure 24.1c) (Wiegmann et al. 2011). Like the Coleoptera, the Diptera have diverse feeding habits, which include phytophagy, saprophagy, hematophagy, and parasitism. Traditionally, Diptera have been divided into two suborders (Nematocera and Brachycera) based on long (>6) or short (<6) antennal segments, respectively. Nematocera is not a monophyletic group (Wiegmann et al. 2011; Lambkin et al. 2013), but Brachycera and many other groups within Brachycera are well-supported monophyletic groups (see groups with arrow in Figure 24.1c). Among all Dipteran families, systematic positions of Tipulidae, Culicidae, Tabanidae, Calliphoridae, Sarcophagidae, Tachinidae, Oestridae, Lauxaniidae, Diopsidae, Sepsidae, and Tephritidae are either weakly supported or differ in various systematic analyses (Wiegmann et al. 2011; Caravas and Friedrich 2013; Singh and Wells 2013). Note that several of these are forensically important taxa, highlighting the need for expertise to correctly identify such species.

Although several families of Diptera are of forensic importance (see Figure 24.1c, with names in bold font), members of Calliphoridae (blow flies) and Sarcophagidae (flesh flies) are encountered more frequently in forensic entomological investigations, and hence it is important to understand phylogenetic relationships within and between these groups. The systematic position of both families within the superfamily Oestroidea is contentious and differs among systematic analyses (Pape 1992; Rognes 1997; Kutty et al. 2010; Wiegmann et al. 2011; Caravas and Friedrich 2013; Lambkin et al. 2013; Singh and Wells 2013).

Traditionally, the Calliphoridae includes 13 subfamilies (Ameniinae, Aphyssurinae, Bengaliinae, Calliphorinae [including Melanomyinae], Chrysomyinae, Helicoboscinae, Luciliinae, Mesembrinellinae, Phumosiinae, Polleniinae, Prosthetosominae, Rhiniinae, and Toxotarsinae), but there is a strong evidence against the inclusion of Polleninae and Mesembrinellinae within Calliphoridae (Singh and Wells 2013). The subfamilies Calliphorinae, Lucillinae, Toxotarsinae, and Chrysomyinae include flies of forensic importance. Except for Chrysomyinae, all other subfamilies of forensic importance form a well-supported monophyletic group (Singh and Wells 2013). Monophyly of the Chrysomyinae is well supported by some (Singh and Wells 2011, 2013), but not all, molecular systematic analyses (Kutty et al. 2010). Chrysomyinae are more closely related to social insects (Rhiniinae and Bengaliinae) than to other blow flies of forensic importance (Singh and Wells 2013). Within Chrysomyinae, *Chrysomya* is more closely related to members of the tribe Phormiini (*Protophormia*, *Phormia*, *Trypocalliphora*, and *Protocalliphora*) than to other members of the tribe Chrysomyiini (*Cochliomyia*, *Compsomyiops*, *Paralucilia*, *Hemilucilia*, *Chloroprocta*, etc.) (Singh and Wells 2013).

The Sarcophagidae family is traditionally divided into three subfamilies Sarcophaginae, Miltogramminae, and Paramacronychinae (Kutty et al. 2010; Pape et al. 2012). Monophyly of all three subfamilies is well supported, but phylogenetic relationships between different subfamilies are unresolved (Kutty et al. 2010). Many members of Sarcophaginae and Paramacronychinae breed on living or dead tissues and are of forensic importance (Wells et al. 2001b; Kutty et al. 2010). In North America, members of genera *Sarcophaga*, *Peckia*, *Blaesoxipha*, *Ravinia*, and *Wohlfahrtia* are of forensic importance in urban areas (Wells et al. 2001b). The monophyly of *Sarcophaga* and *Wohlfahrtia* is well-supported, but monophyly of other genera differs across studies (Kutty et al. 2010; Stamper et al. 2013).

24.2.2 DNA-Based Species Identification

Sperling et al. (1994) first applied DNA-based methods to the identification of insects of forensic importance, and numerous studies have followed these efforts to enable identification of taxonomically and geographically diverse insects of forensic importance (Wells et al. 2001b; Harvey et al. 2008; Meiklejohn et al. 2011; Boehme et al. 2012; DeBry et al. 2013). The most simplistic approach to the identification of a species is to sequence a fragment or locus of the DNA and then compare your unknown sequence to published sequences via a similarity search using the BLAST® tool from GenBank database (Altschul et al. 1997; Parson et al. 2000). However, for forensic casework this practice may not work well mainly because of sequence gaps and errors and the lack of authenticated reference DNA sequences in the database (Dawnay et al. 2007; Wells and Stevens 2008). Other molecular methods to identify insects based on DNA are; (1) polymerase chain reaction–restriction fragment length polymorphism (PCR-RFLP) approaches (Sperling et al. 1994; Ratcliffe et al. 2003; Schroeder et al. 2003), (2) high-resolution melting PCR approaches (Malewski et al. 2010), and (3) phylogenetic reconstruction approaches (Harvey et al. 2003; Wells and Williams 2007). Although the first two methods are comparatively simpler, faster, and cheaper (mainly because sequence variations can be detected without actual sequencing), the phylogenetic reconstruction method is preferred (because sequence data provide the largest amount of information for insect species identification) (Wells and Sperling 2001; Stevens et al. 2002; DeBry et al. 2013).

To use a phylogenetic reconstruction method, it is important that a person performing DNA-based identification is familiar with different phylogenetic methods, as well as the biology and genetics of their target organisms. For accurate species identification, it is essential to have a complete reference dataset for the targeted insect group. Even with a complete reference dataset, reliable identifications are only possible for those species that demonstrate reciprocal monophyly in a phylogeny (Figure 24.1a) (Wells and Williams 2007). If one species (e.g., P) is monophyletic, but another species (e.g., Q) is paraphyletic, then only species P can be correctly identified (Figure 24.1a). If an evidentiary species falls between monophyletic (species P) and paraphyletic (species Q) species clades, it can be identified either as species P or species Q (Figure 24.1a).

Although both mitochondrial and nuclear genes have been used for identifications of forensically important insects (Wells and Sperling 1999; Stevens et al. 2002; Wells and Williams 2007), mitochondrial DNA (mtDNA) loci are ideal (Wells et al. 2001b; Harvey et al. 2003; Wallman et al. 2005). The use of mtDNA markers is preferred because forensic evidence is often degraded or poorly preserved; mtDNA loci can be amplified from degraded samples (as mitochondria occur in multiple copies per cell); mitochondria evolve 5–10 times faster than their nuclear counterparts (which is necessary for the identification of closely related taxa); mitochondria do not recombine (which reduces within species diversity and results in discernible haplotypes); and, practically speaking, universal primers are available for a wide range of taxa (Folmer et al. 1994; Simon et al. 1994). A major concern associated with mtDNA loci is the tendency to construct paraphyletic or polyphyletic species trees in some situations (Stevens et al. 2002; Wells et al. 2007).

For reliable species identification, certain procedures produce the most reliable sequencing and identification results. For example, it is important to amplify and sequence overlapping regions of selected loci. During sequence editing, independent overlapping PCR fragment sequences help in detecting PCR and sequencing errors. Sequences from overlapping regions can be edited and aligned into one contiguous sequence using a software package such as ChromasPro (Technelysium Pty, Ltd., South Brisbane, Australia). A contiguous sequence from the evidentiary sample and reference sequences from the target group are then combined for multiple sequence alignment (MSA) using software packages such as Muscle (Edgar 2004), or ClustalX (Larkin et al. 2007), as implemented in MEGA version 5.2 (Tamura et al. 2011). It is important that the regions used for phylogenetic analyses are represented by a large majority of the taxa (represented ideally by samples from all species likely to be encountered in the location of interest) and aligned DNA regions with low coverage (few overlapping sequences) should be cropped. MSA files can be used for phylogenetic analyses using appropriate distance- (Neighbor-Joining) or character-based (Maximum Parsimony/ Maximum Likelihood/Bayesian) methods using software packages such as MEGA (Tamura et al. 2011), PHYLIP (Felsenstein 1989), PAUP (Swofford 2003), or MrBayes (Huelsenbeck and Ronquist 2001). Evidence samples can be identified from such phylogenetic trees based on their sister-group relationship with reference taxa. Swofford et al. (1996) provide an excellent review on principles underlying different phylogenetic methods, and their advantages and disadvantages.

24.2.3 Identification of Noninsect DNA from the Insect Gut

Insects are well known for their utility in providing information regarding the timing of death, but in some situations DNA found in the digestive tract of an insect can also provide information in criminal cases where insects are left behind at the scene (e.g., the DNA from a perpetrator found in a louse left with the victim), and on the relocation of a corpse (e.g., maggots at a scene where no food source is apparent), the actual food source of a larva (e.g., a body found in a dumpster with viable non-human food options), and the determination of the sex of the decedent (human Y chromosomal DNA in the insect gut) (Clery 2001; Wells et al. 2001a; Campobasso et al. 2005). Given these scenarios, the guts of immature carrion flies, beetles, and hematophagous insects can provide forensically probative information. Identification of the source of meals of insects is possible as their digestive tracts contain a crop that stores food (Clery 2001). Crops typically do not secrete enzymes; thus, consumed tissue is generally undigested, even several hours after ingestion (Clery 2001). This well-preserved carrion tissue yields both insect and noninsect DNA and can help forensic entomologists in the identification of both insects and their vertebrate foods (Wells et al. 2001a). Wells et al. (2001a) provide an excellent review of the details on this methodology and its application in forensic entomology.

24.3 POPULATION GENETICS

While the first section of the chapter dealt with macroevolutionary concepts in forensic entomology, there is a growing body of evidence that microevolutionary processes like local adaptation and genetic drift may also be important to forensics. Essentially, macroevolutionary studies are targeted toward the consistent differences between species to identify evidentiary samples. However, just as humans can span a range of heights, it is well documented that there can be variation in developmental and morphological traits within insect species. Accordingly, researchers in the field are beginning to adapt to these concerns.

To incorporate these concerns into forensic entomology, it is necessary to study populations. Ecological processes, such as the dispersal of insects over time and space, as well as their population structure, are critically important to understanding the dynamics of insect populations. Such understanding is important not only for the control of insect pests but also for the extrapolation of physiological data collected using insects from different geographic areas. Mark and recapture studies are generally employed for these purposes (e.g., the study by Cragg and Hobart [1955]), but they require large teams of collaborators, and generally the information is based on collections from short distances. Population genetics allows the sampling of insects from large geographic areas (and over time) and is useful for describing patterns of dispersal and measuring genetic divergence between local populations.

There is increasing evidence that subpopulations of forensically important insects exist, for one of two reasons. First, local ecological factors can drive the evolution of specific populations toward different developmental optima by selecting for alleles of genes that confer success in that environment. In this case, very specific parts of an insect genome will differ between populations in association with the selected trait. Essentially, this is the same process by which bacteria develop resistance to antibiotics (Pollock 1971; Baquero and Blazquez 1997; Baquero et al. 2009) and insects develop resistance to insecticides (Shanahan 1958; Georghiou and Taylor 1977; Forgash 1984; Ffrench-Constant et al. 2004). However, depending on population dynamics and size there can also be random fluctuations in allele frequencies at loci that affect traits, as well as at loci that are evolutionarily neutral (Glass 1954; Kimura 1955, 1976; Ohta and Kimura 1969). Both selection and drift can affect predictions made with entomological data in casework. Given this knowledge, the current challenge for forensic entomologists is to determine how to best classify a population and the degree to which population level variation affects error in forensic applications with entomological evidence.

In forensic entomology, a biological process of interest is the rate of immature development. This is used to determine the age of an evidentiary sample that is discovered on a corpse, which is used as a clock to estimate the minimum time since death. The age of the sample is determined through the use of published developmental datasets, in which a large number of immature samples were generated in a laboratory environment under controlled settings. The timing of developmental change can be determined with these datasets. In addition, the correlation of morphology (such as length and/or width) of immature insects with their developmental progress can also be informative of age. If a particular biological process differs according to the population structure of the species, then a development dataset compiled from a particular geographic region may not be an accurate model for predicting the age of samples from that population. Most of the published development studies are completed using samples of flies collected over a small period of time or geographic location. For example, published *Phormia regina*, (Meigen) (Diptera: Calliphoridae) development studies used flies collected from a limited number of locations: Washington (Kamal 1958), Florida (Byrd and Allen 2001), Nebraska (Nabity et al. 2006), Texas (Nunez-Vazquez et al. 2013) in the United States, British Columbia, and Canada (Anderson 2000). In this simplified example of development rates from the same species collected from different geographic regions, you can see in Figure 24.2 the variation in accumulated degree hours (ADHs), which essentially reflect heat-adjusted development (as insect development is temperature dependent) given an assumed minimum thermal developmental threshold. Each study was analyzed assuming a base temperature of 6°C as a minimum threshold temperature in this example (which is one possible threshold temperature traditionally used for this species). If the assumption of a developmental threshold is correct, within a population all ADH values should be the same across temperatures. If there were no differences between the populations, then one would expect roughly the same ADH values for a given species regardless of its geographic region of origin. Clearly, neither of these assumptions is met. Each of these studies was conducted under different experimental conditions; therefore, it is possible that these conditions could account for the variation. However, differences in the total amount of time or in the minimum amount of heat required to develop (or both) could explain population differences in this example.

Defining the population structure also has additional applied uses. For example, inferring post-mortem movement of a corpse based on the structure of the insect community on a corpse can be very useful. If a corpse, initially colonized by a local population, is then moved, then the analysis of the insects on the discovered corpse can be used to infer geographic origin of the initial colonization. Though mtDNA sequencing has currently failed to reveal a genetic signature on this spatial scale, other more polymorphic methods may sample the appropriate level of genetic information, thus leading to a fine scale of assays for evidentiary individuals. One example is the use of amplified fragment length polymorphism (AFLP) markers to classify individuals based on their kin structure (Picard and Wells 2012). Using this technology, they have been able to show that not only are flies attracted to a corpse at a given time likely to be more related (Picard and Wells 2010) but also each

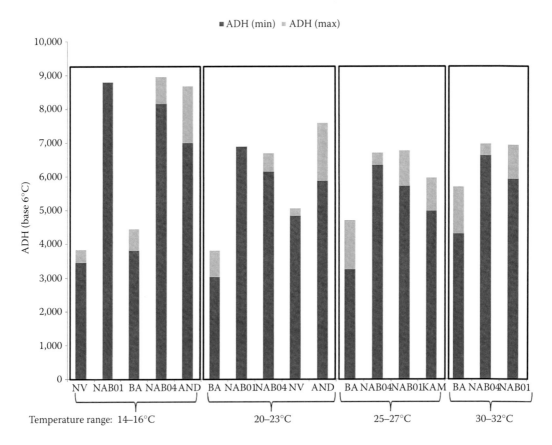

Figure 24.2 Figure highlighting variation in accumulated degree hours (ADHs) values obtained from different developmental datasets for *Phormia regina*. Each dataset was converted from hours to ADH using 6°C as a minimum threshold temperature. If standard deviations were reported, they were used to calculate the minimum and maximum ADH values for total development (to adult). For simplicity, narrow temperature ranges were used to group the data for comparisons. NV, Nunez-Vazquez et al. 2013; NAB01, Nabity et al. 2006 (2001 data); BA, Byrd and Allen 2001; AND, Anderson 2000; NAB04, Nabity et al. 2006 (2004 data); and KAM, Kamal (1958).

fly will deposit hundreds of full sibling larvae. In this study, because a corpse population is made up of hundreds of sets of full siblings, the ability to detect a full sibling larva on a corpse to a larva collected in a separate location can lead to inferences of postmortem movement. This lends further credence that a corpse may be housing a very distinct (in time and space) subpopulation, one that lacks genetic variation due to the kin structure of the flies on the resource.

Another applied use of population genetics is to gain knowledge of the amount of genetic variation that is present in a population, and this is important when attempting to model physiological processes, such as development rate, to a large portion of the population. If each individual developmental dataset was obtained using a small number of flies from a small geographic area, then the variation observed among the different datasets may be, in part, due to genetic differences. In this case, it is imperative that the nature of such variation (selection vs. drift) be understood. If an initial sample is composed of many related individuals, then a study will not contain a random sample of the population, and therefore the developmental information from the strain will be genetically biased. Data suggest that flies maintained in laboratory colonies very quickly lose genetic variation and thus become very homogenous. This has been demonstrated in a few species (McCommas and Bryant 1990; Florin and Gyllenstrand 2002).

Population genetic structures have yielded contradicting results with respect to the presence of subpopulations in blow flies. Using mtDNA sequencing techniques, the population genetic structure of the New World screwworm *Cochliomyia hominivorax* Coquerel (Diptera: Calliphoridae) had distinct populations, and this had direct implications for pest control strategies (Fresia et al. 2011). Meanwhile, other research using mtDNA sequencing techniques suggests that populations of blow flies are panmictic (Gleeson and Sarre 1997; Stevens and Wall 1997; Lyra et al. 2005; Boehme 2006; Torres et al. 2007). In the case of the Old World screwworm fly *Chrysomya bezziana*, (Villeneuve), mtDNA sequence analysis on a small geographic scale (the Gulf region) revealed little genetic differentiation between the samples (Hall et al. 2009; Ready et al. 2009); however, the worldwide population could be subdivided into two geographic races (Hall et al. 2001). Panmixis is further supported in the stable fly, *Stomoxys calcitrans*, (Linnaeus) (Diptera: Muscidae), in which allozyme data suggest a single large interbreeding population (Szalanski et al. 1996). These results suggest that pest management or eradication techniques (e.g., sterile insect technique) may be applied to broad regions, but for some of these species there exists a possibility that population structure would affect management efforts on a large regional or global scale.

Mitochondrial DNA sequencing is a relatively easy and robust method for genotyping; however, the amount of intraspecific variation is sometimes not sufficient to distinguish between population substructures, should they exist. The use of more polymorphic DNA loci, such as randomly amplified polymorphic DNA (RAPD) and microsatellites may have the power to discriminate between small changes in DNA that differentiate a population. RAPDs were used to differentiate between wild samples of *Lucilia sericata*, (Meigen) (Diptera: Calliphoridae) and colony-maintained specimens (Stevens and Wall 1995). By definition colony specimens are an isolated population, and thus this study demonstrated that RAPDs are polymorphic enough to differentiate between different populations. Following mtDNA and RAPDs, additional, perhaps more reliable DNA loci (Black 1993) have been developed and used in carrion insects. One example, AFLP is a powerful method for surveying the population diversity of any species. AFLP was used in determining the population structure of two common North American flies: *L. sericata* and *P. regina* (Picard and Wells 2009; Picard and Wells 2010). These population genetic surveys observed that wild adults caught in a single collection time (<30 minutes) were composed of a greater than expected proportion of related individuals if the sample was a random sample of the population thus demonstrating temporal population structuring. AFLP was also successfully used for differentiating *C. hominivorax* and *Cochliomyia macellaria* Fabricius (Diptera: Calliphoridae) (Alamalakala et al. 2009), as well as the ability to differentiate between specimens collected from different geographic regions. These data further support the purported findings of distinct populations within the *C. hominivorax* community.

Currently, microsatellites are the preferred molecular technique for studying population genetics. For example, these types of loci provide the basis for human forensic identifications. Unfortunately, in the absence of reliable genomic information these can be difficult to isolate and characterize, and developing a genotyping assay can be time consuming and expensive (for a review, refer to the study by Wells and Stevens [2008]). Yet, microsatellite loci were isolated and characterized for several blow fly species. In *Lucilia illustris* Meigen (Diptera: Calliphoridae) and *L. sericata*, microsatellites were isolated and used to assess the genetic variation of lab colonies and wild samples (Florin and Gyllenstrand 2002). These molecular markers also demonstrated that the genetic variation in lab colony populations had been greatly reduced compared to wild samples and that this reduction in genetic variation occurred after a small number of generations. In addition, the wild samples also had lower than expected heterozygosities, which could be attributed to the sample itself (consisting of a large number of related individuals), null alleles, or population structure. Even more recent is a published study on the isolation and characterization of microsatellite loci for *P. regina* using next-generation sequencing technologies (Farncombe et al. 2014). Further studies with microsatellites, once they are developed, will help to clarify our understanding of population structures in other forensically important species.

24.4 QUANTITATIVE AND FUNCTIONAL GENETICS

Section 24.3 was primarily targeted toward understanding the population processes that affect (largely) functionally neutral markers, as a means of understanding relatedness and population structures among flies. However, one of the several motivations for such work was the observations that phenotypes of carrion feeding insects differ among strains of the same species, when raised in the same environment (Gallagher et al. 2010; Tarone et al. 2011; Owings et al. 2014). These observations indicate that functional alleles differ between such strains as well, due to either drift or selection, suggesting the need to identify the genes associated with forensically relevant traits. Unfortunately, many of the most forensically important phenotypes, like development time and body size, are quantitative traits, meaning they are influenced by numerous genetic and environmental factors. To identify (or map) genes associated with variation in phenotypic traits of interest (e.g., development time and size), it is necessary to have access to genetic and genomic tools, such as a sequenced genome or mutant strains for genes of interest, that are not currently available for many insects of forensic interest. However, there is a subfield of quantitative genetics that is already beginning to show promise in forensic entomology: functional genetics.

Functional genetics is focused on the parts of the genome that provide function to organisms (Marchetti et al. 2012). Consequently, this area of genetics is targeted toward the study of mRNA and protein, as these are the regulated parts of the genome that lead to functional changes in an organism. Traditional gene mapping studies will search for alleles that are statistically associated with a phenotype (Falconer 1989; Gold et al. 2008; Atwell et al. 2010), whereas functional genetic studies will search for expression differences between phenotypic, phenological, population, and/or environmental groups (Hatzimanikatis et al. 1999; Lee et al. 2003; Ayroles et al. 2009; Wood 2009; Gase and Baldwin 2012). The distinction between genomic approaches to identifying causal genes can be seen in Figure 24.3, comparing the traditional gene mapping approach versus a gene expression approach, both of which can be used to identify a gene associated with phenotypic variation. Accordingly, because forensic entomologists are interested in a number of insect phenotypes, which are affected both by the environment and by genetics, and the tools currently exist to study them, functional genetic research will be valuable in understanding variation in insect-derived postmortem interval (PMI) estimates. Most research in this area is centered on RNA and protein products (e.g., ecdysone levels, but not expression of ecdysone synthesis proteins), as *de novo* proteomics is notoriously difficult. However, as sequencing technologies advance reference sequences will enable proteomic studies as well. Recently, several studies have engaged in targeted and global approaches toward understanding the molecular ontogeny of flies as they develop throughout the immature portion of their life cycle by studying changes in mRNA levels over developmental time (Tarone et al. 2007; Tarone and Foran 2011; Boehme et al. 2013). In addition, as population differences in phenotypes become apparent in forensically important taxa (Tarone and Foran 2008; Gallagher et al. 2010; Tarone et al. 2011; Owings et al. 2014) it will be more important to link regulatory and genetic differences found in distinct populations to forensically relevant phenotypes. The same can also be done for known environmental influences on insect development (Kaneshrajah and Turner 2004; Clark et al. 2006; Nabity et al. 2006; Tarone and Foran 2006). This section will cover applications of molecular forensic entomology as they relate to the regulation of insect genes, with particular attention on prediction of phenotype and identifying insect age with expression markers.

24.4.1 Age Prediction

Currently, fly age prediction is typically done by evaluating the developmental stage of the sample, and sometimes size-related information like length or width (Byrd and Castner 2010). These data are assessed in conjunction with an estimate of the thermal exposure of that sample (as insect development is temperature dependent) to estimate age. Although this is very useful for early

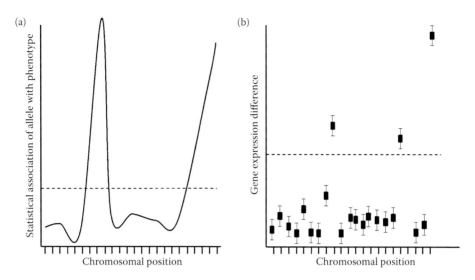

Figure 24.3 A comparison of quantitative genetic approaches to identifying genes associated with phenotypes of interest. (a) In a DNA-based gene mapping approach, such as quantitative trait locus or association mapping, the genome is scanned for alleles that are statistically associated with phenotype scores. These sequences can be linked either due to their causal effects on the trait or due to their physical linkage on the chromosome to the causal genes. While such an approach has been useful for model organisms, tools development is needed to do this in nonmodel insects of forensic importance. (b) In a functional genetic approach, the same loci can be identified by evaluating differences in gene expression (shown here as absolute value in gene expression change where differences in expression are represented by large numbers and no difference represents low expression difference values) that associate with variation in the phenotype, age, environment, or different populations. Functional genetic approaches can also frequently be employed when information about the full genome is not available, allowing for some genomic assessment of the biology of nonmodel organisms like those encountered in forensics. In both examples, genes significantly associated with a trait or environment are found above the dotted lines along a chromosome. The two genomic approaches are complementary and can help provide a more detailed appreciation of phenotypes when used in conjunction. Currently, only functional genetic studies are pursued in forensic entomology.

developmental stages, it is clear that predictions with third instar and pupal samples could be more precise if additional information was incorporated into this process (Tarone and Foran 2008).

The temporal regulation of genes to guide a developing organism through its ontogeny is well established (Arbeitman et al. 2002). There are numerous examples of genes that must be up- and downregulated in a temporally specific fashion to achieve proper developmental progress. In the *Drosophila* (Diptera: Drosophilidae) model, this was appreciated earlier, as it was clear that the temporally regulated hormone ecdysone affected polytene chromosomal structure and the transcription of numerous genes on these chromosomes (Ashburner 1990). Currently, in the *Drosophila* model system developmental information is known for every gene in the genome, at a whole-organism level in typically encountered and nutritionally altered environmental conditions and at a cellular level in response to several developmentally regulated hormones (Beckstead et al. 2005; Gershman et al. 2007; Riddiford et al. 2010).

The forensic entomology community has begun to evaluate gene expression as one means of identifying more precise temporal intervals of blow fly age. In 2007, gene expression changes during embryonic development of the blow fly *L. sericata* were demonstrated (Tarone et al. 2007). By evaluating three genes, they were able to show that early development was punctuated by high expression levels of *bicoid* (*bcd*) and *slalom* (*sll*), which are embryonic patterning genes, and missing or low levels of *chitin synthase* (*chs*). However, as development progressed, *bcd* and *sll* mRNA levels dropped, whereas *chs* levels began to increase as the embryo reached the point where cuticular structures were developing. These results suggested the potential to divide the 10-hour embryonic phase of development into 2 to 3 hour temporal blocks by using just the information provided by the three genes.

This initial observation in embryos was then followed by a study to identify similar changes during larval and pupal development (Tarone and Foran 2011). In an evaluation of the expression levels of nine genes, it was possible to identify differential expression during larval and pupal development. These expression levels could be entered into a statistical model to develop more precise intervals of development in larval and pupal samples. However, with those genes improvements in a blind study were not as obvious (precision increased by 3%–8%). However, even this improvement could reflect a considerable window of time in low temperature conditions. In addition, the larval samples in the blind study yielded improved precision in age prediction compared to the pupal samples, although there was a slight improvement in the prediction of pupal ages too. This study also worked with three populations of *L. sericata* reared in two thermal environments and was therefore able to demonstrate differences in the expression levels across populations and temperatures for several of the genes studied. For the most part, these expression differences were subtle (had little impact on overall statistical model predictions). Interestingly, heat shock protein 90 and acetylcholine esterase exhibited very high expression levels during larval and pupal development at 20°C compared to 33.5°C, yet these effects exerted little influence on overall predictions, presumably due to the information provided by other genes in their prediction model. Overall, this work was able to demonstrate a more realistic possibility for improvements in PMI estimates derived from blow fly samples, by evaluating gene expression levels in a scenario that is likely to be encountered in forensics. The study also indicated that some gene expression profiles may be more appropriate for the identification of populations or thermal environments than for determining the physiological age of an immature blow fly.

While the previous study yielded encouraging results regarding the ability to predict blow fly pupal age based on gene expression, it was limited by the availability of *L. sericata* sequence at the time. The *L. sericata* studies were immediately followed by several evaluations of *Calliphora vicina*, Robineau-Desvoidy (Diptera: Calliphoridae) age-dependent pupal gene expression differences (Boehme et al. 2013). The authors employed a differential enrichment strategy for identifying the most differentially expressed mRNA species between samples. This strategy uses pools of DNA derived from RNA, which are mixed between samples of interest (for instance, young and old pupae) in a manner that enriches sequences that are differentially expressed between them. The enriched samples in that study were then sequenced, and gene expression was evaluated for several identified and unidentified genes that were likely to be differentially expressed in pupae of differing ages. They demonstrated strong differentiation in gene expression levels among pupae, with some support from blind studies (Boehme et al. 2014). These results strongly suggest that improvements in pupal age prediction are possible, if the right loci are chosen for analyses. In addition, this work expanded functional genetic studies of blow flies in forensic entomology beyond the *Lucilia* genus.

24.4.2 Genomics Studies and Phenotype/Environmental Exposure Prediction

As was demonstrated in the work with *L. sericata* (Tarone and Foran 2011), it is possible for any gene to exhibit genetic or environmental differences in expression, with some exhibiting more differences than others. In addition, many of the genes in that same study were clearly correlated (either positively or negatively) with the expression of the *ecdysone receptor* and its cofactor *ultraspiracle*, indicating that they were all ecdysone responsive genes (Tarone and Foran 2011). The addition of each gene, which correlated with the expression of other genes in the study, to the analysis did not add as much power as could have been achieved using a gene whose expression is uncorrelated. These observations are a double-edged sword for forensic entomology. It means that extensive work will need to go into identifying the least variable (across environments and populations) and temporally informative genes in the genome, preferably from groups of genes whose expression levels are minimally correlated with each other. However, once that is done there will be a robust set of genes that can be assayed independent of genetic and environmental conditions. Further, as more is learned about population and environmental effects on gene expression (and other phenotypes), such information will also allow investigators to confirm or refute hypotheses regarding the populations and environments associated with evidentiary samples. However, to conduct such studies it is necessary to have genomic information about forensically important insects. Given the high potential for improved probative value of insect evidence through functional genetic information, it will be important for the forensic entomology community to develop genomic tools.

With the advent of newer high-throughput sequencing technologies, there have been several advances in accruing genomic information for forensically relevant insects. Recently, the genome sizes of numerous forensically important flies were reported, aiding in future efforts to plan the sequencing of those genomes (Picard et al. 2012). Interestingly, the paper by Picard et al. (2012) also demonstrated that within-genus and intersexual differences in genome size could be of forensic value. For example, *L. sericata* larvae demonstrate sex-specific growth patterns and development rates, indicating that sex-specific predictions may be warranted in some species (Picard et al. 2013). The genome size information regarding these species also highlighted the fact that many genomes of forensically important flies are much larger than the genomes of model organisms (some more than a gigabase in size), meaning their assembly may be more difficult. In addition, much of the money spent on sequencing those species genomes will likely go to sequencing the nonexpressed parts of the genomes. However, it is also possible to sequence cDNA derived from mRNA. In these instances, only functional parts of the genome are sequenced (the transcriptome), which provides a better return on sequencing dollars and can also sometimes provide expression-level information. Several projects of this type have been done with forensically important taxa.

In the flesh fly *Sarcophaga crassipalpis*, Macquart (Diptera: Sarcophagidae) which is a model for diapause research, a normalized library of cDNA was sequenced from multiple life history stages of the species (Hahn et al. 2009). Normalization eliminated the chance to evaluate expression levels with the experiment but allowed the capture of rarely expressed genes with minimal sequencing costs. This and similar work was used to pursue functional information regarding the regulation of diapause in that species (Rinehart et al. 2010; Pavlides et al. 2011). Diapause itself is a forensically relevant state (as a diapausing individual will appear younger than its true age), and further research in this area for any forensically important fly is necessary. Molecular markers for diapause in *C. vicina* have just been identified (Fremdt et al. 2013). Such information could be worked into casework to adjust for this physiological state.

In the *Lucilia* genus, two genomic projects have been completed. First, a cDNA library of *L. cuprina* was published (Lee et al. 2011), yielding high-quality information for thousands of genes but again little information regarding age-specific expression. This information has been used to develop high-quality information on the regulation of the sex determination system of this and other flies (Concha et al. 2010, Li et al. 2013), which could aid in future efforts to identify sex-specific growth patterns. However, a second project in *L. sericata* was developed to address age-specific expression directly (Sze et al. 2012), and the study provided information on thousands of potentially differentially expressed genes (many supported by both deep sequencing and subtraction experiments) derived from different covarying gene expression clusters, laying the foundation for more gene-targeted studies of differential gene expression in the species. Finally, several other studies targeting nonforensic questions in forensically relevant flies have been published. In *Chrysomya megacephala* Fabricius (Diptera: Calliphoridae), a project was completed to identify genes that were differentially expressed under variable nutritional conditions (Zhang et al. 2013). In *C. hominivorax*, the transcriptome was sequenced to evaluate genes that may confer insecticide resistance (Carvalho et al. 2010). Although the hypotheses of the articles were of limited immediate use to forensic research, the genomic tools developed for the work will be invaluable for future studies with these, and related species, as forensic indicators.

Recently developed tools can be used to decrease error in forensic entomology through the molecular dissection of fly development. While such an endeavor will likely take time, it is very possible that in the future molecular work in forensic entomology will enable an estimation of the physiological age of an insect sample, the identification of genotypes that may develop differently (i.e., large or small, fast or slow, etc.), and those that may be exposed to numerous environments that could influence their development (for instance, those that induce diapause). The sum of such efforts will be to narrow the window of time in an estimate of the PMI based on insect evidence (improved precision) and the ability to adjust estimates based on genetic and environmental effects on the evidentiary samples (improved accuracy).

24.5 CONCLUSION

Molecular forensic entomology is rapidly changing. While some traditional DNA-based analyses are well established for species identification, there are also newer population and functional genetic techniques that have been demonstrated, but they require further development and/or validation. Population and quantitative genetic studies will aid in determining the proper developmental dataset to apply to evidentiary samples, potentially leading to decreases in error. Functional genetic studies may allow more precise identification of physiological age, as well as more accuracy, through the identification of physiological states (like diapause) and any environments (temperature, humidity, nutrition, light cycle, or disease) or populations associated with differential development rates.

REFERENCES

Alamalakala, L., S. R. Skoda, and J. E. Foster. 2009. Amplified fragment length polymorphism used for inter- and intraspecific differentiation of screwworms (Diptera: Calliphoridae). *Bulletin of Entomological Research* 99: 139–149.

Altschul, S. F., T. L. Madden, A. A. Schäffer, J. Zhang, Z. Zhang, W. Miller, and D. J. Lipman. 1997. Gapped blast and psi-blast: A new generation of protein database search programs. *Nucleic Acids Research* 25: 3389–3402.

Anderson, G. S. 2000. Minimum and maximum development rates of some forensically important Calliphoridae (Diptera). *Journal of Forensic Sciences* 45: 824–832.

Arbeitman, M. N., E. E. Furlong, F. Imam, E. Johnson, B. H. Null, B. S. Baker, M. A. Krasnow, M. P. Scott, R. W. Davis, and K. P. White. 2002. Gene expression during the life cycle of *Drosophila melanogaster*. *Science* 297: 2270–2275.

Ashburner, M. 1990. Puffs, genes, and hormones revisited. *Cell* 61: 1–3.

Atwell, S., Y. S. Huang, B. J. Vilhjálmsson, G. Willems, M. Horton, Y. Li, D. Meng et al. 2010. Genome-wide association study of 107 phenotypes in *Arabidopsis thaliana* inbred lines. *Nature* 465: 627–631.

Ayroles, J. F., M. A. Carbone, E. A. Stone, K. W. Jordan, R. F. Lyman, M. M. Magwire, S. M. Rollmann et al. 2009. Systems genetics of complex traits in *Drosophila melanogaster*. *Nature Genetics* 41: 299–307.

Baquero, F., C. Alvarez-Ortega, and J. L. Martinez. 2009. Ecology and evolution of antibiotic resistance. *Environmental Microbiology Reports* 1: 469–476.

Baquero, F. and J. Blazquez. 1997. Evolution of antibiotic resistance. *Trends in Ecology & Evolution* 12: 482–487.

Barton, P. S., S. A. Cunningham, D. B. Lindenmayer, and A. D. Manning. 2013. The role of carrion in maintaining biodiversity and ecological processes in terrestrial ecosystems. *Oecologia* 171: 761–772.

Baust, J. G. and J. S. Edwards. 1979. Mechanisms of freezing tolerance in an Antarctic midge, *Belgica antarctica Physiological Entomology* 4: 1–5.

Beckstead, R. B., G. Lam, and C. S. Thummel. 2005. The genomic response to 20-hydroxyecdysone at the onset of *Drosophila* metamorphosis. *Genome Biology* 6: R99.

Beutel, R. G. and F. Haas. 2000. Phylogenetic relationships of the suborders of Coleoptera (Insecta). *Cladistics.* 16: 103–141.

Black, W. C. I. 1993. PCR with arbitrary primers: Approach with care. *Insect Molecular Biology* 2: 1–6.

Boehme, P. 2006. Population genetics of forensically important North American (Diptera: Calliphoridae) using the A+T-rich region of the mitochondrial DNA. MS Thesis. Bonn, Germany: Rheinische Friedrich-Wilhelms Universitat.

Boehme, P., J. Amendt, and R. Zehner. 2012. The use of COI barcodes for molecular identification of forensically important fly species in Germany. *Parasitology Research* 110: 2325–2332.

Boehme, P., P. Spahn, J. Amendt, and R. Zehner. 2013. Differential gene expression during metamorphosis: A promising approach for age estimation of forensically important *Calliphora vicina* pupae (Diptera: Calliphoridae). *International Journal of Legal Medicine* 127: 243–249.

Boehme, P., P. Spahn, J. Amendt, and R. Zehner. 2014. The analysis of temporal gene expression to estimate the age of forensically important blow fly pupae: Results from three blind studies. *International Journal of Legal Medicine* 128: 565–573.

Byrd, J. H. and J. C. Allen. 2001. The development of the black blow fly, *Phormia regina* (Meigen). *Forensic Science International* 120: 79–88.

Byrd, J. H. and J. Castner. 2010. *Forensic Entomology: The Utility of Arthropods in Legal Investigations.* 2nd ed. Boca Raton, FL: CRC Press.

Campobasso, C. P., J. G. Linville, J. D. Wells, and F. Introna. 2005. Forensic genetic analysis of insect gut contents. *The American Journal of Forensic Medicine and Pathology* 26: 161–165.

Caravas, J. and M. Friedrich. 2013. Shaking the Diptera tree of life: Performance analysis of nuclear and mitochondrial sequence data partitions. *Systematic Entomology* 38: 93–103.

Carvalho, R. A., A. M. L. Azeredo-Espin, and T. Torres. 2010. Deep sequencing of New World screw-worm transcripts to discover genes involved in insecticide resistance. *BMC Genomics* 11: 695.

Clark, K., L. Evans, and R. Wall. 2006. Growth rates of the blowfly, *Lucilia sericata*, on different body tissues. *Forensic Science International* 156: 145–149.

Clery, J. M. 2001. Stability of prostate specific antigen (Psa), and subsequent Y-Str typing, of *Lucilia (Phaenicia) sericata* (Meigen) (Diptera: Calliphoridae) maggots reared from a simulated postmortem sexual assault. *Forensic Science International* 120: 72–76.

Concha, C., F. Li, and M. J. Scott. 2010. Conservation and sex-specific splicing of the *doublesex* gene in the economically important pest species *Lucilia cuprina. Journal of Genetics* 89: 279–285.

Cragg, J. B. and J. Hobart. 1955. A Study of a field population of the blowflies *Lucilia caesar* (L.) and *L. sericata* (Mg.). *Annals of Applied Biology* 43: 645–663.

Dawnay, N., R. Ogden, R. Mcewing, G. R. Carvalho, and R. S. Thorpe. 2007. Validation of the barcoding gene COI for use in forensic genetic species identification. *Forensic Science International* 173: 1–6.

Debry, R. W., A. Timm, E. S. Wong, T. Stamper, C. Cookman, and G. A. Dahlem. 2013. DNA-based identification of forensically important *Lucilia* (Diptera: Calliphoridae) in the continental United States. *Journal of Forensic Sciences* 58: 73–78.

Edgar, R. C. 2004. Muscle: Multiple sequence alignment with high accuracy and high throughput. *Nucleic Acids Research* 32: 1792–1797.

Falconer, D. S. 1989. *Introduction to Quantitative Genetics.* New York: Longman Wiley.

Farncombe, K. M., D. Beresford, and C. J. Kyle. 2014. Characterization of microsatellite loci in Phormia regina towards expanding molecular approaches in forensic entomology. *Forensic Science International* 240: 122–125.

Felsenstein, J. 1989. PHYLIP – phylogeny inference package (Version 3.2). *Cladistics* 5:164–166.

Ffrench-Constant, R. H., P. J. Daborn, and G. Le Goff. 2004. The genetics and genomics of insecticide resistance. *Trends in Genetics* 20: 163–170.

Fiene, J. G., G. S. Sword, S. VanLaerhoven, and A. M. Tarone. 2014. The role of spatial aggregation in forensic entomology. *Journal of Medical Entomology.* 51:1–9.

Florin, A. B. and N. Gyllenstrand. 2002. Isolation and characterization of polymorphic microsatellite markers in the blowflies *Lucilia illustris* and *Lucilia sericata. Molecular Ecology Notes* 2: 113–116.

Folmer, O., M. Black, W. Hoeh, R. Lutz, and R. Vrijenhoek. 1994. DNA primers for amplification of mitochondrial *Cytochrome C Oxidase Subunit I* from diverse metazoan invertebrates. *Molecular Marine Biology and Biotechnology* 3: 294–299.

Forgash, A. J. 1984. History, evolution, and consequences of insecticide resistance. *Pesticide Biochemistry and Physiology* 22: 178–186.

Fremdt, H., J. Amendt, and R. Zehner. 2013. Diapause-specific gene expression in *Calliphora vicina* (Diptera: Calliphoridae)-a useful diagnostic tool for forensic entomology. *International Journal of Legal Medicine.* In Press.

Fresia, P., M. L. Lyra, A. Coronado, and A. M. De Azeredo-Espin. 2011. Genetic structure and demographic history of new world screwworm across its current geographic range. *Journal of Medical Entomology* 48: 280–290.

Gallagher, M. B., S. Sandhu, and R. Kimsey. 2010. Variation in developmental time for geographically distinct populations of the common green bottle fly, *Lucilia sericata* (Meigen). *Journal of Forensic Sciences* 55: 438–442.

Gase, K. and I. T. Baldwin. 2012. Transformational tools for next-generation plant ecology: Manipulation of gene expression for the functional analysis of genes. *Plant Ecology and Diversity* 5: 485–490.

Georghiou, G. P. and C. E. Taylor. 1977. Genetic and biological influences in evolution of insecticide resistance. *Journal of Economic Entomology* 70: 319–323.

Gershman, B., O. Puig, L. Hang, R. M. Peitzsch, M. Tatar, and R. S. Garofalo. 2007. High-resolution dynamics of the transcriptional response to nutrition in *Drosophila*: A key role for *dfoxo. Physiological Genomics* 29: 24–34.

Glass, B. 1954. Genetic changes in human populations, especially those due to gene flow and genetic drift. *Advances in Genetics* 6: 95–139.

Gleeson, D. M. and S. Sarre. 1997. Mitochondrial DNA variability and geographic origin of the sheep blowfly, *Lucilia cuprina* (Diptera: Calliphoridae), in New Zealand. *Bulletin of Entomological Research* 87: 265–272.

Gold, B., T. Kirchhoff, S. Stefanov, J. Lautenberger, A. Viale, J. Garber, E. Friedman et al. 2008. Genome-wide association study provides evidence for a breast cancer risk locus at 6q22-33. *Proceedings of the National Academy of Sciences of the United States of America* 105: 4340–4345.

Hahn, D. A., G. J. Ragland, D. D. Shoemaker, and D. L. Denlinger. 2009. Gene discovery using massively parallel pyrosequencing to develop ESTs for the flesh fly *Sarcophaga crassipalpis. BMC Genomics* 10: 234.

Hall, M. J., W. Edge, J. M. Testa, Z. J. Adams, and P. D. Ready. 2001. Old world screwworm fly, *Chrysomya bezziana,* occurs as two geographical races. *Medical and Veterinary Entomology* 15: 393–402.

Hall, M. J., A. H. Wardhana, G. Shahhosseini, Z. J. Adams, and P. D. Ready. 2009. Genetic diversity of populations of old world screwworm fly, *Chrysomya bezziana,* causing traumatic myiasis of livestock in the gulf region and implications for control by sterile insect technique. *Medical and Veterinary Entomology* 23 Suppl 1: 51–58.

Harvey, M. L., I. R. Dadour, and S. Gaudieri. 2003. Mitochondrial DNA *Cytochrome Oxidase I* gene: Potential for distinction between immature stages of some forensically important fly species (Diptera) in Western Australia. *Forensic Science International* 131: 134–139.

Harvey, M. L., S. Gaudieri, M. H. Villet, and I. R. Dadour. 2008. A global study of forensically significant calliphorids: Implications for identification. *Forensic Science International* 177: 66–76.

Hatzimanikatis, V., L. H. Choe, and K. H. Lee. 1999. Proteomics: Theoretical and experimental considerations. *Biotechnology Progress* 15: 312–318.

Huelsenbeck, J. P. and F. Ronquist. 2001. Mrbayes: Bayesian inference of phylogenetic trees. *Bioinformatics* 17: 754–755.

Hunt, T., J. Bergsten, Z. Levkanicova, A. Papadopoulou, O. S. John, R. Wild, P. M. Hammond. 2007. A comprehensive phylogeny of beetles reveals the evolutionary origins of a superradiation. *Science* 318: 1913–1916.

Kamal, A. S. 1958. Comparative study of thirteen species of sarcosaprophagous Calliphoridae and Sarcophagidae (Diptera). *Annals of the Entomological Society of America* 51: 261–270.

Kaneshrajah, G. and B. Turner. 2004. *Calliphora vicina* larvae grow at different rates on different body tissues. *International Journal of Legal Medicine* 118: 242–244.

Kimura, M. 1955. Solution of a process of random genetic drift with a continuous model. *Proceedings of the National Academy of Sciences of the United States of America* 41: 144–150.

Kimura, M. 1976. Molecular evolution - random genetic drift prevails. *Trends in Biochemical Sciences* 1: N152–N154.

Kutty, S. N., T. Pape, B. M. Wiegmann, and R. Meier. 2010. Molecular phylogeny of the Calyptratae (Diptera: Cyclorrhapha) with an emphasis on the superfamily Oestroidea and the position of Mystacinobiidae and Mcalpine's fly. *Systematic Entomology* 35: 614–635.

Lambkin, C. L., B. J. Sinclair, T. Pape, G. W. Courtney, J. H. Skevington, R. Meier, D. K. Yeates, V. Blagoderov, and B. M. Wiegmann. 2013. The phylogenetic relationships among infraorders and superfamilies of Diptera based on morphological evidence. *Systematic Entomology* 38: 164–179.

Larkin, M. A., G. Blackshields, N. P. Brown, R. Chenna, P. A. McGettigan, H. McWilliam, F. Valentin et al. 2007. Clustal W and Clustal X version 2.0. *Bioinformatics* 23: 2947–2948.

Lawrence, J. F., A. Ślipiski, A. E. Seago, M. K. Thayer, A. F. Newton, and A. E Marvaldi. 2011. Phylogeny of the coleoptera based on morphological characters of adults and larvae. *Annales Zoologici* 61: 1–217.

Lee, P. S., L. B. Shaw, L. H. Choe, A. Mehra, V. Hatzimanikatis, and K. H. Lee. 2003. Insights into the relation between mRNA and protein expression patterns: II. Experimental observations in *Escherichia coli*. *Biotechnology and Bioengineering* 84: 834–841.

Lee, S. F., Z. Z. Chen, A. Mcgrath, R. T. Good, and P. Batterham. 2011. Identification, analysis, and linkage mapping of expressed sequence tags from the Australian sheep blowfly. *BMC Genomics* 12: 406.

Li, F., S. P. Vensko, E. J. Belikoff, and M. J. Scott. 2013. Conservation and sex-specific splicing of the *transformer* gene in the calliphorids *Cochliomyia hominivorax*. *Cochliomyia macellaria* and *Lucilia sericata*. *PLoS ONE* 8: e56303.

Lyra, M. L., P. Fresia, S. Gama, J. Cristina, L. B. Klaczko, and A. M. L. De Azeredo-Espin. 2005. Analysis of mitochondrial DNA variability and genetic structure in populations of New World screwworm flies (Diptera: Calliphoridae) from Uruguay. *Journal of Medical Entomology* 42: 589–595.

Malewski, T., A. Draber-Monko, J. Pomorski, M. Los, and W. Bogdanowicz. 2010. Identification of forensically important blowfly species (Diptera: Calliphoridae) by high-resolution Melting PCR analysis. *International Journal of Legal Medicine* 124: 277–285.

Marchetti, G., M. Pinotti, B. Lunghi, C. Casari, and F. Bernardi. 2012. Functional genetics. *Thrombosis Research* 129: 336–340.

Mccommas, S. A. and E. H. Bryant. 1990. Loss of electrophoretic variation in serially bottlenecked populations. *Heredity* 64: 315–321.

Meiklejohn, K. A., J. F. Wallman, and M. Dowton. 2011. DNA-Based identification of forensically important Australian Sarcophagidae (Diptera). *International Journal of Legal Medicine* 125: 27–32.

Michaud, J. P., K. G. Schoenly, and G. Moreau. 2012. Sampling flies or sampling flaws? Experimental design and inference strength in forensic entomology. *Journal of Medical Entomology* 49: 1–10.

Nabity, P. D., L. G. Higley, and T. M. Heng-Moss. 2006. Effects of temperature on development of *Phormia regina* (Diptera: Calliphoridae) and use of developmental data in determining time intervals in forensic entomology. *Journal of Medical Entomology* 43: 1276–1286.

National Research Council (U.S.). Committee on DNA forensic science: An update and National Research Council (U. S.). Commission on DNA forensic science: An update. 1996. *The Evaluation of Forensic DNA Evidence*. Washington, DC: National Academy Press.

National Research Council (U.S.). Committee on identifying the needs of the forensic science community, National Research Council (U. S.). Committee on science technology and law policy and global affairs and National Research Council (U. S.). Committee on applied and theoretical statistics. 2009. *Strengthening Forensic Science in the United States: A Path Forward*. Washington, DC: National Academies Press.

Nunez-Vazquez, C., J. K. Tomberlin, M. Cantu-Sifuentes, and O. Garcia-Martinez. 2013. Laboratory development and field validation of *Phormia regina* (Diptera: Calliphoridae). *Journal of Medical Entomology* 50: 252–60.

Ohta, T. and M. Kimura. 1969. Linkage disequilibrium due to random genetic drift. *Genetical Research* 13: 47–55.

Owings, C. G., C. Spiegelman, A. M. Tarone, and J. K. Tomberlin. 2014. Developmental variation among *Cochliomyia macellaria* Fabricius (Diptera: Calliphoridae) populations from three ecoregions of Texas, USA. *International Journal of Legal Medicine* 128: 709–717.

Pape, T. 1987. *The Sarcophagidae (Diptera) of Fennoscandia and Denmark.* Copenhagen, Denmark: E. J. Brill/Scandinavian Science Press Ltd.

Pape, T. 1992. Phylogeny of the Tachinidae family-group (Diptera: Calyptratae). *Tijdschrift Voor Entomologie* 135: 43–86.

Pape, T., G. Dahlem, C. A. de Mello Patiu, and M. Giroux. 2012. The World of Flesh Flies (Diptera: Sarcophagidae), accessed October 25, 2014, http://www.zmuc.dk/entoweb/sarcoweb/sarcweb/sarc_web.htm.

Parson, W., K. Pegoraro, H. Niederstatter, M. Foger, and M. Steinlechner. 2000. Species identification by means of the *Cytochrome B* gene. *International Journal of Legal Medicine* 114: 23–28.

Pavlides, S. C., S. A. Pavlides, and S. P. Tammariello. 2011. Proteomic and phosphoproteomic profiling during diapause entrance in the flesh fly, *Sarcophaga crassipalpis. Journal of Insect Physiology* 57: 635–644.

Picard, C. J., K. DeBlois, F. Tovar, J. L. Bradley, J. S. Johnston, and A. M. Tarone. 2013. Increasing precision in development-based postmortem interval estimates: What's sex got to do with it? *Journal of Medical Entomology* 50: 425–431.

Picard, C. J., J. S. Johnston, and A. M. Tarone. 2012. Genome sizes of forensically relevant Diptera. *Journal of Medical Entomology* 49: 192–197.

Picard, C. J. and J. D. Wells. 2009. Survey of the genetic diversity of *Phormia regina* (Diptera: Calliphoridae) using amplified fragment length polymorphisms. *Journal of Medical Entomology* 46: 664–670.

Picard, C. J. and J. D. Wells. 2010. The population genetic structure of North American *Lucilia sericata* (Diptera: Calliphoridae), and the utility of genetic assignment methods for reconstruction of postmortem corpse relocation. *Forensic Science International* 195: 63–67.

Picard, C. J. and J. D. Wells. 2012. A test for carrion fly full siblings: A tool for detecting postmortem relocation of a corpse. *Journal of Forensic Sciences* 57: 535–538.

Pollock, M. R. 1971. Function and evolution of *Penicillinase. Proceedings of the Royal Society of London, Series B-Biology* 179: 385–401.

Ratcliffe, S. T., D. W. Webb, R. A. Weinzievr, and H. M. Robertson. 2003. PCR-RFLP identification of Diptera (Calliphoridae, Muscidae and Sarcophagidae)—a generally applicable method. *Journal of Forensic Sciences* 48: 783–785.

Ready, P. D., J. M. Testa, A. H. Wardhana, M. Al-Izzi, M. Khalaj, and M. J. Hall. 2009. Phylogeography and recent emergence of the Old World screwworm fly, *Chrysomya bezziana*, based on mitochondrial and nuclear gene sequences. *Medical and Veterinary Entomology* 23 Suppl 1: 43–50.

Riddiford, L. M., J. W. Truman, C. K. Mirth, and Y. C. Shen. 2010. A role for juvenile hormone in the prepupal development of *Drosophila melanogaster. Development* 137: 1117–1126.

Rinehart, J. P., R. M. Robich, and D. L. Denlinger. 2010. Isolation of diapause-regulated genes from the flesh fly, *Sarcophaga crassipalpis* by suppressive subtractive hybridization. *Journal of Insect Physiology* 56: 603–609.

Rognes, K. 1997. The Calliphoridae (blowflies) (Diptera: Oestroidea) are not a monophyletic group. *Cladistics* 13: 27–66.

Saks, M. J. and J. J. Koehler. 2005. The coming paradigm shift in forensic identification science. *Science* 309: 892–895.

Schroeder, H., H. Klotzbach, S. Elias, C. Augustin, and K. Pueschel. 2003. Use of PCR-RFLP for differentiation of calliphorid larvae (Diptera, Calliphoridae) on human corpses. *Forensic Science International* 132: 76–81.

Shanahan, G. J. 1958. Resistance to dieldrin in *Lucilia cuprina* Wied., the Australian sheep blowfly. *Nature* 181: 860–861.

Simon, C., F. Frati, A. Beckenbach, B. Crespi, H. Liu, and P. Flook. 1994. Evolution, weighting, and phylogenetic utility of mitochondrial gene sequences and a compilation of conserved polymerase chain reaction primers. *Annals of the Entomological Society of America* 87: 651–701.

Singh, B. and J. D. Wells. 2011. Chrysomyinae (Diptera: Calliphoridae) is monophyletic: A molecular systematic analysis. *Systematic Entomology* 36: 415–420.

Singh, B. and J. D. Wells. 2013. Molecular systematics of the Calliphoridae (Diptera: Oestroidea): Evidence from one mitochondrial and three nuclear genes. *Journal of Medical Entomology* 50: 15–23.

Song, H., N. C. Sheffield, S. L. Cameron, K. B. Miller, and M. F. Whiting. 2010. When phylogenetic assumptions are violated: Base compositional heterogeneity and among-site rate variation in beetle mitochondrial phylogenomics. *Systematic Entomology* 35: 429–448.

Sperling, F. A. H., G. S. Anderson, and D. A. Hickey. 1994. A DNA-based approach to the identification of insect species used for postmortem interval estimation. *Journal of Forensic Sciences* 39: 418–427.

Stamper, T., G. A. Dahlem, C. Cookman, and R. W. Debry. 2013. Phylogenetic relationships of flesh flies in the subfamily Sarcophaginae based on three mtDNA fragments (Diptera: Sarcophagidae). *Systematic Entomology* 38: 35–44.

Stevens, J. R. and R. Wall. 1995. The use of random amplified polymorphic DNA (RAPD) analysis for studies of genetic variation in populations of the blowfly *Lucilia sericata* (Diptera: Calliphoridae) in Southern England. *Bulletin of Entomological Research* 85: 549–555.

Stevens, J. R. and R. Wall. 1997. Genetic variation in populations of the blowflies *Lucilia cuprina* and *Lucilia sericata* (Diptera:Calliphoridae). Random amplified polymorphic DNA analysis and mitochondrial DNA sequencing. *Biochemical and Systematic Ecology* 26: 81–97.

Stevens, J. R., R. Wall, and J. D. Wells. 2002. Paraphyly in Hawaiian hybrid blowfly populations and the evolutionary history of anthropophilic species. *Insect molecular biology* 11: 141–148.

Swofford, D. L. 2003. PAUP*. Phylogenetic analysis using parsimony (*and other methods). Version 4. Sunderland, MA: Sinauer Associates, Inc.

Swofford, D. L., G. J. Olsen, P. J. Waddell, and D. M. Hillis. 1996. Phylogenetic inference. *Molecular Systematics*, eds. D. M. Hillis, C. Moritz, and B. K. Mable, 407–514. Sunderland, MA: Sinauer Associates.

Szalanski, A. L., D. B. Taylor, and R. D. Peterson, 2nd. 1996. Population genetics and gene variation of stable fly populations (Diptera:Muscidae) in Nebraska. *Journal of Medical Entomology* 33: 413–420.

Sze, S. H., J. P. Dunham, B. Carey, P. L. Chang, F. Li, R. M. Edman, C. Fjeldsted, M. J. Scott, S. V. Nuzhdin, and A. M. Tarone. 2012. A *de novo* transcriptome assembly of *Lucilia sericata* (Diptera: Calliphoridae) with predicted alternative splices, single nucleotide polymorphisms and transcript expression estimates. *Insect Molecular Biology* 21: 205–221.

Tamura, K., D. Peterson, N. Peterson, S. Stecher, M. Nei, and S. Kumar. 2011. Mega5: Molecular evolutionary genetics analysis using maximum likelihood, evolutionary distance, and maximum parsimony methods. *Molecular Biology and Evolution* 28: 2731–2739.

Tarone, A. M. and D. R. Foran. 2006. Components of developmental plasticity in a Michigan population of *Lucilia sericata* (Diptera: Calliphoridae). *Journal of Medical Entomology* 43: 1023–1033.

Tarone, A. M. and D. R. Foran. 2008. Generalized additive models and *Lucilia sericata* growth: Assessing confidence intervals and error rates in forensic entomology. *Journal of Forensic Sciences* 53: 942–948.

Tarone, A. M. and D. R. Foran. 2011. Gene expression during blow fly development: Improving the precision of age estimates in forensic entomology. *Journal of Forensic Sciences* 56 Suppl 1: S112–122.

Tarone, A. M., K. C. Jennings, and D. R. Foran. 2007. Aging blow fly eggs using gene expression: A feasibility study. *Journal of Forensic Sciences* 52: 1350–1354.

Tarone, A. M., C. J. Picard, C. Spiegelman, and D. R. Foran. 2011. Population and temperature effects on *Lucilia sericata* (Diptera: Calliphoridae) body size and minimum development time. *Journal of Medical Entomology* 48: 1062–1068.

Tomberlin, J. K., M. E. Benbow, A. M. Tarone, and R. M. Mohr. 2011a. Basic research in evolution and ecology enhances forensics. *Trends in Ecology & Evolution* 26: 53–55.

Tomberlin, J. K., R. Mohr, M. E. Benbow, A. M. Tarone, and S. Vanlaerhoven. 2011b. A roadmap for bridging basic and applied research in forensic entomology. *Annual Review of Entomology* 56: 401–421.

Torres, T. T., M. L. Lyra, P. Fresia, and A. M. L. Azeredo-Espin. 2007. Assessing genetic variation in the New World screwworm *Cochliomyia hominivorax* population from Uruguay. In *Area-Wide Control of Insect Pests*, eds. M. J. B, Vreysen, A. S. Robinson, and J. Hendrichs, 183–91. Netherlands: Springer Netherlands.

Wallman, J. F., R. Leys, and K. Hogendoorn. 2005. Molecular systematics of Australian carrion-breeding blowflies (Diptera: Calliphoridae) based on mitochondrial DNA. *Invertebrate Systematics* 19: 1–15.

Wells, J. D., F. Introna, G. Di Vella, C. P. Campobasso, J. Hayes, and F. A. Sperling. 2001a. Human and insect mitochondrial DNA analysis from maggots. *Journal of Forensic Sciences* 46: 685–687.

Wells, J. D., T. Pape, and F. A. Sperling. 2001b. DNA-based identification and molecular systematics of forensically important Sarcophagidae (Diptera). *Journal of Forensic Sciences* 46: 1098–1102.

Wells, J. D. and F. A. Sperling. 1999. Molecular phylogeny of *Chrysomya albiceps* and *C. rufifacies* (Diptera: Calliphoridae). *Journal of Medical Entomology* 36: 222–226.

Wells, J. D. and F. A. Sperling. 2001. DNA-based identification of forensically important Chrysomyinae (Diptera: Calliphoridae). *Forensic Science International* 120: 110–115.

Wells, J. D. and J. R. Stevens. 2008. Application of DNA-based methods in forensic entomology. *Annual Review of Entomology* 53: 103–120.

Wells, J. D., R. Wall, and J. R. Stevens. 2007. Phylogenetic analysis of forensically important *Lucilia* flies based on *Cytochrome Oxidase I* sequence: A cautionary tale for forensic species determination. *International Journal of Legal Medicine* 121: 229–233.

Wells, J. D. and D. W. Williams. 2007. Validation of a DNA-based method for identifying Chrysomyinae (Diptera: Calliphoridae) used in a death investigation. *International Journal of Legal Medicine* 121: 1–8.

Wiegmann, B. M., M. D. Trautwein, I. S. Winkler, N. B. Barr, J. K. Kim, C. Lambkin, M. A. Bertone et al. 2011. Episodic radiations in the fly tree of life. *Proceedings of the National Academy of Sciences of the United States of America* 108: 5690–5695.

Wood, T. K. 2009. Insights on *Escherichia coli* biofilm formation and inhibition from whole-transcriptome profiling. *Environmental Microbiology* 11: 1–15.

Zhang, M., H. Yu, Y. Yang, C. Song, X. Hu, and G. Zhang. 2013. Analysis of the transcriptome of blowfly *Chrysomya megacephala* (Fabricius) larvae in responses to different edible oils. *PLoS ONE* 8: e63168.

CHAPTER 25

Engineering and Forensic Entomology

Glen C. Rains, Jeffery K. Tomberlin, and Robin Fencott

CONTENTS

25.1 INTRODUCTION

Engineering can be defined as the creative application of scientific principles in the design and development of structures, machines, apparatus, or manufacturing processes to improve or develop intended functions that make tasks more efficient, safe, or practical. As applied to forensic entomology, it can mean the correct choice of measurement instruments, or the development and design of new ones. It can also be development of computer programs, computer apps, and/or creative applications of behavioral sciences. This chapter focuses on tools and devices for measuring climatic, weather, soil, and volatile data at the scene of the forensic entomologist's investigation. The potential for future engineering solutions applied to forensic entomological investigations is also explored.

25.2 CURRENT MEASUREMENTS WITH FORENSIC ENTOMOLOGY APPLICATIONS

25.2.1 Environmental Measurements (North America)

Ambient temperature, relative humidity (RH), and precipitation from the location of a cadaver discovery are crucial to an accurate estimation of the postmortem interval (PMI). Estimations based on extrapolation from nearest weather station data lead to uncertainty, especially as the nearest temperature information gets further removed from the location of interest (Archer 2004; Dourel et al. 2010; Johnson et al. 2012).

Localized weather information as well as soil temperature, moisture, and hydrological data are available to varying degrees in all states, Canada, and Mexico. The best source of information is the hydrometeorological network web page: www.eol.ucar.edu/projects/hydrometnet/, which is managed by the U.S. National Center for Atmospheric Research. This website lists most meteorological and hydrological networks in the United States and Canada. Currently, a summary Mexico's network is listed as "in development." Even if some links are broken, the names of the networks are available and can be searched by Internet to find the new link or a contact person to find more information. This resource lists multiple government agencies and even local meteorological clubs that keep data in specific geographical locations. Local meteorological clubs should not be dismissed as a potential weather source. Weather records from these sources could be close to the investigation. Another good source is the weather network (www.theweathernetwork.com). It provides historical air temperature and precipitation from all over North America. Within that site is the farmzone that covers the Canadian provinces only (www.farmzone.com). Mainly a farming resource, it provides historical data for free as well, but includes more details such as ultraviolet index, dew point temperature, humidity, and hours of sunshine.

25.2.2 Environmental Measurements (World)

Each country has meteorological and agricultural data resources, most connected to the country's government. The meteorological data will cover the ambient temperature, precipitation, and other common measures of climate. The world meteorological organization (WMO) has the representative country organizations on their website: http://www.wmo.int/pages/index_en.html. The agricultural department for each country may be useful in providing data on soil parameters such as temperature and moisture. Many countries and states in northern latitudes use a road weather information system composed of environmental sensor stations to monitor and disseminate information on road conditions (Eriksson and Norman 2001). These stations record, among other things, temperature and precipitation. In the United States, the Federal Highway Administration is the home to the Road Weather Management Program and would have information on where stations are located. The appropriate transportation administration of countries in Northern Europe, such as Sweden, would need to be contacted to access their roadway weather data.

There may also be local weather clubs and organizations that collect their own data. Check with local extension offices in the United States and similar organizations in other countries to find local sources of information. As stated, local clubs may provide data that are extremely close to the scene of an investigation. In any instance, the challenge is obtaining data close to the scene of the investigation and at the dates of interest to an investigation.

25.2.3 Temperature Measurements

Temperature measurements at the site of the investigation can be made using multiple instruments, depending on the needs of the case. Point measurements are the most common and easiest to make on the cadaver, between the cadaver and soil interface, soil and larval mass of the colonizing insects. Surface measurements and measurements inside the remains can each be made using a probe containing the temperature measuring device. The primary thermometer measuring systems are the analog mercury/alcohol thermometer, resistance temperature detector (RTD), thermistor, and thermocouple.

For each of these methods, it is important that the temperature measuring device has a recent calibration, particularly digital thermometers using RTD, thermistor, and thermocouple technology; often these calibrations are performed by the company. You can perform a test of the accuracy of a device using ice water, which should measure 0°C. For all thermometers, ambient temperature measurements should be made where there is no direct sunlight and little to no air movement. In either case, the temperature reading would be affected if not protected. Sunlight causes radiative heating of the sensor that would artificially raise the ambient temperature measurement. Air movement around the sensor also causes convective cooling, which will artificially lower the ambient temperature measurement. For further recommendations, measurement standards can be found at ASTM International (www. ASTM.org).

25.2.3.1 Mercury/Alcohol Thermometer

The most commonly used temperature measuring device is the mercury/alcohol bulb thermometer. It can be used to measure the temperature of the cadaver, soil, air, insect egg, larvae, and adults. Currently, mercury thermometers are mainly used in scientific laboratories for their accuracy and broad range of temperature measurement. However, mercury is toxic and the glass is easily breakable. Many bulb thermometers are now alcohol based. Alcohol-based bulb thermometers are considered as accurate as mercury thermometers at the temperature range expected for forensic investigations. Typical accuracy is ±1°–2°C. However, care should be used when using such thermometers with cadavers as their use risks damage to the remains.

25.2.3.2 Bimetallic

Bimetallic thermometer's primary application is as a meat thermometer. Two metals are joined in a long strip and wrapped into a coil. As the temperature changes, the difference in expansion and contraction of the two separate metals causes the coil to move a dial that is calibrated to a temperature gauge reading. This device can be used as a probe to determine the temperature below the surface of the target, such as below the surface or inside a corpse. Typically they are accurate to within 1°C.

25.2.3.3 Resistive Temperature Sensors

The two primary resistive types of temperature sensors are the RTD and the thermistor. Each uses a sensing element (RTD is a fine coiled metal and thermistor is a semiconductor) that changes resistance with a change in temperature. Temperature is correlated to changes in voltage across the sensors when given a constant excitation current. However, each uses a slightly different principle of operation. The thermistor has a more narrow operating range and is less expensive than the RTD, but the RTD is more common among digital temperature probes. Newer technology and computer processing has made it possible to get cheaper thermistors with accuracy as high as or higher (<1°C) than RTDs and with minimal drift over time.

25.2.3.4 Thermocouples

A thermocouple consists of two dissimilar metals joined at one end. A change in temperature at the joint creates a voltage potential across the two metals called the Seebeck voltage. This means thermocouples do not require an external voltage source. Electronic cold junction compensation is required to convert the voltage into an accurate temperature. There are multiple combinations of metals that cover different ranges of temperatures. For example, Type J thermocouples use an iron–constantan combination of metals and have a temperature range of –40°C to 750°C and are accurate to within ±1.5°C. Thermocouples produce an analog signal that can be amplified and read directly with a gauge, or more generally, it is digitized and read as a digital thermometer. Thermocouples' accuracy is about the same as resistive temperature sensors.

25.2.3.5 Infrared Noncontact Thermometer

Infrared thermometers make noncontact and hard-to-reach temperature measurement possible. An Infrared noncontact gun uses a low power (<1 mW) laser beam for aiming and an array of sensors to measure the emitted, reflected, and transmitted radiant energy from an object at which the gun is pointing. The size of the object and distance to the object affect the measurement accuracy. The spot size that the detector "sees" is proportional to the distance from the object. The spot size must be smaller than the object aimed at. Also, an assumed emissivity of the object is programmed into many thermal thermometers. This can lead to a false reading for glossy and/or shiny objects. Emissivity adjustments are possible with many instruments. The instruments manual will provide specific instructions for objects that have an emissivity that requires some type of compensation for an accurate measurement. Typical accuracy is ±1% of reading or ±1°C, whichever is greater.

25.2.3.6 Infrared Noncontact Camera (2D Picture)

Infrared cameras are useful if you want the temperature gradients over an area instead of a spot and you want to see how the temperature gradient changes within the field of view. These cameras detect radiation in the infrared range of the electromagnetic spectrum, just like the gun, except they have an array of sensors to produce a 2D image. These infrared thermal imagers are much more expensive than the gun and are of limited use in forensic investigations.

25.2.4 Data Logging

Data logging ambient air temperature, soil temperature, and other environmental measurements can be accomplished using several off-the-shelf data logging systems (Table 25.1). The Onset Hobo datalogger comes with internal temperature measurement and RH for some models as well as connections for external sensors to measure temperature, rainfall, and soil moisture. They also supply the sensor with the datalogger. Spectrum, Campbell, and Thermodata each provide similar systems. Dataloggers can be left unattended and programmed to collect data at intervals of the users

Table 25.1   Datalogger Companies, Products, Software, and Websites

Company	Products	Software	Website
Onset	Hobo datalogger, external sensors	HOBOware, HOBOlink	www.onsetcomp.com
Spectrum	WatchDog datalogger, external sensors	SpecWare	www.specmeters.com
Campbell Scientific	CRxx dataloggers, external sensors	LoggNet and PCxx datalogger	www.campbellsci.com
Thermodata	TLxx thermologger	Thermodata viewer	www.store.thermodata.us
Extech	Temp & Humidity dataloggers	Windows compatible	www.Extech.com

Note: xx refers to different model numbers.

choosing, such as every 15, 30, or 60 minutes. The dataloggers can then be connected to a computer to download the collected data. This can provide diurnal data on temperature and/or humidity at the location of interest.

25.3 VOLATILE CHEMICALS

Volatile chemicals are released to some extent from almost all chemical compounds. For forensic investigations, they can be produced through aerobic and anaerobic decomposition by invertebrates and microbes. In cases involving a corpse, the decomposition by invertebrate and microbial metabolic processes leads to specific volatile chemical production. Vass et al. (2008) cataloged volatile compounds for 4 years of corpse decomposition in a shallow grave at the body farm in Tennessee. Their study showed the breakdown of compounds that are (1) continuously released, (2) released during the first year, and (3) present as long as soft tissue remains with the body. The value of this study is the breakdown of chemicals present during the decomposition process with the accompanying field conditions, such as RH, temperature, and local precipitation. More recent studies have begun to unravel the mechanisms of what decomposition processes produce certain compounds and how these may be regulated between inter-kingdom relationships. For example, volatile compounds were found to vary by a bacteria species commonly present on human remains depending on their ability to swarm (Tomberlin et al. 2012). These shifts in volatile production have been shown to affect the egg-laying (oviposition) of blow flies. Volatiles can also be used to find contraband, most notably illegal drugs, explosives, and accelerants from arson cases, as is commonly done using trained canines.

A burgeoning area of detection technology is the development of insects as biosensors for the purpose of identifying and detecting chemicals of interest (Rains et al. 2008; Tomberlin et al. 2008; Frederickx et al. 2011). The following sections examine common electronic and biological volatile detection systems.

25.3.1 Electronic Volatile Detection

The electronic nose uses sensors to detect volatile compounds that it can be trained to recognize. Much like animal olfaction, multiple electrical sensors (receptors) can detect overlapping arrays of chemicals (Pearce 2003). The resulting response of each sensor creates a pattern, stored as a specific scent and is recognized when it is subsequently detected. These devices are best used when contaminating background odors are minimal, which is rarely the case outside of a laboratory or other controlled environments. The best commercial example of this type of electronic nose is the Cyranose 320 (Intelligent Optical Systems). It uses 32 electrically conducting polymers (chemresistors) that each adsorbs different chemicals like a sponge, changing the electrical resistance. Another device, called the Z-nose (Electronic Sensor Technology), uses a surface acoustic wave (SAW) sensor along with a gas chromatograph column to detect individual compounds from a sample as they are separated by the column. Consequently, the odor pattern is created by the SAW response to the compounds and this response can be converted to a "smellprint" similar to other electronic nose devices. Another device has been developed at the Oak Ridge National Laboratory called the LABRADOR (Light-weight Analyzer for Buried Remains and Decomposition Odor Recognition). It uses 12 unique metal oxide chemicals sensors and onboard electronics to detect 30 classes of chemicals produced by a corpse (Vass et al. 2010). It is portable and is currently licensed by Agile Technologies, who is developing it for commercial use. Each e-nose relies on statistical or other methods for pattern recognition to classify each odor pattern and then identify it when exposed during testing with unknown samples.

Portable gas detection and monitoring equipment are also available that are designed to detect only specific chemical compounds. These are generally used to detect dangerous and/or combustible gases such as carbon monoxide, chlorine, or ammonia in industrial settings. Typically, these devices use electrochemical sensors and also photoionization detectors (PID) and infrared sensors to detect multiple chemicals. Some of the primary manufacturers of these sensors are Dräger and RAE instruments.

There are other e-nose sensing technologies, as well as chemical detection based on quantum dots, DNA/RNA aptamers, and surface plasmon resonance (SPR), to name a few. Most chemical sensor development is focused on high sensitivity to one specific chemical.

25.3.2 Whole-Organism Invertebrate Volatile Detection

Although dogs are the most well-developed method of search and detection of odors, studies have shown that insect olfaction can be used for cadaver detection and discovery (Tomberlin et al. 2008). Insects can be conditioned to link chemical odors to food or other resources through associative learning techniques, and then exhibit a specific behavioral response to that odor. This is accomplished by classically conditioning the insect to recognize a conditioned stimulus (CS) (the target odor) in association with an unconditioned stimulus (US) (food, host, or other resource). The training protocol requires an exposure between the CS and US for a specific period of time, repetition of training, and interval between training repetitions. It also requires some control of the physiological state of the insect. Generally, insects are starved for 1–2 days before training in the case of food–odor associative learning. In the case of the parasitic wasp, *Microplitis croceipes* (Cresson) (Hymenoptera: Braconidae), the entire training routine to classically condition an individual wasp to an odor is approximately 5–10 minutes (Tertuliano et al. 2005). Once conditioned, insects can respond to the odor they associate with food with a conditioned response (CR). Honey bee, *Apis mellifera* L. (Hymenoptera: Apidae), studies of classical conditioning usually rely on the proboscis extension response (PER) (Bitterman et al. 1983; Giurfa and Sandoz 2012). Wasps have been shown to exhibit a food-searching or coiling response that includes specific body movements and antennation of the odor source. The hawk moth, *Manduca sexta* (Linnaeus) (Lepidoptera: Sphingidae), feeding response has been monitored by electromyography, detecting a feeding signal from the muscles of the moth (King et al. 2004). In each of these cases, the insect is confined to a small arena, or harnessed for observation of the behavior.

The parasitic wasp, *M. croceipes*, has also been shown to exhibit a context-dependent CR when conditioned to two different CS (Olson et al. 2003). Wasps were conditioned to one odor with food and another odor with larval feces (frass). When exposed to the odors after conditioning, the wasps exhibited unique behaviors for each odor, one a food-searching behavior and the other a coiling behavior. This study demonstrated the ability to use the insect as a sensor for at least two target odors by observing the appropriate CR.

To provide an objective measurement of the behavioral response of insects, instruments have been developed that measure those responses and indicate the presence of a target chemical. The Wasp Hound was developed to detect volatile compounds using the trained parasitic wasp, *M. croceipes* (Rains et al. 2006). Five wasps, conditioned to the odor or specific volatile compound of interest, are placed in a small cartridge mounted inside the device. Once the cartridge is placed inside the instrument, a fan pulls sample air into the cartridge and a web camera takes real-time video of the wasp's behavior (Figure 25.1). The camera is connected to a laptop computer running a

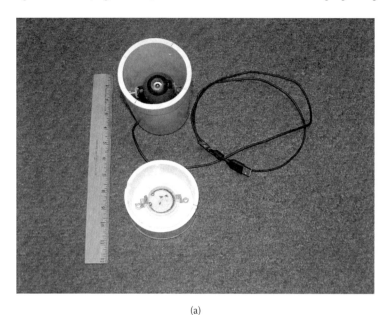

(a)

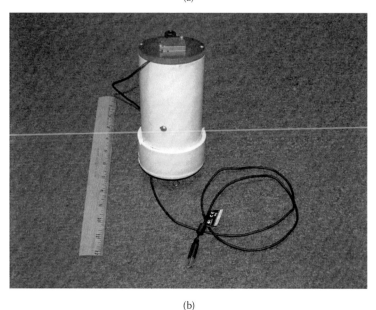

(b)

FIGURE 25.1 **(See color insert.)** (a) Wasp Hound assembled showing fan on top for pulling sample air into the instrument. (b) Wasp Hound chemical detector showing the top removed to expose the camera and wasp cartridge, where odor enters into the instrument. Air enters through the bottom. USB connector plugs into computer running real-time behavioral analysis program.

software program that analyzes the behavior and indicates when the behavior has changed and the odor has been detected (Utley et al. 2007). The device determines the presence of the target odorant within 20–30 seconds. The Wasp Hound has been used to detect multiple chemicals and odors including putrescine and cadaverine (Tomberlin et al. 2008).

The company Inscentinel (www.Inscentinel.com) is developing chemical detection devices using immobilized conditioned honey bees that measures the PER. Honey bees are harnessed so that only their proboscis extends when the target odor is detected. In each case, sampled air is pulled over the insect's antennae and the instrument provides the user with a reading to determine if the target odor is present.

Further tests with the Wasp Hound and the wasp *M. croceipes* indicate that through improved analysis of the behavioral response (Zhou et al. 2012) or modification of the conditioning routine (Olson et al. 2012) an indication of odor concentration can also be determined. The ability to measure odor concentration could lead to a method of tracking an odor to its source, as in finding clandestine gravesites or contraband locations (Rains et al. 2008).

Another method relies on sensing movements of conditioned insects in free flight as they search and find target odorants. Conditioning in this system is typically done in-mass and the bees are allowed to freely forage. LiDAR is used to detect the wing beat frequency of the bees as they fly near an odor of interest (Hoffman et al. 2007; Carlsten et al. 2011). Although free-flying insect tracking has been studied for land mine detection, it is also possible that such a system could be used in detecting clandestine gravesites, drugs, arson evidence, and other odors of forensic interest.

25.3.3 Insect Antennal Sensor

Insect antennae remain biologically active when severed from the insect for more than an hour (Park et al. 2002) and have olfactory receptor neurons (ORN) that can depolarize when exposed to volatile chemical signals. A 4–8 channel electroantennogram was developed as a biosensor to detect chemical odorants using antennae excised from different insects (Park and Baker 2002). Using multiple insect antennae allowed for a wider range of reaction to multiple chemicals. The device has been updated and used with Global Positioning System (GPS) and wind anemometer to also track pheromones to their source when using moths harnessed inside the device and the electroantennogram attached to the antennae. By keeping the insect intact, the antennae do not deteriorate and several hours of use can be achieved as demonstrated: they found that the moth pheromone was tracked upwind to within 0.2 m of the source (Myrick and Baker 2011). Another study used a similar approach using the whole blow fly and electroantennogram to detect odors (Huotari 2000). The action potential from a single ORN was monitored when exposed to 1,4-diaminobutane, 1-hexanol, and butanoic acid. The ORN was found to be responsive to each chemical to varying degrees depending on concentration levels.

25.3.4 Behavior versus Electroantennogram Methods

The two mentioned insect detection mechanisms are based on examining the behavioral response of an insect or multiple insects, or reading the antennae electrical impulses, referred to as spikes, directly using an electroantennogram. Each has advantages and disadvantages. The behavioral method uses the "brains" of the insect for associative learning and a behavioral response, or feedback, to indicate when the odor has been detected. Insects in these systems are classically conditioned to odors of interest and the insect's behavioral response is objectively interpreted by a system that reads that behavior electronically. In essence, the brain is programmed to the odors of interest through the associative learning technique of classical conditioning.

The electroantennogram method reads the signals from the antennae as they respond to chemicals interacting with the ORNs. In this case, the interpretation of the signal must be done electronically through some process to recognize the odor when it is detected. The antennae may respond to many odors, but the process of deciphering what the signals mean requires the development of hardware and/or software filters and recognition programs.

In behavioral systems, the insect can be used for a finite number of positive detections, after which the behavioral response to the odor is attenuated by the lack of a positive reinforcement, such as food. On the other hand, the electroantennogram can be used as long as the antennae are viable, which can be for several hours to days depending on whether the whole insect is kept intact. There are currently no studies that compare the accuracy or sensitivity of the two approaches or of different insects.

25.4 MOBILE COMPUTING AND DATA COLLECTION

Mobile phones have made computing, web browsing, photography, videography, voice recognition, and detection available in a small handheld platform. The most prominent feature for forensic investigations is the camera phone. Although these phones may not have all the features of standard camera or single-lens reflex cameras, their advancement over the last few years means they will rival professional photography in the near future. Lacking on phone cameras and a necessity for entomologist scene investigators is a macro lens for pictures of small insects. However, there are multiple third-party vendors, such as Photojojo (www.photojojo.com), that provide an add-on macro lens to camera phones.

25.4.1 There's an App for That?

"Apps" are programs (applications) built to run on handheld phones and tablets. They are generally cheap to buy and many program developers are becoming proficient in mobile software. Currently, the two predominant platforms for apps are the Google Android and Apple iOS operating systems. Android is freely available open operating system, which runs on a wide range of devices, whereas Apple is the sole user of their proprietary operating system, iOS. Microsoft and Blackberry are currently the two other major operating systems for phones. These operating system platforms are frequently updated with newer versions; check the operating system required for the app for best results. An app must be written for a specific operating system platform. To work on multiple platforms, the app must be rewritten for the platform.

Apps for forensic investigations will usually make use of one or more of the onboard sensors that come with most smartphones. Currently, these sensors are GPS, magnetometer, accelerometer, thermometer, camera, RH sensor, voice recorder, and barometric sensor. Using these sensors, apps are available that can calculate how level the camera/tablet is, the location in GPS coordinates, temperature, RH, and barometric pressure.

Apps for general forensic investigations, such as a blood spatter and forensic medicine app, are currently on the market. As of this writing, two forensic entomology specific apps were found. SmartInsects (University of Western Australia, Centre for Software Practice) provides a guide for collecting and preserving entomological samples, primarily aimed at nonentomologists such as coroners, medical examiners, and pathologists. This app is available on iPhone and android devices and currently four versions in English, Italian, Chinese (Mandarin), Brazilian–Portuguese. iFly (Stamper and Strong 2013) (Purdue University) is an iOS app for digitally recording codified crime scene information. By encouraging all data collection to take place in a single app, iFly addresses many issues associated with pen- and paper-based note taking (illegibility, transcription errors, disassociation from other project files) and allows audio, video, and photography to be collated with written notes taken at the crime scene. iFly uses an SQLite database to store data internally, and allows data to be transferred to a desktop computer. At the time of writing, the iFly was in beta development and not yet not available on the Apple App Store. In addition, ForenSeek is a recent software developed for calculating egg laying times, with additional details about this software described in Chapter 12.

25.4.2 Mobile Computing Concerns

Smartphones and tablet computers are items of consumer electronics, and may not be as reliable, rugged, or practical as purpose-built devices. Screen visibility in bright sunlight, waterproofing, data signal availability, and battery life are all important considerations, whereas their onboard sensors may not be as accurate, sensitive, or fault tolerant as those found in scientific instruments.

Furthermore, due to the vast array of devices on the market, the behavior of an app might vary between different devices, operating systems, and software versions, potentially leading to discrepancies in measurements and problems of repeatability in experimental work. Should mobile devices become commonplace in investigative work, it may be important to evaluate their performance, reliability, and accuracy. Evaluations could for instance focus on the differences between mobile device models, or provide a comparison of mobile app performance to results obtained through more established techniques.

25.4.3 Web Apps

In comparison to mobile apps discussed earlier, Web Apps run on remote servers (in "the cloud"), and are typically accessed by visiting a web page using a standard Internet browser. Google Drive (www.drive.google.com), Amazon Cloud Drive (www.amazon.com/clouddrive), and DropBox (www.dropbox.com) are popular cloud applications, allowing multiple people to share and collaboratively edit documents. Because all cloud data are stored online, users accessing a shared document always see the most up-to-date version, and changes are instantly shared with all other users.

Cuttiford and Fencott have developed the "ADH Monitor" Web App (http://spiderylue.wordpress.com/adh-monitor/) to assist in the conduct of forensic entomology experimental work. The ADH Monitor tracks real-time temperature data near an investigation site, and automatically sends e-mail alerts when user-specified accumulated degree hour (ADH) values are reached. The ADH Monitor can potentially gather temperature data from any live temperature data source on the Internet. Currently, data are collected from the Wunderground weather station network (http://www.wunderground.com/), and from a datalogger at the Texas State University Forensic Anthropology Facility. The ADH Monitor automatically checks these sources at regular intervals and updates its ADH calculations when new data become available. The system can also send an e-mail warning if a data source becomes unavailable.

By automating repetitive data collection tasks, the ADH Monitor frees up investigator time, and is capable of performing ADH calculations for hundreds of simultaneous experiments. Furthermore, because multiple researchers can log into the system simultaneously, the ADH Monitor can be viewed as an online collaborative research tool, which provides up-to-date status information about ongoing experimental work.

At the time of writing, the ADH Monitor is under active development, and is being used in experimental work by researchers at Texas A&M University. The ADH Monitor is written in Java and hosted on the Google App Engine.

25.5 FUTURE TECHNOLOGIES FOR FORENSIC ENTOMOLOGY

25.5.1 Innate Insect Behavior

Previously, insects trained to detect target chemicals were discussed as potential instruments for clandestine corpse discovery. However, insects that have an innate attraction to corpse odors may be more sensitive and require less or no conditioning to be used as sensors (Rains et al. 2008). Several insect species that colonize a corpse could be potential sensor elements in a detection device used to search and find clandestine gravesites and/or cadavers after natural disasters, fires, and other events that require finding bodies in debris. Multiple carrion colonizing beetle and fly species could be candidates as detectors of cadavers. Behavioral changes would be monitored inside a detection device, similar to the Wasp Hound. Another potential use is as a verification tool for cadaver dogs. Once a cadaver dog positively identifies a cadaver location, the insects could be used to verify they also detect the odor. This could reduce lost time and effort from false-positive hits by the cadaver dog.

25.5.2 Mobile Sensing and Data Sharing

Apps are developed continuously, but one potentially useful concept, and others like it to come, is the PressureNet App. Currently barometric pressure is collected from mobile phones running the app automatically and logged with the GPS location of the phone. Their website includes a graph of the data for specific regions on a Google map. PressureNet developers are updating the software continually and users can also log local weather conditions. This and other similar apps designed to log temperature and RH could be useful in pinpointing localized weather conditions in urban areas if there is sufficient public interest. Access to historical data would be required and is currently available on PressureNets website (pressurenet.cumulonimbus.ca).

Instruments for improved investigations are also likely to emerge. Handheld DNA analysis at the crime scene is connected to a network database that could link DNA to a person and potentially reduce time for suspect apprehension or victim identification. Further, lab-on-a-chip technology could conduct blood testing (may be extracted from mosquitoes or other blood-sucking insect) for drugs or test soil samples and soil organisms.

25.5.3 Developing Apps for Forensic Entomology

A fruitful approach toward the development of new software for forensic entomology might be to initiate collaboration between software developers and practicing forensic entomologists. A dialog between these groups could provide practitioners with insights into the technological possibilities from mobile devices, web services, and modern programming languages, whereas software developers stand to gain by understanding in greater detail the process and workflows of practicing investigators. This could lead to new and innovative design concepts for mobile apps, web services, and other forms of computer technology.

Although there is at present a paucity of information concerning the design of software for forensic entomology investigation (see Chapter 12 for discussion of the ForenSeek program), the field of Human–Computer Interaction (HCI) (Dix et al. 2004) has contributed a wealth of methodologies and "user-centric" design (Norman and Draper 1986) practices to aid in the design of computing systems. Of particular note is the practice of "Participatory Design" (Sears and Jacko 2009), which involves engagement by both designers and end users at all stages of the design process to produce systems, which meet their requirements and are usable by the target end users. Workplace observation (Heath et al. 2000) is another common HCI research technique for studying the working practices of teams and individuals. A variety of established ethnographic methodologies exist to provide a structured approach to these forms of observational studies. Within the context of forensic entomology, observations of field work and laboratory activities could lead to a more thorough understanding of how and where new technologies might be developed.

25.5.4 Software Development Practices

Software design and implementation and maintenance is a skilled process. Poorly written software may at best prove to be frustrating to use, however, at its worst may result in wasted time, miscalculations, and data loss. A variety of software development practices exist to counteract these problems. Of particular note is the practice of Test Driven Development (TDD) (Beck 2003) that can be used to empirically validate the functionality of all computer code within the software. For example, TDD can be used to identify "bugs," to confirm that mathematical calculations produce the correct results for a given set of test cases, and ensure that all input to the software is correctly validated to reduce margins for user error.

25.5.5 User Testing and Iterative Development

In addition to validating the functionality of all computer code, it is essential for software to be evaluated and tested in use by those for whom the software is intended to be used. Once again, there are a variety of HCI methodologies for testing and evaluating the success of user interfaces and software designs. Finally, once software is "in the wild," problems, ideas, and new requirements invariably arise. At this stage, the design and implementation of the software may need to be revised so that new iterations of the software can be developed and released.

REFERENCES

Archer, M.S. 2004. The effect of time after body discovery on the accuracy of retrospective weather station ambient temperature corrections in forensic entomology. *Journal of Forensic Sciences* 49 (3): 553–559.

Beck, K. 2003. *Test-Driven Development by Example*. Boston, MA: Addison Wesley.

Bitterman, M.E., R. Menzel, A. Fietz, and S. Schafer. 1983. Classical-conditioning of proboscis extension in honeybees (*Apismellifera. Journal of Comparative Psychology* 97 (2): 107–119.

Carlsten, E.S., G.R. Wicks, K.S. Repasky, J.L. Carlsten, J.J. Bromenshenk, and C.B. Henderson. 2011. Field demonstration of a scanning lidar and detection algorithm for spatially mapping honeybees for biological detection of land mines. *Applied Optics* 50 (14): 2112–2123.

Dix, A., J. Finlay, G.D. Abowd, and R. Beale. 2004. *Human-Computer Interaction*, 3rd ed. Harlow, England: Prentice Hall.

Dourel, L., T. Pasquerault, E. Gaudry, and B. Vincent. 2010. Using estimated on-site ambient temperature has uncertain benefit when estimating postmortem interval. *Psyche (Cambridge)* 2010: 1–7.

Eriksson, M. and J. Norman. 2001. Analysis of station locations in a road weather information system. *Meteorological Applications* 8: 437–448.

Frederickx, C., F.J. Verheggen, and E. Haubruge. 2011. Biosensors in forensic sciences. *Biotechnologie Agronomie Societe et Environnement* 15 (3): 449–458.

Giurfa, M. and J.C. Sandoz. 2012. Invertebrate learning and memory: Fifty years of olfactory conditioning of the proboscis extension response in honeybees. *Learning & Memory* 19 (2): 54–66.

Heath, C., H. Knoblauch, and P. Luff. 2000. Technology and social interaction: The emergence of 'workplace studies'. *British Journal of Sociology* 51 (2): 299–320.

Hoffman, D.S., A.R. Nehrir, K.S. Repasky, J.A. Shaw, and J.L. Carlsten. 2007. Range-resolved optical detection of honeybees by use of wing-beat modulation of scattered light for locating land mines. *Applied Optics* 46 (15): 3007–3012.

Huotari, M.J. 2000. Biosensing by insect olfactory receptor neurons. *Sensors and Actuators B-Chemical* 71 (3): 212–222.

Johnson, A.P., J.F. Wallman, and M.S. Archer. 2012. Experimental and casework validation of ambient temperature corrections in forensic entomology. *Journal of Forensic Sciences* 57 (1): 215–221.

King, T.L., F.M. Horine, K.C. Daly, and B.H. Smith. 2004. Explosives detection with hard-wired moths. *IEEE Transactions on Instrumentation and Measurement* 53 (4): 1113–1118.

Myrick, A.J. and T.C. Baker. 2011. Locating a compact odor Source using a four-channel insect electroantennogram sensor. *Bioinspiration & Biomimetics* 6 (1): 016002.

Norman, D.A. and S.W. Draper, eds. 1986. *User Centered System Design: New Perspectives on Human-Computer Interaction*. Hillsdale, NJ: Erlbaum Associates.

Olson, D., F. Waeckers, and J.E. Haugen. 2012. Threshold detection of boar taint chemicals using parasitic wasps. *Journal of Food Science* 77 (10): S356–S361.

Olson, D.M., G.C. Rains, T. Meiners, K. Takasu, M. Tertuliano, J.H. Tumlinson, F.L. Waeckers, and W.J. Lewis. 2003. Parasitic wasps learn and report diverse chemicals with unique conditionable behaviors. *Chemical Senses* 28 (6): 545–549.

Park, K.C. and T.C. Baker. 2002. Improvement of signal-to-noise ratio in electroantennogram responses using multiple insect antennae. *Journal of Insect Physiology* 48 (12): 1139–1145.

Park, K.C., S.A. Ochieng, J.W. Zhu, and T.C. Baker. 2002. Odor discrimination using insect electroantennogram responses from an insect antennal array. *Chemical Senses* 27 (4): 343–352.

Pearce, T.C. 2003. *Handbook of Machine Olfaction Electronic Nose Technology*. Weinheim, Germany: Wiley-VCH.

Rains, G.C., J.K. Tomberlin, and D. Kulasiri. 2008. Using insect sniffing devices for detection. *Trends in Biotechnology* 26 (6): 288–294.

Rains, G.C., S.L. Utley, and W.J. Lewis. 2006. Behavioral monitoring of trained insects for chemical detection. *Biotechnology Progress* 22 (1): 2–8.

Sears, A. and J. A. Jacko, eds. 2009. *Participatory Design: The Third Space in HCI*. Boca Raton, FL: CRC Press.

Stamper, T. and C. Strong. 2013. iFly: an iPad® program for recording forensic entomology field research and student instruction. *North American Forensic Entomology Association*, July 2013.

Tertuliano, M., J.K. Tomberlin, Z. Jurjevic, D. Wilson, G.C. Rains, and W.J. Lewis. 2005. The ability of conditioned *Microplitis croceipes* (Hymenoptera: Braconidae) to distinguish between odors of aflatoxigenic and non-aflatoxigenic fungal strains. *Chemoecology* 15 (2): 89–95.

Tomberlin, J.K., T.L. Crippen, A.M. Tarone, B. Singh, K. Adams, Y.H. Rezenom, M.E. Benbow et al. 2012. Interkingdom responses of flies to bacteria mediated by fly physiology and bacterial quorum sensing. *Animal Behaviour* 84 (6): 1449–1456.

Tomberlin, J.K., G.C. Rains, and M.R. Sanford. 2008. Development of *Microplitis croceipes* as a biological sensor. *Entomologia Experimentalis et Applicata* 128 (2): 249–257.

Utley, S.L., G.C. Rains, and W.J. Lewis. 2007. Behavioral monitoring of *Microplitis croceipes*, a parasitoid wasp, for detecting target odorants using a computer vision system. *Transactions of the Asabe* 50 (5): 1843–1849.

Vass, A., C.V. Thompson, and M. Wise. 2010. *A New Forensics Tool: Development of an Advanced Sensor for Detecting Clandestine Graves*. Oak Ridge, TN: Oak Ridge National Laboratory.

Vass, A.A., R.R. Smith, C.V. Thompson, M.N. Burnett, N. Dulgerian, and B.A. Eckenrode. 2008. Odor analysis of decomposing buried human remains. *Journal of Forensic Sciences* 53 (2): 384–391.

Zhou, Z., G.C. Rains, and D. Kulasiri. 2012. Development of a behavior parameter in classically conditioned parasitic wasps that detect changes in odor intensity. *Biological Engineering* 5 (1): 19–31.

CHAPTER 26

Behavioral Ecology and Forensic Entomology

Michelle R. Sanford, Jeffery K. Tomberlin, and Sherah L. VanLaerhoven

CONTENTS

26.1 INTRODUCTION

What is insect evidence? It is certainly any insect alive or dead, or associated debris from insects' developmental stages such as shells of hatched eggs, shed exoskeletons left behind from the molt of one immature stage to another, or empty puparia left behind when the adults emerge. Equally vital as evidence are signs of insect feeding, bite marks, defecation, as well as the timing of adult arrival to the body, oviposition, and location of oviposition (egg laying). Thus, it is important to recognize that it is not just the insects themselves but also their behavior that can be of vital evidentiary value in providing information regarding the time frame or conditions of the body prior to discovery.

Behavioral ecology is a field of scientific inquiry that asks what is the relationship between the environment in which an organism has evolved and its behavior. Specifically, it determines the function of behavior, the evolutionary context of behavior, and what role behavior has in determining population dynamics and community patterns. Thus, behavior of insects is a function of their genetic composition and their interaction with the environment in which they find themselves. At the most basic level, the ability for insects to be active, fly, locate mates and food resources, and produce successful offspring depends on the genetic expression of their physiological tolerances at specific temperature and humidity conditions, as well as their behavior in selecting specific micro-habitats that provide the abiotic conditions required for successful development and reproduction. Acquisition of these sites and resources is likely to depend on the insect's ability to compete with conspecifics and heterospecifics, avoid or defend against predators and parasitoids, and maximize their use of the resource. The specific behavioral strategies employed by different insect species and populations, plasticity in which they modify their strategies under different abiotic and biotic conditions, and limits to which they can endure are all a function of the interaction of genetics with the environment.

The evidentiary value of insects in forensic investigations is based on the premise that insect behavior is predictable, with insects arriving and colonizing dead bodies in a predictable manner (Figure 26.1). The speed at which insects arrive and lay their eggs, signaling use of the body as a suitable resource for their offspring, depends on several biotic factors such as what location on the body is an acceptable food source for their offspring; the diet breadth of the species, which influences whether the insect is a specialist, feeding only on flesh, or a more generalist decomposer, feeding on a range of decomposing vegetative and animal material; and how the presence of other species either facilitate colonization or inhibit and potentially competitively exclude colonization. There are other abiotic factors that influence insect behavior such as the time of day, season, or temperature.

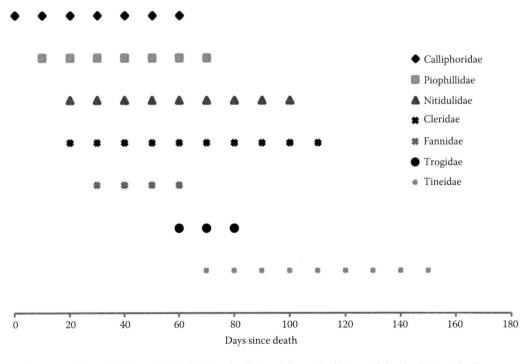

Figure 26.1 Generalized succession of insect families arriving, colonizing, and developing on a body.

Thus, understanding the behavioral ecology of these insects commonly associated with decomposition is critical to predicting where, when, and under what circumstances they will colonize bodies. Although there is some understanding of how various factors relate to insect behavior, a much more thorough understanding of the mechanisms involved is required.

26.1.1 Scope of Chapter

In this chapter, the study of how the behavioral ecology of forensically important insects can benefit the application of insect data to death investigations is explored. Forensic entomology relies heavily on key behaviors such as oviposition, larval migration, and competition to determine an estimate of colonization time or minimum postmortem interval (min-PMI). Colonization and succession, at the heart of modern forensic entomology practice, are at the interface of behavior and the environment. The study of behavioral ecology in forensically important insects is early, and many of the methods and analyses that have been developed in other taxa (e.g., mosquitoes and *Drosophila*) have not yet been applied to forensic entomology. For the purpose of this chapter, other entomology examples from taxa not commonly encountered in a forensic setting have been integrated to provide greater explanation of a given principle. These examples are meant to exemplify methods and analyses that exist but have not yet been applied in forensic entomology casework.

26.2 GENETICS IN BEHAVIORAL ECOLOGY OF FORENSICALLY IMPORTANT INSECTS

Genetic information is one component underlying the expression of behavior in insects. At its basic level, it is the directly measurable information content of the insect. The genetic sequence of an individual insect can be analyzed for information about genes associated with certain traits and behaviors. The application of genetic information to insects of forensic importance is really in its infancy. Thus, in this section three examples are presented of both how scientists are currently using genetic information content in forensic entomology and how examples from other insects and study systems may be applicable in death investigations. The examples presented here are merely a brief overview. For a thorough review of how genetic tools are being used and developed in forensic entomology, see the following recent reviews: reviews by Wells et al. (2007), Wells and Stevens (2007), and Wells and Stevens (2010), as well as Chapter 24 in this book.

26.2.1 Genetic Information for Species Identification

Species identification is the critical first step to any forensic entomology casework as it relates the developmental dataset to the species collected from the body. It is of equal importance when one initiates the process of collecting temperature-dependent developmental data as it can be significantly different across species. Among the true flies (Diptera), there are many examples of species that share morphological characteristics yet differ in their ecological, behavioral, and distributional characteristics. In insects that are highly morphologically similar, practitioners have been left with few tools to determine the species, until the recent developments and more widespread accessibility of genetic tools such as the polymerase chain reaction (PCR) for species identification.

Genetic tools are widely used to differentiate insect species complexes that have indistinguishable morphology but significant differences in behavior. Among other fly groups, mosquitoes (Diptera: Culicidae) significantly impact humans and animals due to their ability to serve as vectors for a number of pathogens. In sub-Saharan Africa, the *Anopheles gambiae* s.l. species complex comprises at least six morphologically indistinguishable species (Coetzee et al. 2000). Within this complex, there are a range of behaviors and physiological differences in each mosquito species including susceptibility to infection with *Plasmodium falciparum*, the most deadly form of malaria. Of these species, *A. gambiae* s.s. Giles is considered the most important vector because of its very strong preference for feeding on humans (della Torre et al. 2002), whereas *A. arabiensis* Patton has a very flexible feeding preference and will feed on both humans and animals, thus reducing the risk for transmission of malaria. In addition, *A. quadriannulatus* Theobald is not considered a vector of malaria parasites because it feeds exclusively on cattle. As all of these species are morphologically inseparable, the use of a PCR diagnostic (Scott et al. 1993) has become the standard tool for identifying vectors to implement control procedures that are based on the individual vector's behavior.

The most well-known examples of morphologically problematic insects in forensic entomology are probably the flesh flies (Sarcophagidae) (Byrd and Castner 2000). These flies are morphologically similar in both the immature and adult stages, making it difficult to distinguish among species and hence to conclude that species have different niches, behaviors, and distributions. Flesh flies are important flies in the process of decomposition and have the potential to be very important in developing min-PMI estimations as they are often among the first fly families to colonize a body. Genetic tools have become an attractive option for differentiating flesh fly species as the costs and availability of sequencing and PCR have lowered in recent years.

26.2.2 Genetic Information and Species Boundaries

The greenbottle flies, *Lucilia cuprina* (Wiedemann) and *Lucilia sericata* (Meigen) (Diptera: Calliphoridae), are among the most widely distributed blow fly species in the world. They are considered good biological species throughout most of their known distribution sharing ecological niches and overlapping distributions, yet there is genetic evidence of both ancient and recent hybridization events (Williams and Villet 2013). By examining the relationship between genetic and phenotypic traits important to forensic entomology, such as oviposition preference in closely related species, an understanding can begin of how genes influence these behaviors and determine how gene flow and behavior interact at the population level.

One of the most striking examples of behavioral differentiation in this *Lucilia* species pair has been in parts of their distribution where both species are considered significant pests in sheep myiasis (invasion of living tissues by fly larvae, also known as fly-strike) and large economic losses. In Australia *L. cuprina* is the species most often implicated in fly-strike of sheep, whereas in nearby New Zealand *L. sericata* is responsible for about 50% of the fly-strike incidence (Heath and Bishop 2006). Waterhouse and Paramonov (1950) noted a significant difference in oviposition preference between the two species, observing that *L. cuprina* was more attracted directly to sheep while *L. sericata* laid eggs after *L. cuprina*. In addition to this strong behavioral difference, they were also able to obtain hybrids in the laboratory that morphologically resembled *L. cuprina*; however, the oviposition preference of the hybrid was not tested. This information could prove vital when determining if colonization of a decedent occurred prior to or after death. For example, on a decedent found with diabetic or pressure ulcers that might be associated with natural disease processes found deceased without knowledge about this history, larvae collected from these wounds may be erroneously used to estimate a time of colonization that does not match the PMI (Sanford et al. 2014).

The morphological similarity between *L. cuprina* and *L. sericata* is strong but not complete and morphologically intermediate hybrids between the two species are rarely encountered, suggesting that hybridization is rare in nature. However, the recent work by Williams and Villet (2013) demonstrated how hybrids might go unnoticed when examining morphology or even by using only a small set of genetic markers. They showed that when only using nuclear genes the difference between the two species seems very clear; however, when also examining a mitochondrial gene the evidence of introgression and hybridization becomes evident. Their data suggest the action of a sterilizing bacterial endosymbiont, such as a *Wolbachia* sp., which commonly shifts populations of Diptera and has a similar maternal inheritance pattern to mitochondrial DNA (mtDNA), may be the driving force behind their differentiation. The action of a sterilizing endosymbiont would also explain the unidirectional pattern of laboratory-derived hybrids and morphology observed by Waterhouse and Paramonov (1950). These data suggest caution should be used when assigning a given behavior, such as myiasis, to a single species within a given region as hybridization and incomplete species boundaries could result in behavioral shifts between species and even among populations of the same species.

26.2.3 Genetic Information beyond the Sequence

With continuous new developments in next-generation sequencing (NGS), more is being learned about the genomic sequence of forensically important insects. As researchers build enormous data-sets of information, the realization that there is more to describing an organism than what can be gleaned based solely on its genetic sequence becomes pronounced. One of the most widely adopted NGS techniques to examine patterns of development and behavior is RNAseq, which is a technique for examining the expressed parts of the genome (exome). RNAseq at its basic level involves submitting the organism to some stimulus or capturing its RNA transcripts at a known level of development, converting the collected RNA to DNA, and sequencing the DNA to determine what genes are expressed and hence related to whatever stimulus or stage the organism was undergoing. The identification of genes related to stimuli has so far only been applied to developmental events, but the potential to apply it to behavioral stimuli and use it to identify genes underlying specific behaviors is on the horizon and could have profound application in understanding behavior related to forensic entomology.

RNAseq has already seen adoption in forensic entomology as a tool with potential application in determining the developmental age of a fly larva. The whole exome of *L. sericata* has been sequenced (Sze et al. 2012) and includes information about variation and transcripts that might be differentially expressed and thus potentially used to determine the age of a larva or pupa found during the course of a forensic investigation. Our current ability to determine age of collected larval specimens is limited to known morphological indicators of stage (the number of slits on the postspiracular plate) or to larval length, which is not linearly related to age. Thus, the development of new genetic tools such as the analysis of genome-wide expression patterns (Tarone and Foran 2011) (discussed in Chapter 24) and markers such as those proposed by Boehme et al. (2013) are exciting developments for determining insect age within a particular life stage. Boehme et al. (2013) examined expression of several genes using quantitative PCR over the course of pupal development in the blow fly *Calliphora vicina* Robineau-Desvoidy (Diptera: Calliphoridae) and found that some of these genes could be used to indicate age within the pupal stage at a given temperature. Once methods are established for associating genes in blow flies with life-history events, the prospect of associating genes with behavioral cues and traits becomes even more plausible.

In the *A. gambiae* s.l. species complex, the information being obtained about differentiation between the morphologically inseparable, yet biologically differentiating molecular forms provides clues as to the underlying processes at the molecular level that may be transferred to our knowledge of forensically important Diptera. As discussed previously, *A. gambiae* s.s comprises two molecular forms, M and S, that are thought to be undergoing a sympatric speciation event in sub-Saharan Africa (della Torre et al. 2002). One of the mechanisms thought to be responsible for the genetic differentiation has been hypothesized to be the action of chromosomal inversions. Chromosomal inversions can protect specific genes and those closely surrounding them from recombination events and allow for selection to act on these genes. In certain parts of the *A. gambiae* distribution, the pattern of inversion polymorphism is clearly related to environmental patterns such as aridity where particular karyotypes (patterns of chromosomal inversions) are more commonly encountered and associated with a particular molecular form (Touré et al. 1998). The application of microarray technology has suggested that the chromosomal inversions are indeed areas where molecular differentiation is maximized (Turner et al. 2005; Lee et al. 2013), yet this does not appear to explain all the basis for genetic differentiation and researchers are still in the process of discovering the specific genes that underlie the phenotypic divergence observed in the field. Using NGS approaches to determine the extent and basis of differentiation between the molecular forms and their associated karyotypes will allow researchers to not only answer basic speciation questions but also apply this knowledge in the development of genetically modified mosquito control of this species. The use of NGS in closely related species and populations of forensically important insects will allow for more specific determination of the genes underlying development and perhaps even those genes underlying speciation itself.

In addition to the adoption of newly developing NGS methods, there is great potential to mine the growing databases in the future for information about noncoding DNA, structural elements, epigenetic interactions, and many other rapidly developing areas that are only just now beginning to develop. Understanding the interactions among genetic elements has the potential to move our understanding from the DNA sequence to explaining complex traits and behaviors. The use of genetic tools in forensic entomology is just beginning and has the potential to bring a rigorous scientific foundation to such plastic behaviors as oviposition and the resulting development.

26.3 ENVIRONMENT

The behavior of an insect is tightly aligned with its ecology (Thomas 1993) and relative environment. The success of an individual insect hinges on its ability to operate within a given environment and secure the essential materials for successful transmittance of its genetic material to the next generation: food, mates, and resources for its offspring. The expression of particular behaviors associated with these essentials may have a basis in the genetic makeup of the individual or even the species, but such behaviors are expressed within the larger context of the environment. The purpose of this section is to discuss behavior within the context of the environment and the specific factors that can regulate it.

26.3.1 Defining the Environment

In most natural settings, the environment is stochastic, necessitating constant adjustments by the individual experiencing these changes. Some changes are observable on a very short time scale (e.g., temperature fluctuations), whereas others are not (e.g., immigration of a predator at a given location). Large-scale temporal changes, such as urban sprawl into an agricultural setting, can take decades to occur. Within this temporal scale two additional sets of factors can be defined that regulate and influence behavior including abiotic factors, which are tied to nonbiological components of the environment (e.g., temperature and humidity), and biotic factors, which are tied to interactions with other organisms (e.g., competitors and parasites).

26.3.2 Abiotic Environmental Factors

When considering decomposing human remains, the environmental abiotic conditions that influence insect behavior include factors that affect the ability of a forensic entomologist to accurately determine colonization and developmental progression. These abiotic factors can have direct implications on larval growth such as the use of narcotics by the individual prior to death (Goff and Lord 1994; Gagliano-Candela and Aventaggiato 2001; Murthy and Mohanty 2010) or the type of tissue that the larvae consume for growth and development (Clark et al. 2006; Day and Wallman 2006; Boatright and Tomberlin 2010).

One of the most well-documented abiotic factors affecting larval development is temperature. Temperature has long been recognized as a factor regulating behavior of forensically important insects. Some blow fly species are active during cooler weather, whereas others are the opposite. However, cold or heat tolerance can vary between populations within a species, which is related to gene-by-environment interactions, as further discussed in Section 26.4. For example, *Phormia regina* (Meigen) (Diptera: Calliphoridae) activity in Texas typically occurs from January through April, whereas this species is active in more northern climates throughout the year (Tenorio et al. 2003). This difference in activity within a single species illustrates the importance of assessing the mechanisms driving succession patterns globally. Failing to account for the locality of seasonal behavioral patterns could lead to false assumptions being made with respect to species activity within a given location simply due to reliance on data produced in a different locale.

In addition to adult activity patterns, the effect of temperature is a critical factor in determining larval development and age. Every published larval development dataset available is recorded with a set temperature regime to compare and apply it to collected data from the scene (Byrd and Butler 1996; Byrd 1998; Byrd and Allen 2001; Boatright and Tomberlin 2010). Temperature is relatively easy to measure but can be drastically impacted by the spatial scale of the environment. Temperature also appears to be important at the temporal scale as well, with long-term selection apparently acting on populations of flies that live in different climates (Gallagher et al. 2010).

Development based on temperature is important to document as larval behavior can change as temperature shifts within a given environment. Larvae can regulate their microhabitat, to some extent, through the formation of aggregate larval masses to generate additional heat through metabolic heat production, or cool, depending on the optimal temperature needed to maximize that particular species' own development. At this time, little is known about how temperature regulates larval mass formation and the significance of this behavior on the development of fly larvae. Understanding the mechanisms that regulate this behavior could allow for greater refinements to min-PMI estimates based on insect development.

Other abiotic factors are known to affect larval development. Agitation and disruption of larval feeding can lead to premature dispersal from a resource (De Jong and Chadwick 1999). Rain is also thought to result in premature dispersal (Lewis and Benbow 2011). In contrast, handling larvae could prolong larval feeding. These behaviors are most important to document when attempting to rear larvae for age determination at the time of collection from the remains. Interpretation of such behavioral activity could lead to the erroneous conclusion that the larvae have entered the postfeeding stage in the case of larvae disturbed by rain, or some other form of moisture, and are much older than they really are or result in delayed development due to agitation as in the case of handling larvae collected from a scene.

26.3.3 Biotic Environmental Factors

Biotic factors consist of interactions between organisms and can have significant impacts on behaviors such as arrival and colonization of decomposing remains by forensically important insects. These factors may include other species of insect present on the remains (Carter et al. 2007; Yang and Shiao 2012), vertebrate scavengers (DeVault et al. 2003; Wilson and Wolkovich 2011), and microbes (Chaudhury et al. 2002; Carter et al. 2007; Pechal 2013).

Interactions between species can significantly affect the behaviors of all species involved and is discussed in the following examples. *Chrysomya rufifacies* (Macquart) (Diptera: Calliphoridae), a species introduced to North America from Asia, is a predator on the native *Cochliomyia macellaria* (Fabricius) (Diptera: Calliphoridae) in the western hemisphere and comprises a model example of predator–prey interactions (Wells and Greenberg 1992). *Chrysomya rufifacies* larvae tend to prey on third instar *C. macellaria* larvae. Interestingly, a blow fly that co-occurs with *Ch. rufifacies* in its native range, *Chrysomya megacephala* (Fabricius) (Diptera: Calliphoridae), has evolved larval behaviors that enable it to escape an attack by *Ch. rufifacies*. To date, such behaviors have not been observed in *C. macellaria*.

However, *C. macellaria* may be altering its behavior in other ways to reduce this negative interaction with the introduced species. Current data have demonstrated that this predator–prey interaction has selected for *C. macellaria* that respond to the mere presence of *Ch. rufifacies* eggs (Brundage and Tomberlin 2009; Brundage 2012) and larvae (Flores 2013). These interactions are known as nonconsumptive effects. Behaviorally, *C. macellaria* adults avoid resources colonized with *Ch. rufifacies*, which would be indicative of reverse aggregation or avoiding sites where another species aggregates. This response appears to be temporally based with *C. macellaria* adults avoiding *Ch. rufifacies* eggs (Brundage 2012) and being attracted to third instar *Ch. rufifacies* larvae (Flores 2013). This response by *C. macellaria* could be due to avoidance of the stage most likely to induce predation. Thus, when analyzing the succession pattern of these co-occurring species, it is important to note that this temporal colonization pattern might not be immediately obvious from the life stages based on morphological characters in the collected samples, but rather this would be related to the behavioral timing of colonization: *Ch. rufifacies* larvae would be similar in age to *Cochliomyia macellaria* larvae, resulting in predation when they both reach the third instar. In instances when *Ch. rufifacies* colonizes substantially earlier than *C. macellaria*, there is less risk of predation, as *Ch. rufifacies* is not known to predate on eggs or first instar *C. macellaria* larvae (Wells and Greenberg 1992).

Furthermore, the presence of *Ch. rufifacies* larvae also impacts larval behavior indirectly, as related to development and the predictions of dispersal of third instar larvae from the remains (Flores 2013). Third instar *C. macellaria* larvae will disperse from remains once *Ch. rufifacies* colonizes. In addition, development of *C. macellaria* is accelerated in the third instar stage in the presence of *Ch. rufifacies* (Flores 2013). These shifts in behavior and development could impact succession patterns of insects on vertebrate carrion and human remains and, consequently, the ability of a forensic entomologist to use such information to predict min-PMI. In the case of succession data, studies conducted prior to the introduction of *Ch. rufifacies* might not be applicable to current succession patterns where local populations have adapted to the presence of this introduced species. If one were not to know that *C. macellaria* is a possible primary colonizer and *Ch. rufifacies* is a secondary colonizer, estimates of the time of colonization of human remains due solely to the presence of *Ch. rufifacies* could be a significant underestimate, based on the suspected colonization behaviors of these species.

Volatile organic compounds (VOCs) play a major role in regulating insect behavior and attendance at a carrion resource, and many VOCs are related to the microbes as well as the insects present (Dekeirsschieter et al. 2009). Research into this relationship has determined that the presence of specific insect species influences the behavioral responses of con- and heterospecific species, as previously described with *C. macellaria* and *Ch. rufifacies*. Another example of such interactions is with *P. regina*, which is thought to feed predominately during the later successional stages of decay long after initial colonization has occurred (Denno and Cothran 1976). Similarly, *Ch. megacephala* will avoid ovipositing on pork liver colonized by the predator *Ch. rufifacies* (Yang and Shiao 2012). However, *Ch. megacephala* showed no preference when presented with resources with or without conspecific larvae or larvae of the nonpredator *Hemipyrellia ligurriens* (Wiedemann) (Diptera: Calliphoridae) (Yang and Shiao 2012). While with each of these examples researchers suggest that the insects present serve as the mechanism regulating subsequent colonization by other insects, microbes also play a role in regulating the behaviors exhibited by later colonizers (Janzen 1977; Chaudhury et al. 2002; Burkepile et al. 2006). See Chapter 21 for a more detailed discussion on microbial–insect interactions in decomposition.

Parasitoids, which are insects that deposit their offspring in or on another insect resulting in the death of the host, also regulate behaviors exhibited by the insects colonizing and developing on human remains. Blow flies are no different than any other insect as they have a number of specialized parasitoids. Because of this close association, blow flies have evolved behavioral strategies to avoid parasitism. In the case of *L. sericata* larvae, like most other blow fly larvae that burrow into soil, the depth at which they pupate can vary depending on soil compaction and the presence of parasitoids. Parasitoids can shift the burrowing behaviors exhibited by the blow fly larvae including the distance dispersed from the remains on which they developed and the depth burrowed for pupation (Cammack et al. 2010). Understanding these relationships could enhance methods currently employed for collecting entomological evidence from a scene. For example, in cases where pupae are parasitized the forensic entomologist might take deeper soil cores if in fact larvae attempting to escape parasitism burrow deeper.

26.3.4 Spatial Scale of the Environment

The environment has impacts on behavior along multiple spatial scales. At the center of this spatial scale of the environment is the individual (Figure 26.2). When one extrapolates out beyond the individual, the microhabitat experienced by that individual consists of the very small aspect of the environment directly impacting that individual. Environments occurring on a microhabitat level can often be overlooked and consequently not taken into consideration when assessing behavioral output from insects on deceased or living victims.

Of course, defining a microhabitat environment can be difficult as it depends on the observer, scale, and method of measure. Most often than not, the microhabitat is restricted temporally and by location when entomological evidence is encountered. For example, the microhabitat could be blow fly larval interactions on a decomposing body (e.g., species vs. species) or even within an individual larval mass. One of the best examples of this scale is the observation that larval mass temperature often differs from ambient temperature. Larval fly aggregations, often called maggot masses, can generate their own heat, therefore drastically changing their microhabitat and the subsequent developmental progression (Cianci and Sheldon 1990) even when the body is stored under refrigeration (Huntington et al. 2007). This ability can have a significant impact on the age or developmental rate of the larvae and hence the length of the estimated time that the body has been colonized.

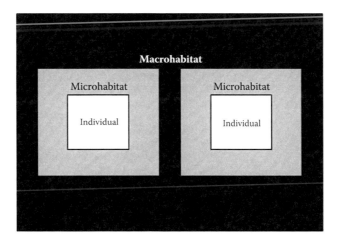

Figure 26.2 Describing the nested environmental spatial scale.

The microhabitat has been emphasized in most of the standard operating procedures developed for forensic entomology. In fact, Amendt et al. (2007) have a section specifically addressing microclimatic conditions with specific reference to describing the condition of the corpse, the scene, and temperature. Forensic entomology texts, including those by Byrd and Castner (2010) and Haskell and Williams (2008), also provide an assessment of similar information to be collected with respect to an investigation. Variables experienced at the micro- and macrohabitat levels influence behaviors exhibited by insects and could influence their colonization, development, and succession on vertebrate carrion.

Environments occurring at a macrohabitat level are less often taken into consideration during an investigation, as its connection seems more removed and to have less obvious influence than what the insects experience in the microhabitat. The abiotic and biotic variables previously listed also apply to the macrohabitat level. However, this level is more complex and contains other variables that must be considered. Blow fly species in particular can be specific in their habitat preference. For example, Brundage et al. (2011) demonstrated in Santa Clara County, California (San Francisco area), that of the seven blow fly species collected, six were influenced by habitat with their occurrence being different between urban, rural, and riparian environments. They also demonstrated that season (i.e., spring, summer, fall, and winter) influenced the distribution of all seven species. Furthermore, the habitat and season interaction was statistically significant with respect to the distribution of six of the seven species collected. Although these significant interactions have been determined, the mechanisms explaining their occurrence has not been well defined to date.

26.3.5 Environmental Integration and Colonization

Insect attraction and colonization of human remains are dependent on a host of environmental factors. State of decomposition of remains has been related to occurrence of specific insects on decomposing vertebrate remains. Blow flies are typically more attracted to remains early in the decomposition process (Early and Goff 1986; Tomberlin and Adler 1998; Apichat et al. 2007), whereas most coleopteran larvae are present on remains in later stages of decomposition (Grassberger and Frank 2004). These colonization behaviors are partly due to the resources utilized by the insects in question. Blow fly larvae feed on soft tissue of the remains and, in contrast, many beetle larvae are more generalists, feeding on fly larvae, soft tissue, skin, cartilage, and hair. However, there is a tremendous amount of variation in the arrival of different insect species depending on locality, climate, and population-regulated factors, yet why it occurs is not fully appreciated and thus at this time cannot be included in time of colonization estimates based on insect succession.

As previously mentioned, the initial attraction of insects to decomposing flesh is predominately regulated by associated VOCs and interest in the chemical ecology aspects of this process is growing. Studies are beginning to characterize the odor plumes associated with vertebrate carrion decomposition (Dekeirsschieter et al. 2009), which is providing the framework allowing for the examination of semiochemicals in regulating blow fly (Frederickx et al. 2012a,b) and other insect attraction and colonization of remains. Researchers are now discovering that attraction and oviposition by blow flies to larval resources are partly governed by VOCs that are released from bacteria in a given environment (Chaudhury et al. 2010). Although this is only one example of a group of insects associated with carrion, the same principle most likely holds true for other groups including Coleoptera (Kalinová et al. 2009). Furthermore, it is now being discovered that these compounds are related to quorum sensing (communication) by bacteria (Ma et al. 2012). Disruption of the quorum sensing ability of the bacteria decreases blow fly attraction and oviposition on associated resources (Tomberlin et al. 2012). In essence, these studies demonstrate that bacteria strongly influence the behaviors exhibited by the insects colonizing decomposing bodies.

26.4 GENE-BY-ENVIRONMENT INTERACTION IN BEHAVIORAL ECOLOGY OF FORENSICALLY IMPORTANT INSECTS

Neither genetics nor environment alone can account for the phenotypes observed in a living organism. Like physically measurable phenotypes, such as length or weight, behavior is a plastic phenotype that is subject to the same selection pressures. The interaction between genetics and environment is fundamental to the fitness of the individual and ultimately to the action of selection on a population. The importance of this interaction makes examples of genes and environment affecting the behavior of forensically important insects abundant. Discussion on this topic has been limited to three examples, which have important implications for forensic entomology casework.

26.4.1 Oviposition Preference and Resource Quality

A female fly making a critical decision when choosing where to lay her eggs reflects one of the most direct gene-by-environment interaction examples in forensic entomology. The type of tissue that fly larvae develop on can have significant effects on the size and fitness of the resulting adult fly. The size of the adult is impacted by larval nutrition (Day and Wallman 2006; Ireland and Turner 2006), which affects fitness through mating success as smaller males are less successful at mating with females (Stoffolano et al. 2000) and through fecundity as smaller females can carry fewer eggs (Saunders and Bee 1995). Thus, colonization of a suboptimal resource could result in variation in larval size and development, two measures critical to using insect evidence for estimating min-PMI of human remains.

In addition to tissue type, the stage of decomposition and the related bacterial communities have an impact on the success of immature development. The insect succession patterns observed during decomposition reflect each individual species' optimal resource exploitation with respect to bacterial populations, conspecifics, and the whole suite of other environmental factors (see Section 26.3) that impact larval development. Bacterial symbionts can regulate oviposition cues via density dependence in some species such as the housefly, *Musca domestica* L. (Diptera: Muscidae) (Lam et al. 2007). Changes in the bacterial community can regulate oviposition on the broader fly community as well. In general, some species are more attracted to bacterially rich media over sterile oviposition media (Eisemann and Rice 1987). In addition, some bacterially produced chemicals may be cues used by flies to determine oviposition substrate qualities (Tomberlin et al. 2012).

26.4.2 Nocturnal Oviposition

Circadian rhythms are an excellent example of gene-by-environment interactions. Genes regulate the activity patterns of the organism, but they are tuned directly by the environment. Light cycles and the associated circadian clock genes are one of the most well-studied gene-by-environment interactions affecting insect behavior. Light is one of the most recognizable and easiest to manipulate environmental cues that insects can use to set their "circadian clocks." From a forensic entomology perspective, the nighttime is often considered a time when no oviposition will occur because the flies are not active. However, recent studies (Baldridge et al. 2006; Berg and Benbow 2013) in conjunction with the fact that humans often have artificial light in urban and suburban areas make this subject of particular interest for forensic investigations because making the assumption that no activity will occur at night may not always be appropriate in these settings.

One well-recognized example of how insect behavior is tuned by environmental cues can be observed in the cricket (Orthoptera). Only in the presence of a defined light:dark cycle will crickets synchronize their mating calls to a narrow 2-hour window of time (Loher 1972) and hence increase their chances of finding a suitable mate. Much of what is known about the genetic component of circadian rhythms comes from the study of the vinegar fly, *Drosophila melanogaster* Meigen (Diptera: Drosophilidae). Identification of the genes responsible for circadian activity such as *period*, *timeless*, and *doubletime* was made possible through the use of genetic dropout mutant fruit flies (Konopka and Benzer 1971). Flies with mutations in these genes were observed to have significantly different activity patterns from control flies.

Research in blow fly circadian rhythms is not yet to the level of knowledge of that for *Drosophila* sp.; however, there has been a lot of recent research in this area due to its importance in forensic casework and min-PMI estimation. The literature is conflicted as to whether blow flies can fly at night (Wooldridge et al. 2007; Zurawski et al. 2009); whether they will lay eggs at night (Baldridge et al. 2006); and whether nocturnal oviposition, when it occurs, is an artifact of environmental factors like artificial lighting (Greenberg 1990). If blow flies can lay eggs at night, even if only under very specific conditions, the implications for casework are significant because the current practice is to assume that flies are not active during the night.

26.4.3 Developmental Variation at the Population Level

The environment varies on many different scales (e.g., microhabitat vs. macrohabitat) in a given geographic area. Thus, it is not surprising that the species that live in a specific geographic area are capable of local adaptation and different levels of plasticity in response to local conditions. If one considers developmental variation to be a reflection of the behavioral events of the precolonization period (sensu Tomberlin et al. 2011), culminating in the developmental response of the insect, population-level differences in development likely reflect behavioral differences in such events as oviposition, which is another example of an event reflecting the sum of several behavioral steps. This is important in forensic entomology casework because it raises important considerations when applying species-specific temperature-based developmental datasets collected from a population of a species in one part of the distribution and comparing it to evidence collected in another location.

Population-level differences are probably the most noticeable in species with wide distributions, such as the greenbottle fly, *L. sericata*. Gallagher et al. (2010) found that populations of this species collected in Sacramento, California; San Diego, California; and Easton, Massachusetts, responded differently in terms of rate of development when reared at the same temperatures. The study showed that each population had a different development time and although it was not directly tied to a temperature-based hypothesis, it demonstrated the need for local population data for the most accurate data and min-PMI estimates, thus reinforcing the need for locally obtained population-specific temperature-based development data for widely distributed forensically important insect species.

26.5 CONCLUSIONS

The behavioral ecology of forensically important insects is connected to a wide variety of internal and external mechanisms that influence the data obtained during forensic death investigations and its interpretation. Although it is known that certain factors influence the behaviors associated with colonization and development, it is not yet understood exactly how these factors can be accounted for practically in applied forensic entomology. At present, the forensic entomologist can document those factors and data that are deemed important and work to better understand how those data can be implemented as corrections and adjustments for retrospective analysis of trends and development of new tools in forensic entomology in death investigation.

REFERENCES

Amendt, J., C.P. Campobasso, E. Gaudry, C. Reiter, H.N. LeBlanc, and M.J.R. Hall. 2007. Best practice in forensic entomology—standards and guidelines. *International Journal of Legal Medicine* 121 (2): 90–104.

Apichat, V., P. Wilawan, T. Udomsak, P. Chanasorn, and N. Saengchai. 2007. A preliminary study on insects associated with pig (*Sus scrofa*) carcasses in Phitsanulok, northern Thailand. *Tropical Biomedicine* 24 (2): 1–5.

Baldridge, R.S., S.G. Wallace, and R. Kirkpatrick. 2006. Investigation of nocturnal oviposition by necrophilous flies in central Texas. *Journal of Forensic Sciences* 51 (1): 125–126.

Berg, M.C. and M.E. Benbow. 2013. Environmental factors associated with *Phormia regina* (Diptera: Calliphoridae) oviposition. *Journal of Medical Entomology* 50 (2): 451–457.

Boatright, S.A. and J.K. Tomberlin. 2010. Effects of temperature and tissue type on the development of *Cochliomyia macellaria* (Diptera: Calliphoridae). *Journal of Medical Entomology* 47 (5): 917–923. doi:10.1603/ME09206.

Boehme, P., P. Spahn, J. Amendt, and R. Zehner. 2013. Differential gene expression during metamorphosis: A promising approach for age estimation of forensically important *Calliphora vicina* pupae (Diptera: Calliphoridae). *International Journal of Legal Medicine* 127 (1): 243–249.

Brundage, A. 2012. Fitness effects colonization time of *Chrysomya rufifacies* and *Cochliomyia macellaria*, and their response to Intra- and Inter-specific eggs and egg-associated microbes. Doctoral dissertation, Texas A&M University, TX, available at http://hdl.handle.net/1969.1/ETD-TAMU-2012-05-10803.

Brundage, A., S. Bros, and J.Y. Honda. 2011. Seasonal and habitat abundance and distribution of some forensically important blow flies (Diptera: Calliphoridae) in Central California. *Forensic Science International* 212 (1–3): 115–120.

Brundage, A.L. and J.K. Tomberlin. 2009. Attraction of two forensically important fly species: *Chrysomya rufifacies* (Macquart) and *Cochliomyia macellaria* (Fabricius) to inter- and intraspecific eggs. *Proceedings of American Academy of Forensic Sciences* 15: 271.

Burkepile, A.L., J.D. Parker, C.B. Woodson, H.J. Mills, J. Kubanek, P.A. Sobecky, and M.E. Hay. 2006. Chemically mediated competition between microbes and animals: Microbes as consumers in food webs. *Ecology* 87 (11): 2821–2831.

Byrd, J.H. 1998. *Temperature dependent development and computer modeling of insect growth: Its application to forensic entomology*. PhD diss., University of Florida, Gainesville, FL.

Byrd, J.H. and J.C. Allen. 2001. The development of the black blow fly, *Phormia regina* (Meigen). *Forensic Science International* 120: 79–88.

Byrd, J.H. and J.F. Butler. 1996. Effects of temperature on *Cochliomyia macellaria* (Diptera: Calliphoridae) development. *Journal of Medical Entomology* 33 (6): 901–905.

Byrd, J.H. and J.L. Castner. 2000. Insects of Forensic Importance. In *Forensic Entomology: The Utility of Arthropods in Legal Investigations*, edited by J.H. Bryd and J.L. Castner, 43–78. Boca Raton, FL: CRC Press.

Byrd, J.H. and J.L. Castner. 2010. *Forensic Entomology: The Utility of Arthropods in Legal Investigations*. Boca Raton, FL: CRC press.

Cammack, J., P.H. Adler, J.K. Tomberlin, Y. Arai, and W.C. Jr., Bridges. 2010. Influence of parasitism and soil compaction on pupation of the green bottle fly, *Lucilia sericata*. *Entomologia Experimentalis et Applicata* 136 (2): 134–141.

Carter, D., D. Yellowlees, and M. Tibbett. 2007. Cadaver decomposition in terrestrial ecosystems. *Naturwissenschaften* 94 (1): 12–24.

Chaudhury, M.F., S.R. Skoda, A. Sagel, and J.B. Welch. 2010. Volatiles emitted from eight wound-isolated bacteria differentially attract gravid screwworms (Diptera: Calliphoridae) to oviposit. *Journal of Medical Entomology* 47 (3): 349–354.

Chaudhury, M.F., J.B. Welch, and L.A. Alvarez. 2002. Response of fertile and sterile screwworm (Diptera: Calliphoridae) flies to bovine blood inoculated with bacteria originating from screwworm infested animal wounds. *Journal of Medical Entomology* 39 (1): 130–134.

Cianci, T.J. and J.K. Sheldon. 1990. Endothermic generation by blow fly larvae *Phormia regina* developing in pig carcasses. *Bulletin of the Society for Vector Ecology* 15 (1): 33–40.

Clark, K., L. Evans, and R. Wall. 2006. Growth rates of the blowfly, *Lucilia sericata*, on different body tissues. *Forensic Science International* 156 (2–3): 145–149.

Coetzee, M., M. Craig, and D. le Sueur. 2000. Distribution of African malaria mosquitoes belonging to the *Anopheles gambiae* complex. *Parasitology Today* 16 (2): 74–77.

Day, D.M. and J.F. Wallman. 2006. Influence of substrate tissue type on larval growth in *Calliphora augur. Lucilia cuprina* (Diptera: Calliphoridae). *Journal of Forensic Sciences* 51 (3): 657–663.

De Jong, G.D. and J.W. Chadwick. 1999. Decomposition and arthropod succession on exposed rabbit carrion during summer at high altitudes in Colorado, USA. *Journal of Medical Entomology* 36 (6): 833–845.

Dekeirsschieter, J., F.J. Verheggen, M. Gohy, F. Hubrecht, L. Bourguignon, G. Lognay, and E. Haubruge. 2009. Cadaveric volatile organic compounds released by decaying pig carcasses (*Sus domesticus* L.) in different biotopes. *Forensic Science International* 189 (1): 46–53.

della Torre, A., C. Costantini, N.J. Besansky, A. Caccone, V. Petrarca, J.R. Powell, and M. Coluzzi. 2002. Speciation within *Anopheles gambiae*—the glass is half full. *Science* 298 (5591): 115–117.

Denno, R.F. and W.R. Cothran. 1976. Competitive interactions and ecological strategies of sarcophagid and calliphorid flies inhabiting rabbit carrion. *Annals of the Entomological Society of America* 69 (4): 109–113.

DeVault, T.L., O.E. Rhodes, Jr., and J.A. Shivik. 2003. Scavenging by vertebrates: Behavioral, ecological, and evolutionary perspectives on an important energy transfer pathway in terrestrial ecosystems. *Oikos* 102 (2): 225–234.

Early, M. and M.L. Goff. 1986. Arthropod succession patterns in exposed carrion on the island of Oahu, Hawaiian Islands, USA. *Journal of Medical Entomology* 23 (5): 520–531.

Eisemann, C.H. and M.J. Rice. 1987. The origin of sheep blowfly, *Lucilia cuprina* (Wiedemann) (Diptera: Calliphoridae), attractants in media infested with larvae. *Bulletin of Entomological Research* 77: 287–294.

Flores, M. 2013. Life-history traits of *Chrysomya rufifacies* (Marquart) (Diptera: Calliphoridae) and its associated non-consumptive effects on *Cochliomyia macellaria* (Fabricius) (Diptera: Calliphoridae) behavior and development. PhD Diss., Texas A&M University. http://hdl.handle.net/1969.1/151308

Frederickx, C., J. Dekeirsschieter, Y. Brostaux, J.P. Wathelet, F.J. Verheggen, and E. Haubruge. 2012a. Volatile organic compounds released by blowfly larvae and pupae: New perspectives in forensic entomology. *Forensic Science International* 219 (1): 215–220.

Frederickx, C., J. Dekeirsschieter, F.J. Verheggen, and E. Haubruge. 2012b. Responses of *Lucilia sericata* Meigen (Diptera: Calliphoridae) to cadaveric volatile organic compounds. *Journal of Forensic Sciences* 57 (2): 386–390.

Gagliano-Candela, R. and L. Aventaggiato. 2001. The detection of toxic substances in entomological specimens. *International Journal of Legal Medicine* 114 (4–5): 197–203.

Gallagher, M.B., S. Sandhu, and R. Kimsey. 2010. Variation in developmental time for geographically distinct populations of the common green bottle fly, *Lucilia sericata* (Meigen). *Journal of Forensic Sciences* 55 (2): 438–442.

Goff, M.L. and W.D. Lord. 1994. Entomotoxicology: A new area for forensic investigation. *The American Journal of Forensic Medicine and Pathology* 15 (1): 51–57.

Grassberger, M. and C. Frank. 2004. Initial study of arthropod succession on pig carrion in a central European urban habitat. *Journal of Medical Entomology* 41 (3): 511–523.

Greenberg, B. 1990. Nocturnal oviposition behavior of blow flies (Diptera: Calliphoridae). *Journal of Medical Entomology* 27 (5): 807–810.

Haskell, N.H. and R.E. Williams. 2008. *Entomology & Death: A Procedural Guide*. 2nd ed. Clemson, SC: Joyce's Print Shop.

Heath, A.C.G. and D.M. Bishop. 2006. Flystrike in New Zealand: An overview based on a 16-year study, following the introduction and dispersal of the Australian sheep blowfly, *Lucilia cuprina* Wiedemann (Diptera: Calliphoridae). *Veterinary Parasitology* 137 (3–4): 333–344.

Huntington, T.E., L.G. Higley, and F.P. Baxendale. 2007. Maggot development during morgue storage and its effect on estimating the post-mortem interval. *Journal of Forensic Sciences* 52 (2): 453–458.

Ireland, S. and B. Turner. 2006. The effects of larval crowding and food type on the size and development of the blowfly, *Calliphora vomitoria*. *Forensic Science International* 159 (2–3): 175–181.

Janzen, D.H. 1977. Why fruits rot, seeds mold, and meat spoils. *American Naturalist* 111: 691–713.

Kalinová, B., H. Podskalská, J. Růžička, and M. Hoskovec. 2009. Irresistible bouquet of death—how are burying beetles (Coleoptera: Silphidae: *Nicrophorus*) attracted by carcasses. *Naturwissenschaften* 96 (8): 889–899.

Konopka, R.J. and S. Benzer. 1971. Clock mutants of *Drosophila melanogaster*. *Proceedings of the National Academy of Sciences of the United States of America* 68 (9): 2112–2116.

Lam, K., D. Babor, B. Duthie, E. Babor, M. Moore, and G. Gries. 2007. Proliferating bacterial symbionts on house fly eggs affect oviposition behaviour of adult flies. *Animal Behaviour* 74 (1): 81–92.

Lee, Y., T.C. Collier, M.R. Sanford, C.D. Marsden, A. Fofana, A.J. Cornel, and G.C. Lanzaro. 2013. Chromosome inversions, genomic differentiation and speciation in the African malaria mosquito *Anopheles gambiae*. *PLOS ONE* 8 (3): e57887.

Lewis, A.J. and M.E. Benbow. 2011. When entomological evidence crawls away: *Phormia regina* en masse larval dispersal. *Journal of Medical Entomology* 48: 1112–1119.

Loher, W. 1972. Circadian control of stridulation in the cricket *Teleogryllus commodus* Walker. *Journal of Comparative Physiology* 79 (2): 173–190.

Ma, Q., A. Fonseca, W. Liu, A.T. Fields, M.L. Pimsler, A.F. Spindola, A.M. Tarone, T.L. Crippen, J.K. Tomberlin, and T.K. Wood. 2012. *Proteus mirabilis* interkingdom swarming signals attract blow flies. *The ISME Journal* 6(7): 1356–1366.

Murthy, V.C.R. and M. Mohanty. 2010. Entomotoxicology: A review. *Journal of Indian Academy of Forensic Medicine* 32 (1): 82–84.

Pechal, J. 2013. The importance of microbial and primary colonizer interactions on an ephemeral resource. PhD Diss., Texas A&M University, TX. http://hdl.handle.net/1969.1/ETD-TAMU-2012-05-11016

Sanford, M., T. Whitworth, and D. Phatak. 2014. Human wound colonization by Lucilia eximia and Chrysomya rufifacies (Diptera: Calliphoridae). Myiasis, perimortem or postmortem colonization? *Journal of Medical Entomology*. 51: 716–719.

Saunders, D.S. and A. Bee. 1995. Effects of larval crowding on size and fecundity of the blow fly, *Calliphora vicina* (Diptera: Calliphoridae). *European Journal of Entomology* 92: 615–622.

Scott, J.A., W.G. Brogdon, and F.H. Collins. 1993. Identification of single specimens of the *Anopheles gambiae* complex by the polymerase chain reaction. *The American Journal of Tropical Medicine and Hygiene* 49 (4): 520–529.

Stoffolano, J.G., E.Y. Gonzalez, M. Sanchez, J. Kane, K. Velazquez, A.L. Oquendo, G. Sakolsky, P. Schafer, and C. Yin. 2000. Relationship between size and mating success in the blow fly *Phormia regina* (Diptera: Calliphoridae). *Annals of the Entomological Society of America* 93 (3): 673–677.

Sze, S.H., J.P. Dunham, B. Carey, P.L. Chang, F. Li, R.M. Edman, C. Fjeldsted, M.J. Scott, S. V Nuzhdin, and A.M. Tarone. 2012. A de novo transcriptome assembly of *Lucilia sericata* (Diptera: Calliphoridae) with predicted alternative splices, single nucleotide polymorphisms and transcript expression estimates. *Insect Molecular Biology* 21 (2): 205–221.

Tarone, A.M. and D.R. Foran. 2011. Gene expression during blow fly development: Improving the precision of age estimates in forensic entomology. *Journal of Forensic Sciences* 56 (Suppl 1): S112–S122.

Tenorio, F.M., J.K. Olson, and C.J. Coates. 2003. Decomposition studies, with a catalog and descriptions of forensically important blow flies (Diptera: Calliphoridae) in Central Texas. *Southwestern Entomologist* 28: 267–272.

Thomas, D.B. 1993. Behavioral aspects of screwworm ecology. *Journal of the Kansas Entomological Society* 66: 13–30.

Tomberlin, J., R. Mohr, M.E. Benbow, A.M. Tarone, and S. VanLaerhoven. 2011. A roadmap for bridging basic and applied research in forensic entomology. *Annual Review of Entomology* 56: 401–421.

Tomberlin, J.K. and P.H. Adler. 1998. Seasonal colonization and decomposition of rat carrion in water and on land in an open field in South Carolina. *Journal of Medical Entomology* 35 (5): 704–709.

Tomberlin, J.K., T.L. Crippen, A.M. Tarone, B. Singh, K. Adams, Y.H. Rezenom, M.E. Benbow et al. 2012. Interkingdom responses of flies to bacteria mediated by fly physiology and bacterial quorum sensing. *Animal Behaviour* 84 (6): 1449–1456.

Toure, Y.T., V. Petrarca, S.F. Traore, A. Coulibaly, H.M. Maiga, O. Sankare, M. Sow, M.A. Di Deco, and M. Coluzzi. 1998. The distribution and inversion polymorphism of chromosomally recognized taxa of the *Anopheles gambiae* complex in Mali, West Africa. *Parassitologia* 40 (4): 477–511.

Turner, T.L., M.W. Hahn, and S.V. Nuzhdin. 2005. Genomic islands of speciation in *Anopheles gambiae*. *PLoS Biology* 3 (9): e285.

Waterhouse, D.F. and S.J. Paramonov. 1950. The status of the two species of *Lucilia* (Diptera, Calliphoridae) attacking sheep in Australia. *Australian Journal of Scientific Research Series B* 3: 310–336.

Wells, J.D. and B. Greenberg. 1992. Laboratory interaction between introduced *Chrysomya rufifacies* and native *Cochliomyia macellaria* (Diptera: Calliphoridae). *Environmental Entomology* 21 (3): 640–645.

Wells, J.D. and J.R. Stevens. 2007. Application of DNA-based methods in forensic entomology. *Annual Review of Entomology* 53 (1): 103–120.

Wells, J.D. and J.R. Stevens. 2010. Molecular methods for forensic entomology. In *Forensic Entomology: The Utility of Arthropods in Legal Investigations*, edited by J.H. Byrd and J.L. Castner, 2nd ed., 437–452. Boca Raton, FL: CRC Press.

Wells, J.D., R. Wall, and J.R. Stevens. 2007. Phylogenetic analysis of forensically important *Lucilia* flies based on cytochrome oxidase I sequence: A cautionary tale for forensic species determination. *International Journal of Legal Medicine* 121 (3): 229–233.

Williams, K. and M.H. Villet. 2013. Ancient and modern hybridization between *Lucilia sericata* and *Lu. cuprina* (Diptera: Calliphoridae). *European Journal of Entomology* 110 (2): 187–196.

Wilson, E.E. and E.M. Wolkovich. 2011. Scavenging: How carnivores and carrion structure communities. *Trends in Ecology and Evolution* 26: 129–135.

Wooldridge, J., L. Scrase, and R. Wall. 2007. Flight activity of the blowflies, *Calliphora vomitoria* and *Lucilia sericata*, in the dark. *Forensic Science International* 172 (2–3): 94–97.

Yang, S.T. and S.F. Shiao. 2012. Oviposition preferences of two forensically important blow fly species, *Chrysomya megacephala* and *Ch. rufifacies* (Diptera: Calliphoridae), and implications for postmortem interval estimation. *Journal of Medical Entomology* 49: 424–435.

Zurawski, K.N., M.E. Benbow, J.R. Miller, and R.W. Merritt. 2009. Examination of nocturnal blow fly (Diptera: Calliphoridae) oviposition on pig carcasses in mid-Michigan. *Journal of Medical Entomology* 46 (3): 671–679.

CHAPTER 27

Community Ecology

Jennifer L. Pechal and M. Eric Benbow

CONTENTS

27.1 INTRODUCTION

A community is a temporal and spatial assemblage of organisms occupying the same geographic area or region. In forensic entomology, one may consider a community as those organisms associated with decomposing vertebrate remains (e.g., human cadavers), or what has been defined as the necrobiome (Benbow et al. 2013a). The necrobiome consists of microbes (e.g., bacteria, fungi, and protists), arthropods, and vertebrate consumers that utilize carrion as a resource patch for feeding, for breeding, and as a habitat. Thus, species of the necrobiome should follow the general rules of community ecology principles such as species sorting, succession, and aggregation.

Each member of the necrobiome has a functional role in the decomposition process ranging from microbial processes acting as interkingdom (or interdomain) signaling cues for blow fly (Diptera: Calliphoridae) arrival (Ma et al. 2012; Tomberlin et al. 2012) to vultures consuming and dispersing large quantities of biomass into the landscape (Wilson and Wolkovich 2011). Forensic practitioners typically collect evidence that can be seen with the naked eye such as blow fly and beetle (Coleoptera) larvae and/or adults (Greenberg 1991; Byrd and Castner 2001; Amendt et al. 2004), but there is new evidence demonstrating promise for the use of bacterial communities in estimating a minimum postmortem interval (PMI_{min}) (Pechal et al. 2013, 2014a) and, thus, expanding the utility of necrophagous communities to include microbes.

With such a diversity of organisms using these unpredictable, ephemeral, and limited resources, it is becoming increasingly evident that the community ecology of carrion and its potential application in forensics is highly complex and in need of greater study. In this chapter, an ecological framework developed from basic community ecology principles and how they can be applied in general and more specifically to forensic entomology is presented. Several questions related to community ecology that bridge basic and applied science are addressed, followed by a discussion on how such concepts can be used to refine estimates of the PMI_{min} and the minimum period of insect activity (PIA_{min}) in forensic applications (Amendt et al. 2007; Tomberlin et al. 2011). A summary and recommendations for future research conclude the chapter.

27.2 NECROBIOME ECOLOGY APPLICATIONS IN FORENSICS

The necrobiome of carrion and human remains can be divided into three trophic levels: microbial decomposers (Burkepile et al. 2006), arthropod primary consumers (Norris 1965; Payne et al. 1968; Putman 1978), and arthropod and vertebrate secondary consumers (DeVault et al. 2003). Forensic entomologists have historically used data derived from arthropods (primary and secondary consumers) that have evolved to colonize and consume decomposing animal remains (Byrd and Castner 2010; Benbow et al. 2013b). Data on insect development rates and necrophagous taxa succession have been used to estimate the time of initial insect colonization during a forensic investigation because blow flies typically lay eggs within minutes to hours after death (Byrd and Castner 2001). This estimate of initial insect colonization is a reasonable surrogate for PMI_{min} and has been argued to affect the overall sequence and timing of species assembly (succession) patterns on carrion (Tomberlin et al. 2011). In many forensic cases, the arthropod evidence and resulting entomologically based PMI_{min} estimates are a product of initial colonization and subsequent community succession (Byrd and Castner 2010; Wells and Lamotte 2010). Many factors can affect these entomologically based estimates of PMI_{min} (Tomberlin et al. 2011) including temperature (Ames and Turner 2003); physical disturbance to the body (De Jong and Hoback 2006); chemicals (Goff 2010); weather (Catts and Goff 1992); nocturnal oviposition (Berg and Benbow 2013); and biotic interactions such as competition on, or for, the resource (Denno and Cothran 1976).

Necrobiome succession consists of the sequential arrival of different species of arthropods to a set of remains throughout the decomposition process (see Section 27.3 for a detailed definition). Because insects and other arthropods can have predictable life histories, habitats, known distributions, and developmental rates, the presence/absence and size of a specific species at a crime scene, such as a homicide, can provide important information about when, where, and how a particular crime occurred (Merritt and Benbow 2009; Byrd and Castner 2010; Tomberlin et al. 2011).

Arthropods play a natural role in carrion decomposition in the environment, consuming the decomposing organic material and recycling energy and nutrients as part of their development and life cycle. When a vertebrate organism dies, endogenous bacteria that were once held in equilibrium by the immune system begin to digest proteins, lipids, and carbohydrates as energy sources, creating both gaseous and liquid by-products that act as olfactory cues (i.e., a smell) for colonization of the remains by multicellular organisms such as flies (Diptera) and beetles (Vass 2001; Vass et al. 2002; Carter et al. 2007). In most instances, the initial arthropod colonizers are adult blow flies that feed and lay eggs on the remains (Byrd and Castner 2010). Within a few hours to days, the decomposing remains act as food resource patches for newly eclosed (hatched) larvae, which grow and develop through their life stages at temperature-dependent rates. The presence of blow fly larvae attracts predators and parasites such as beetles, mites (Acari), ants (Hymenoptera), wasps (Hymenoptera), and spiders (Arachnida), which feed on or parasitize the eggs, larvae, pupae, or adults of other insects present at the remains. This is followed by another wave or sere of arthropod species that feed on previously unconsumed or conditioned (e.g., dry skin) remains such as hide beetles (Coleoptera: Dermestidae), which ultimately reduces the carrion to dry bones and hair (Payne 1965; Schoenly 1992; Byrd and Castner 2010). Incidental arthropods such as grasshoppers (Orthoptera) and butterflies (Lepidoptera) can occur at the remains throughout any period of the decomposition process but are typically not used in forensic cases. Figure 27.1 illustrates data on

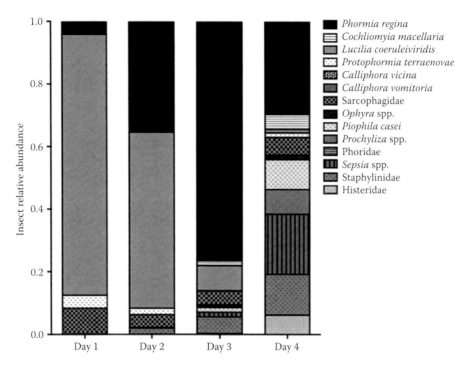

Figure 27.1 Adult insect succession pattern for fresh swine remains placed in the field within 2 hours of death. Blow flies were predominate in the early stages of decomposition (Days 1–3) with lesser flies such as Piophilidae and beetles arriving during later stages of decomposition (Day 4). (Adapted from Figure 2 of Pechal et al., *Ecosphere*, 5, art45, 2014.)

necrophagous arthropod succession that occurs during carrion decomposition in a single geographic area (Pechal et al. 2014). Community succession on vertebrate carrion regulates the decomposition rates based on species accessibility, arrival, and composition, thus becoming important to forensic investigations. In Section 27.3, conceptual aspects of metacommunity and landscape ecology are applied to community succession both within and among carcasses to highlight how ecological concepts can inform forensic entomology.

27.3 CARRION METACOMMUNITY AND LANDSCAPE ECOLOGY

Carrion resources are discrete patches of space, nutrients, and energy that are made available soon after death and in many ecosystems are considered food-falls that act as resource subsidies to the local habitat, often with functional impact on the surrounding ecosystem (Burkepile et al. 2006; Yang et al. 2008; Parmenter and MacMahon 2009). The process by which these resources become available to other parts of the ecosystem is limited by the natural succession of organisms that occupy and modify the patch over time: succession is the sequential colonization of a resource patch by organisms, where the first colonizers modify the patch through colonization of space, feeding, and occupation in ways that make it suitable for a second group of organisms, with the second group further refining that patch for the next group, and so on (Clements 1936; Muller 1940; Connell and Slatyer 1977). The fate of carrion resources is defined by these biotic succession patterns, as species modify the resources in ways that limit or potentially facilitate the additional use by other species, thereby affecting the rate and permanence of a metacommunity assembly among these patches; a metacommunity is the collective set of communities that are close enough for dispersal, interactions, and consequently detectable gene flow among the populations of those communities (Leibold et al. 2004; Logue et al. 2011). How these metacommunities are structured and how carrion food-falls are colonized and used is an active area of theoretical ecology (Logue et al. 2011); but this line of research also has important applications related to natural resources management, conservation ecology, and forensics. The biological interactions that occur during the colonization of carrion food-falls are complex and often ecosystem dependent (Yang et al. 2008), a product of varying climate and other abiotic conditions that define specific ecosystems (Chase and Bengtsson 2010).

Abiotic variables affect metacommunities by habitat modification and defined ambient conditions that limit species distribution, competitive ability, and persistence of certain species in a landscape. Understanding how abiotic factors interact with biotic communities can be important for predicting metacommunity assembly and nutrient cycling of ecosystems (Verhoef and Morin 2010). Additional understanding of how abiotic and biotic factors interact to structure metacommunities is important for describing ecological systems and food web dynamics in decomposing organic material systems (Petchey et al. 2010). These areas of theoretical and applied ecology would benefit from focused hypothesis testing using model systems that can be practically manipulated in the field (Logue et al. 2011). And because carrion resources, or remains of decedents at potential crime scenes, are a resource pulse of habitat, nutrients, and energy, their decomposition process is presumably driven by these processes described by landscape ecology, patch dynamics, and island biogeography, all of which can be applied to forensic entomology.

Landscape ecology is defined as the natural setting or habitat of multiple communities and how organisms function within that habitat (Forman 1995). Landscape dynamics conjure ideas of human influence on the environment (landscape); this is true in forensics as humans are modifying the environment and in the case of a homicide can alter landscape dynamics by placing the victim in an environment purposefully, thus causing an unnatural or unexpected resource.

Patch dynamics are important to community assembly and succession on the remains because the structure of patch mosaics affects community heterogeneity (Wu and Loucks 1995). They also drive metacommunity processes, such as species sorting, occurring at the metacommunity level that is defined by the potential interaction of species among communities (Leibold et al. 2004). Specifically, landscape patch configurations affect ecological processes and are scale dependent from the perspective of the consumer. Decaying human remains and their associated necrobiome community assembly become part of this configuration.

Patch heterogeneity reduces competition both within and among patches (Atkinson and Shorrocks 1981). For example, landscape effects on arthropods (e.g., carrion beetles and flies) have been documented in fragmented forest habitats where fly abundance increased by 150% while beetle abundance reduced by 66% as habitat fragmentation increased (Gibbs and Stanton 2001). Better understanding of these ecological interactions that influence community changes and species-specific development rates of arthropods on decomposing remains are paramount for forensic entomology, as these interactions ultimately influence PIA_{min} or PMI_{min} estimates. Here, the concepts of metacommunity ecology that are an integration of landscape ecology and patch dynamics are further discussed. This integrated perspective may inform future studies that address how studying ecological processes can be used to develop new models for entomologically based PIA_{min} or PMI_{min} estimates.

Landscape ecology and patch dynamics are conceptually linked and scale dependent in a way that determines the processes and outcomes of metacommunity ecology, and both are important for answering several questions relevant to forensic entomology. These questions depend on the scale of inquiry and can be roughly divided into (1) how does the landscape or ecoregion scale influence species presence in a given region? and (2) how does the local habitat or ecosystem scale influence species distributions and mediate metacommunity interactions (Hanski 1998; Leibold et al. 2004)?

At the landscape and ecoregion scale (~100 to > 1000 km²), several questions (Table 27.1) are derived from landscape and metacommunity ecology relevant to forensic entomology. Answering the first four questions provides the necessary large-scale information for determining the spatial extent and periods of necrophagous species distribution, essentially providing information on the

Table 27.1   Landscape and Ecoregion and Local-Scale Questions Derived from Landscape and Metacommunity Ecology Relevant to Forensic Entomology

Spatial Scale (Area)	Question	Relevant Ecological Concepts	Significance for Forensic Entomology
Landscape and ecoregion (100 to >1000 km²)	1. What species of necrophagous arthropods are residents of a given geographic area or ecoregion during specific times of the year?	Spatial extent and periods of necrophagous species distribution by habitat modification limits, which can affect species distributions, competitive ability, and persistence of certain species.	Identifying the potential species pool available in a geographic area, which can potentially be a crime scene location.
	2. What abiotic and biotic environmental, such as long-term climate variation and predominate vegetation, characteristics are important factors that influence which necrophagous species occur in a given geographic area or ecoregion?	The characteristics of a resource patch within a given geographic area or ecoregion such as heterogeneity and/or quality can influence the species composition (community assembly) and niche utilization (species sorting) of an ecoregion.	Identify the abiotic factors that affect the potential species pool within a given geographic area.
	3. What are the large-scale geographic boundaries for necrophagous arthropod and scavenger species?	Large-scale geographic boundaries such as mountain ranges and oceans affect the dispersal capabilities of species and can mitigate species composition for a given location.	Provides data for making probability estimates of the occurrence of individual species and communities of species within a geographic area.
	4. How well do models based on ecoregion-specific variables estimate large-scale necrophagous arthropod or scavenger species distribution?	Species have different responses to environmental conditions and different dispersal, colonization, and extinction rates or the mass effect that affect species distributions at an ecoregion scale.	Provides analytical confidence in making probability estimates of species occurrence within the ecoregion.
	5. Can ecoregion-scale variables of a crime scene be used to determine the probability that a species is from a given geographic location during a specific time of year?	Including the previous four ecological concepts, one must consider the population turnover rate at a given location(s) and how this affects the species pool within and among years.	Estimates the likelihood that a given species (or multiple species) collected at a crime scene, or similar legally relevant location, is expected to be from that region; this is valuable for making inferences about the location(s) of remains, and probability that a certain species will colonize a set of remains.

(Continued)

Table 27.1   Landscape and Ecoregion and Local-Scale Questions Derived from Landscape and Metacommunity Ecology Relevant to Forensic Entomology (*Continued*)

Spatial Scale (Area)	Question	Relevant Ecological Concepts	Significance for Forensic Entomology
Local (1–10 km²)	1. Do specific local or habitat-scale environmental characteristics "select" for a subset of species from the regional species pool that will potentially colonize or use (e.g., scavenge) remains or other material (e.g., stored products) of legal importance, and do these factors influence the arrival time and colonization sequence in a given habitat?	Species sorting results from favorable environmental conditions and heterogeneity (e.g., niche availability) within a given area. These differences in availability can affect the species arrival and successional patterns to the resource.	Determining what "subset" community of necrophagous species from the landscape and ecoregion species pool will arrive and potentially colonize a resource in a specific habitat, or patch, within the larger ecoregion or geographic area; and understanding how variable the arrival and colonization process is within a local environment and how this process may be affected by the location, size, and configuration of similar and different patches of habitat in the region.
	2. How do local environmental conditions such as daily variation in temperature, humidity, precipitation, and vertebrate scavenging affect the arrival time and colonization sequence of a necrophagous insect species from the regional pool?	The tolerance of species to abiotic conditions (e.g., temperature) will affect species composition and, thus, species sorting is based on environmental gradients and individual species thresholds to abiotic conditions.	Identify the abiotic factors that affect the potential species pool within a given local environment.
	3. Does the location, in both space and time, of other necrophagous resources in the surrounding habitat influence the arrival time and colonization sequence of a necrophagous species from the regional pool?	Dispersal depends on both proximity and configuration of species to adjacent patches (edge effects), patch heterogeneity (fragmentation), and configuration.	Provides data for making probability estimates of the occurrence of individual species and communities of species within a geographic area and when interacting with other species in the environment.
	4. How well do models based on local environmental variables estimate habitat-scale necrophagous arthropod species arrival time and colonization sequence in one or more locations and among seasons?	Succession models taken into account competition (intra- and interspecific) and aggregation influences on population structures. However, these models are dependent on local biotic and abiotic conditions.	Provides analytical confidence in making probability estimates of species occurrence on the remains within a specific geographic area.
	5. Can local-scale variables of a crime scene be used to determine the probability that a certain species, or several species, will arrive and colonize a resource during over a specific period, and does that process depend on season?	Invertebrate arrival and succession patterns are dependent on patch (e.g., carrion) availability and quality. Furthermore, priority effects and competition between and among trophic levels (e.g., blow flies and vertebrate scavengers) will dictate those species present on the remains at any given time.	Estimates the likelihood that a given species or multiple species collected at crime scene arrived and colonized a resource in a given amount of time under specific habitat environmental conditions, which has potential to provide an entomological time line that can be used in estimating a PIA_{min} or PMI_{min}.

most important factors for identifying the potential species pool available in that geographic area (Figure 27.2). For the fifth question, understanding and having the ability to estimate the likelihood that a given species or multiple species collected at a crime scene, or similar legally relevant location, are expected to be from that area is valuable for making inferences about the location and potential movement of the remains from one location to another.

At the local scale (~1–10 km²), several additional questions (Table 27.1) are derived from the concepts of patch dynamics, metacommunity ecology, and island biogeography that are relevant to forensic entomology. Answering the first four questions provides local-scale information for determining which "community subset" of necrophagous species from the regional species pool will arrive and potentially colonize a resource for a specific habitat (patch) within a larger ecoregion (geographic area). Answers to the aforementioned questions also provide information for understanding how variable the arrival and colonization process is within a local environment and how this process may be affected by the location, size, and configuration of similar and different patches of habitat within the region. For the fifth question, understanding and having the ability to estimate the likelihood that a species or multiple species collected from a crime scene arrived and colonized a resource in a given amount of time under specific habitat environmental conditions has the potential to provide a more refined entomological time line, which can be used to estimate a PIA_{min} or PMI_{min}. There are several community ecology paradigms that can be used to test how and when subsets of species will colonize a resource patch, and these are summarized in Figure 27.2.

The consideration and answers to these questions begin to bridge the basic understanding of carrion landscape and metacommunity ecology, patch dynamics, and island biogeography with application in forensics; however, there have been limited studies to empirically consider the importance of such environmental contexts for making statistical or mathematical estimates of necrophagous insect detection, acceptance, and colonization time frames (Schoenly 1992; Schoenly et al. 2007;

Paradigm

Patch dynamics

- Assumes homogeneous patches
- Local dynamics occur faster than dispersal to new patches
- Competition-colonization trade-offs important

Species sorting

- Assumes heterogeneous patches
- Patch quality and dispersal affect community composition
- Dispersal tracks changes in environmental conditions

Mass effects

- Assumes heterogeneous patches
- Dispersal influences local dynamics
- Different quality patches connected by dispersal lead to source-sinks relations

Neutral

- Environmental context irrelevant
- All species are demographically similar
- Species composition, a product of metacommunity size

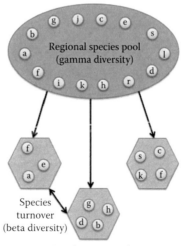

Figure 27.2 Community ecology paradigms that can be used to identify the potential species pool (e.g., necrophagous arthropods) available for patchy habitats (e.g., cadavers) in any given geographic area. (Modified from Figure 1 of Mihaljevic, J.R, *Trends Ecol. Evol.*, 27, 323–329, 2012.)

Michaud and Moreau 2009; Michaud et al. 2012), which have been argued to define a portion of the PIA_{min} (Tomberlin et al. 2011). There have been some excellent papers describing the conceptual importance of ecological theory to forensic entomology (Schoenly and Reid 1987; VanLaerhoven 2010), but very few have tested explicit hypotheses related to how these factors can affect estimates of PIA_{min} or PMI_{min}. This may be because most forensic entomology studies have been conducted at small spatial scales of < 1 km^2 (VanLaerhoven 2010; Michaud et al. 2012), which are scales considered within the adult dispersal range of at least some necrophagous fly species (Braack and De Vos 1990).

27.4 A REVIEW OF CARRION COMMUNITY ECOLOGY

The patterns of carrion necrophagous arthropod community assembly vary by geographic region, and the mechanisms that influence the arrival sequence, colonization, and replacement of specific species are not well known. Blow flies are spatially and temporally distributed throughout the world (Roy and Dasgupta 1975; Whitworth 2006; Sukontason et al. 2007; Harvey et al. 2008). Spatial patterns of blow flies depend on environmental and abiotic conditions (Levin 1976), while habitat heterogeneity influences the species sorting and composition of necrophagous insects on the remains. Species sorting is how community assembly is regulated by the local (e.g., competition) and regional factors (e.g., dispersal, immigration events) within a geographic area (Levin 1976). The local community is influenced by the abundance of taxa at the regional scale and the more widespread a taxon at the regional scale, the more abundant the taxon at the local scale, as discussed earlier. Barriers separating populations can be large geographic structures such as mountains and oceans or extensive urban development. However, genetic analyses of blow fly populations (see Chapter 24) show that the temporal, and not the spatial, scale is more important in determining population relatedness (Picard and Wells 2009, 2010). Recently, Barton et al. (2013) decoupled spatial aggregation and coexistence theory from succession theory, recognizing that both are needed for a unified theory of carrion ecology.

Necrophagous insects are ubiquitous in the most terrestrial environments, yet two questions remain: from what distance can a blow fly detect carcasses or human remains? Further, what signals are the insects using to detect the remains? Adult necrophagous insects have traits that allow them to effectively use unpredictable and ephemeral resources. Many insects are generalist, opportunistic consumers with the capability to utilize a wide breadth of resources and patchy resources (e.g., carcass, manure, and garbage) until a suitable resource is found (Archer and Elgar 2003). However, there are specialists or highly mobile insects that can cover larger spatial areas to detect the necessary substrate or habitat without detrimental costs to overall fitness. But, the distances adult blow flies travel between resource patches remain understudied (Cragg and Hobart 1955; Norris 1965; Smith and Wall 1998). Blow flies have traits of both generalists and specialists that allow them to utilize a wide breadth of resources when carrion is not available and yet can cover larger spatial distances to find a suitable resource (Tessmer and Meek 1996; Smith and Wall 1998). Blow fly mechanisms for detecting carrion have previously been studied by characterizing behavioral responses to a single volatile organic compound (VOC) or simple blends (e.g., two to three VOCs) collected from the headspace of carrion (LeBlanc 2008; Frederickx et al. 2012). Yet, the biological and ecological mechanisms governing blow fly behavioral responses to carrion-associated complex microbial communities, or the epinecrotic microbial communities, remain unexplored (Tomberlin et al. 2012; Benbow et al. 2013a; Pechal et al. 2013, 2014a).

Once insects have detected remains there must be acceptance of the remains, which is typically a female ovipositing on the resource (Tomberlin et al. 2011). Observations have documented oviposition events occurring within 1 minute of carcass placement in the field up to 24 hours later (personal observations). Yet, could the microbial communities (see Chapters 21 and 22) and VOCs emitted from the carcass regulate the arrival patterns and subsequent oviposition events? Adult oviposition preference and acceptance of the resource depends on the female reproductive status (Barton et al. 1987) and exposure of the remains; oviposition has been documented to initially occur in the mouth followed by oviposition in skin folds of the remains approximately 24 hours after the initial oviposition event (Archer and Elgar 2003). Subsequent arthropod succession could also be influenced by the timing and magnitude of the initial oviposition events.

Succession patterns are governed by niche dynamics and environmental factors, whereas biodiversity is determined from spatial and temporal heterogeneity (Grinnell 1917; Elton 1927; Hutchinson 1961). Succession can also be influenced by neutral dynamics or the immigration and extinction rates of a species within a habitat (Hubbell 2001). Necrophagous arthropod community variation has been documented during colonization of vertebrate resources in similar habitats (Watson and Carlton 2003). Annual variation of insect communities could be attributed to differences in resource size (Braack 1987; Benbow et al. 2013a) or priority effects of initial colonizers altering subsequent arthropod community structure (Chase 2010; Fukami and Nakajima 2011). For example, asynchronous blow fly succession was described for alligator carcasses (Nelder et al. 2009), which suggested niche theory mechanisms may explain colonization patterns. However, there is inconsistency in the literature as to whether niche or neutral theory dictates succession on carrion.

Early colonizers of remains, and thus potential competitors, may also influence acceptance of the remains as a suitable resource by later arrivals. The invasive species *Chrysomya rufifacies* (Macquart) (Diptera: Calliphoridae) is a secondary colonizer of remains and can exhibit delayed colonization by up to 24 hours (Wells and Greenberg 1992). Yet, larvae from both native blow fly, *Cochliomyia macellaria* (Fabricius) (Diptera: Calliphoridae), and the invasive *Ch. rufifacies* can occupy the same resource and develop at a similar rate until the invasive species drives the native species from the carcass and forces early pupation of the native (Wells and Greenberg 1992). Finally, delayed insect access to remains can result from abiotic factors (e.g., temperatures below the arthropod activity threshold) and physical barriers (e.g., wrappings and burial). The precolonization interval as defined in the work of Tomberlin et al. (2011) is an area of decomposition that has been widely understudied by forensic practitioners, as there is limited insect evidence with only the adults being present at the remains. However, a recent study suggested the potential use of epinecrotic microbial community succession as a means to estimate decomposition time in lieu of insect evidence (Pechal et al. 2014a).

Multiple species can use a single resource of variable quality at a given time point (Levin 1976), specifically carrion (Braack 1987). Resource quality is determined by carbon and nutrient availability, microbial community activity, and water content of the remains (Carter et al. 2007). Equal resource quality of the remains only occurs at the beginning and end of decomposition (Payne 1965). Depending on resource size, a single set of remains can be used by consumers in multiple at times during the decomposition process because of the varying degrees of resource quality (Braack 1987). For example, tri-trophic competition for carrion is common with microbes competing with higher trophic level consumers during initial decomposition, whereas interspecific insect competition and vertebrate competition with lower trophic level consumers can occur throughout the decomposition process. Microbe-laden carcasses in aquatic habitats can deter higher level consumers; scavenging rates in marine systems were reported to vary from 66% when carrion microbial communities were allowed to proliferate undisturbed to 89% in the absence of mature microbial communities (Barlocher 1979; Burkepile et al. 2006). It has yet to be determined what, and if, there is a microbial threshold for outcompeting higher trophic levels (e.g., insects and vertebrate scavengers) in terrestrial systems.

Yet, some necrophagous taxa have evolved mechanisms to reduce competition from the microbial communities as documented for calliphorid larvae and silphid beetles (Coleoptera: Silphidae). For instance, a waste by-product of blow fly larvae is ammonia, which increases the pH of the surrounding area and inhibits the growth of many bacteria species while providing favorable conditions for proteolytic enzyme activity (Nigam et al. 2006; Barnes and Gennard 2011). Nicrophorus beetles care for their brood balls by coating them with oral and anal secretions, which are thought to maintain antibiotic molecules (Hoback et al. 2004). Although not all carrion beetles use antibiotics in such a direct manner, some beetles are not attracted to carrion until a species that utilizes antibiotics has been at the same resource (Hoback et al. 2004).

Arrival patterns of blow flies are influenced by initial insect colonization (Spivak et al. 1991). Early colonization may allow better access to, and acquisition of, nutrients on fresh carrion for some consumers, whereas others may require some degree of resource processing by pioneer microbial communities before consumption (Barlocher 1979; Burkepile et al. 2006). Aggregation models, or the use of a resource by organisms within a limited spatial area, have been proposed to describe how competitors (e.g., blow flies and microbes) coexist on resource pulses (Woodcock et al. 2002). Coexistence of blow flies on carrion resources is common (Hanski 1987; Kouki and Hanski 1995) and can be explained by aggregations of individual species consuming a resource patch and not others, thus providing opportunities for another species (Hartley and Shorrocks 2002; Abos et al. 2006) or community assemblages to utilize the same resource (MacArthur and Wilson 1967). Competition among necrophagous taxa determines community composition and subsequent resource consumption (Denno and Cothran 1975), having impact on decomposition rates at any given time. For example, adult silphid or histerid beetles are functional predators consuming the developing dipteran larvae on carrion (Summerlin et al. 1982; Carlton et al. 1996). This competition is important as calliphorid larvae development is used most often for PMI_{min} estimates (Amendt and Hall 2007). The effect of behavioral interactions and outcomes of competition are discussed in Chapter 26.

The dispersal ability of insects is affected by spatial attributes (Levin 1976). In outdoor crime scenes, larval dispersal can be affected by soil type, precipitation, land gradient, and vegetation. Larvae associated with carrion have restricted dispersal ability because of their morphology unlike the adult stage, which can disperse great distances (Hanski 1987). Previous recommendations in collecting postfeeding third instars or pupae have been restricted to 10 m (Catts and Haskell 1990); however, recent evidence has demonstrated larval en masse dispersal capabilities of greater than 20 m (Lewis and Benbow 2011).

When insects are collected at a scene, it is rarely a single species but rather a community of individuals from several species. Describing entire communities could refine PIA_{min} estimates in addition to development datasets from individual species. The gravid status of females arriving to the remains (Mohr and Tomberlin 2014) and adult community composition of adults (Pechal et al. 2014b; Benbow et al. 2013a) have the potential to estimate decomposition time. Better understanding insect dynamics, including adult interactions, on remains will refine PMI_{min} estimates.

27.5 CONCLUSIONS AND FUTURE OF COMMUNITY ECOLOGY IN FORENSIC ENTOMOLOGY

Community structure on carrion over time has often been described as a process of competition between or among species for a resource (Janzen 1977; Polis and Strong 1996; DeVault et al. 2003). Necrophagous insect taxa of the necrobiome such as blow flies, flesh flies (Diptera: Sarcophagidae), and carrion beetles have been the most studied in terms of their life history traits and community interactions when using carrion. There remains a paucity of information about microbial communities and vertebrate scavengers, both of which are important consumers of carrion with the potential to enhance the knowledge of the necrobiome ecological interactions and further refine PMI_{min} estimates. Furthermore, the spatial and temporal scale influences (landscape and patch mosaics) on community interactions and the ability of necrophagous insects to utilize unpredictable, patchy habitats require further investigation. Understanding how arthropods interact with their habitat at both landscape and local levels may improve the use of arthropod community dynamics within forensic applications. Ecological models for succession have also gone unused/underused by carrion ecologists, such as Markov chains (including backward directed versions; mentioned in Chapter 19). The future of forensic entomology should include additional studies that use community ecology theory and data to improve PMI_{min} estimates (Schoenly et al. 2006).

REFERENCES

Abos, C. P., F. Lepori, B. G. McKie, and B. Malmqvist. 2006. Aggregation among resource patches can promote coexistence in stream-living shredders. *Freshwater Biology* 51: 545–553.

Amendt, J., C. P. Campobasso, E. Gaudry, C. Reiter, H. N. LeBlanc, and M. Hall. 2007. Best practice in forensic entomology—Standards and guidelines. *International Journal of Legal Medicine* 121: 90–104.

Amendt, J. and M. Hall. 2007. Forensic entomology—Standards and guidelines. *Forensic Science International* 169: S27.

Amendt, J., R. Krettek, and R. Zehner. 2004. Forensic entomology. *Naturwissenschaften* 91: 51–65.

Ames, C. and B. Turner. 2003. Low temperature episodes in development of blowflies: Implications for post-mortem interval estimation. *Medical and Veterinary Entomology* 17: 178–186.

Archer, M. S. and M. A. Elgar. 2003. Female breeding-site preferences and larval feeding strategies of carrion-breeding Calliphoridae and Sarcophagidae (Diptera): A quantitative analysis. *Australian Journal of Zoology* 51: 165–174.

Atkinson, W. D. and B. Shorrocks. 1981. Competition on a divided and ephemeral resource: A simulation model. *Journal of Animal Ecology* 50: 461–471.

Barlocher, F. 1979. On trophic interactions between microorganisms and animals. *The American Naturalist* 114: 147–148.

Barnes, K. M. and D. E. Gennard. 2011. The effect of bacterially-dense environments on the development and immune defences of the blowfly *Lucilia sericata*. *Physiological Entomology* 36: 96–100.

Barton, B. L., A. Van Gerwen, and P. Smith. 1987. Relationship between mated status of females and their stage of ovarian development in field populations of the Australian sheep blowfly, *Lucilia cuprina* (Wiedemann) (Diptera: Calliphoridae). *Bulletin of Entomological Research* 77: 609–615.

Benbow, M. E., A. J. Lewis, J. K. Tomberlin, and J. L. Pechal. 2013a. Seasonal necrophagous insect community assembly during vertebrate carrion decomposition. *Journal of Medical Entomology* 50: 440–450.

Benbow, M. E., R. W. Merritt, and J. L. Pechal. 2013b. Entomology. In *Wiley Encyclopedia of Forensic Science*. West Sussex, United Kingdom: Wiley.

Berg, M. C. and M. E. Benbow. 2013. Environmental factors associated with *Phormia regina* (Meigen) (Diptera: Calliphoridae) oviposition. *Journal of Medical Entomology* 50: 451–457.

Braack, L. and V. De Vos. 1990. Feeding habits and flight range of blow-flies (Chrysomyia spp.) in relation to anthrax transmission in the Kruger National Park, South Africa. Onderstepoort Journal Veterinary Research 57: 141–142.

Braack, L. E. O. 1987. Community dynamics of carrion-attendant arthropods in tropical African woodlands. *Oecologia* 72: 402–409.

Burkepile, D. E., J. D. Parker, C. B Woodson, H. J. Mills, J. Kubanek, P. A. Sobecky, and M. E. Hay. 2006. Chemically mediated competition between microbes and animals: Microbes as consumers in food webs. *Ecology* 87: 2821–2831.

Byrd, J. H. and J. L. Castner. 2001. *Forensic Entomology: The Utility of Arthropods in Legal Investigations*. Boca Raton, FL: CRC Press.

Carlton, C., R. Leschen, and P. Kovarik. 1996. Predation on adult blow flies by a Chilean hister beetle, *Euspilotus bisignatus* (Erichson) (Coleoptera: Histeridae). *The Coleopterists' Bulletin* 50: 154.

Carter, D. O., D. Yellowlees, and M. Tibbett. 2007. Cadaver decomposition in terrestrial ecosystems. *Naturwissenschaften* 94: 12–24.

Catts, E. P. and M. L. Goff. 1992. Forensic entomology in criminal investigations. *Annual Review of Entomology* 37: 253–272.

Catts, E. P. and N. H. Haskell. 1990. *Entomology & Death: A Procedural Guide*. Clemson, SC: Joyce's Print Shop.

Chase, J. M. 2010. Stochastic community assembly causes higher biodiversity in more productive environments. *Science* 328: 1388–1391.

Chase, J. M. and J. Bengtsson. 2010. Increasing spatio-temporal scales: Metacommunity ecology. In *Community Ecology: Processes, Models, and Applications*, edited by H. A. Verhoef and P. J. Morin, pp. 57–68. Oxford, United Kingdom: Oxford University Press.

Clements, F. E. 1936. Nature and structure of the climax. *Journal of Ecology* 24: 252–284.

Connell, J. H. and R. O. Slatyer. 1977. Mechanisms of succession in natural communities and their role in community stability and organization. *The American Naturalist* 111: 1119–1144.

Cragg, J. and J. Hobart. 1955. A study of a field population of the blowflies *Lucilia caesar* (L.) and *L. sericata* (Mg.). *Annals of Applied Biology* 43: 645–663.

De Jong, G. D. and W. W. Hoback. 2006. Effect of investigator disturbance in experimental forensic entomology: Succession and community composition. *Medical and Veterinary Entomology* 20: 248–258.

Denno, R. F. and W. R. Cothran. 1975. Niche relationships of a guild of necrophagous flies. *Annals of the Entomological Society of America* 68: 741–754.

Denno, R. F. and W. R. Cothran. 1976. Competitive interactions and ecological strategies of sarcophagid and calliphorid flies inhabiting rabbit carrion. *Annals of the Entomological Society of America* 69: 109–113.

Devault, T. L., O. E. Rhodes, and J. A. Shivik. 2003. Scavenging by vertebrates: Behavioral, ecological, and evolutionary perspectives on an important energy transfer pathway in terrestrial ecosystems. *Oikos* 102: 225–234.

Elton, C. 1927. *Animal Ecology*. London, England: Sidgwick and Jackson.

Forman, R. T. 1995. Some general principles of landscape and regional ecology. *Landscape Ecology* 10: 133–142.

Frederickx, C., J. Dekeirsschieter, F. J. Verheggen, and E. Haubruge. 2012. Responses of *Lucilia sericata* Meigen (Diptera: Calliphoridae) to cadaveric volatile organic compounds. *Journal of Forensic Sciences* 57: 386–390.

Fukami, T. and M. Nakajima. 2011. Community assembly: Alternative stable states or alternative transient states? *Ecology Letters* 14: 973–984.

Gibbs, J. P. and E. J. Stanton. 2001. Habitat fragmentation and arthropod community change: Carrion beetles, phoretic mites, and flies. *Ecological Applications* 11: 79–85.

Goff, M. L. 2010. Early postmortem changes and stages of decomposition. In *Current Concepts in Forensic Entomology*. New York, NY: Springer.

Greenberg, B. 1991. Flies as forensic indicators. *Journal of Medical Entomology* 28: 565–577.

Grinnell, J. 1917. The niche-relations of the California Thrasher. *Auk* 427–433.

Hanski, I. 1987. Carrion fly community dynamics: Patchiness, seasonality and coexistence. *Ecological Entomology* 12: 257–266.

Hanski, I. 1998. *Metapopulation Ecology*. Oxford, United Kingdom: Oxford University Press.

Hartley, S. and B. Shorrocks. 2002. A general framework for the aggregation model of coexistence. *Journal of Animal Ecology* 71: 651–662.

Harvey, M. L., S. Gaudieri, M. H. Villet, and I. R. Dadour. 2008. A global study of forensically significant calliphorids: Implications for identification. *Forensic Science International* 177: 66–76.

Hoback, W. W., A. A. Bishop, J. Kroemer, J. Scalzitti, and J. J. Shaffer. 2004. Differences among antimicrobial properties of carrion beetle secretions reflect phylogeny and ecology. *Journal of Chemical Ecology* 30: 719–729.

Hubbell, S. 2001. *The Unifed Neutral Theory of Biodiversity and Biogeography*. Princeton, NJ: Princeton University Press.

Hutchinson, G. E. 1961. The paradox of the plankton. *The American Naturalist* 95: 137–145.

Janzen, D. H. 1977. Why fruits rot, seeds mold, and meat spoils. *The American Naturalist* 111: 691–713.

Kouki, J. and I. Hanski. 1995. Population aggregation facilitates coexistence of many competing carrion fly species. *Oikos* 72: 223–227.

LeBlanc H. N. 2008. Olfactory stimuli associated with the different stages of vertebrate decomposition and their role in the attraction of the blowfly Calliphora vomitoria (Diptera: Calliphoridae) to carcasses, The University of Derby, Derby, United Kingdom.

Leibold, M. A., M. Holyoak, N. Mouquet, P. Amarasekare, J. M. Chase, M. F. Hoopes, M. F. Holt. 2004. The metacommunity concept: A framework for multi-scale community ecology. *Ecology Letters* 7: 601–613.

Levin, S. A. 1976. Population dynamic models in heterogeneous environments. *Annual Review of Ecology and Systematics* 7: 287–310.

Lewis, A. J. and M. E. Benbow. 2011. When entomological evidence crawls away: *Phormia regina en masse* larval dispersal. *Journal of Medical Entomology* 48: 1112–1119.

Logue, J. R. B., N. Mouquet, H. Peter, and H. Hillebrand. 2011. Empirical approaches to metacommunities: A review and comparison with theory. *Trends in Ecology & Evolution* 26: 482–491.

Ma, Q., A. Fonseca, W. Liu, A. T. Fields, M. L. Pimsler, A. F. Spindola, A. M. Tarone, T. L. Crippen, J. K. Tomberlin, T. K. Wood. 2012. *Proteus mirabilis* interkingdom swarming signals attract blow flies. *The ISME Journal: Multidisciplinary Journal of Microbial Ecology* 6: 1356–1366.

MacArthur, R. H. and E. O. Wilson. 1967. *The Theory of Island Biogeography*. Princeton, NJ: Princeton University Press.

Merritt, R. and M. Benbow. 2009. Forensic entomology. In *Wiley Encyclopedia of Forensic Science*, edited by A. Jamieson and A. Moenssens. Hoboken, NJ: Wiley.

Michaud, J.-P., and G. Moreau. 2009. Predicting the visitation of carcasses by carrion-related insects under different rates of degree-day accumulation. *Forensic Science International* 185: 78–83.

Michaud, J.-P., K. G. Schoenly, and G. Moreau. 2012. Sampling flies or sampling flaws? Experimental design and inference strength in forensic entomology. *Journal of Medical Entomology* 49: 1–10.

Mihaljevic J. R. 2012. Linking metacommunity theory and symbiont evolutionary ecology. *Trends in Ecology & Evolution* 27: 323–329.

Mohr R. M. and J. K. Tomberlin. 2014. Environmental Factors Affecting Early Carcass Attendance by Four Species of Blow Flies (Diptera: Calliphoridae) in Texas. Journal of Medical Entomology 51: 702–708.

Muller, C. H. 1940. Plant succession in the Larrea-Flourensia climax. *Ecology* 21: 206–212.

Nelder, M. P., J. W. Mccreadie, and C. S. Major. 2009. Blow flies visiting decaying alligators: Is succession synchronous or asynchronous? *Psyche* 2009: 1–7.

Nigam, Y., A. Bexfield, S. Thomas, and N. A. Ratcliffe. 2006. Maggot therapy: The science and implication for cam—Part II—Maggots combat infection. *Evidence-based Complementary and Alternative Medicine* 3: 303–308.

Norris, K. R. 1965. The bionomics of blow flies. *Annual Review of Entomology* 10: 47–68.

Parmenter, R. and J. Macmahon. 2009. Carrion decomposition and nutrient cycling in a semiarid shrub-steppe ecosystem. *Ecological Monographs* 79: 637–661.

Payne, J. A. 1965. A summer carrion study of the baby pig *Sus scrofa* Linnaeus. *Ecology* 46: 592–602.

Payne, J. A., E. W. King, and G. Beinhart. 1968. Arthropod succession and decomposition of buried pigs. *Nature* 219: 1180–1181.

Pechal J. L., M. E. Benbow, T. L. Crippen, A. M. Tarone, J. K. Tomberlin. 2014. Delayed insect access alters carrion decomposition and necrophagous insect community assembly. Ecosphere 5: art45.

Pechal, J. L., T. L. Crippen, M. E. Benbow, A. M. Tarone, S. Dowd, and J. K. Tomberlin. 2014a. The potential use of bacterial community succession in forensics as described by high throughput metagenomic sequencing. *International Journal of Legal Medicine* 128: 193–205.

Pechal, J. L., T. L. Crippen, A. M. Tarone, A. J. Lewis, J. K. Tomberlin, and M. E. Benbow. 2013. Microbial community functional change during vertebrate carrion decomposition. *PLoS ONE* 8: e79035.

Petchey, O. L., P. J. Morin, and H. Olff. 2010. The topology of ecological interaction networks: The state of the art. In *Community Ecology: Processes, Models, and Applications*, edited by H. A. Verhoef and P. J. Morin, pp. 7–21. Oxford, United Kingdom: Oxford University Press.

Picard, C. J. and J. D. Wells. 2009. Survey of the genetic diversity of *Phormia regina* (Diptera: Calliphoridae) using amplified fragment length polymorphisms. *Journal of Medical Entomology* 46: 664–670.

Picard, C. J. and J. D. Wells. 2010. The population genetic structure of North American *Lucilia sericata* (Diptera: Calliphoridae), and the utility of genetic assignment methods for reconstruction of postmortem corpse relocation. *Forensic Science International* 195: 63–67.

Polis, G. A. and D. R. Strong. 1996. Food web complexity and community dynamics. *The American Naturalist* 147: 813–846.

Putman, R. J. 1978. The role of carrion-frequenting arthropods in the decay process. *Ecological Entomology* 3: 133–139.

Roy, P. and B. Dasgupta. 1975. Seasonal occurrence of muscid, calliphorid and sarcophagid flies in Siliguri, West Bengal, with a note on the identity of *Musca domestica* L. *Oriental Insects* 9: 351–374.

Schoenly, K. G. 1992. A statistical analysis of successional patterns in carrion-arthropod assemblages: Implications for forensic entomology and determination of the postmortem interval. *Journal of Forensic Sciences* 37: 1489–1513.

Schoenly, K. G., N. H. Haskell, R. D. Hall, and J. R. Gbur. 2007. Comparative performance and complementarity of four sampling methods and arthropod preference tests from human and porcine remains at the forensic anthropology center in Knoxville, Tennessee. *Journal of Medical Entomology* 44: 881–894.

Schoenly, K. G., N. H. Haskell, D. K. Mills, C. Bieme-Ndi, K. Larsen, and Y. Lee. 2006. Using pig carcasses as model corpses to teach concepts of forensic entomology and ecological succession. *The American Biology Teacher* 68: 402–410.

Schoenly, K. G. and W. Reid. 1987. Dynamics of heterotrophic succession in carrion arthropod assemblages: Discrete seres or a continuum of change? *Oecologia* 73: 192–202.

Smith, K. E. and R. Wall. 1998. Estimates of population density and dispersal in the blowfly *Lucilia sericata* (Diptera: Calliphoridae). *Bulletin of Entomological Research* 88: 65–73.

Spivak, M., D. Conlon, and W. J. Bell. 1991. Wind-guided landing and search behavior in fleshflies and blowflies exploiting a resource patch (Diptera: Sarcophagidae, Calliphoridae). *Annals of the Entomological Society of America* 84: 447–452.

Sukontason, K., P. Narongchai, C. Kanchai, K. Vichairat, P. Sribanditmongkol, T. Bhoopat, H. Kurahashi et al. 2007. Forensic entomology cases in Thailand: A review of cases from 2000 to 2006. *Parasitology Research* 101: 1417–1423.

Summerlin, J., D. Bay, R. Harris, D. Russell, and K. Stafford. 1982. Predation by four species of histeridae (Coleoptera) on horn fly (Diptera: Muscidae). *Annals of the Entomological Society of America* 75: 675–677.

Tessmer, J. W. and C. L. Meek. 1996. Dispersal and distribution of Calliphoridae (Diptera) immatures from animal carcasses in southern Louisiana. *Journal of Medical Entomology* 33: 665–669.

Tomberlin, J. K., M. E. Benbow, A. M. Tarone, and R. Mohr. 2011. Basic research in evolution and ecology enhances forensics. *Trends in Ecology and Evolution* 26: 53–55.

Tomberlin, J. K., T. L. Crippen, A. M. Tarone, B. Singh, K. Adams, Y. H. Rezenom, M. E. Benbow et al. 2012. Interkingdom responses of flies to bacteria mediated by fly physiology and bacterial quorum sensing. *Animal Behaviour* 84: 1449–1456.

Tomberlin, J. K., R. Mohr, M. E. Benbow, A. M. Tarone, and S. Vanlaerhoven. 2011. A roadmap for bridging basic and applied research in forensic entomology. *Annual Review of Entomology* 56: 401–421.

VanLaerhoven, S. 2010. Ecological theory and its application in forensic entomology. In *Forensic Entomology: The Utility of Arthropods in Legal Investigations*, edited by J. H. Byrd and J. L. Castner, pp. 493–518. Boca Raton, FL: CRC Press.

Vass, A. A. 2001. Beyond the grave—Understanding human decomposition. *Microbiology Today* 28: 190–192.

Vass, A. A., S. A. Barshick, G. Sega, J. Caton, J. T. Skeen, J. C. Love, J. A. Synstelien. 2002. Decomposition chemistry of human remains: A new methodology for determining the postmortem interval. *Journal of Forensic Sciences* 47: 542–553.

Verhoef, H. A. and P. J. Morin. 2010. *Community Ecology: Processes, Models, and Applications*. Oxford, United Kingdom: Oxford University Press.

Watson, E. J. and C. E. Carlton. 2003. Spring succession of necrophilous insects on wildlife carcasses in Louisiana. *Journal of Medical Entomology* 40: 338–347.

Wells, J. D. and B. Greenberg. 1992. Interaction between *Chrysomya rufifacies* and *Cochliomyia macellaria* (Diptera: Calliphoridae): The possible consequences of an invasion. *Bulletin of Entomological Research* 82: 133–137.

Wells, J. D. and L. R. Lamotte. 2010. Estimating the postmortem interval. In *Forensic Entomology: The Utility of Arthropods in Legal Investigations*, edited by J. H. Byrd and J. L. Castner, pp. 367–388. Boca Raton, FL: CRC Press.

Whitworth, T. 2006. Keys to the genera and species of blow flies (Diptera: Calliphoridae) of America north of Mexico. *Proceedings of the Entomological Society of Washington* 108: 689–725.

Wilson, E. E. and E. M. Wolkovich. 2011. Scavenging: How carnivores and carrion structure communities. *Trends in Ecology & Evolution* 26: 129–135.

Woodcock, B. A., A. D. Watt, and S. R. Leather. 2002. Aggregation, habitat quality and coexistence: A case study on carrion fly communities in slug cadavers. *Journal of Animal Ecology* 71: 131–140.

Wu, J. and O. L. Loucks. 1995. From balance of nature to hierarchical patch dynamics: A paradigm shift in ecology. *Quarterly Review of Biology* 439–466.

Yang, L. H., J. L. Bastow, K. O. Spence, and A. N. Wright. 2008. What can we learn from resource pulses? *Ecology* 89: 621–634.

CHAPTER 28

Surface Hydrocarbons as min-PMI Indicators. Fit for Purpose?

Hannah E. Moore and Falko P. Drijfhout

CONTENTS

28.1 INTRODUCTION

28.1.1 New Developments in Forensic Entomology

The first crucial step a forensic entomologist has to undertake when presented with a crime scene is to identify the insects present. Closely related arthropods that colonize vertebrate carrion can have substantially different development rates, and a misidentification can lead to inaccurate minimum postmortem interval (min-PMI) estimations. The current method used for their identification is dichotomous keys, which are based on morphological features within different insect species. Using these keys require analysts to be very familiar with the anatomical characteristics of the insects examined. However, most identification keys are for adult, and not immature (e.g., egg, larva, pupa) flies. In investigations where immatures are collected, the investigator is asked to rear them to the adult stage. This process can be time consuming and is not guaranteed to result in adults. One way to compliment current identification methods is DNA-based analyses (Sperling et al. 1994). These techniques have been applied in the field of forensic entomology for over two decades. Sperling et al. (1994) were the first to test the concept in forensic entomology and much more has been done since then. For examples, refer to the following: Malgorn and Coquoz 1999, Vincent et al. 2000, Stevens and Wall 2001, Wallman and Donnellan 2001, Wells and Sperling 2001, Harvey et al. 2003, Ratcliffe et al. 2003, Schroeder et al. 2003, Ames et al. 2006a, Cainé et al. 2009, Oliveira et al. 2011). DNA is a sound tool for identifying immature blow flies (Diptera: Calliphoridae), as well as showing great potential for ageing some of the life stages. However, DNA-based techniques have a few disadvantages, which could limit their use in the field. Most studies are using DNA barcoding to identify species, but the clear difference between the number of intra- and interspecific nucleotide variation can be an artifact of insufficient sampling across taxa (Fremdt et al. 2012). Also, the analyst must be a technical expert, and the process can be time consuming and expensive.

28.1.2 Cuticular Hydrocarbons

The cuticle (surface) of insects is covered by a layer of epicuticular waxes, consisting of hydrocarbons, fatty acids, alcohols, waxes, glycerides, phospholipids, and glycolipids. This cuticular lipid layer prevents desiccation as well as penetration of microorganisms (Gibbs and Crockett 1998). Hydrocarbons predominate this layer in many insect species (Gibbs and Pomonist 1995) and are very stable (Drijfhout 2010). A vast number of different hydrocarbons are present in many combinations, indicating that potentially each species of insect could have its own unique hydrocarbon profile, often referred to as a fingerprint (Everaerts et al. 1997; Ye et al. 2007; Martin et al. 2008a,b).

One technique that may have the potential to give the same accuracy as DNA-based techniques for species identification as well as a potential ageing tool is cuticular hydrocarbon or surface hydrocarbon analysis. Hydrocarbons are long linear molecules and are observed in their saturated and unsaturated form in insects and can have one or more methyl groups attached to the chain length. In the saturated form, *n*-alkanes (or paraffins) consist of all the carbons being joined together with single bonds but they may have one or more methyl groups present. In the unsaturated form (olefins), one (alkene), two (alkadiene), or three (alkatriene) double bonds maybe present along the length of the chain. Olefins can be in the form of two isomeric structures, referred to as *cis*-alkenes (Z-alkenes) or *trans*-alkenes (E-alkenes) (Drijfhout 2010). Insect hydrocarbons are usually made up of *n*-alkanes, Z-alkenes, and methyl branched alkanes, of which examples are given in Figure 28.1.

These epicuticular lipids serve a variety of roles in different insect species (e.g., in ants [Hymenoptera: Formicidae]). Often they are used for communication and recognition of colony nest mates (Martin et al. 2008a,c), but their primary function is to protect against desiccation (Blomquist and Bagnères 2010). This is achieved by the structure of the *n*-alkanes, which allows them to be

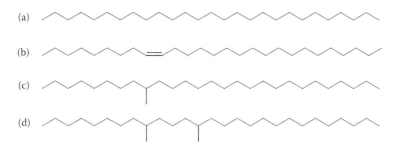

Figure 28.1 Examples of linear long-chain cuticular hydrocarbons typically found in *Calliphora vicina*. (a) Heptacosane, (b) Z(9)-heptacosene, (c) 9-methylheptacosane, and (d) 9,13-dimethylheptacosane.

closely packed (Gibbs 1998). The reader is referred to reviews on the general role and function of cuticular hydrocarbons in insects by Howard and Blomquist (2005) and more to insects related to forensic entomology by Drijfhout (2010).

28.1.3 Cuticular Hydrocarbons in Forensic Entomology

Hydrocarbons as a means of species identification has been demonstrated for several groups of insects: ants (Hymenoptera: Formicidae) (Tissot et al. 2001; Akino et al. 2004; Martin et al. 2008b); termites (Dictyoptera: Blattodea) (Haverty et al. 1997); cockroaches (Dictyoptera: Blattodea) (Everaerts et al. 1997); mosquitos (Diptera: Culicidae) (Anyanwu et al. 2000); grasshoppers (Orthoptera: Acrididae) (Chapman and Espelies 1995; Tregenza et al. 2000); wasps (Hymenoptera: Pteromalidae) (Bernier et al. 1998); honeybees (Hymenoptera: Apinae) (Lavine and Vora 2005); beetles (Coleoptera: Tenebrionidae) (Lockey 1991, 1992; Page et al. 1997); and flies (Diptera: Muscidae) (Urech et al. 2005). Guillem et al. (2012) proposed to use chemotaxonomy to address the difficulty of morphological identification of two ant species, *Myrmica sabuleti*, Meinert (Hymenoptera: Formicidae) and *Myrmica scabrinodis*, Nylander (Hymenoptera: Formicidae), that are morphologically similar. Their hydrocarbon profiles were examined to determine if they could provide a quick solution for differentiating between the two species. They discovered that the presence or absence of two species-specific compounds (one for each species) allowed for these species to be distinguished.

Cuticular hydrocarbons have also been used in the field of entomology for ageing (Chen et al. 1990; Brown et al. 1992; Desena et al. 1999a,b; Brown et al. 2000; Tregenza et al. 2000) and gender identification (Jackson and Bartelt 1986; Mpuru et al. 2001; Marican et al. 2004; Steiner et al. 2006). Trabalon et al. (1992) examined the cuticular hydrocarbons of adult *Calliphora vomitoria* (Linnaeus) (Diptera: Calliphoridae) with relation to age and sex. Using principal component analysis (PCA), they were able to distinguish between different groups based on relative proportions of male and female hydrocarbons. Young (3–6 hours) adult flies could be distinguished from older (24–120 hours) individuals, and males could be distinguished from females, with the sex pheromones (alkenes) being the main class of compounds allowing for this distinction.

Recent studies on flies have also shown that hydrocarbons can be used for taxonomic purposes (Brown et al. 1992; Urech et al. 2005; Ye et al. 2007) as well as a potential ageing tool (Zhu et al. 2006, 2007; Roux et al. 2008). Roux et al. (2008) were the first to carry out a complete ontogenetical study of three forensically important blow flies: *Calliphora vicina* Robineau-Desvoidy (Diptera: Calliphoridae), *C. vomitoria*, and *Protophormia terraenovae* Robineau-Desvoidy (Diptera: Calliphoridae) using gas chromatography with a flame ionization detector (GC-FID). Results showed the potential to use hydrocarbons as an application in forensic entomology. The discrimination analysis they carried out allowed for clear differentiations between the life stages (i.e., larvae, postfeeding larvae, puparia, and adult fly).

28.1.4 Cuticular Hydrocarbons to Solve Complex PMI Estimations

A major problem in the field of forensic entomology is the complexity of min-PMI estimations involving many different factors such as species identification, ageing the oldest life stage present, environmental conditions before the discovery of the body, condition of the body (e.g., clothed or not), geographic location of the body, and potential intake of toxins, which can alter the rate of development for larvae (Amendt et al. 2004). If any of these steps are factored incorrectly, the min-PMI calculation could be misleading and therefore unreliable. Species identification alone is a complex procedure that must be carried out by a highly experienced taxonomist and forensic entomologist and even then, some insects can be impossible to identify to species level, especially in the larval stages.

A potential application of hydrocarbon analysis is to identify and age forensically important blow flies. When the hydrocarbon profiles are found to be species-specific and distinguishable at different ages, this could provide a reliable method that can complement current methods used for taxonomic and ageing purposes. This approach has the potential to enhance the precision of min-PMI estimations. The technique of hydrocarbon analysis also could reduce time invested in criminal investigations as it could eliminate the need to rear the larvae to adult flies for species identification. One chromatogram would be sufficient to facilitate the age of the insect, alongside the species identification.

The main aim of this chapter is, therefore, to review findings from studies on the cuticular hydrocarbons found in Calliphoridae that allow for the identification and age to be determined for these forensically important blow flies. The chapter begins with an overview of the basic methodology of the technique followed by more detailed discussion of how it is used for species identification and fly age determination.

28.2 METHODOLOGY

28.2.1 Extractions

Before cuticular hydrocarbons can be analyzed they need to be removed (extracted) from the surface (cuticle) of the insect. Extracting the hydrocarbons can be done in two different ways: liquid extraction or solid-phase microextraction (SPME). In liquid extraction, the hydrocarbons are removed from the cuticle by dissolving them in an organic solvent such as hexane or dichloromethane. SPME is based on the affinity of compounds to an absorbent (e.g., a liquid stationary phase) (Pawliszyn 1997). During sampling the compounds of interest are absorbed on the stationary phase and the compounds can then be thermally desorbed within the injector of the GC or gas chromatography–mass spectrometry (GC-MS). Alternatively, the extracted liquid can be analyzed by liquid chromatography coupled to mass spectrometry (LC-MS). With SPME, sampling is even possible by rubbing the SPME needle over the body part of an insect, to absorb only those compounds found on the specific body part (Lommelen et al. 2006).

Extraction should be performed carefully. If the method is too aggressive, many of the internal compounds from the insect will be coextracted. Adult insects collected from plants or within their nests usually give "clean" extracts, but if material (e.g., larvae, eggs, larvae) is collected from corpses, this could potentially lead to more "dirty" or complex extracts as some of the corpse tissue may have accumulated on the insects being examined. In such cases, the extract may need to be cleaned as described by Roux et al. (2008) and Moore et al. (2012). Millar and Sims (1998) also give a good review on the preparation, cleanup, and fractionation of various extractions.

28.2.2 Chemical Analysis

The number of studies on cuticular chemistry has greatly increased over the past 20 years, especially concerning the hydrocarbons of social insects including ants (Martin et al. 2008a,b), bees (Lavine and Vora 2005), termites (Haverty et al. 1997), and wasps (Bernier et al. 1998). The majority of studies carried out on cuticular hydrocarbons has benefitted from the widespread availability and popularity of GC-MS, a well-established technique.

28.2.2.1 Gas Chromatography–Mass Spectrometry

The result of combining the two analytical techniques of GC-MS produces a very powerful and user-friendly analytical instrument. Over the years, it has become the primary analysis in many branches of science owing to its ability to separate, detect, and identify a large number of chemical compounds (McMaster and McMaster 1998).

The sample is injected into the GC via the injector port and the vaporized samples are carried into the column of the GC by the mobile phase. The hydrocarbons being analyzed will interact strongly with the column that sits in an oven, and a linear temperature program is applied to aid separation of the analytes. The lower molecular weight hydrocarbons (and therefore lower boiling point) will elute off the column first with the high boiling point compounds eluting last. The temperature program is gradually ramped up and then held at a high temperature, ensuring all analytes and impurities are removed from the column, before returning to ambient temperature. The compounds remain in the gas phase as they leave the GC oven, via the heated transfer line. These separated analytes pass into the MS where they are ionized by electron ionization. The ions are pushed through the system into the quadrupole analyzer where they are measured according to their mass-over-charge (m/z) ratio. The ions are then detected and plotted into a mass spectrum showing the masses against the peak intensities. The end result is of a GC chromatogram with each peak detected in the GC displaying its own mass spectrum enabling identification and structural information.

Because of the complex mixtures extracted from the cuticles of blow flies, GC is needed to separate the analytes followed by MS to identify the compounds. For a more detailed description on GC-MS, the reader is referred to general textbooks (e.g. Heath and Dueben [1998] and the review by Drijfhout [2010]).

28.2.2.2 Direct Analysis Real-Time Mass Spectrometry

An alternative to GC-MS is an analytical technique capable of producing rapid taxonomic information for blow fly species. It is the recently developed ion source, direct analysis in real time (DART), which was developed in 2005 by Laramée and Cody (2005). When coupled to high-resolution MS, it allows for rapid qualitative and quantitative analysis of a wide variety of samples (Cody et al. 2007). DART's potential is being exploited in a broad range of applications from food science (Hajslova et al. 2011; Kim et al. 2011) to drug analysis (Steiner and Larson 2009; Grange and Sovocool 2011). It has also been used for nonpolar compound analysis, which is of particular interest when investigating its potential to analyze insect cuticular hydrocarbons. One disadvantage of DART analysis is that it does not separate components in time (like chromatography) for sample mixtures. However, when this technique is coupled with a high-resolution MS, the relatively simple mass spectra (mostly $[M+H]^+$ for positive ion, $[M-H]^-$ for negative ion) provide high mass accuracy measurements that can be used to calculate the elemental compositions for each peak observed in the DART mass spectra. In addition, if chromatography is necessary, the DART source can be interfaced with the GC column output (Cody 2009). However, this technique adds time to the overall analysis, which undercuts the speed that DART offers without chromatography.

Work has also been published previously on hydrocarbon analysis (without chromatography) using DART, where Yew et al. (2008) analyzed adult fly species of the *Drosophila* (Diptera: Drosophilidae). They were able to determine differences in the chemical composition between male and female profiles based on unsaturated hydrocarbons profiles. Besides the rapid analysis DART offers, unlike GC-MS, it does not require the insect to be killed for hydrocarbon analysis to be undertaken, so a larva/pupa/fly can be analyzed several times and followed through its life cycle.

28.2.3 Data Analysis

28.2.3.1 Multivariate Analysis

The complexity of the hydrocarbon profiles of insects, which can consist of up to 100 hydrocarbons, makes it very difficult to visually observe differences between two samples of different species or ages. Problems encountered in the analysis of hydrocarbon profiles include the following: (1) log-level differences in the percentage values of individual hydrocarbons, (2) presence of large number of zeroes, (3) auto-correlations between data points, and (4) the "constant-sum constraint" property of compositional data (Ranganathan and Borges 2011). Added to this complexity within the chemical profile is the fact that the total cuticular hydrocarbon abundance usually shows a wide spread of values for each individual. It is not uncommon to find two samples whose total abundance differs by a factor of two. Because of this issue, normalization is often applied to the data set. In most analyses, the absolute abundance of each compound (peak area) is divided by the total abundance of all compounds (total peak area) giving a relative abundance, minimizing effects of the high within-treatment variation. The relative abundance for these compounds varies from <0.5% to 40% of the total profile. Compounds with a very low abundance (<0.5%–1%), usually have a high coefficient of variance and are therefore often omitted from most analyses (Kent et al. 2007; Moore 2013). However, as these compounds may only be present in low relative abundance during a specific developmental stage, while more abundant in other stages (e.g., young larvae vs. old larvae), some researchers still omit them, whereas others decided to include those in their analyses. Clearly there is much variation in the analysis of compounds with very low relative abundance (i.e., low concentration) on the surface of blow flies. Other authors use a log-contrast method in which the logarithm of the ratio of a compound of interest to another compound is used (Blows and Allan 1998). Many studies use the method developed by Aitchison (1986), using square root transformation or log transformation with the addition of an arbitrary constant (varying from 0.01 to 0.00001) to accommodate the presence of zero values.

Classification of the data obtained in the steps highlighted above can be done in various ways by clustering methods, of which PCA and multidimensional scaling (MDS) are often used. A quick search on Google Scholar (using "cuticular hydrocarbons" + "GC-MS" + "PCA" as keywords) indicated that 178 articles used PCA to classify the samples, whereas 47 articles used MDS or nonmetric multidimensional scaling (NMDS) (using "cuticular hydrocarbons" + "GC-MS" + "MDS" OR "NMDS" as keywords). These multivariate analysis techniques are used for reducing correlated data sets from large numbers of samples, making them more manageable and allowing for any trends or similarities within the data to be revealed. This is done by grouping data according to statistically significant similarities using either principal components in PCA or dimensions in MDS (Ozturk et al. 2009; Suinyuy et al. 2013). Because of the common use of PCA, a brief summary of PCA in clustering samples based on their hydrocarbon profile is given in this section; however, for further details, readers are referred to Brereton (2003). The chromatogram of the surface extract (referred to the chemical profile) of any insect consists of several data pairs (retention time and peak area). All relative peak areas (or relative abundance, see above) in the chromatogram from each sample can be described by a linear combination of principal components, each of which is weighted by a specific set of numbers, known as the loadings (or scores). The first principal component (PC1) describes the largest amount of variation within a dataset whereas the second and sequential principal components reveal additional variation, each of successively less significance (Adam et al. 2008). Each principal component has a corresponding eigenvalue that displays the value of variance within a dataset as a percentage (Miller and Miller 2005). The eigenvalue is used to determine the number of principal components that need to be applied to the dataset. It is favorable when data can be described by the least number of principal components, because the principal components decrease in significance, leading to the exclusion of some data. Four to six principal components are usually sufficient to describe the main variations, thus greatly reducing the datasets from the original site of the input data from up to 100 data pairs (retention time and peak area) to 4-6 principal components. The output of a PCA calculation is compiled into a scatter-plot using the loadings (or scores) of two selected principal components. This enables any clustering to be easily visualized, signifying samples that are similar to each other. The Euclidean distance can be calculated to measure the distance between two points, given by Pythagoras' theorem (Brereton 2003). This technique is calculated using the loading values allowing for comparison based on all principal components.

The importance of multivariate analysis in conjunction with analytical techniques such as chromatographic data or even MS to reduce the number of variables is clear. Therefore it is not surprising that coinventor of the DART has also designed software allowing for PCA and linear discriminant analysis (LDA) to be applied directly to datasets obtained from DART spectra. This allows for trends to be visualized of different species cluster in different areas of the PCA/LDA plots.

Although PCA can be very useful, it is not without any problems (Martin and Drijfhout 2009). Log transformations or square-root transformations carried out in PCA as well as the addition of a constant to accommodate for zero values in a dataset are some of the difficulties using PCA-based analysis. Much of these procedures are based on assuming a particular data distribution, independent of data points, as well as absence of interaction between data points. However, the nonparametric approach, used in NMDS, does not require the same parametric assumptions as applied in PCA. Another nonparametric approach at the researcher's disposal is Random Forests (Ranganathan and Borges 2011), although not yet applied to the analysis of hydrocarbons. Random Forests is a data-mining algorithm with many features making it suitable for analyzing complex data sets. Two features of Random Forests are of particular interest in hydrocarbon analysis: (1) no implicit assumptions on the structure of the data points and (2) accommodation of any interactions and/or correlations between data points. As Random Forests is a nonparametric method, it can also address issues with data points varying in log scales and with hydrocarbons having zero values.

28.2.3.2 Artificial Neural Networks

Gaining a PCA plot can be time consuming as the analyst has to plot all of the combinations of PCs to visualize the best possible combination. Only two or three dimensions can be plotted, meaning information that may be important in excluded PCs can be lost. The process of running the PCA and plotting the numerous combinations of PCs has to be rerun when new data are collected. Artificial neural networks (ANNs) (Haykin 1999) are an attractive alternative to the PCA or NMDS techniques. ANNs are modeled on the functionality of biological neurons in the brain (Butcher et al. 2009). They have been widely studied and used in the field of computational intelligence because of their ability to learn and recognize characteristic features within datasets that they are exposed to. Similar to PCA, they are capable of identifying trends within datasets, but they work on a training and learning basis. Once trained, ANNs can reduce the amount of analysis time required by clarifying novel data based on their knowledge of the domain that they have acquired during training. There are numerous characteristics that make the application of ANNs appealing in many branches of science, but for the purpose of this chapter, the main advantage they hold is the ability to accommodate data with significant noise (Butcher et al. 2013). For further, detailed information on ANN, see references Haykin (1999, 2008) and Butcher et al. (2013).

There are many types of supervised and unsupervised ANNs (Butcher et al. 2009), but one of the most common types of unsupervised ANNs is self-organizing maps (SOMs) (Kohonen 1990). SOMs have been applied with a great deal of success to a broad range of applications from detecting defects in reinforced concrete (Butcher et al. 2009) to DNA classifications (Nacnna et al. 2003). A typical SOM consists of an input layer that receives input data and an output layer, onto which the input data are mapped according to its underlying characteristics (Butcher et al. 2013).

The great potential ANNs hold for data analysis has led them to be applied in the field of forensic science from the analysis of human skeletal remains (Prescher et al. 2005) to digital forensic investigation (Fei et al. 2006). They have also been used in the field of entomology. A study by Bianconi et al. (2010), applied three different types of ANNs on datasets obtained from the blow fly species *Chrysomya megacephala* Fabricius (Diptera: Calliphoridae) to predict the number of larvae reaching the adult stage based on food quantity, larval density, and duration of the immature stage. ANNs have also been used for insect hydrocarbon data in a study by Bagnéres et al. (1998) from four termite species (Blattaria: Rhinotermitidae), with the overall objective of trying to classify the caste for each insect. This approach enhanced performance in comparison to multivariate techniques. The successful use of ANNs in ageing *Lucilia sericata* Meigen, 1826 (Diptera: Calliphoridae) larvae using their hydrocarbon profiles is described in the work of Butcher et al. (2013).

28.3 SPECIES IDENTIFICATION

One of the first steps in establishing the min-PMI is to identify insects collected from the human remains (Amendt et al. 2004), as growth rate is dependent on the species. Identification is therefore needed to use the correct developmental data. The identification of each developmental stage is possible (e.g., morphological or molecular techniques), but it can be challenging (Amendt et al. 2004) and typically requires a trained entomologist or molecular biologist to carry out the procedures. The earlier highlighted studies on using hydrocarbons in chemotaxonomy offer a means to compliment the current identification techniques and have the potential to be applied to unknown specimens (Guillem et al. 2012), or to confirm uncertain species identifications. Hence, hydrocarbon analysis has the potential to complement current identification methods used for Calliphoridae, especially for the more challenging life stages, such as eggs and empty egg shell cases, first-instar larvae, and empty puparial cases, all of which are difficult to morphologically characterize. It could also be developed to assist with the identification of more difficult insect families, such as the Sarcophagidae (Diptera). It is not possible to identify the larvae of flesh flies using morphological criteria alone, and the only way to successfully identify them is to rear them to adulthood, which may cause delays in criminal investigations, or through molecular techniques (Amendt et al. 2004). Sarcophagidae females are notoriously difficult to identify, as many keys refer to male genitalia only. Therefore, if established, hydrocarbon analysis may form a useful identification tool for specimens at the larval stages, as well as adult flies.

28.3.1 Eggs

If eggs are the only insect evidence collected at a crime scene, it is likely they will be reared to third instar or ideally to adult to confirm their identification. Otherwise, the level of taxonomic certainty will be very limited. However, confident early identification of this life stage can provide valuable information and allow for more timely min-PMI estimations (Mendonça et al. 2008). They can also give an indication of whether the body has been moved because the presence of eggs of a Diptera specific to a certain geographical location, differing from where the body was discovered, might suggest the victim was killed in a different location and was moved following death (Benecke 2001; Mendonça et al. 2008).

Currently, the best method for identifying an insect egg of forensic relevance is with the aid of a Scanning Electron Microscope (SEM) (Sukontason et al. 2007a; Mendonça et al. 2008). Another technique is staining the egg with potassium permanganate and examining it with a light microscope (Sukontason et al. 2004). This technique has the advantage of being rapid while also using microscopy is less expensive method. However, up to the present time, little information is available on the validity of this technique.

There are currently very few studies that look into the surface chemistry of eggs to identify them. Nelson and Leopold (2003) published findings of the hydrocarbons extracted from the vitelline membrane surface of dechorionated eggs of the house fly, *Musca domestica* L. (Diptera; Muscidae), the New World screwworm, *Cochliomyia hominivorax* Coquerel (Diptera: Calliphoridae), the secondary screwworm, *Cochliomyia macellaria* Fabricius (Diptera: Calliphoridae), the green bottle fly, *Phaenicia (Lucilia) sericata* Meigen (Diptera: Calliphoridae), the sheep blow fly, *Lucilia cuprina* Wiedemann (Diptera: Calliphoridae), and the Mexican fruit fly, *Anastrepha ludens* Loew (Diptera: Anastrepha). This study revealed quantitative and qualitative differences within the profiles. Although there was no forensic entomology application, the work demonstrated that the chemical profiles of the eggs are species-specific. Roux et al. (2008) also examined eggs when profiling the hydrocarbons taken from three forensically important species, *C. vomitoria, C. vicina*, and *P. terraenovae*, however, they did not specify if the profiles of all the three species were characteristically different.

In depth, chemical analysis carried out by Moore (2012) examined the cuticular hydrocarbon profiles extracted from eggs and empty egg cases of *C. vicina, C. vomitoria*, and *L. sericata* and found them to be distinguishable from each other. The profiles were analyzed using GC-MS followed by PCA. As expected, all the three species yielded profiles containing a wide range of hydrocarbons, consisting of alkanes (linear, mono- and di-methyl branched) and alkenes. However, the dominating compounds were the branched hydrocarbons. The differing peak ratios as well as some unique branched compounds for the three individual species allowed for identification to be established.

28.3.2 Larvae

Larvae are the life stage that is thought of in association with decomposing remains and provide the most useful information in the estimation of the min-PMI. Many taxonomic keys have been published to allow for their identification using morphological techniques (Szpila 2010). These morphological traits are most visible in third-instar larvae and therefore this larval life stage is commonly used when applying morphological techniques (Szpila 2010). Although it is possible to identify species of Luciliinae (Szpila et al. 2013b), Chrysomyinae (Szpila et al. 2008, 2013a), and Calliphoridae (Szpila et al. 2008) in the first-instar larval stage, in general early stage larvae can be difficult to identify using morphological traits and in some cases DNA-based analyses (Ames et al. 2006a,b; Reibe et al. 2009; Mazzanti et al. 2010) are the only techniques that can be successfully applied. Moore et al. (2014) showed that hydrocarbon analysis offers a complimentary technique, as they show that species identification of *C. vicina, C. vomitoria*, and *L. sericata* can be obtained from first-instar larvae using hydrocarbon analysis (with GC-MS) when combined with PCA. Therefore, although it is possible for blow fly larvae to be correctly identified by a taxonomic expert, this technique complements the current morphological identifications and shows potential to be applied to Sarcophagidae, as it is still not possible to gain confident identification as to species in the larval stage of this family without confirmation to adult (Amendt et al. 2004; Saigusa et al. 2005).

28.3.3 Puparia

The pupa represents the longest developmental life stage of the fly life cycle (≈50% [Zehner et al. 2006]), accurately ageing them would be beneficial, but as with the other life stages, to do this the identity of any collected puparia must first be established (Sukontason et al. 2007b) and this is further complicated because of the fact that blow fly puparia are similar across species (Siriwattanarungsee et al. 2005). Current techniques in identification use microscopy to visualize morphological features (Amorim and Ribeiro 2001; Siriwattanarungsee et al. 2005; Sukontason et al. 2006a,b), or more recently, DNA-based methods (Amendt et al. 2004; Wells and Stevens 2008). DNA techniques are being successfully used (Siriwattanarungsee et al. 2005), and because the puparia is the hardened skin of the larvae, it is covered with hydrocarbons. Therefore, hydrocarbon analysis could offer a rapid and complementary method to the current identification techniques used in the field.

Ye et al. (2007) carried out a cuticular hydrocarbon study on the pupal exuviae of six necrophagous flies for taxonomic differentiation. GC-MS analysis was used to analyze the extracted hydrocarbons. Ye et al. (2007) noted the absence of alkenes in the profile of *L. sericata* pupal exuviae, which is inconsistent with results presented by Moore (2013) where six alkenes were detected for this species. Consistent results were found with regard to the methyl branched hydrocarbons. However, Moore (2012) found that using only the *n*-alkanes in subsequent PCA provided sufficient information to distinguish between *L. sericata* and *C. vicina* (the profile of *C. vomitoria* was not examined for this particular life stage). Interestingly, puparia yielded significantly fewer hydrocarbons in their profiles than larvae. Notably fewer methyl branched alkanes are present and the majority of the profile comprises *n*-alkanes, with the heavier molecular weight hydrocarbons (in particular heptacosane [C27:H], nonacosane [C29:H], and hentriacontane [C31:H]) significantly dominating the lower molecular weight *n*-alkanes. Methyl branched hydrocarbons (and alkenes) are believed to reduce the overall melting point of the compound and they are therefore required to keep the cuticular surface flexible (Morgan 2010). If the abundance of long chain *n*-alkanes is observed, it is likely the number of methyl branched alkanes/alkenes will also increase. However, puparia are rigid and do not require the need for a soft or flexible cuticle, which could explain why a simple profile made up of mainly straight chain *n*-alkanes are observed.

Armold and Regnier (1975) analyzed hydrocarbon biosynthesis in *Sarcophaga bullata* at pupariation and there are a small number of studies examining the cuticular hydrocarbon variations that occur during diapause. Yoder et al. (1992) examined the hydrocarbons of *Sarcophaga crassipalpis* Macquart, 1839 (Diptera: Sarcophagidae). They reported that puparia of diapausing pupae contained twice as many hydrocarbons in comparison to nondiapausing pupae, and hypothesized that the abundance of hydrocarbons present on diapausing puparia were either a consequence of increased deposition of select hydrocarbons or from an overall increase in deposition (Yoder et al. 1995).

28.3.4 Remnants

Puparial cases are often the only entomological evidence present in criminal investigations involving decomposed remains (Zhu et al. 2007). Many studies have been published using larvae and pupal stages for min-PMI estimations (Adams and Hall 2003; Ames et al. 2006a; Donovan et al. 2006; Greenberg 1991; Wang et al. 2008). However, puparial cases are rarely used with any great significance in criminal investigations because of the difficulty in identification. In the past decade, studies have suggested that invaluable information could be extracted from these puparial cases and hence new identification methods are being developed (Ye et al. 2007; Zhu et al. 2007). Hydrocarbon analysis can offer an accurate identification of puparial cases with the main advantage that species identification can not only be established on young puparial cases, but also on old cases (because of the stability of hydrocarbons) that have been crushed or deteriorated because of weathering, making the usual morphological characteristics difficult or impossible to visualize under a microscope.

The profiles of the empty puparial cases closely resemble that of the adult fly. The puparia extractions are only from the surface of the cuticle, whereas empty puparial case extractions will remove compounds residing within the case, which could be left from the cuticle of the emergent fly. This is believed to be the case as compounds extracted from the surface of the empty puparial case should be identical to those extracted from the surface of the puparia. However, with an increase in the number of methyl branched alkanes observed for the empty puparial cases, it is assumed these hydrocarbons are from the cuticle of the adult fly. In a study by Moore (2012), the hydrocarbons extracted from the empty puparial cases of *C. vicina*, *C. vomitoria*, and *L. sericata* were examined. PCA was applied using the *n*-alkanes alone, which gave excellent separation of species. As the species can be distinguished using only the *n*-alkanes, in this case it was not necessary to carry out full profile identification. This allows a novice in analytical chemistry to identify species using this technique, whereas an experienced forensic entomologist/taxonomist would be required if using morphological criteria.

Cuticular hydrocarbons of *L. sericata*, *Cochliomyia macellaria* (Diptera: Calliphoridae), and *Ch. rufifacies* Macquart (Diptera: Calliphoridae) were extracted from the empty puparial cases. Results showed clustering within the LDA plot, allowing for the species to be differentiated from each other. Samples from *C. macellaria* were examined from Ohio and Texas, and interestingly, distinctions within the chemical profiles were also observed within the same species potentially indicating geographic differences (unpublished).

28.4 AGE DETERMINATION

Once the identity of the insect has been established, the next stage in determining the min-PMI is based on the age of the insects present. An indication of age for the larvae is usually estimated using temperature data in combination with development data (Grassberger and Reiter 2001) for the species in question (Amendt et al. 2007).

Recent studies on Calliphoridae have shown that hydrocarbons have the potential as a possible ageing tool (Zhu et al. 2006; Roux et al. 2008). Different hydrocarbon profiles at distinguishable ages provide a reliable method complementary to the current methods used for ageing all of the stages of the blow flies life cycle.

28.4.1 Larvae

A common issue forensic entomologist are often faced with when presented with a crime scene is the dominance of third-instar larvae, leading to a large time window corresponding to larval age at discovery. Being able to age third-instar larvae is vital for an accurate min-PMI to be established.

Recent papers have suggested that hydrocarbons could be used to establish the age of forensically important blow fly larvae as their profiles change over time (Zhu et al. 2006; Roux et al. 2008). Obtained results showed that hydrocarbon analysis could potentially complement the current ageing techniques, allowing for more accurate min-PMI estimations.

Roux et al. (2008) examined the cuticular hydrocarbons of three forensically important blow flies (*C. vicina*, *C. vomitoria*, and *P. terraenovae*). They examined the ontogenetic study of these three species from egg through to 8-day-old adult flies. Short-chain hydrocarbons present in the profiles of larvae and postfeeding evolved into long-chain compounds for the pupae and adult flies. The methyl branched alkanes were also seen to be more abundant in the immature stages of the larvae, with a substantial decrease as they became postfeeding. Zhu et al. (2006) presented the hydrocarbon profiles from the developing larvae of *Ch. rufifacies* with the overall aim of establishing if this technique could be used to determine the larval age. The hydrocarbons were analyzed using GC-FID and GC-MS. The statistical analysis consisted of using the *peak ratios* of nonacosane (C29:H) and another eight selected peaks, which appeared to increase with larval age. They were then able to model their age using exponential or power functions.

However, results presented by Moore et al. (2013) revealed that high odd-chain-length alkanes (e.g., nonacosane, hentriacontane, and tritriacontane) significantly increased with age. Hence, the method of relating peak areas to nonacosane, used as a reference compound or marker, was not suitable for this study, and hence PCA was applied to the data set instead. Recently published work by Butcher et al. (2013) highlighted the great potential ANN analysis holds for this type of data after applying this artificial intelligence technique to hydrocarbon data to age *L. sericata* larvae (in comparison to PCA) (Moore et al. (2013). They presented enhanced results in comparison to PCA, with where the ANNs correctly classified the data with accuracy scores of 80.8% and 87.7%. Further inspection of these results shows the ANNs to confuse two consecutive days that are of the same life stage (final 2 days of postfeeding) and as a result, their chemical profiles are very similar. The grouping of these 2 days into one class further improved results where accuracy scores 89% and 97.5% were obtained.

The four main factors believed to be influential for the composition of hydrocarbon pools are development, genetics, physiological state, and environmental conditions (Blomquist et al. 1987; Espelie and Payne 1991; Roux et al. 2008). The changes observed during blow fly development may be affected by the environment they are exposed to. Larvae develop in warm, humid conditions (decomposing remains) and in this stage of their life cycle (pre- and postfeeding), they yield profiles consisting of a mixture of low and high molecular weight hydrocarbons. Moore (2013) showed that the higher molecular weight hydrocarbons become more abundant as the larvae age. When the larvae become older and gradually move into the postfeeding stage of the life cycle they move away from the source of food and seek a site for pupation, exposing them to a drier environment. Therefore, they have a greater need for conserving water than when they are at a younger age, where they are usually submerged within their food source that is warm and moist (Toolson and Kupersimbron 1989; Tregenza et al. 2000; Ferveur 2005). The display of higher boiling point *n*-alkanes mixed in with alkenes and methyl branched hydrocarbons has also been linked to flexibility of the cuticle (Morgan 2010). To help flexibility in the larvae's cuticle, it will need a composition of methyl branched alkanes and alkenes, which have lower melting points compared to the straight-chain alkanes.

28.4.2 Pupae

Little work has been carried out on ageing forensically important puparia in comparison to larvae, but this is of great importance as they comprise 50% of the immature developmental cycle (Zehner et al. 2006). Therefore, this life stage may give valuable information when estimating the min-PMI. Techniques are being developed for ageing this life stage using gene expression (Zehner et al. 2006, 2009; Tarone et al. 2011; Boehme et al. 2013; Davies and Harvey 2013). There is also ongoing work using low-energy X-rays to image developing flies through the outer case (Martin Hall et al., pers. comm.). This technique has enabled pupae to be imaged multiple times during their development with no adverse effects.

There are studies on hydrocarbon profiles of puparia and it has not yet been used for ageing this life stage. Roux et al. (2008) examined pupae of *C. vicina* and reported that they had difficulty in detecting hydrocarbons for this species in the puparia life stage. They therefore had to extract hydrocarbons from the nymph membrane after the puparium was opened. Results from Moore (2013) appeared to be in agreement, as very few hydrocarbons were present in comparison to the other life stages but *n*-alkanes were detectable and although it was not possible to establish the age of the puparia, enough compounds were extracted from the profile, allowing for identification to be established between three species of Calliphoridae.

Given the developments in gene expression analysis and the results it gives on the potential for ageing puparia, hydrocarbon analysis may be better applied to other life stages, where other ageing techniques have been less successful.

28.4.3 Adults

When a forensic entomologist is presented with an indoor crime scene where adult flies have accumulated (e.g., by a window), it would be advantageous if the age of the flies could be established. This could determine the age of the older flies that are potentially the initial colonizers of the body or the younger flies that have developed on the body and hatched at the scene.

The only morphological changes that occur in a fly occur within the first few hours after it has emerged. As it emerges from the puparial case, flies have an unusual appearance. There is a protruding region of the head of the fly called the ptilinum, which becomes inflated with hemolymph and enables the emerging fly to push their way out of the case. This feature retracts into the head a few hours after emergence forming the ptilinal suture. The wings of the fly are also crumpled and the body is brown/gray. However, once the ptilinum sinks back into the facial structure, the wings are fully formed and the body gains color.

Hydrocarbon analysis on mature insects such as grasshoppers (Orthoptera) (Tregenza et al. 2000) and cockroaches (Brown et al. 2000) has been carried out to establish the age and sex. Hydrocarbon studies have also been carried out on several Diptera (Mpuru et al. 2001; Urech et al. 2005; Hugo et al. 2006) and Calliphoridae (Trabalon et al. 1992), but few have been carried out in relation to its importance in forensic entomology (Roux et al. 2008). Trabalon et al. (1988) was the first to study cuticular hydrocarbons of a forensically important blow fly in relation to age and sexual behavior. Using female *C. vomitoria* adults as a model, they were able to detect considerable differences within the profiles of young (up to 12 hours after eclosion) and older (after 24 hours) individuals. These results were the first to show differences in age, sexual attractiveness, and receptivity for this blow fly species. A later paper by Trabalon et al. (1992) examined the same species of blow fly in relation to age and sex, comparing individuals from 3 to 120 hours postemergence. The profiles of both sexes contained *n*-alkanes, mono- and dimethyl alkanes ranging from 20 to 31 carbon atoms. However, alkenes were only detected in females, with males holding fewer hydrocarbons in their profiles. Roux et al. (2008) carried out cuticular hydrocarbon analysis on three forensically important blow flies (*C. vicina*, *C. vomitoria*, and *P. terraenovae*), using GC-FID, producing results that have shown the potential to use hydrocarbons for a forensic entomological application. The discrimination analysis they carried out allowed for clear differentiations between the life stages (larvae, postfeeding larvae, puparia, and adult fly), as well as differences within the chemical profiles of adult flies up to 8 days old, with the main differences being observed between young (<24 hour) and old (>24 hour) individuals. Moore (2012) observed similar results to Trabalon et al (1992) and Roux et al (2008) who investigated the changes within the hydrocarbon profiles at varies ages for *L. sericata*, *C. vicina*, and *C. vomitoria*. In all the three species, the chemical profiles changed drastically between the young (day 1) and older (<day 5) adults. Further examination at older ages (days 10, 20, and 30) showed small differences within the profiles, allowing the adults of *C. vicina* and *C. vomitoria* to be aged up to 20 days old. Therefore, a differentiation between flies that have developed on the cadaver and are newly emerged, to older flies that have been attracted by odor can be established (Roux et al. 2008).

Promising results were obtained when the hydrocarbons were extracted from days 1, 5, 10, 20, and 30 for female individuals of both *C. macellaria* and *Ch. rufifacies*. Both species clustered into the separate age groups, allowing for days 1, 5, and 10 to be established, and *C. macellaria* clustering days 20 and 30 (Jennifer Pechal, pers. comm.). Further work is in the process to establish if field caught females of both species can be added to the PCA model to establish ageing.

28.5 CONCLUSION

Overall, a wealth of information can be obtained from cuticular hydrocarbon analysis of forensically important Calliphoridae. The field of forensic entomology could greatly benefit from a technique that facilitates identification as well as age, from one chemical profile of an insect. The methodology is simple to execute, allowing a nonexpert to follow and perform the extraction procedures. This chapter highlighted the potential hydrocarbons hold of being used to determine the age of the insect in each of the life stages. The combination of using the analytical technique of GC-MS (for chemical identification using cuticular hydrocarbons) and statistical analysis means the identity and an indication of age can be established for the majority of the life stages of some forensically important Calliphoridae.

The next stage to validate hydrocarbon analysis fully in forensic entomology is to obtain some field-based samples to test the practicalities of the technique. All of the experimental data reviewed in this chapter were obtained under standard laboratory conditions. Future work should examine the stability of the hydrocarbons under various temperatures (stable and fluctuating) (Niederegger et al. 2010), and the effects of food sources, as this will vary when larvae feed on a cadaver as they tend to feed from the head downward (if no wounds are present), therefore ingesting various bodily tissues and organs. Most studies published in the field of forensic entomology use liver as a food source to rear the larvae, but studies have shown their growth can be accelerated when feeding on other tissues such as heart, lung, kidney, and brain (Kaneshrajah and Turner 2004). It would also be useful to study the effects that toxicological substances may have on the hydrocarbon profiles. If larvae feed on a cadaver that ingested drugs before death, this can have an effect on their development and can either accelerate or retard growth (Bourel et al. 1999; Introna et al. 2001), which could have potential implications on reliable min-PMI estimations.

A hydrocarbon database of all forensically important insects could be created and, for the wider picture, a chemotaxonomy database of insects in the United Kingdom and elsewhere would be highly beneficial for identification purposes. More work, both in chemical analysis (e.g., using DART-MS) and data analysis (e.g., ANN), is needed especially using closely related blow flies and for the flesh flies.

REFERENCES

Adam, C.D., S.L. Sherratt, and V.L. Zholobenko. 2008. Classification and individualization of black ballpoint pen inks using principal component analysis of UV-vis absorption spectra. *Forensic Science International* 174: 16–25.

Adams, Z. and M.J.R. Hall. 2003. Methods used for the killing and preservation of blowfly larvae, and their effect on post-mortem larval length. *Forensic Science International* 138: 50–61.

Aitchison, J. 1986. *The Statistical Analysis of Compositional Data. Monographs in Statistics and Applied Probability*. London: Chapman & Hall.

Akino, T., K. Yamamura, S. Wakamura, and R. Yamaoka. 2004. Direct behavioral evidence for hydrocarbons as nestmate recognition cues in *Formica japonica* (Hymenoptera: Formicidae). *Applied Entomology and Zoology* 39: 381–387.

Amendt, J., C.P. Campobasso, E. Gaudry, C. Reiter, H.N. LeBlanc, and M.J.R. Hall. 2007. Best practice in forensic entomology—Standards and guidelines. *International Journal of Legal Medicine* 121: 90–104.

Amendt, J., R. Krettek, and R. Zehner. 2004. Forensic entomology. *Naturwissenschaften* 91: 51–65.

Ames, C., B. Turner, and B. Daniel, 2006a. Estimating the post-mortem interval (I): The use of genetic markers to aid in identification of Dipteran species and subpopulations. *International Congress Series* 1288: 795–797.

Ames, C., B. Turner, and B. Daniel. 2006b. The use of mitochondrial cytochrome oxidase I gene (COI) to differentiate two UK blowfly species: *Calliphora vicina Calliphora vomitoria Forensic. Science International* 164: 179–82.

Amorim, J.A. and O.B. Ribeiro. 2001. Distinction among the puparia of three blowfly species (Diptera: Calliphoridae) frequently found on unburied corpses. *Memórias do Instituto Oswaldo Cruz* 96: 781–4.

Anyanwu, G.I., D.H. Molyneux, and A. Phillips. 2000. Variation in cuticular hydrocarbons among strains of the *Anopheles gambiae* sensu stricto by analysis of cuticular hydrocarbons using gas liquid chromatography of larvae. *Memórias do Instituto Oswaldo Cruz* 95: 295–300.

Armold, M.T. and F.E. Regnier. 1975. Stimulation of hydrocarbon biosynthesis by ecdysterone in the flesh fly *Sarcophaga bullata*. *Journal of Insect Physiology* 21: 1581–1586.

Bagneres, A.G., G. Riviere, and J.L. Clement. 1998. Artificial neural network modeling of caste odor discrimination based on cuticular hydrocarbons in termites. *Chemoecology* 8: 201–209.

Benecke, M. 2001. A brief history of forensic entomology. *Forensic Science International* 120: 2–14.

Bernier, U.R., D.A. Carlson, and C.J. Geden. 1998. Analysis of the cuticular hydrocarbons from parasitic wasps of the genus *Muscidifurax*. *Journal of the American Society for Mass Spectrometry* 9: 320–332.

Bianconi, A., C.J. Von. Zuben, A.B. de S. Serapião, and J.S. Govone. 2010. Artificial neural networks: A novel approach to analysing the nutritional ecology of a blowfly species, *Chrysomya megacephala Journal of Insect Science* 10: 1–18.

Blomquist, G.J. and A.G. Bagnères. 2010. *Insect Hydrocarbons: Biology, Biochemistry, and Chemical Ecology*. Cambridge, United Kingdom: Cambridge University Press.

Blomquist, G.J., D.R. Nelson, and M. de Renobales. 1987. Chemistry, biochemistry, and physiology of insect cuticular lipids. *Archives of Insect Biochemistry and Physiology* 6: 227–265.

Blows, M.W. and R.A. Allan, 1998. Levels of mate recognition within and between two Drosophila species and their hybrids. *The American Naturalist* 152: 826–837.

Boehme, P., P. Spahn, J. Amendt, and R. Zehner. 2013. The analysis of temporal gene expression to estimate the age of forensically important blow fly pupae: Results from three blind studies. *International Journal of Legal Medicine* 128: 565–573. doi 10.1007/s00414-013-0922-8.

Bourel, B., V. Hedouin, L. Bouyer-Martin, A. Becart, G. Tournel, M. Deveaux, and D Gosset. 1999. Effects of morphine in decomposing bodies on the development of *Lucilia sericata* (Diptera: Calliphoridae). *Journal of Forensic Sciences* 44: 354–358.

Brereton, R.G. 2003. *Chemometrics: Data Analysis for the Laboratory and Chemical Plant*. Chichester, United Kingdom: Wiley.

Brown, W.V., R. Morton, and J.P. Spradbery. 1992. Cuticular hydrocarbons of the Old World screw-worm fly, *Chrysomya bezziana* Villeneuve (Diptera: Calliphoridae). Chemical characterization and quantification by age and sex. *Comparative Biochemistry and Physiology. Part B: Comparative Biochemistry* 101: 665–671.

Brown, W.V., H.A. Rose, M.J. Lacey, and K. Wright. 2000. The cuticular hydrocarbons of the giant soil-burrowing cockroach *Macropanesthia rhinoceros* Saussure (Blattodea: Blaberidae: Geoscapheinae): Analysis with respect to age, sex and location. *Comparative Biochemistry and Physiology. Part B: Biochemistry and Molecular Biology* 127: 261–277.

Butcher, J.B., M. Lion, C.R. Day, and P.W. Haycock. 2009. A low frequency electromagentic probe for detection of corrosion in steel-reinforced concrete. In *Concrete Solutions*, edited by M. Grantham, C. Majorana, and V. Salomoni, pp. 417–424. Leiden, The Netherlands: CRC Press.

Butcher, J.B., H.E. Moore, C.R. Day, C.D. Adam, and F.P. Drijfhout. 2013. Artificial neural network analysis of hydrocarbon profiles for the ageing of *Lucilia sericata* for post mortem interval estimation. *Forensic Science International* 232: 25–31.

Cainé, L.M., F.C. Real, M.I. Saloña-Bordas, M.M. de Pancorbo, G. Lima, T. Magalhães, and F. Pinheiro. 2009. DNA typing of Diptera collected from human corpses in Portugal. *Forensic Science International* 184: 21–23.

Chapman, R.F. and K.E. Espelies. 1995. Use of cuticular lipids in grasshopper taxonomy: A study of variation in *Schistocerca shoshone* (Thomas). *Biochemical Systematics and Ecology* 23: 383–398.

Chen C.S., M.S. Mulla, R.B. March, and J.D. Chaney. 1990. Cuticular hydrocarbon patterns in *Culex quinquefasciatus* as influenced by age and sex, and geography. *Bulletin of the Society for Vector Ecology* 15: 129–139.

Cody, R.B. 2009. Observation of molecular ions and analysis of nonpolar compounds with the Direct Analysis in Real Time ion source. *Analytical Chemistry* 81: 1101–1107.

Cody, R.B. and J.A. Laramee. 2005. Versatile new ion source for the analysis of materials in open air under ambient conditions. *Journal of Analytical Chemistry* 77: 2297–2302.

Cody, R.B., J.A. Laramee, J.M. Nilles, and H.D. Durst. 2007. *Direct Analysis in Real Time (DART) Mass Spectrometry. Applications Notebook.* Peabody, MA: JEOL.

Davies, K. and M.L. Harvey. 2013. Internal morphological analysis for age estimation of blow fly pupae (Diptera: Calliphoridae) in postmortem interval estimation. *Journal of Forensic Sciences* 58: 79–84.

Desena M.L., J.M. Clark, J.D. Edman, S.B. Symington, T.W. Scott, G.G. Clark, and T.M. Peters. 1999. Potential for aging female *Aedes aegypti* (Diptera: Culicidae) by gas chromatographic analysis of cuticular hydrocarbons, including a field evaluation. *Journal of Medical Entomology* 36: 811–823.

Desena, M.L., J.D. Edman, J.M. Clark, S.B. Symington, and T.W. Scott. 1999. *Aedes aegypti* (Diptera: Culicidae). Age determination by cuticular hydrocarbon analysis of female legs. *Journal of Medical Entomology* 36: 824–830.

Donovan, S.E., M.J.R. Hall, B.D. Turner, and C.B. Moncrieff. 2006. Larval growth rates of the blowfly, *Calliphora vicina*, over a range of temperatures. *Medical and Veterinary Entomology* 20: 106–114.

Drijfhout, F.P. 2010. Cuticular hydrocarbons: A new tool in forensic entomology? In *Current Concepts in Forensic Entomology*, edited by J. Amendt, C.P. Campobasso, M.L. Goff, and M. Grassberger, pp.179–204. Dordrecht, The Netherlands: Springer.

Espelie, K.E. and J.A. Payne. 1991. Characterization of the cuticular lipids of the larvae and adults of the pecan weevil, *Curculio caryae. Biochemical Systematics and Ecology* 19: 127–132.

Everaerts, C., J.P. Farine, and R. Brossut. 1997. Changes of species specific cuticular hydrocarbon profiles in the cockroaches *Nauphoeta cinerea* and *Leucophaea maderae* reared in heterospecific groups. *Behavioral Ecology and Sociobiology* 85: 145–150.

Fei, B.K.L., J.H.P. Eloff, M.S. Olivier, and H.S. Venter. 2006. The use of self-organising maps for anomalous behaviour detection in a digital investigation. *Forensic Science International* 162: 33–37.

Ferveur, J.F. 2005. Cuticular hydrocarbons: Their evolution and roles in Drosophila pheromonal communication. *Behavior Genetics* 35: 279–295.

Fremdt, H., K. Szpila, J. Huijbregts, A. Lindstrom, R. Zehner, and J. Amendt. 2012. *Lucilia silvarum* Meigen, 1826 (Diptera: Calliphoridae): A new species of interest for forensic entomology in Europe. *Forensic Science International* 222: 335–339.

Gibbs, A.G. 1998. Water-proofing properties of cuticular lipids. *American Zoologist* 38: 471–482.

Gibbs, A.G. and E.L. Crockett. 1998. The biology of lipids: Integrative and comparative perspectives. *Integrative and Comparative Biology* 38: 265–267.

Gibbs, A. and J.G. Pomonist. 1995. Physical properties of insect cuticular hydrocarbons: The effects of chain length, methyl-branching and unsaturation. *Comparative Biochemistry and Physiology* 112B: 243–249.

Grange, A.H. and G.W. Sovocool. 2011. Detection of illicit drugs on surfaces using Direct Analysis in Real Time (DART) time-of-flight mass spectrometry. *Rapid Communications in Mass Spectrometry* 25: 1271–1281.

Grassberger, M. and C. Reiter. 2001. Effect of temperature on *Lucilia sericata* (Diptera: Calliphoridae) development with special reference to the isomegalen- and isomorphen-diagram. *Forensic Science International* 120: 32–36.

Greenberg, B. 1991. Flies as forensic indicators. *Journal of Medical Entomology* 28: 565–577.

Guillem, R.M., F.P. Drijfhout, and S.J. Martin. 2012. Using chemo-taxonomy of host ants to help conserve the large blue butterfly. *Biological Conservation* 148: 39–43.

Hajslova, J., T. Cajka, and L. Vaclavik. 2011. Challenging applications offered by Direct Analysis in Real Time (DART) in food-quality and safety analysis. *Trends in Analytical Chemistry* 30: 204–218.

Harvey, M.L., I.R. Dadour, and S. Gaudieri, 2003. Mitochondrial DNA cytochrome oxidase I gene: Potential for distinction between immature stages of some forensically important fly species (Diptera) in Western Australia. *Forensic Science International* 131: 134–139.

Haverty, M.I., M.S. Collins, L.J. Nelson, and B.L. Thorne. 1997. Cuticular hydrocarbons of termites of the British Virgin Islands. *Journal of Chemical Ecology* 23: 927–964.

Haykin, S. 1999. *Neural Networks: A Comprehensive Foundation.* Upper Saddle River, NJ: Prentice Hall.

Haykin, S. 2008. *Neural Networks and Learning Machines.* Upper Saddle River, NJ: Prentice Hall.

Heath, R.R. and D. Dueben. 1998. Analytical and preparative gas chromatography. In *Methods in Chemical Ecology*, edited by J.G. Millar and K.F. Haynes, pp.85–126. Dordrecht, The Netherlands: Kluwer.

Howard, R.W. and G.J. Blomquist. 2005. Ecological, behavioral and biochemical aspects of insect hydrocarbons. *Annual Review of Entomology* 50: 371–393.

Hugo, L.E., B.H. Kay, G.K. Eaglesham, N. Holling, and P.A. Ryan. 2006. Investigation of cuticular hydro-carbons for determining the age and survivorship of Australasian mosquitoes. *The American Journal of Tropical Medicine and Hygiene* 74: 462–474.

Introna, F., C.P. Campobasso, and L.M. Goff. 2001. Entomotoxicology. *Forensic Science International* 120: 42–47.

Jackson, L.L. and R.J. Bartelt. 1986. Cuticular hydrocarbons of sex. *Insect Biochemistry* 16: 433–439.

Kaneshrajah, G. and B. Turner. 2004. *Calliphora vicina* larvae grow at different rates on different body tissues. *International Journal of Legal Medicine* 118: 242–244.

Kent, C., R. Azanchi, B. Smith, A. Chu, and J. Levine. 2007. A model-based analysis of chemical and temporal patterns of cuticular hydrocarbons in male *Drosophila melanogaster*. *PLoS One* e962: 1–21.

Kim, H.J., W.S. Baek, and Y.P. Jang. 2011. Identification of ambiguous cubeb fruit by DART-MS-based finger-printing combined with principal component analysis. *Food Chemistry* 129: 1305–1310.

Kohonen, T. 1990. Self-organising map. *Proceedings of the IEEE* 78: 1464–1480.

Lavine, B.K. and M.N. Vora. 2005. Identification of Africanized honeybees. *Journal of Chromatography A* 1096: 69–75.

Lockey, K. 1991. Insect hydrocarbon classes: Implications for chemotaxonomy. *Insect Biochemistry* 21: 91–97.

Lockey, K. 1992. Insect hydrocarbon chemotaxonomy: Cuticular hydrocarbons of adult and larval epiphysa spe-cies blanchard and adult *Onymacris unguicularis* (HAAG) (Tenebrionidae: Coleoptera). *Comparative Biochemistry and Physiology. Part B: Comparative Biochemistry* 102: 451–470.

Lommelen, E., C.A. Johnson, F.P. Drijfhout, J. Billen, T. Wenseleers, and B. Gobin. 2006. Cuticular hydrocar-bons provide reliable cues of fertility in the ant *Gnamptogenys striatula*. *Journal of Chemical Ecology* 32: 2023–2034.

Malgorn, Y. and R. Coquoz. 1999. DNA typing for identification of some species of Calliphoridae. An interest in forensic entomology. *Forensic Science International* 102: 111–119.

Marican, C., L. Duportets, S. Birman, and J.M. Jallon. 2004. Female-specific regulation of cuticular hydrocar-bon biosynthesis by dopamine in Drosophila melanogaster. *Insect Biochemistry and Molecular Biology* 34: 823–830.

Martin, S.J. and F.P. Drijfhout. 2009. How reliable is the analysis of complex cuticular hydrocarbon profiles by multivariate statistical methods? *Journal of Chemical Ecology* 35: 375–382.

Martin, S.J., H. Helanterä, and F.P. Drijfhout. 2008a. Colony-specific hydrocarbons identify nest mates in two species of Formica ant. *Journal of Chemical Ecology* 34: 1072–1080.

Martin, S.J., H. Helanterä, and F.P. Drijfhout. 2008b. Evolution of species-specific cuticular hydrocarbon pat-terns in Formica ants. *Biological Journal of the Linnaean Society* 95: 131–140.

Martin, S.J., E. Vitikainen, H. Helanterä, and F.P. Drijfhout. 2008c. Chemical basis of nest-mate discrimination in the ant *Formica exsecta*. *Proceedings of the Royal Society B-Biological Sciences* 275: 1271–1278.

Mazzanti, M., F. Alessandrini, A. Tagliabracci, J.D. Wells, and C.P. Campobasso. 2010. DNA degradation and genetic analysis of empty puparia: genetic identification limits in forensic entomology. *Forensic Science International* 195: 99–102.

McMaster, M., and C. McMaster. 1998. *GC/MS: A Practical User's Guide*. Hoboken, NJ: Wiley-VCH.

Mendonça, P.M., J.R. dos Santos-Mallet, R.P. de Mello, L. Gomes, and M.M. de Carvalho Queiroz. 2008. Identification of fly eggs using scanning electron microscopy for forensic investigations. *Micron* 39: 802–807.

Millar, J.G. and J.J. Sims. 1998. Preparation cleanup and preliminary fractionation of extracts. In *Methods in Chemical Ecology*, edited by J.G. Millar and K.F. Haynes, pp. 85–126. Dordrecht, The Netherlands: Kluwer.

Miller, J.N. and J.C. Miller. 2005. *Statistics and Chemometrics for Analytical Chemistry*. Harlow, United Kingdom: Pearson.

Moore, H.E. 2012. Analysis of cuticular hydrocarbons in forensically important blowflies using mass spec-trometry and its application in post mortem interval estimations. Thesis. Staffordshire, United Kingdom, Keele University.

Moore, H.E., C.D. Adam, and F.P. Drijfhout. 2013. Potential use of hydrocarbons for aging *Lucilia sericata* blowfly larvae to establish the postmortem interval. *Journal of Forensic Sciences* 58: 404–412.

Moore, H.E., C.D. Adam, and F.P. Drijfhout. 2014. Identifying 1st instar larvae for three forensically important blow-fly species using "fingerprint" cuticular hydrocarbon analysis. *Forensic Science International* 240: 48–53.

Morgan, D. 2010. *Biosynthesis in Insects*. Cambridge, United Kingdom: The Royal Society of Chemistry.

Mpuru, S., G.J. Blomquist, C. Schal, M. Roux, M. Kuenzli, G. Dusticier, J.L. Clément, and A.G. Bagnères. 2001. Effect of age and sex on the production of internal and external hydrocarbons and pheromones in the housefly, *Musca domestica. Insect Biochemistry and Molecular Biology* 31: 139–155.

Nacnna, T., R.A. Bress, and M.J. Embrechts. 2003. DNA classifications with self-organizing maps (SOMs), pp. 151–154. *IEEE International Workshop on Soft Computing in Industrial Applications*, Binghamton, NY.

Nelson, D. and R. Leopold. 2003. Composition of the surface hydrocarbons from the vitelline membranes of dipteran embryos. *Comparative Biochemistry and Physiology. Part B: Biochemistry and Molecular Biology* 136: 295–308.

Niederegger, S., J. Pastuschek, and G. Mall. 2010. Preliminary studies of the influence of fluctuating temperatures on the development of various forensically relevant flies. *Forensic Science International* 199: 72–78.

Oliveira, A.R., A. Farinha, M.T. Rebelo, and D. Dias. 2011. Forensic entomology: Molecular identification of blowfly species (Diptera: Calliphoridae) in Portugal. *Forensic Science International: Genetics Supplement Series* 3: e439–e440.

Ozturk, I., H. Orhan, and Z. Dogan. 2009. Comparison of principal component analysis and multidimensional scaling methods for clustering some honey bee genotypes. *Journal of Animal and Veterinary Advances* 8: 413–419.

Page, M., L.J. Nelson, G.J. Blomquist, and S.J. Seybold. 1997. Cuticular hydrocarbons as chemotaxonomic characters of pine engraver beetles (Ips spp.) in the grandicollis Subgeneric Group. *Journal of Chemical Ecology* 23: 1053–1099.

Pawliszyn, J. 1997. *Solid Phase Microextraction: Theory and Practice*. New York: Wiley-VCH.

Prescher, A., A. Meyers, and D.G.V. Keyserlingk. 2005. Neural net applied to anthropological material: A methodical study on the human nasal skeleton. *Annals of Anatomy–Anatomischer Anzeiger* 187: 261–269.

Ranganathan, Y. and R.M. Borges.2011. To transform or not to transform: That is the dilemma in the statistical analysis of plant volatiles. *Plant Signalling and Behaviour* 6: 113–116.

Ratcliffe, S.T., D.W. Webb, R.A. Weinzievr, and H.M. Robertson. 2003. PCR-RFLP identification of Diptera (Calliphoridae, Muscidae and Sarcophagidae): A generally applicable method. *Journal of Forensic Sciences* 48: 783–785.

Reibe, S., J. Schmitz, and B. Madea. 2009. Molecular identification of forensically important blowfly species (Diptera: Calliphoridae) from Germany. *Parasitology Research* 106: 257–261.

Roux, O., C. Gers, and L. Legal. 2006. When, during ontogeny, waxes in the blowfly (Calliphoridae) cuticle can act as phylogenetic markers. *Biochemical Systematics and Ecology* 34: 406–416.

Roux, O., C. Gers, and L. Legal. 2008. Ontogenetic study of three Calliphoridae of forensic importance through cuticular hydrocarbon analysis. *Medical and Veterinary Entomology* 22: 309–317.

Saigusa, K., M. Takamiya, and Y. Aoki. 2005. Species identification of the forensically important flies in Iwate prefecture, Japan based on mitochondrial cytochrome oxidase gene subunit I (COI) sequences. *Legal Medicine* 7: 175–178.

Schroeder, H., H. Klotzbach, S. Elias, C. Augustin, and K. Pueschel. 2003. Use of PCR–RFLP for differentiation of calliphorid larvae (Diptera, Calliphoridae) on human corpses. *Forensic Science International* 132: 76–81.

Siriwattanarungsee, S., K.L. Sukontason, B. Kuntalue, S. Piangjai, J.K. Olson, and K. Sukontason. 2005. Morphology of the puparia of the housefly, *Musca domestica* (Diptera: Muscidae) and blowfly, *Chrysomya megacephala* (Diptera: Calliphoridae). *Parasitology research* 96: 166–170.

Sperling, F.A., G.S. Anderson, and D.A. Hickey. 1994. A DNA-based approach to the identification of insect species used for postmortem interval estimation. *Journal of Forensic Sciences* 39: 418–427.

Steiner, R.R. and R.L. Larson. 2009. Validation of the direct analysis in real time source for use in forensic drug screening. *Journal of Forensic Sciences* 54: 617–622.

Steiner, S., N. Hermann, and J. Ruther. 2006. Characterization of a female-produced courtship pheromone in the parasitoid *Nasonia vitripennis. Journal of Chemical Ecology* 32: 1687–1702.

Stevens, J. and R. Wall. 2001. Genetic relationships between blowflies (Calliphoridae) of forensic importance. *Forensic Science International* 120: 116–123.

Suinyuy, T.N., J.S. Donaldson, and S.D. Johnson. 2013. Variation in the chemical composition of cone volatiles within the African cycad genus Encephalartos. *Phytochemistry* 85: 82–91.

Sukontason, K.L., N. Bunchu, T. Chaiwong, B. Kuntalue, and K. Sukontason. 2007. Fine structure of the egg-shell of the blow fly, *Lucilia cuprina*. *Journal of Insect Science* 7: 1–8.

Sukontason, K.L., C. Kanchai, S. Piangjai, W. Boonsriwong, N. Bunchu, D. Sripakdee, T. Chaiwong, B. Kuntalue, S. Siriwattanarungsee, and K. Sukontason.2006. Morphological observation of puparia of *Chrysomya nigripes* (Diptera: Calliphoridae) from human corpse. *Forensic Science International* 161: 15–19.

Sukontason, K.L., P. Narongchai, C. Kanchai, K. Vichairat, S. Piangjai, W. Boonsriwong, N. Bunchu et al. 2006. Morphological comparison between *Chrysomya rufifacies* (Macquart) and *Chrysomya villeneuvi* Patton (Diptera: Calliphoridae) puparia, forensically important blow flies. *Forensic Science International* 164: 230–234.

Sukontason, K.L., R. Ngern-Klun, D. Sripakdee, and K. Sukontason. 2007. Identifying fly puparia by clearing technique: Application to forensic entomology. *Parasitology Research* 101: 1407–1416.

Sukontason, K., K.L. Sukontason, S. Piangjai, N. Boonchu, H. Kurahashi, M. Hope, and J.K. Olson. 2004. Identification of forensically important fly eggs using a potassium permanganate staining technique. *Micron* 35: 391–395.

Szpila, K. 2010. Key for the identification of third instars of European blowflies (Diptera: Calliphoridae) of forensic importance. In *Current Concepts in Forensic Entomology*, edited by J. Amendt, C.P. Campobasso, M. Goff, and M. Grassberger, pp. 43–56. Dordrecht, The Netherlands: Springer.

Szpila, K., M.J.R. Hall, T. Pape, and A Grzywacz. 2013a. Morphology and identification of first instars of the European and Mediterranean blowflies of forensic importance. *Part II. Luciliinae. Medical and Veterinary Entomology* 27: 349–366.

Szpila, K., M.J.R. Hall, K.L. Sukontason, and T.I. Tantawi. 2013b. Morphology and identification of first instars of the European and Mediterranean blowflies of forensic importance. *Part I: Chrysomyinae. Medical and Veterinary Entomology* 27: 181–193.

Szpila, K., T. Pape, and A Rusinek. 2008. Morphology of the first instar of *Calliphora vicina* and *Phormia regina Lucilia illustris* (Diptera, Calliphoridae). *Medical and Veterinary Entomology* 22: 16–25.

Tarone, A.M. and D.R. Foran. 2011. Gene expression during blow fly development: Improving the precision of age estimates in forensic entomology. *Journal of Forensic Sciences* 56: S112–S122.

Tissot, M., D.R. Nelson, and D.M. Gordon. 2001. Qualitative and quantitative differences in cuticular hydro-carbons between laboratory and field colonies of *Pogonomyrmex barbatus. Comparative Biochemistry and Physiology. Part B: Biochemistry and Molecular Biology* 130: 349–358.

Toolson, E.C. and R. Kuper-Simbron. 1989. Laboratory evolution of epicuticular hydrocarbon composition and cuticular permeability in *Drosophila pseudoobscura*: Effects on sexual dimorphism and thermal-acclimation ability. *Evolution* 43: 468–473.

Trabalon, M., M. Campan, J.L. Clement, C. Lange, and M.T. Miquel. 1992. Cuticular hydrocarbons of *Calliphora vomitoria* (Diptera): Relation to age and sex. *General and Comparative Endocrinology* 85: 208–216.

Trabalon, M., M. Campan, J.L. Clément, B. Thon, C. Lange, and J. Lefevre. 1988. Changes in cuticular hydro-carbon composition in relation to age and sexual behaviour in the female *Calliphora vomitoria* (Dipetra). *Behavioural Processes* 17: 107–115.

Tregenza, T., S.H. Buckley, V.L. Pritchard, and R.K. Butlin. 2000. Inter- and Intra-population effects of sex and age on epicuticular composition of meadow grasshopper, *Chorthippus parallelus. Journal of Chemical Ecology* 26: 257–278.

Urech, R., G.W. Brown, C.J. Moore, and P.E. Green. 2005. Cuticular hydrocarbons of buffalo fly, *Haematobia exigua*, and chemotaxonomic differentiation from horn fly, *H. irritans. Journal of Chemical Ecology* 31: 2451–2461.

Vincent, S., J.M. Vian, and M.P. Carlotti. 2000. Partial sequencing of the cytochrome oxidase b subunit gene I: A tool for the identification of European species of blow flies for postmortem interval estimation. *Journal of Forensic Sciences* 45: 820–823.

Wallman, J.F. and S.C. Donnellan. 2001. The utility of mitochondrial DNA sequences for the identification of forensically important blowflies (Diptera: Calliphoridae) in southeastern Australia. *Forensic Science International* 120: 60–67.

Wang, J., Z. Li, Y. Chen, Q. Chen, and X. Yin. 2008. The succession and development of insects on pig carcasses and their significances in estimating PMI in south China. *Forensic Science International* 179: 11–18.

Wells, J.D. and F.A.H. Sperling. 2001. DNA-based identification of forensically important Chrysomyinae (Diptera: Calliphoridae). *Forensic Science International* 120: 110–115.

Wells, J.D. and J.R. Stevens. 2008. Application of DNA-based methods in forensic entomology. *Annual Review of Entomology* 53: 103–120.

Ye, G., K. Li, J. Zhu, G. Zhu, and C. Hu. 2007. Cuticular hydrocarbon composition in pupal exuviae for taxonomic differentiation of six necrophagous flies. *Journal of Medical Entomology* 44: 450–456.

Yew, J.Y., R.B. Cody, and E.A. Kravitz. 2008. Cuticular hydrocarbon analysis of an awake behaving fly using direct analysis in real-time time-of-flight mass spectrometry. *Proceedings of the National Academy of Sciences of the United States of America* 105: 7135–7140.

Yoder, J.A., G.J. Blomquist, and D.L. Denlinger. 1995. Hydrocarbon profiles from puparia of diapausing and nondiapausing flesh flies (*Sarcophaga crassipalpis*) reflect quantitative rather than qualitative differences. *Archive of Insect Biochemistry and Physiology* 28: 377–385.

Yoder, J.A., D.L. Denlinger, M.W. Dennis, and P.E. Kolattukudy.1992. Enhancement of diapausing flesh fly puparia with additional hydrocarbons and evidence for alkane biosynthesis by a decarbonylation mechanism. *Insect Biochemistry and Molecular Biology* 22: 237–243.

Zehner, R., J. Amendt, and P. Boehme. 2009. Gene expression analysis as a tool for age estimation of blowfly pupae. *Forensic Science International: Genetics Supplement Series* 2: 292–293.

Zehner, R., S. Mösch, and J. Amendt. 2006. Estimating the postmortem interval by determining the age of fly pupae: Are there any molecular tools? *International Congress Series* 1288: 619–621.

Zhu, G.H., X.H. Xu, X.J. Yu, Y. Zhang, and J.F. Wang. 2007. Puparial case hydrocarbons of *Chrysomya megacephala* as an indicator of the postmortem interval. *Forensic Science International* 169: 1–5.

Zhu, G.H., G.Y. Ye, C. Hu, X.H. Xu, and K. Li. 2006. Development changes of cuticular hydrocarbons in *Chrysomya rufifacies* larvae: Potential for determining larval age. *Medical and Veterinary Entomology* 20: 438–444.

CHAPTER 29

Standard Practices

Jens Amendt, Gail Anderson, Carlo P. Campobasso, Ian Dadour, Emmanuel Gaudry,
Martin J. R. Hall, Thiago C. Moretti, Kabkaew L. Sukontason, and Martin H. Villet

CONTENTS

29.1 INTRODUCTION

Approved standards and practices are necessary for forensic entomology (Amendt et al. 2007; CINFSC 2009; Disney 2011; Villet and Amendt 2011; Tomberlin et al. 2012) because its application must be reliable in court (Disney 2011). Forensic entomology requires a template for collecting, analyzing, and reporting its evidence using common minimum standards and best practices that can be defended in court.

This chapter provides a protocol modeled on medicocriminal cases that is easily modified for urban, stored-product, or environmental forensic cases. It is based on the forensic process, starting with collection of evidence at an investigation site, and culminating in the documentation of the forensic findings.

29.2 COLLECTION OF ENTOMOLOGICAL EVIDENCE

Trained forensic entomologists should be called to investigation scenes and autopsy rooms to collect any insect-related evidence. This maximizes the expertise available, allows the person interpreting the samples to see their full context, helps to ensure that the insect-related evidence is collected, preserved, labeled, and transported correctly so that it is admissible and useful in court, and reduces the time that live insects are in uncontrolled conditions, which minimizes confounding effects (Huntington et al. 2007; Dourel et al. 2010; Villet et al. 2010). Because it is not always possible for a forensic entomologist to attend a scene, crime scene technicians and forensic pathologists should be trained to perform insect sampling (Gaudry et al. 2001). Personnel can contact a forensic entomologist from an investigation site to request instructions about sampling the insects and send photographs electronically, for example, via e-mail to receive real-time advice. This chapter provides baseline guidance for training; more detailed accounts can be found in the literature (Catts and Haskell 1990; Bishop 2008; Haskell and Williams 2008a; Byrd et al. 2010). It is strongly recommended that forensic entomologists attend the autopsy because they may find evidence that was not accessible at the original scene (Haskell and Schoenly 2008).

29.2.1 Preparation

Forensic entomologists may be summoned at very short notice, so they should prepack a clean, solid, swiftly accessible toolbox with the following recommended equipment (Pasquerault et al. 2006; Amendt et al. 2007; Haskell and Williams 2008b; Byrd et al. 2010; Hall et al. 2012):

1. Sterile gloves.
2. A small paint brush— to collect eggs.
3. A small disposable plastic spoon—to collect larvae.
4. Entomological forceps made of spring steel—to collect larvae, pupae and adults.
5. A source of near-boiling water—to kill fly larvae (Adams and Hall 2003).
6. Leak-proof screw-topped vials containing ≥80% ethanol—to preserve insects.
7. Blank labels that will fit inside the vials—to record collection data.
8. An insulated box: to transport live samples. This should contain a temperature data logger (e.g., an iButton or Dallas Key) for monitoring its internal temperature.
9. Ventilated rearing vials (Figure 29.1) containing crumpled tissue paper: to transport live specimens. Screw-tops are preferable because fly larvae can force open other lids.
10. A lead pencil—to write labels. Inks dissolve in solvents like water and ethanol.
11. Blank labels—to identify and annotate the ventilated vials.
12. A shovel or trowel—to take soil and leaf-litter samples and search for buried insects.
13. Zip-lock® resealable plastic or robust (double-bagged) paper bags—to transport soil and leaf litter samples.
14. Standard entomological evidence forms—to record the origin of each sample and the associated conditions.
15. Standard chain-of-custody form—to record the chain of custody of insect evidence.
16. Scene indicator flags—to indicate the location of insect evidence in photographs.
17. A high definition video camera—to record general site conditions and specific details.
18. A digital thermometer—to measure temperatures within masses of fly larvae.
19. A geographical positioning system (GPS) receiver or its equivalent—to establish accurate location data that may help to establish the aspect of exposure of the corpse, times of local sunrise and sunset, and the distance(s) to the weather station(s) used in the investigation.

20. Two (one for backup) electronic data loggers, preset to record temperatures to match local weather station recording intervals (usually at hourly intervals, on the hour)—to determine temperatures on site after body recovery, which can be used to estimate temperatures before body discovery. Such loggers must be protected from rain or direct sunlight (as at a weather station) to get comparable readings.

Forensic entomologists called to a scene should request information that will affect what additional equipment they bring (Nystrom 2008), such as the size of the death scene, whether it is terrestrial or aquatic, and whether special equipment might be necessary, for example, a soil temperature probe at a burial site. An infrared-sensitive camera is useful to record the whole thermal environment of the corpse (Ridgeway et al. 2014). Although this is not yet standard equipment, it is recommended for relating ambient temperatures to those actually experienced by the insects.

29.2.2 Examining the Scene

Before sampling, forensic entomologists should introduce themselves to the investigator in charge of the scene, obtain permission to collect evidence, and obtain a case number (Bishop 2008; Nystrom 2008). Forensic entomologists and forensic pathologists should interact similarly at mortuaries (Campobasso and Introna 2001; Haskell and Williams 2008b).

When possible, forensic entomologists should assess the insects present on site before other personnel have disturbed them (Bishop 2008) because, once disturbed, adult insects may leave the scene and the distribution of insect larvae is quickly disrupted. This is a good time to take photographs or video recordings, ideally including a certified, identifiable measurement scale, calibrated in millimeters. Such images may be useful to confirm details later, but are a poor substitute for examining actual insect specimens. The basics of collecting insect evidence from the body itself are well established (Catts and Haskell 1990; Pasquerault et al. 2006; Amendt et al. 2007; Haskell and Schoenly 2008; Haskell and Williams 2008b; Hall et al. 2012). When collecting any stage of insect, it is always important to establish whether this is the oldest stage on the remains. Once larvae have been collected from the body, search for insects that have reached the wandering stage of the third instar and have left the body, then determine if any of these have pupariated and whether the puparia are still intact or are empty, indicating adult emergence.

In indoor scenarios, wandering larvae and pupae may be hidden, for example, under carpets, clothes, pillows, skirting boards, and even heavy furniture. Specimens can even be found over 20 m from a body, in adjacent rooms and downstairs from where the body is discovered (Anderson 2011; Lewis and Benbow 2011).

In outdoor scenarios, larvae of some species disperse and pupate in soil or leaf litter or under objects like logs and stones. Sample to a depth of about 15 cm. Soil can be sieved and searched on a plastic sheet at the scene or placed in paper bags for inspection in the laboratory. Paper bags forestall the problems caused by condensation. Species like *Chrysomya albiceps* (Wiedemann), *Chrysomya rufifacies* (Macquart), and *Protophormia terraenovae* Robineau-Desvoidy (Diptera: Calliphoridae) tend to pupate on or beneath bodies and in clothing. Search for dispersed stages at all cardinal points to a minimum distance of 5–20 m from the source. Potential contamination by insects, for example, in dead animals or garbage, should be considered (Archer and Ranson 2005). Control samples to indicate the background level of insect activity in soil and leaf litter should be taken in comparable conditions outside the decomposition zone.

ASTM International's Committee E30.11 on Interdisciplinary Forensic Science Standards, a subcommittee of the technical Committee E30 on Forensic Sciences Standards, maintains a consensus standard for labeling of evidence (ASTM International 2006), which is currently under review. All soil and insect samples must have labels written in pencil and placed where they will not get damaged or separated from the sample.

29.2.3 Examining the Body and Collecting Samples

Specimens from different body regions should be stored in separate containers, labeled in detail, and recorded on a sampling form (Amendt et al. 2007) to associate them with the site of infestation and its temperature and any photographs. Inconsistencies may be significant, for example, when the oldest immature stages are not in the head orifices, where infestations usually start (Baumjohann et al. 2011; Ridgeway et al. 2014).

Samples should be divided into suitable batches for four purposes:

1. Samples killed (see Section 29.2.3.1) and preserved immediately in twice their volume of ≥80% ethanol provide a record of their sizes at the time of collection (for estimating their ages), and voucher specimens for confirming identifications. Insects can be preserved in the field, but it may be easier to do this in the laboratory provided that the transportation of samples will not delay preservation by more than approximately 2 hours.
2. Samples killed (see Section 29.2.3.1) and preserved immediately in at least twice their volume of ≥80% ethanol or frozen (–20°C) in the laboratory for studies of either insect or human DNA or RNA (Campobasso et al. 2005; Stuyt et al. 2010; Boehme et al. 2013). For gene-expression analysis using RNA where specimens cannot be kept alive, samples should be stored at –80°C.
3. Samples ($n \geq 30$) frozen (–20°C) in the laboratory for toxicological analysis (where needed), without any preserving liquid. Standard practices of forensic toxicology should be followed (Penders and Verstraete 2006; Society of Forensic Toxicologists, Toxicology Section of the American Academy of Forensic Sciences 2006; American Board of Forensic Toxicology 2011) to avoid compromising evidence.
4. Samples kept alive and reared in the laboratory to the next developmental landmark for estimating their ages, or to adulthood to confirm identifications.

29.2.3.1 Preserving Samples

Killing and preserving specimens when they are collected enables the forensic entomologist to estimate the age of the insects collected at the scene without needing to take into account the temperatures of the couriering vehicle and any intervening places where the specimens may have been kept before analysis, thus avoiding a reason for uncertainty and questioning in court. Ideally, samples should be preserved on site and maintained in this state indefinitely so that they can be submitted for a second opinion.

It is recommended that ≥80% ethanol is used as a preservative because when specimens are added, their internal fluids will dilute it in proportion to the amount of specimens (Martin 1977). Ethanol becomes an ineffective preservative if it is diluted below 70%; specimens should initially be placed in at least twice their volume of ≥80% ethanol and the ethanol drained off the following day and replaced. All specimens intended for DNA analysis should also be refrigerated.

Eggs are fragile. Collect them carefully by using a fine paint brush or, when they occur in large masses, with fine forceps. The stage of development of embryos may be informative. Eggs can be preserved immediately in ≥80% ethanol.

Larvae are collected using pliable forceps or a spoon. Collect a sample that reflects the abundance and diversity on the body, for example, 50–100 fly larvae from each significant, distinct region of colonization. It is important to collect a subset of all larvae and not only the biggest and presumably oldest. Differences in size may reflect age differences or feeding-site-specific accelerated development due to the presence of drugs (Lord 1990), differences in species-specific growth characteristics, or the presence of precocious larvae (Villet et al. 2010; Davies and Harvey 2012).

Fly larvae (maggots) should be killed by dropping them into freshly boiled water (>90°C) for about 30 seconds before being preserved in ≥80% ethanol (Tantawi and Greenberg 1993; Adams and Hall 2003), to maximally extend them (to a condition comparable with larvae measured in the production of published tables of development) and prevent their decomposition. Specimens placed alive into ethanol may subsequently shrink and die in unstandardized, curled postures that can obscure diagnostic characters and complicate measuring them to estimate their age. It can sometimes be difficult to obtain sufficiently hot water in the field, in which case the insects can be killed on return to the laboratory if it is less than 2 hours away.

Beetle (Coleoptera) larvae should be placed in ≥80% ethanol and not boiled like fly larvae because their unevenly thickened skeletons make them curl in hot water, which makes measuring them difficult (Midgley and Villet 2009).

Pupae can be collected using pliable forceps or a spoon. They should be pricked, taking care to avoid injury to personnel, to enable preservative to enter the tissues promptly when placed into ≥80% ethanol (Singh and Greenberg 1994; Davies and Harvey 2013; Cameron Richards, pers. comm.).

Eclosed puparia provide evidence that at least one generation of flies has completed development. They may also contain evidence of the presence of drugs or poisons (Lopes de Carvalho 2010; Gosselin et al. 2011). If possible, collect specimens with the cephalic valves attached because the larval mouth hooks are usually attached to the inside surfaces and are important for identification. They should be stored dry in vials labeled on the inside.

Adults of beetles or other invertebrate groups should be killed and preserved according to their eventual purpose in the investigation, as outlined at the start of this section. Live adult insects collected at the scene can be killed in the laboratory by freezing them overnight. Dead specimens left in sealed vials will decompose, so after thawing they must be pinned, dried, and stored in a dry environment or preserved in ≥80% ethanol. Although it is a standard entomological killing agent, ethyl acetate should not be used in forensic cases because it severely damages DNA, making specimens useless for DNA analysis.

Dead adult flies might be found indoors in high numbers on window sills or in front of patio doors. There are already some tools to estimate the age of such specimens (Villet and Amendt 2011), and they can be collected and preserved in ≥80% ethanol as they might be relevant to the investigation.

29.2.3.2 Collecting and Transporting Live Samples

Various types of container can be used for transporting live insect samples, but absorbent paper tissue should be placed inside them to soak up liquids, excretions, and respiratory condensation. Live insects require air, so they must be placed in ventilated vials (Figure 29.1). One can replace the vial's lid with two layers of paper towel held in place with an elastic band, as long as it remains dry. Punching holes in an existing lid rarely provides sufficient air and allows larvae to escape.

Eggs for rearing should be placed on moist tissue paper in ventilated vials to maintain humidity because they are very susceptible to lethal desiccation.

Larvae should not fill a vial more than one layer deep, or they may suffocate, especially in warmer weather and if the larvae are contaminated with decaying material. Live larvae from separated sites should be kept separate where species such as *C. albiceps* or *C. rufifacies* occur as these may eat other species. Commonly, these species do not feed together with other species on the same corpse (Villet et al. 2010).

Pupae for rearing should be placed in damp tissue paper in a ventilated vial.

Live samples should be delivered to the laboratory promptly because while they are not well fed and under controlled environmental conditions, the effects of confounding influences accumulate (Huntington et al. 2007; Villet et al. 2010). Live insects should be placed in a carrier, insulated against heat but not chilled, for example, not left in a fridge or the boot of a car on a warm day. The so-called Kaufmann effect compromises analytical models when the immature stages are at

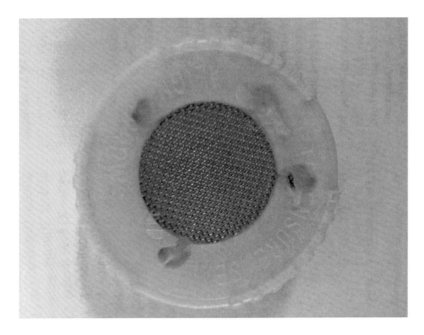

Figure 29.1 Plastic cap for ventilated vial with stainless steel mesh (0.28 mm) (Courtesy of I. Dadour, UWA).

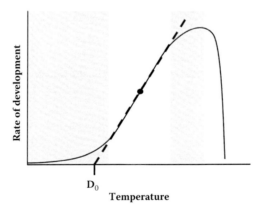

Figure 29.2 The growth rate of insect larvae usually has a sigmoidal response (solid line) to temperature below the upper lethal limit. Linear models used to estimate the growth rate of insects (dashed line) are fitted to the nearly linear part of the sigmoid (Higley and Haskell 2010). The Kauffman effect or rate summation effect (Worner 1992; Ikemoto and Egami 2013) occurs in the gray regions, where the fit of the linear model become exponentially worse at temperatures further from the inflection in the sigmoid (marked with a black dot), which is the developmental optimum. The lower developmental threshold (D_0) falls within the affected region. (Courtesy of M.H. Villet, RU.)

temperatures near to their lower developmental threshold (see Figure 29.2 for explanation) (Worner 1992; Ikemoto and Egami 2013). Until nonlinear models of development (Baqué and Amendt 2013; Ikemoto and Egami 2013) become standard, specimens should be transported at temperatures least affected by the Kaufmann effect and a reliable record of the temperature should be kept, for example, by a data logger placed in the container. The temperature record should be printed out in the laboratory and archived.

29.2.4 Reconstructing Field Temperatures

Soon after arriving on site, the forensic entomologist should set up temperature data loggers to record environmental temperatures at the scene, which are important for two reasons. First, environmental temperatures affect how quickly insects arrive at the scene (Michaud and Moreau 2009, 2011; Matuszewski 2012) to lay eggs/larvae and start the biological "clock" of development. Second, temperature affects the rates both of various decomposition processes that may be used as biological clocks (Heaton et al. 2010; Simmons et al. 2010; Vass 2011; Zhou and Byard 2011), and of insect development (Higley and Haskell 2010; Villet et al. 2010).

To estimate the temperature at the site before the discovery of the body, one can use data from the nearest weather station (Dourel et al. 2010), although poor correlation between different locations can undermine this approach (Archer 2004; Amendt et al. 2007; Hall et al. 2012). If the evidence is presented at trial, a lack of cross-validating measurements from the scene can raise difficult questions about the applicability of weather station data to the conditions at the scene, which can be used to cast doubt on the evidence and the analyst. Regression analysis of the temperatures at the scene and the weather station will calibrate the relationship between the two sets of data and allow the temperatures at the site to be reconstructed for the period in question (Huntington and Higley 2008). Multiple regression models may be used if more than one weather station is available.

Temperatures at the scene should ideally be recorded with two calibrated devices to ensure against equipment failure, and provide cross-validation. To facilitate regression analysis, the recording frequency should match the local weather station's recording interval, which is usually at hourly intervals, on the hour. Temperatures should be recorded for at least 5 days, preferably 10 days or more, to provide sample sizes that are adequate for the statistical analysis used to produce convincing and reliable prediction intervals. The data should be printed out and archived. To ensure admissibility in court, weather station data for comparative analysis should only be taken from certified facilities.

29.3 CHAIN OF CUSTODY

There are consensus technical standards for documenting the chain of custody (ASTM International 2007, 2011a).

A chain-of-custody form should be used to record when samples change hands, and should accompany the specimens at all times (Haskell and Catts 2008). For quality assurance (QA), record the numbers and types of samples, and all of the events dealing with their sampling, preserving, and processing to establish an auditable timeline of events. This record should include a printout of the data from the temperature loggers that accompanied the specimens from the field to the laboratory. All relevant communications with officials (names, dates, and case relevant information) from the initial contact should accompany the processing of samples. Most of this information is covered by the request for forensic analysis, the autopsy report, the table of evidence of the Crime Scene Unit, and related documents. Incidental communications, for example, by telephone, should also be recorded—these notes could be helpful when writing the report and later at trial.

On arrival of the samples at the laboratory, the date and time of arrival is registered against a specific laboratory case number, which should be related to the case number of the police investigation. A table of evidence must be compiled, which includes the number of specimens, referring to the labels of the respective vials. All identifications and measurements must be reported in a related table.

All living samples must be processed immediately (see Section 29.4) and the time when rearing begins must be recorded. If samples are subdivided for toxicological, DNA, or other analyses, a system should be in place for using the case number of the laboratory to backtrack to the original field sample. Every step related to the rearing of live samples should be reported in the case notes (e.g., time of checking the samples, time of reaching certain landmarks of development) as well as the temperature record of the rearing incubator.

All dead samples must be preserved, registered, identified, and measured promptly to forestall the risk of physical modifications because of poor preservation in the field or laboratory (Midgley and Villet 2009). At the end of the examination process and after finishing the forensic report, the samples must be stored for a period defined by local legislation before disposal or returning to the client.

29.4 REARING INSECTS

There are several benefits in rearing insects collected from a death scene to adulthood. Several taxa such as flesh flies (Diptera: Sarcophagidae) and muscid flies (Diptera: Muscidae) are easier to identify morphologically as adults, and there are still few resources to enable their reliable identification by molecular means. A soundly designed rearing program can provide quantitative data for the statistical estimation of the ages of the insects (and therefore the minimum postmortem interval or PMI_{min}) with a statistically rigorous measure of the associated uncertainty (Richards and Villet 2008). Developmental landmarks such as hatching, ecdysis, pupation, pupariation, or eclosion of the adult insect provide an additional reference point or, sometimes, the only way to calculate the age of the specimens at the time of sampling. Finally, rearing the next generation of that "corpse-specific population" will provide the opportunity to rear them under controlled conditions that replicate the death scene. The latter approach could be helpful if knowledge of the development of the species in question is scanty or to minimize problems in using published reference data, for example, due to differences caused by geographical variation (Richards et al. 2008) or inconsistent rearing substrates (Richards et al. 2013).

Insects should be reared to adulthood in a climate chamber at a known constant and optimal temperature that is near the inflection in the sigmoidal curve relating growth rate and temperature, which is around 18°C–23°C for temperate species of blow fly and 22°C–26°C for more tropical species. Insects develop faster at temperatures above this range, which decreases the precision and increases the relative error with which their development can be monitored, while at very low temperatures, development becomes excessively attenuated (Figure 29.2). A temperature data logger should be placed inside the incubator to accurately document the temperature, and the data record should be printed out and archived.

As soon as larvae hatch, they should be placed into small rearing containers (e.g., 50 mL disposable plastic cups) on a food substrate. Approximately 2 g of food (see later for type of food) per fly larvae should be sufficient for completing development, and about 20 fly larvae should be kept together because this allows for optimal growth without generating appreciable amounts of confounding metabolic heat (Goodbrod and Goff 1990). Fly larvae are gregarious and grow suboptimally at low densities (Goodbrod and Goff 1990; cf. Davies and Ratcliffe 1994).

The rearing containers should be placed into larger containers (Figure 29.3) with sawdust or sterilized sand or soil, into which they can disperse to pupariate. It is important to ensure that all rearing containers are clearly labeled with their original data, for example, case and sample number, and that insects cannot move between rearing containers. Insects being reared to adulthood should be checked twice a day and a record of visit times and pupation or eclosion rates noted.

Once emerged, the adults should be removed from the samples and a record kept of the numbers emerging each day. To avoid any confusion over emergence date, puparia can be kept in individual storage vials. Alternatively, a group of puparia can be kept in standard commercial or custom-built

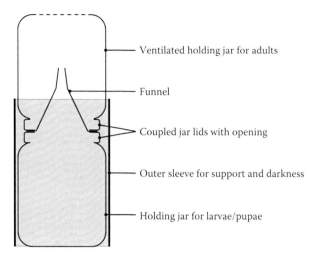

Figure 29.3   Plan for an emergence cage for flies. (Courtesy of M.H. Villet, RU.)

emergence traps (Figure 29.3). Emerged adults can be easily removed by taking advantage of the fact that they will fly readily to light. They can be killed by freezing and then pinned or stored in ≥80% ethanol before identification. Each specimen should be labeled with the appropriate information, including sample number and date and time of adult emergence.

Ideally the rearing process should be standardized, because there is a wide variation in the type of substrate used for development studies in research and casework (Richards et al. 2013). Until there is standardization, best practice would be to rear specimens on the same media as used to rear the samples from which published databases were established.

29.5 IDENTIFYING INSECTS

Because of the species-specific nature of insect development, accurate identification of all specimens is essential in any entomological investigation.

Adults are most easily identified by their physical characteristics, and even very young fly larvae of several species can usually be distinguished (Szpila et al. 2008; Szpila and Villet 2011; Szpila et al. 2013a,b). The make and model of instruments used to examine specimens and the citations for published identification keys should be provided in the investigation report.

If morphological identifications are not possible, molecular biological methods can support the identification of all life stages (Meiklejohn et al. 2013). These analyses should be conducted by experts who are familiar with QA precautions for laboratories performing DNA-based identifications of individuals (Salas et al. 2005; Schneider 2007). Sequencing one mitochondrial gene (e.g., *COI, ND4, 16S rRNA*) has generally been considered a standard procedure for insect identification until now, but because of potential complications such as hybridization, introgression, incomplete lineage sorting, paralogy, and primer failure (Debry et al. 2013; Williams and Villet 2013), it might be necessary in future to recommend the sequencing of at least one nuclear gene (e.g., *28S rRNA, ITS2*) in addition to one mitochondrial gene. Identifications can be made using online tools like BLAST and its successors (Edgar 2010), to compare the sequence of the samples with those of specimens published on the internet; however, a final check should be made by phylogenetic analysis, because not all data on those freely accessible sources are correct due to the potential for an initially wrong (morphological) identification leading to the database being fed with the sequence of a misidentified taxa. All of the primers and reagents and online sources (databases) should be recorded and the associated trace files and sequences should be archived, for example, on an online database.

Unfortunately, there are still no reliable developmental data for many flies like, for example, Sarcophagidae (Villet et al. 2006) and beetles, so establishing their identity may not lead to a precisely estimated PMI_{min}, but will delay matters and increase costs (Gaudry et al. 2009). Here, identification to genus or even family level might be all that is possible.

29.6 ESTIMATING AGES OF INSECTS

When estimating the age of preserved larvae, they must be measured promptly after killing to minimize errors due to shrinkage or swelling (Midgley and Villet 2009). Estimating the age of insects requires not only reliable quantitative methods but also extensive qualitative consideration (Villet et al. 2010). There are many approaches and new ones evolve regularly, along with standards (Villet and Amendt 2011; Baqué and Amendt 2013).

There are important standards for the age estimates themselves, however they are made. The estimates should be based on credible, publicly available benchmark data sources, such as peer-reviewed scientific publications (cf. the *Frye*, *Daubert* and *Kumho Tire* criteria for U.S. admissibility of expert evidence: Villet and Amendt [2011]). They should also be framed in terms of a window of opportunity rather than a specific time, and should be qualified by considering confounding variables that were not included in the quantitative analysis (Villet et al. 2010). Several factors might trigger developmental plasticity, and a certain amount of variability is inevitable, which is still not covered adequately in the literature, nor is the problem of introducing probabilities and confidence intervals (Lagnado et al. 2013).

As a consequence, age estimation can, at best, only narrow down an event to a calendar date (Villet et al. 2010), but sometimes the estimates might have to be extended, for example, to within 1–2 weeks in cooler weather. An exception could be the first stage of development, the eggs, where it is possible to estimate a window of time of just a few hours (Bourel et al. 2003; Anderson 2004).

The age of an insect does not estimate the actual time of death but rather a PMI_{min} (Villet and Amendt 2011). This is because the time when insects first colonized a body is not necessarily when the person died. The first colonization could occur within hours or even minutes of death if the body is outdoors in summer conditions (Anderson 2011; Matuszewski 2012), but colonization can be delayed if, for example, the body is buried, wrapped (Goff 1992; Kelly et al. 2008, 2009, 2011; Gaudry 2010), stored indoors (Reibe and Madea 2010; Anderson 2011), first exposed during cold or rainy weather (Baumjohann et al. 2011), or during a season of low insect activity (Wetzel et al. 2009).

29.7 ENTOMOLOGICAL REPORT

The forensic entomologist's final report, often submitted as a Witness Statement, documents all of the evidence and analysis that has led to the answer to the commissioning person's question. It is a legal document that will be served in court, and therefore needs to be clear, concise, and comprehensive (ASTM International 2011b). A well-written report can assure the admissibility of the evidence, forestall cross-examination, and avoid any question about the professionalism of the forensic expert.

Terminology has been coined to indicate what period the entomological evidence is being used to estimate (Villet and Amendt 2011), and many of these terms have been given acronyms. Many of the terms are inexact and some of the acronyms are confusingly similar (Villet and Amendt 2011).

In the interests of clear and effective communication with nonexperts, it is recommended that authors refer directly, explicitly, and in full to the duration or event that is being estimated and avoid ill-defined TLAs[*] (Villet and Amendt 2011).

The following standard elements should appear in forensic entomological reports about medico-criminal cases (modified from Haskell and Catts 2008):

1. Investigation reference number and, if known, the name of the dead person.
2. Name and contact details of the commissioning authority.
3. Details of notification of case (e.g., time when the call for the report arrived) and background details that introduce the inquiry, for example, details of deceased, time of discovery, and descriptions of the scene and remains.
4. Purpose of the entomological report: the report should deal with this and not other aspects of the case.
5. List of material and samples to be analyzed, their catalogue numbers, and approximate descriptions of the contents as shown on the exhibit label.
6. Outline of initial chain of custody: name of who sampled the material, when and where and what time it arrived at the laboratory.
7. Any temperature data (summarized graphically if possible) used in the report, and its source.
8. The identity, numbers, origin, measurements, description, rearing conditions, and other details of the specimens and relevant information about their biology.
9. The estimated age of the specimens and the source of the calibration data.
10. Reason for not considering any specimens, for example, no calibration data and no relevance for investigation.
11. An interpretation of the evidence in terms of the purpose of the report. This may include a timeline of events. It should be clear that the person could have been dead for a period longer than the insects were present. *Always* give this PMI_{min}, even if the victim appears to have been missing for much longer. The time since death indicated by the insects should be addressed, not the more general aspects of the case. The focus should be on interpreting the insect evidence and its limitations.
12. Summarize the highlights of your report (purpose, material examined, insect species, insect age, PMI_{min}).
13. References that were used and mentioned in the report.
14. Appendices (e.g., temperature data of the meteorological station).
15. A brief statement of your professional qualifications, which enable you to provide this report.

Depending on the relevant national legal system, the report should be signed on the front page by the author and signed or initialed on every page. Copies should be provided to the commissioning person(s), and one should be archived by the author.

29.8 QUALITY ASSURANCE AND ACCREDITATION OF STANDARD PRACTICES

The setting of standards for methods in forensic science is currently not legislated in any country (Lentini 2009), but there are consensus technical standards that cover the general forensic investigation process (ASTM international 2006, 2007, 2011a,b, 2013a,b), that aspire to be internationally relevant. Specific standards for forensic entomology do not exist yet, but the existing general standards are relevant to entomological investigations. These standards have been generated and revised by a consensus process involving professional organizations, and hold authority only for parties that voluntarily subscribe to them. The report of the U.S. National Research Council's Committee on Identifying the Needs of the Forensic Sciences Community (CINFSC 2009) has provided an impetus to formalize and accredit forensic methods and techniques.

---

[*] http://en.wikipedia.org/wiki/Three-letter_acronym.

It is possible for forensic laboratories to seek the International Organization for Standardization's (ISO) accreditation for their analytical work under the ISO 17025 and ISO 17020 standards and certification of their quality management systems under the ISO 9001 standard (Villet and Amendt 2011). Recently, the British Home Office created the post of the U.K. Forensic Regulator (see Hall et al., Chapter 9), who has in turn produced an article, "A Review of the Options for the Accreditation of Forensic Practitioners" and recommended that all providers of forensic science services should gain accreditation from the United Kingdom Accreditation Service (UKAS). Because of the European Union's decision, aiming to reinforce police and judicial cooperation in criminal matters, common standards for forensic service providers will soon be in place in the fields of DNA and fingerprints (Council of European Union 2009).

The debate around these documents is still in progress, but it is clear that accreditation of institutions and certification of practitioners both require a benchmark against which to test proficiency, and therefore rely on the existence of agreed technical standards.

The implications for the standardization of methods in forensic entomology and their international coordination are not yet clear. Some methods in forensic entomology are more reliable than others, they are evolving continuously, and the newest methods are likely to be contentious for a period after their introduction. The past two decades have seen the introduction of new working habits and new rules to comply with in the field of forensic sciences. QA, accreditation, certification of individual experts, and testing (forensic) laboratories are current challenges. Forensic entomology will have to embrace such trends in the near future, because all forensic work, from the collection of evidence to the release of the forensic report, must comply with local legislation and local police forces' jurisdictions.

The impact of forensic sciences on criminal investigations and court proceedings is growing (Greenberg and Kunich 2002; Schneider 2007). Forensic experts are expected by judges, investigators, lawyers, and victims to work in compliance with legal requirements. They must also comply with reliable codes of practice (both laboratory and field standards).

Crime investigation laboratories are progressively imposing a principle of QA management for their forensic departments. Their objective is to obtain accreditation (i.e., general requirements of testing and calibration laboratories), in compliance with standards defined by the International Organization for Standardization, especially NF EN ISO/CEI 17025 standard (ENFSI 2003; Norm EN ISO/IEC 17025 2005; Vanden-Berghe 2009; Gaudry and Dourel 2013).

Forensic entomology is a constantly developing scientific field (Tomberlin et al. 2012). It does not follow the analytical method of other forensic specialities derived from genetics, chemistry, or physics (DNA analyses, drug detection, etc.). It derives from the naturalist branch of biology, which is dependent on an individual's expertise in interpreting evidence that demonstrates biological variation. Forensic entomologists around the world have various backgrounds, working habits, and writing techniques combined with specific and sometimes unique national legal requirements. Nevertheless, forensic entomology must demonstrate good working practice and also provide guarantees of its analytical rigor and technical reliability to satisfy the claimants, the court, and the scientific community.

The last decade saw a significant rise in the routine use of insect evidence in criminal investigations. A sound QA management has to maintain the chain of custody of samples, data management, and to eliminate any risk of PMI_{min} miscalculations (Arnaldos et al. 2006; Pasquerault et al. 2006; Amendt et al. 2007; Haskell and Williams 2008a; Byrd and Castner 2010). There are existing references such as Guidelines for Forensic Science laboratories (DNA Advisory Board 1998, ILAC 2002, ENFSI 2003) and some forensic entomology laboratories have put together standardized forms to normalize sampling. There are several guidelines relating to forensic entomology (collecting, processing, and reporting entomological evidence) published (Catts and Haskell 1990; Anderson and Cervenka 2001; Greenberg and Kunich 2002; Arnaldos et al. 2006; Pasquerault et al. 2006; Amendt et al. 2007; Hall et al. 2012), which could provide a useful basis for the establishment of laboratory QA management systems.

29.8.1 Accreditation of Methods

Accreditation aims to assure a technical competency, which is based on several criteria. Such an assessment is partly based on dedicated documentation complying with specific norms (ISO standards). A forensic entomology quality management system should be in compliance to meet the ISO/CEI 17025 standard and various ASTM International standards (ASTM International 2006, 2007, 2011a,b, 2013a,b). Developing a quality plan encompasses several areas, from documentation to analytical process and proficiency testing.

It should be made up of several sections detailing requirements applicable to forensic entomology:

- Personnel (organization, technical ability, education, and training)
- Accommodation and environmental conditions
- Testing systems, calibration methods, and methods validation
- Measurement traceability: handling of test and calibration items, assuring the quality of testing and calibration results
- Reporting the results

Such a plan should also be made up of technical guidelines and users' instructions, for example, critical equipment (climatic chambers, data loggers), laboratory maintenance, and supply of chemicals and products.

Different testing methods, detailing the processing of entomological samples, identifying and controlling the most critical parameter used in the PMI_{min} calculation (i.e., temperature) should be written: sampling of entomological evidence on site (use of kit and data logger, etc.); preserving, packaging, and storing entomological evidence; insect rearing (protocol, temperature monitoring, etc.); preparing insect samples before identification of species of forensic interest; analyzing insect development; and estimating the PMI.

Accreditation audits are generally carried out by an independent agency originating from the same country (e.g., COFRAC [Comité Français d'accréditation] in France, which carried out the accreditation audits for the entomological department of the French Gendarmerie). They focus on every aspect of the quality plan of the forensic entomology laboratory, especially the testing methods. Testing methods are crucial points, helping to assess and ratify a competence in testing and calibration, in compliance with ISO/CEI 17025 standard. Their revisions, showing a dynamic approach through a system of maintenance and monitoring, are also checked.

29.8.2 Accreditation of Laboratories

Applying global standards to forensic entomology will help to raise the confidence of judges, prosecutors, and investigators and future fit it to the changes that will occur in forensic science generally.

QA in forensic entomology needs constant monitoring alongside the regular update of controlled documentation. Accreditation of laboratories encourages visible and defined working practices, and also aims to reduce the number of mistakes due to sample processing; it also protects experts by proving that their work has been carried out in compliance with best practice (Chaturvedi et al. 2009; Bleay et al. 2012).

29.8.3 Accreditation of Experts

Discussions about the accreditation of experts are on the increase (http://www.newlawjournal.co.uk/nlj/content/no-get-out-experts). Many experts do not work within a laboratory that is seeking accreditation and so, working alone, do not benefit from a quality service support. Laboratory accreditation that complies with the EN ISO/IEC 17025 norm does not assess the expert directly, but only the system in which they work. In these conditions, an assessment of the activity can be difficult. The American Board of Forensic Entomology (ABFE) specifies the certification procedure of individuals to demonstrate competence in forensic entomology (general requirements, education, professional experience, examinations, assessment, renewal of certification, and revocation). There is currently no general consensus practice among the global forensic entomology community on this topic.

The forensic entomology community must adopt common practices complying with international standards, agreed by this community, scientists, forensic practitioners and, most importantly, the judicial system to provide confidence in work done by experts in this field. A recent decision of the Council of the European Union (2009) will soon make it mandatory for every forensic service provider to comply with common standards for two fields: DNA profiling and fingerprints. It is likely that such a decision will be extended to other aspects of forensic science, including forensic entomology, and be adopted by regulators on other continents.

REFERENCES

Adams, Z.J.O. and M.J.R. Hall. 2003. Methods used for the killing and preservation of blowfly larvae, and their effect on post-mortem larval length. *Forensic Science International* 138: 50–61.

Amendt, J., C.P. Campobasso, E. Gaudry, C. Reiter, H.N. Leblanc, and M.J.R. Hall. 2007. Best practice in forensic entomology—Standards and guidelines. *International Journal of Legal Medicine* 121: 90–104.

American Board of Forensic Toxicology. 2011. ABFT forensic toxicology laboratory accreditation manual. American Board of Forensic Toxicology, [Online]: 40. Available at http://www.abft.org/files/ABFTLaboratoryManual.pdf.

Anderson, G.S. 2004. Determining time of death using blow fly eggs in the early postmortem interval. *International Journal of Legal Medicine* 118: 240–241.

Anderson, G.S. 2011. Comparison of decomposition rates and faunal colonization of carrion in indoor and outdoor environments. *Journal of Forensic Sciences* 56: 136–142.

Anderson, G.S. and V.J. Cervenka. 2001. Insects associated with the body: Their use and analyses. In *Advances in Forensic Taphonomy. Methods, Theory and Archeological Perspectives*, eds. W. Haglund and M. Sorg, 174–200. Boca Raton, FL, CRC Press.

Archer, M.S. 2004. The effect of time after body discovery on the accuracy of retrospective weather station ambient temperature corrections in forensic entomology. *Journal of Forensic Sciences* 49: 553–559.

Archer, M.S. and D.L. Ranson. 2005. Potential contamination of forensic entomology samples collected in the mortuary: A case report. *Medicine Science and the Law* 45: 89–91.

Arnaldos, M.I., A. Luna, J.J. Presa, E. Lopez-Gallego, and M.D. Garcia. 2006. Entomologia forense en España: Hacia una buena practica profesional. *Ciencia Forense* 8: 17–38.

ASTM International. 2006. Standard E1020–1996 (2006). Standard practice for reporting incidents that may involve criminal or civil litigation. ASTM International, West Conshohocken, PA. Available at www.astm.org.

ASTM International. 2007. Standard E860–2007. Standard practice for examining and preparing items that are or may become involved in criminal or civil litigation. ASTM International, West Conshohocken, PA. Available at www.astm.org.

ASTM International. 2011a. Standard E1188–2011. Practice for collection and preservation of information and physical items by a technical investigator. ASTM International, West Conshohocken, PA. Available at www.astm.org.

ASTM International. 2011b. Standard E620–2011. Standard practice for reporting opinions of scientific or technical experts. ASTM International, West Conshohocken, PA. Available at www.astm.org.

ASTM International. 2013a. Standard E1459–2013. Standard guide for physical evidence labeling and related documentation. ASTM International, West Conshohocken, PA. Available at www.astm.org.

ASTM International. 2013b. Standard E678–2007 (2013). Standard practice for evaluation of scientific or technical data. ASTM International, West Conshohocken, PA. Available at www.astm.org.

Baqué, M. and J. Amendt. 2013. Strengthen forensic entomology in court–The need for data exploration and the validation of a generalised additive mixed model. *International Journal of Legal Medicine* 127: 213–223.

Baumjohann, K., K.H. Schiwy-Bochat, and M. Rothschild. 2011. Maggots reveal a case of antemortal insect infestation. *International Journal of Legal Medicine* 125: 487–492.

Bishop M.R. 2008. *Insect Evidence: Basic Collection Procedures at the Death Scene.* Bloomington, IN: Xlibris, 1–112.

Bleay, S.M., V.G. Sears, H.L. Bandey, A.P. Gibson, V.J. Bowman, R. Downham, L. Fitzgerald, and T. Ciuksza. 2012. Chapter 3: Finger mark development techniques within scope of ISO 17025. In *Fingerprint Source Book*, Home Office: Centre for Applied Science and Technology (CAST), [Online]: 233–289. http://www.homeoffice.gov.uk/publications/science/cast/crime-investigation/fingerprint-source-book-2012/fsb-chap3-sec1to3-development?.pdf.

Boehme, P., P. Spahn, J. Amendt, and R. Zehner. 2013. Differential gene expression during metamorphosis: A promising approach for age estimation of forensically important *Calliphora vicina* pupae (Diptera: Calliphoridae). *International Journal of Legal Medicine* 127: 243–249.

Bourel, B., B. Callet, V. Hédouin, and D Gosset. 2003. Flies eggs: A new method for the estimation of short-term post-mortem interval? *Forensic Science International* 135: 27–34.

Byrd, J.H. and J.L. Castner. 2010. *Forensic Entomology: The Utility of Arthropods in Legal Investigations.* London, United Kingdom: CRC Press.

Byrd, J.H., W.D. Lord, J.R. Wallace, and J.K. Tomberlin. 2010. Collection of entomological evidence during legal investigations. In *Forensic Entomology: The Utility of Arthropods in Legal Investigations*, 2nd ed., eds. J.H. Byrd and J.L. Castner, 127–175. London, United Kingdom: CRC Press.

Campobasso, C.P. and F. Introna. 2001. The forensic entomologist in the context of the forensic pathologist's role. *Forensic Science International* 120: 132–139.

Campobasso, C.P., J.G. Linville, J.D. Wells, and F. Introna. 2005. Forensic genetic analysis of insect gut contents. *American Journal of Forensic Medicine and Pathology* 26: 161–165.

Carvalho, L.M.L. 2010. Toxicology and forensic entomology. In *Current Concepts in Forensic Entomology*. 1st ed., eds. J. Amendt, C.P. Campobasso, M.L. Goff, and M. Grassberger, 163–178. Heidelberg, Germany: Springer.

Catts, E.P. and N.H. Haskell. 1990. *Entomology and Death: A Procedural Guide*, 1st ed. Clemson, SC: Joyce's Print Shop.

Chaturvedi, A.K., K.J. Craft, P.S. Cardona, P.B. Rogers, D.V. Canfield. 2009. The FAA's postmortem forensic toxicology self-evaluated proficiency test program: The second seven years. *Journal of Analytical Toxicology* 33: 229–236.

Committee on Identifying the Needs of the Forensic Sciences Community, National Research Council. 2009. *Strengthening Forensic Science in the United States: A Path Forward.* Washington, DC: The National Academies Press.

Council of European Union. 2009. Council Framework Decision on accreditation of forensic service providers carrying out laboratory activities. Legislative Acts and Other Instruments, 15905/09, JAI 930, ENFOPOL 290, DGH3A, Brussels.

Davies, K. and M.L. Harvey. 2013. Internal morphological analysis for age estimation of blow fly pupae (Diptera: Calliphoridae) in postmortem interval estimation. *Journal of Forensic Sciences* 58: 79–84.

Davies, L. and G.G. Ratcliffe. 1994. Development rates of some pre-adult stages in blowflies with reference to low temperatures. *Medical and Veterinary Entomology* 8: 245–254.

Debry, R.W., A.E. Timm, E.S. Wong, T. Stamper, C. Cookman, and G.A. Dahlem. 2013. DNA-based identification of forensically important *Lucilia* (Diptera: Calliphoridae) in the continental United States. *Journal of Forensic Sciences* 58: 73–78.

Disney, R.H.L. 2011. Forensic science is not a game. *Pest Technology* 5: 16–22.

DNA Advisory Board. 1998. Quality Assurance Standards for Forensic DNA Testing Laboratories, Standard 13. Available at www.fbi.gov/about-us/lab/biometric-analysis/codis/qas_testlabs.

Dourel, L., T. Pasquerault, E. Gaudry, and B. Vincent. 2010. Using estimated on-site ambient temperature has uncertain benefit when estimating postmortem interval. *Psyche* Article ID 610639.

Edgar, R.C. 2010. Search and clustering orders of magnitude faster than BLAST. *Bioinformatics* 26: 2460–2461.

ENFSI. 2003. Guidance on the production of best practice manuals within ENFSI. QCC-BPM-001.

Gaudry, E. 2010. The insects colonization on buried remains. In *Current Concepts in Forensic Entomology*, 1st ed., eds. J. Amendt, M.L. Goff, C.P. Campobasso, and M. Grassberger, 273–312. Dordrecht, the Netherlands: Springer.

Gaudry, E. and L. Dourel. 2013. Forensic entomology: Implementing quality assurance for expertise work. *International Journal of Legal Medicine* 127: 1031–1037.

Gaudry, E., J.B. Myskowiak, B. Chauvet, T. Pasquerault, F. Lefebvre, and Y. Malgorn. 2001. Activity of forensic entomology department of the French Gendarmerie. *Forensic Science International* 120: 68–71.

Gaudry, E., T. Pasquerault, B. Chauvet, L. Dourel, and B. Vincent. 2009. L'entomologie légale: Une identification ciblée pour une réponse adaptée. *Mémoires de la SEF/Société Entomologique de France* 8: 85–92.

Goff, M.L. 1992. Problems in estimation of postmortem interval resulting from wrapping of the corpse: A case study from Hawaii. *Journal of Agricultural Entomology* 9: 237–243.

Goodbrod, J.R. and M.L. Goff. 1990. Effects of larval population density on rates or development and interactions between two species of *Chrysomya* (Diptera: Calliphoridae) in laboratory culture. *Journal of Medical Entomology* 27: 338–343.

Gosselin, M., S.M.R. Wille, M.M.R. Fernandez, V. Di Fazio, N. Samyn, G. De Boeck, and B. Bourel. 2011. Entomotoxicology, experimental set-up and interpretation for forensic toxicologists. *Forensic Science International* 208: 1–9.

Greenberg, B. and J.C. Kunich. 2002. *Entomology and the Law: Flies as Forensic Indicators*, 1st ed. Cambridge, United Kingdom: Cambridge University Press.

Hall, M.J.R., A.P. Whitaker, and C.S. Richards. 2012. Forensic entomology. In *Forensic Ecology Handbook: From Crime Scene to Court*, 1st ed., eds. N. Márquez-Grant and J. Roberts, 111–140. Chichester, United Kingdom: Wiley-Blackwell.

Haskell, N.H. and E.P. Catts. 2008. The paper trail: Case records and reports. In *Entomology and Death: A Procedural Guide*, 2nd ed., eds. N.H. Haskell and R.E. Williams, 160–170. Clemson, SC: Forensic Entomology Partners.

Haskell, N.H. and K.G. Schoenly. 2008. Entomological collection techniques at autopsy and for other environments in the terrestrial setting. In *Entomology and Death: A Procedural Guide*, 2nd ed., eds. N.H. Haskell and R.E. Williams, 102–113. Clemson, SC: Forensic Entomology Partners.

Haskell, N.H. and R.E. Williams. 2008a. *Entomology and Death: A Procedural Guide*. Clemson, SC: Forensic Entomology Partners.

Haskell, N.H. and R.E. Williams. 2008b. Collection of entomological evidence at the death scene. In *Entomology and Death: A Procedural Guide*, 2nd ed., eds. N.H. Haskell and R.E. Williams, 85–101. Clemson, SC: Forensic Entomology Partners.

Heaton, V., A. Lagden, C. Moffatt, and T. Simmons. 2010. Predicting the postmortem submersion interval for human remains recovered from U.K. waterways. *Journal of Forensic Sciences* 55: 302–307.

Higley, L.G. and N.H. Haskell. 2010. Insect development and forensic entomology. In *Forensic Entomology: The Utility of Arthropods in Legal Investigations*, 2nd ed., eds. J.H. Byrd and J.L. Castner, 389–405. London, United Kingdom: CRC Press.

Huntington, T.E. and L.G. Higley. 2008. Collection and analysis of climatological data. In *Entomology and Death: A Procedural Guide*. 2nd ed., eds. N.H. Haskell and R.E. Williams, 144–159. Clemson, SC: Forensic Entomology Partners.

Huntington, T.E., L.G. Higley, and F.P. Baxendale. 2007. Maggot development during morgue storage and its effect on estimating the post-mortem interval. *Journal of Forensic Sciences* 52: 453–458.

ILAC. 2002. Guideline for Forensic Science Laboratories ILAC-G19:2002.

Ikemoto, T. and C. Egami. 2013. Mathematical elucidation of the Kaufmann effect based on the thermodynamic SSI model. *Applied Entomology and Zoology* 48: 313–323.

Kelly, J.A., T.C. Van Der Linde, and G.S. Anderson. 2008. The influence of clothing and wrapping on carcass decomposition and arthropod succession: A winter study in central South Africa. *Canadian Society of Forensic Science Journal* 41: 135–147.

Kelly, J.A., T.C. Van Der Linde, and G.S. Anderson. 2009. The influence of clothing and wrapping on carcass decomposition and arthropod succession during the warmer seasons in central South Africa. *Journal of Forensic Sciences* 54: 1105–1112.

Kelly, J.A., T.C. Van Der Linde, and G.S. Anderson. 2011. The influence of wounds, severe trauma, and clothing, on carcass decomposition and arthropod succession in South Africa. *Canadian Society of Forensic Science Journal* 44: 144–157.

Lagnado, D.A, N. Fenton, and M. Neil. 2013. Legal idioms: A framework for evidential reasoning. *Argument & Computation* 4: 46–63.

Lentini, J.J. 2009. Forensic science standards: Where they come from and how they are used. *Forensic Science Policy & Management* 1: 10–16.

Lewis, A.J. and M.E. Benbow. 2011. When entomological evidence crawls away: *Phormia regina en masse* larval dispersal. *Journal of Medical Entomology* 48: 1112–1119.

Lord, W.D. 1990. Case histories of the use of insects in investigations. In *Entomology and Death: A Procedural Guide*. 1st ed., eds. E.P. Catts and N.H. Haskell, 9–37. Clemson, SC: Joyce's Print Shop.

Martin, J.E.H. 1977. The insects and arachnids of Canada. Part 1: Collecting, preparing, and preserving insects, mites, and spiders. Publication 1643, Research Branch, Canada Department of Agriculture, Hull, Québec, Canada. 182 pp.

Matuszewski, S. 2012. Estimating the preappearance interval from temperature in *Creophilus maxillosus* L. (Coleoptera: Staphylinidae). *Journal of Forensic Sciences* 57: 136–145.

Meiklejohn, K.A., J.F. Wallman, and M. Dowton. 2012. DNA barcoding identifies all immature life stages of a forensically important flesh fly (Diptera: Sarcophagidae). *Journal of Forensic Sciences* 58: 184–187.

Michaud, J.P. and G. Moreau. 2009. Predicting the visitation of carcasses by carrion-related insects under different rates of degree-day accumulation. *Forensic Science International* 185: 78–83.

Michaud, J.P. and G. Moreau. 2011. A statistical approach based on accumulated degree days to predict decomposition-related processes in forensic studies. *Journal of Forensic Sciences* 56: 229–232.

Midgley, J.M. and M.H. Villet. 2009. Effect of the killing method on post-mortem change in length of larvae of *Thanatophilus micans* (Fabricius, 1794) (Coleoptera: Silphidae) stored in 70% ethanol. *International Journal of Legal Medicine* 123: 103–108.

NORM EN ISO/IEC 17025. 2005. General requirements for the competence of testing and calibration laboratories (ISO/IEC 17025:2005).© 2005 CEN/CENELEC BRUSSELS.

Nystrom, J.C. 2008. Standard evidence recovery protocols and procedures at the crime scene. In *Entomology and Death: A Procedural Guide*. 2nd ed., eds. N.H. Haskell and R.E. Williams, 71–84. Clemson, SC: Forensic Entomology Partners.

Pasquerault, T., B. Vincent, L. Dourel, B. Chauvet, and E. Gaudry. 2006. Los muestreos entomologicos: De la escena del crimen a la peritacion. *Ciencia Forense* 8: 39–56.

Penders, J. and A. Verstraete. 2006. Laboratory guidelines and standards in clinical and forensic toxicology. *Accreditation and Quality Assurance: Journal of Quality, Comparability and Reliability in Chemical Measurement* 11: 284–290.

Reibe, S. and B. Madea. 2010. How promptly do blowflies colonise fresh carcasses? A study comparing indoor with outdoor locations. *Forensic Science International* 195: 52–57.

Richards, C.S., I.D. Paterson, and M.H. Villet. 2008. Estimating the age of immature *Chrysomya albiceps* (Diptera: Calliphoridae), correcting for temperature and geographical latitude. *International Journal of Legal Medicine* 122: 271–279.

Richards, C.S., C.C. Rowlinson, L. Cuttiford, R. Grimsley, and M.J.R. Hall. 2013. Decomposed liver has a significantly adverse affect [sic] on the development rate of the blowfly *Calliphora vicina*. *International Journal of Legal Medicine* 127: 259–262.

Richards, C.S. and M.H. Villet. 2008. Factors affecting accuracy and precision of thermal summation models of insect development used to estimate postmortem intervals. *International Journal of Legal Medicine* 122: 401–408.

Ridgeway, J., J.M. Midgley, I.J. Collett, and M.H. Villet. 2014. Advantages of using development models of the carrion beetles *Thanatophilus micans* (Fabricius) and *T. mutilatus* (Castelneau) (Coleoptera: Silphidae) for estimating minimum post mortem intervals, verified with case data. *International Journal of Legal Medicine* 128: 207–220.

Salas, A., A. Carracedo, V. Macaulay, M. Richards, and H.J. Bandelt. 2005. A practical guide to mitochondrial DNA error prevention in clinical, forensic, and population genetics. *Biochemical and Biophysical Research Communications* 335: 891–899.

Schneider, P.M. 2007. Scientific standards for studies in forensic genetics. *Forensic Science International* 165: 238–243.

Simmons, T., R.E. Adlam, and C. Moffatt. 2010. Debugging decomposition data comparative taphonomic studies and the influence of insects and carcass size on decomposition rate. *Journal of Forensic Sciences* 55: 8–13.

Singh, D. and B. Greenberg. 1994. Survival after submergence in the pupae of five species of blow flies (Diptera: Calliphoridae). *Journal of Medical Entomology* 31: 757–759.

Society of Forensic Toxicologists, Toxicology Section of the American Academy of Forensic Sciences. 2006. *SOFT/AAFS forensic toxicology laboratory guidelines.* Society of Forensic Toxicologists Inc. and American Academy of Forensic Sciences, Toxicology Section. 24 pp.

Stuyt, M., A.R. Ursic-Bedoy, D. Cooper, N.R. Huitson, G.S. Anderson, and C. Lowenberger. 2010. Identification of host material from crops and whole bodies of *Protophormia terraenovae* (R-D) (Diptera) larvae, pupae, and adults, and the implications for forensic studies. *Canadian Society of Forensic Science Journal* 43: 97–107.

Szpila, K., M.J.R. Hall, T. Pape, and A. Grzywacz. 2013b. Morphology and identification of first instars of the European and Mediterranean blowflies of forensic importance. Part II: Luciliinae. *Medical and Veterinary Entomology* 27: 349-66.

Szpila, K., M.J.R. Hall, K. Sukontason, and T. Tantawi. 2013a. Morphology and identification of first instars of the European and Mediterranean blowflies of forensic importance. Part I: Chrysomyinae. *Medical and Veterinary Entomology* 27: 181-193.

Szpila, K., T. Pape, and A. Rusinek. 2008. Morphology of the first instar of *Calliphora vicina. Phormia regina* and *Lucilia illustris* (Diptera, Calliphoridae). *Medical and Veterinary Entomology* 22: 16–25.

Szpila, K. and M.H. Villet. 2011. Morphology and identification of first instar larvae of African blowflies (Diptera: Calliphoridae) commonly of forensic importance. *Medical and Veterinary Entomology* 48: 738–752.

Tantawi, T.I. and B. Greenberg. 1993. The effect of killing and preservative solutions on estimates of maggot age in forensic cases. *Journal of Forensic Sciences* 38: 702–707.

Tomberlin, J.K., J.H. Byrd, J.R. Wallace, and M.E. Benbow. 2012. Assessment of decomposition studies indicates need for standardized and repeatable research methods in forensic entomology. *Journal of Forensic Research* 3: 147.

Vanden-Berghe, B. 2009. En France, L'Institut de recherche criminelle de la gendarmerie nationale obtient l'accréditation ISO/CEI 17025, *ISO Management Systems,* www.iso.org/ims. P. 18.

Vass, A.A. 2011. The elusive universal post-mortem interval formula. *Forensic Science International* 204: 34–40.

Villet, M.H. and J. Amendt. 2011. Advances in entomological methods for estimating time of death. In *Forensic Pathology Reviews*, ed. E.E. Turk, 213–238. Heidelberg, Germany: Humana Press.

Villet, M.H., B. Mackenzie, and W.J. Muller. 2006. Larval development of the carrion-breeding flesh fly *Sarcophaga (Liosarcophaga) tibialis* Macquart (Diptera: Sarcophagidae) at constant temperatures. *African Entomology* 14: 357–366.

Villet, M.H., C.S. Richards, and J.M. Midgley. 2010. Contemporary precision, bias and accuracy of minimum post-mortem intervals estimated using development of carrion-feeding insects. In *Current Concepts in Forensic Entomology*, 1st ed., eds. J. Amendt, C.P. Campobasso, M.L. Goff, and M. Grassberger, 109–137. Heidelberg, Germany: Springer.

Wetzel, W., S. Reibe, and B. Madea. 2009. An entomological case report during the winter months: Estimation of the post-mortem interval considering the influence of cold temperatures on the development of the forensically important blowfly *Calliphora vomitoria. Archiv für Kriminologie* 223: 123–130.

Williams, K.A. and M.H. Villet. 2013. Ancient and modern hybridization between *Lucilia sericata* and *Lucilia cuprina* (Diptera: Calliphoridae). *European Journal of Entomology* 110: 187–196.

Worner, S.P. 1992. Performance of phenological models under variable temperature regimes: Consequences of the Kaufmann or rate summation effect. *Environmental Entomology* 21: 689–699.

Zhou, C. and R.W. Byard. 2011. Factors and processes causing accelerated decomposition in human cadavers: An overview. *Journal of Forensic and Legal Medicine* 18: 6–9.

CHAPTER 30

International Collaborations and Training

Beryl Morris, Michelle Harvey, and Ian Dadour

CONTENTS

30.1 INTRODUCTION

With the number of insect species far outnumbering the rest of the animal species presently on Earth, the surprising aspect of forensic entomology is not that biological information derived from insects is a useful tool for assisting in legal matters, but rather, so few resources are devoted to forensic entomological education, training, and research. An examination of the traditional and emerging uses of insects in forensic situations amply shows the breadth of knowledge required by entomologists, opportunities for further development of the field, and in parallel, the potential intellectual capital available from scientific studies of insects and their arthropod relatives for applications to the courts.

Traditionally, both civil and criminal law have benefited from the entomologist's knowledge of insect taxonomy, physiology, and behavior. Typical uses (Catts and Goff 1992) include:

- Urban entomology, which relates to civil actions arising from disputes about insects and human-built structures, as may occur with termites (Isoptera) and buildings
- Stored product entomology, for instance in civil actions related to insect infestations of commodities such as food
- Medicolegal entomology, which most commonly relates to criminal cases involving the estimate of time since death for decomposing remains of humans or animals

Although entomological expertise continues to be used by the legal profession mainly in these three areas, potential applications of forensic entomology are continually evolving, particularly where informed by interdisciplinary research collaborations and innovative use of scientific apparatus largely developed for other scientific fields. A range of special legal cases has shown that knowledge about insects has aided in the detection of the following:

- Toxins, drugs, and gunshot residues (Beyer et al. 1980; Crosby et al. 1986; Gunatilake and Goff 1989; Catts and Goff 1992; Roeterdink et al. 2004)
- Traffic accidents with no immediately obvious cause (Dadour and Harvey 2008)
- Determining the site of human death (Byrd et al. 2009)
- Criminal misuse of insects (Leclercq 1969)
- Movement of vehicles and transport of remains (Smith 1986)
- Injuries after death (Haglund et al. 1989)
- Movement of people as determined by bites or infestations (Prichard et al. 1986)
- Neglect of the elderly (Benecke 2003)
- Child abuse (Goff et al. 1991)
- Neglect and abuse of animals in veterinary and wildlife forensics (Anderson 1999; Merck 2007; Tomberlin et al. 2012)

The breadth of potential uses for insects in legal cases and the capacity to utilize the accuracy of information derived from such applications offer seemingly boundless opportunities for forensic entomologists. If we accept science as a communal and globalizing activity, then increased international collaboration is certainly an aspirational goal for our widespread community in realizing the opportunities and challenges available in this field. Indeed, collaborators, regardless of language, time differences, location, and economic background are able to use a spectrum of technologies to combine diverse forms of expertise for the benefit of bringing novel mixes of knowledge, products, and solutions to problems (Walsh and Maloney 2002).

Such a conceptualization of collaboration surely allows for more than coauthorship in scholarly publications. As Katz and Martin (1997) envisage, collaboration necessitates a demonstration of active involvement in research projects, requiring an intense form of interaction among the participating knowledge producers with every effort being made to achieve effective communication within the group so that there can be an equitable sharing of skills, competencies, and resources until a final report or outcome is generated together.

Information and communications technology are readily available in most countries and are central to facilitating long-distance collaborative work (Ynalvez et al. 2009). Indeed, it has been long acknowledged that researchers prefer to collaborate internationally rather than with their geographically proximate colleagues (Godin and Ippersiel 1996), a not-surprising observation given that scientific competition is foremost international rather than national. Current funding sources complement the tendency to work with colleagues at a distance rather than those closer to home, as evidenced by the recent explosion in dispersed collaboration spawned by funding agencies such as the National Science Foundation (NSF) in the United States, the Framework Programmes in the European Union, and the Australian Research Council, all of which aim for diverse organizational representation in research funding applications. The trade-off to such innovation opportunities is the coordination costs to the researchers in overseeing communication and data interchange, an important issue for forensic entomologists who generally operate by themselves or in small teams at the best of times (Magni et al. 2013).

At an informal level, the digital age has made possible the proliferation of networks at the local, country, and international level, which has provided the framework for the current knowledge-based mode of development of forensic entomology. This framework is exemplified by the many public and subscription-only websites that provide a record of publications and describe the history and other aspects of the field. Links to forensic departments or personal websites dealing with forensic entomology are available from the website of the European Association of Forensic Entomology. Three examples of public websites are as follows:

1. American Board of Forensic Entomology (ABFE)
   http://www.forensicentomologist.org/
   The ABFE site provides a short overview of the science and history of forensic entomology, as well as case studies in forensic entomology.
2. Forensic Entomology Pages, International
   http://folk.uio.no/mostarke/forens_ent/forensic_entomology.html
   Created by Morten Stærkeby, an independent consultant in forensic entomology in Norway, this website provides an overview of the many uses of insects and other arthropods as evidence and case histories.
3. Forensic Entomology: Insects in Legal Investigations
   http://www.forensicentomology.com/index.html
   Created by Dr. J. H. Byrd, formerly of Virginia Commonwealth University, this site includes definitions, death scene procedures, life cycles, information on entomological collection equipment, a PDF entomological field notes death scene form, and further links.

Many factors contribute to the unique nature of forensic entomology. Among these factors are the lack of a universally recognized forensic entomology education pathway, few stand-alone employment opportunities, and scarcity of funding for education and research. In addition local specialization is necessitated by geography, legal jurisdictions, language, regulatory systems, politics, and faunal taxonomy. Despite such specialization, regional specialists find collaboration opportunities through the universality of scientific methodology and scientific apparatus, membership of specialist, multi-country associations, and shared knowledge of insect behavior and physiology.

With widespread availability of information technology, it is increasingly possible to see evidence of its use in an expanding level of collaboration in forensic entomology with two or more forensic entomologists working together on research projects, sharing their knowledge, skills, and resources for achievement of a specific goal or new insight, regardless of their respective home base. An example of this type of collaboration is exhibited between the University of Western Australia (Centre for Forensic Science) and Italian Law Enforcement to produce the first forensic entomology mobile phone app, the multilingual SmartInsects (discussed in Chapter 14), designed as an instructional and training tool to assist non-entomologists with the collection and preservation of insects at a crime scene. Other signs of collaboration are the upsurge over the past decade in books with an international cast of editors and chapter authors and an increasing number of scholarly papers published with authors from multiple countries. Even with such achievements, forensic entomology remains a subject discipline rather than a profession. Of necessity, its practitioners largely obey the norms of independence and individualism rather than teamwork and collaboration. The main contributors to this phenomenon are expanded on in the remainder of this chapter.

30.2 CAREERS

Almost everyone is aware of some level of detail of forensic entomology—its public profile has been quite prominent for a number of years. Newspaper and television reporting of evidence provided in high profile murder trials, fame of the "Body Farm" at the University of Tennessee, Knoxville (UTK) Anthropology Research Facility, Tennessee, United States (Rodriguez and Bass 1983), and the continuing extraordinary popularity of the crime-writing genre have guaranteed frequent use of insect-based storylines in books and high profile television crime dramas, ranging from *Quincy* in the 1970s to the more contemporary long-running American television series, *CSI—Crime Scene Investigation*.

Given such widespread exposure and general public awareness of forensic entomology, it is a surprise for students looking at career options to learn that forensic entomologists are generally not employed in forensic science crime laboratories (Magni et al. 2013). Indeed, the total number of practicing forensic entomologists worldwide is relatively small, no more than perhaps a hundred, and career opportunities are limited. In some cases, individuals may function as part-time forensic entomology consultants but in general, forensic entomologists are most often university based with small numbers working in museums, law enforcement agencies, or in private employment (Magni et al. 2013). These entomologists are often called on to assist with cases on an "as required" basis by medical examiners, coroners, law enforcement agencies, or attorneys, depending on the practices of their respective local situation. The remainder of their time is generally devoted to teaching; research; conducting workshops to teach detectives, crime scene field officers, and others on the use of insects in crime scene investigations; serving on committees; and preparing graduate students for roles in forensic entomology or other related disciplines.

There are also entomologists who are rarely involved in a forensic case and so are not known in the guise of a practicing forensic entomologist, except to regional police and forensic laboratories. These are likely to be the entomologists who work in museums, universities, or agricultural departments and who are called on due to their ability to identify an insect that has been found in an investigation. They may also be called on due to their role in medical entomology, perhaps in a hospital, or veterinary parasitology, or entomology department in a university. Following an initial unplanned brush with crime scene investigation and given the fascination of the field, there are known instances of such individuals changing their insect species specialization and career direction to become more formally recognized as contributing to the field of forensic entomology.

30.3 FORENSIC ENTOMOLOGY ASSOCIATIONS

Even when counting entomologists, who occasionally provide assistance with legal cases along with those practitioners who have studied forensic science, conducted research purposefully in the forensic entomological field, and who have more substantial courtroom experience, the number of forensic entomologists in the world is relatively small. This is evidenced to some extent by results of two surveys, the first conducted in 1986, and the second, 23 years later, in 2009. In the first survey, carried out by Lord and Stevenson (1986) to compile a database of practitioners, just 62 forensic entomologists from six countries were recorded. Later, in 2009, 70 forensic entomologists from 24 countries responded to a survey designed to determine who in the world is practicing forensic entomology and in what capacity (Magni et al. 2013).

Identifying who is involved with forensic entomology has been made easier by use of the membership lists of forensic entomology associations. These entities exist as national organizations, which as yet do not make decisions on the practice of forensic entomology but instead, play a constructive role in disseminating news about emerging technologies and publications. Examples are the North American Forensic Entomology Association (NAFEA), the European Association of Forensic Entomology (EAFE), Gruppo Italiano per l'Entomologia Forense (GIEF), and the Malaysian Association of Forensic Entomology (MAFE). The beneficial role of such groups flows beyond their membership as they also assist in disseminating credible information through their websites to those with an occasional need to touch base with forensic entomology, such as lawyers, and to curious members of the general public.

In addition to the publicly accessible forensic entomology associations, there are subscription-based groups that distribute information to their member forensic entomologists and students of forensic entomology quickly and informally, as is the practice of the Forensic Entomology Yahoo Group. A final category of forensic entomology entities is the scientific working group (SWG). This category generally occurs within broader scientific assemblages, such as forensic science associations

Table 30.1 European Association of Forensic Entomology (EAFE), North American Forensic Entomology Association (NAFEA), and Other Forensic Entomology Association Member Countries (2013)

Algeria	Argentina	Austria	Australia	Belgium	Bosnia Herzegovina
Brazil	Bulgaria	Cameroon	Canada	Chile	Colombia
Croatia	Czech Republic	Egypt	Estonia	France	Germany
Greece	Hungary	India	Iran	Ireland	Italy
Kuwait	Lithuania	Malaysia	Netherlands	Norway	Poland
Portugal	Qatar	Romania	Serbia	South Africa	Spain
Sudan	Sweden	Switzerland	Turkey	United Kingdom	United States

and largely at a national level. For example, the Entomology SWG in Australia, which formed under the Senior Managers Australia and New Zealand Forensic Laboratories (SMANZFL) reports to the Australia and New Zealand Policing Advisory Agency National Institute of Forensic Science Forum (ANZPAA NIFS 2012). In the Australian and New Zealand instance, the Entomology SWG is not an information dissemination group or representational of any practitioners in these countries, although it does operate to influence decisions at the national level, which potentially impact on practitioners of forensic entomology in the region.

Members of most of the aforementioned forensic entomology associations and technical communities are not true indicators of the number of forensic entomology practitioners as their membership lists typically include forensic entomologists as well as biologists, medicolegal practitioners, other forensic experts, and people with a general interest in the topic. Indicative of their personal levels of interest and collaboration, some individuals are members of multiple associations. Of these, EAFE, and to some extent NAFEA, have assumed the roles of global associations, having a broad geographic base, which at this time collectively spans over 40 countries, as shown in Table 30.1.

Operating as global professional associations and providing current information on their websites has attracted a critical mass of participants for EAFE and NAFEA, with the highest number of members coming from the United States and United Kingdom. Thus, NAFEA and EAFE give their members access to the largest networks. In turn, network size translates to greater diversity of collaborative opportunities and variety of resources. These associations also usefully parallel the history of forensic entomology, showing that no country has a monopoly on it. Although the national associations continue to grow, there are, however, insufficient numbers of forensic entomologists to sustain a network of local chapters, leaving the more generalist entomological societies and clubs or forensic science societies in local jurisdictions to cater for entomologists' need for a local forum to which they can reasonably travel for the purpose of networking and seeking professional support. The membership of such groups is generally broad and attendance at meetings large enough to allow inclusion of talks and posters on matters related to forensic entomology, alongside a multitude of other specialty fields, which now exist among the diverse group of entomological subdisciplines.

Forensic entomology associations serve the valuable role of providing continuing education, which is a primary benefit of membership. Continuing education is available in many formats: at conferences, meetings, online, or in journal articles. The associations help their student members to further their education through the availability of scholarships and research grants. Furthermore, through their annual research conferences, the associations provide avenues for undergraduate and graduate students to hone project and presentation skills, which assist in assuring the future publication of their work. The associations also allow student members to access practitioners with whom they can negotiate internships to help them establish their own careers and from whom they can learn subtle values and priorities not easily communicated elsewhere.

A less tangible benefit of forensic entomology associations is the "glue" that they provide to a membership that is relatively low in number, widely distributed geographically around the world and otherwise, professionally isolated. Interacting with other forensic entomologists through the associations provides a sense of belonging, a collective identity, and a broader perspective is gained with the realization that other entomologists face and overcome the same obstacles. Members can also draw the attention of those in the law enforcement and legal professions to their participation in one or more of the forensic entomology associations as a means of showing their credentials as a practitioner able to provide expert evidence.

30.4 GLOBAL APPLICATION OF FORENSIC ENTOMOLOGY

The world's forensic entomologists have much in common. They face similar research challenges, apply the same scientific methodology, study the same kinds of evidence, and access global research databases. Nevertheless, some regional heterogeneity inevitably exists. History, language, culture, and economic situations are obviously variable in the world and to a large degree, persist in resource availability, training, research quality, and organization. For those collaborating with distant places and different cultures, recognition of this variation is vital to understanding the heterogeneity in training, research, and deployment of forensic entomology.

The contrast between forensic entomology practice in Europe and the United States with some other parts of the world shows the global dissimilarity of forensic entomological practice. Forensic entomology has been beneficial in crime scene investigations for more than a century in some parts of Europe where several of the countries shown in Table 30.1 pioneered the routine use of the discipline, as discussed in other chapters. In comparison, for a growing number of countries, for example, Cameroon, Chile, Colombia, Iran, Sudan, and Turkey, the first applications of entomology to legal situations have been undertaken and published only this century and in others, such as some Middle Eastern countries, entomology is yet to be used in legal investigations.

Differences in the maturity of forensic entomology knowledge and application in different countries are not necessarily correlated to levels of regional economic development. Today, with the pervasiveness and accessibility of the Internet, researchers and students pioneering forensic entomology in their region are not professionally isolated and are able to quickly be part of the larger entomological community, both nationally and internationally, accessing scientific support, mentoring, resources, training, casework, and perhaps travel scholarships through the forensic entomology associations and informal networks such as the Forensic Entomology Yahoo Group. Collaboration and cooperation among forensic entomologists is critical in promoting quality outcomes, which in turn enhances the credibility and wider uptake of the science, to the benefit of all forensic entomologists.

For most countries, and particularly those that have complex and diverse ecosystems, such as India and Malaysia, the current priority is to stimulate use of forensic entomology by establishing databases on distribution, niche preferences, life cycle, and identification characteristics for the key regional species (Sumodan 2002). Even in countries where forensic entomology is routinely used as a tool in reconstructing the history of corpses found on crime scenes in accordance with principles found in the rapidly developing body of literature, there is still much to learn about forensically useful insects. Examples of regional gaps include the taxonomy of lesser known carrion insect species, seasonal occurrence of carrion species in the country's various geographical regions, and rates of development of the local species modeled in the many site situations of forensic interest.

Known paradigms for forensically important insects are also under challenge as insect populations move beyond human-established geographical borders. The ease of international transportation and increasing global trade allow the convenient dispersal of insects and even the invasion of new pests and associated diseases into new environments (Ward et al. 2006). Some species will flourish in ways we cannot predict whereas others will inevitably not adapt to change (Hill et al. 2011; Dunn and Fitzpatrick 2012). The outcome is likely to impact on the insect ecology of cadavers in each region. Climate change is already causing global changes in species diversity and distributions (Turchetto and Vanin 2010). These changes unleash endless opportunities for forensic entomological research and the forging of collaborations, which will be made easier by the Internet and open access publication sources making possible synchronous and asynchronous communication on a global basis about entomological problems and solutions.

30.5 EDUCATION

The formal education required for a forensic entomologist is broad and is derived from two postsecondary categories: undergraduate and graduate level study. To date, there is no specifically named university bachelor degree awarded through a university or college program, which allows graduation as a forensic entomologist. Thus, most forensic entomologists have developed their expertise over a considerable time following their initial education. At the undergraduate level, individuals make their own course choices, selecting according to personal preference and aptitude from among courses such as biology, chemistry, genetics, taxonomy, medical entomology, biochemistry, forensic law, parasitology, statistics, and, naturally, entomology. Most practicing forensic entomologists have completed a unit in entomology at undergraduate degree level, graduating with a bachelor degree generally in science, applied science, agricultural science, or forensic science. More recent graduates have most likely also taken a molecular biology course due to the increasing importance of this field to forensic science and to resolving problems in the area of insect identification. Figure 30.1 shows an undergraduate class undertaking molecular biology experiments as part of a forensic entomology course.

Figure 30.1 Molecular biology classes are taken by entomology students (Courtesy of M.L. Harvey).

Ultimately, the success and credibility of forensic entomology relies on the entomological evidence being analyzed by a person qualified to provide an opinion on how the evidence fits the facts of a particular case. Whether or not an individual is qualified to deliver an expert opinion in the court of law regarding a case involving entomological evidence depends on the court's approval, his or her education, experience, and personal attributes. The expectation that the more credentialed expert will be best received by the jury militates toward choosing the entomologist with the highest relevant academic degrees and certainly if both prosecution and defense call experts, the entomologist with the higher degrees will most likely be perceived as the more authoritative and credible. Thus, most practicing forensic entomologists have postgraduate qualification, which may be a PhD, masters, or MD in entomology or a related field and may be board certified in the discipline (see Section 30.7).

In the United States, there are at least five universities offering masters coursework programs related to forensic entomology: University of California Davis, Michigan State University, Purdue University, Texas A&M University, and University of Nebraska Lincoln. In Australia, since 2002 it has been possible at the University of Western Australia to undertake a combined master/PhD of forensic science, a program that allows students to undertake the coursework master's program, covering all the basic skills for a technical career in forensic science, in parallel to carrying out independent PhD research in a discipline of study such as entomology. The result is a graduate program delivering strong laboratory and research components.

The doctoral programs offered around the world are based on research and analysis. Entry requirements for PhD research on insects will vary by both country and institutions. In some cases, a university may require completion of a 1-year honors program at the end of an undergraduate degree during which a project is undertaken to show an individual's capacity for independent research. In other universities, the requirement is a master's degree in entomology by research. The unique study and research infrastructure of the institution in which the study occurs will dictate to a large extent the nature of the research that is undertaken. Examples of resources and infrastructures, which play a role in the scope of a forensic entomology research project, include the availability of insect-rearing facilities, a field site for carrion studies, and a molecular biology laboratory. Depending on whether a student is enrolled fulltime or part-time, PhD programs may take up to 7 years to complete.

Postgraduate education and training often present opportunities for early development of international networks useful for future collaborations. In some cases, students benefit from the networks of their supervisor, particularly if they lead to time spent in overseas laboratories acquiring new skills, perhaps funded by scholarships or collaborative research programs. Alternatively, the graduate student may undertake their degree abroad, possibly acquiring additional language skills in addition to the degree. Provided the students are able to gain the initial funding from their government or competitive scholarships, graduate training abroad, which can last anything from 2 to 5 years, is an attractive path for students from countries where forensic entomology is not well developed and hence, potential supervisors are scarce. In addition to the prospect of lifelong networks and future collaboration on scholarly publications and grant applications, international students have an opportunity to acquire and develop their technical skills and familiarize themselves with cutting-edge research. Ynalvez and Shrum (2009) highlighted the impact on future collaboration of the type of study undertaken and the mentoring style on the likelihood of students undertaking collaborations in the future with researchers from the country in which they studied. For example, a PhD involving some course work dilutes time spent with a supervisor and research colleagues. In comparison, a PhD completely focused on research and strongly mentored within the supervisor's laboratory for the duration of the degree will involve total immersion with a supervisor and there is a strong probability that the graduate student will collaborate with members of that laboratory over the remainder of his or her career. Regardless of whether a graduate student undertakes their research project locally or internationally, it is probable that when they later establish their own laboratories, they will conduct research using the same ideas, principles, and similar interests to those from their original research groups and hence, the impact of alma mater on internationalization of forensic entomology is significant.

30.6 PROFESSIONAL SKILLS AND PERSONAL ATTRIBUTES

No matter in which country forensic entomologists work, results of their efforts are destined for use in court. This is where a number of soft skills and personal attributes are as important as qualifications and experience in determining the ultimate effectiveness of individuals against such a formidable backdrop. Given the absence of a formal registration process as a forensic entomologist, persons preparing themselves for recognition in the field should ensure they develop or acquire such attributes, in addition to the very specific technical skills and entomological knowledge arrived at through their undergraduate and graduate degrees.

Individuals who are expert witnesses in the area of forensic entomology need more than skills in insect identification. They need to be experienced in identification of all life stages of the insect in question and evaluating the factors that affect the rate of growth of those insects, skilled in the quantitative methodologies to estimate a minimum postmortem interval (mPMI), and have a broad and thorough understanding of insect ecology. Of great importance is the need to understand factors involved in controlling insect growth rates on animal carcasses and a preparedness to adapt this knowledge to the human situation. Experience in collecting specimens from cadavers, or an understanding of the difficulties involved, so that appropriate allowances can be made for influences associated with collection and preservation by non-entomologists, is essential. Participants in the legal and court processes should be mindful that although an individual may be an expert in one of these aspects, such as insect identification, it would be prudent to engage someone who has a broad range of entomological skills and, importantly, experience in the application of such skills to forensic situations.

Customarily forensic entomologists must show independence and resilience due to their professional isolation, but there will be times when they need to work with police and other forensic scientists in examining physical evidence. Rapport with others is therefore a necessary attribute in such teamwork. Further, it assists in establishing working relationships with forensic science laboratories, delivering workshops to forensic field officers, and is a particular asset to the entomologist throughout questioning in court. Here, a sense of camaraderie assists in maintaining calm and patience and preventing future hostility with any other entomologist who may be called to give expert evidence in the same case, presumably for the opposing legal team.

When looking at a crime scene or physical evidence, forensic entomologists must deploy the excellent problem-solving and critical-thinking skills that helped earn them a graduate qualification. In an investigation, the entomologist must not only have an eye for detail but be curious and able to pose questions, such as "which insects should be here at this time of year; is something missing and if so, what might be the explanation?" Unanswered questions may even become excellent problems for future graduate student research projects.

A forensic entomologist also needs organizational skills, particularly with respect to maintaining continuity of evidence provided to them for legal cases and for structuring reports. Such skills are important too for maintaining robust administrative systems over a lengthy time span. Effective administrative systems are invaluable for retrieving information and specimens for court, perhaps after the passage of years from the time of initial receipt of material until the legal profession deems there is sufficient evidence to take a case to court. Organizational skills are an aid in court as well, especially for fluently referring to supporting paperwork or specimens while under question.

For expert opinion to add value to the judicial process, it must provide the court with information that is likely outside the court or jury's knowledge and experience but which nevertheless gives the court the help it needs in forming evidence-based conclusions. To succeed in the role of expert, an entomologist therefore requires well-developed written and oral communication skills and an ability to deploy these without calling upon unnecessary technical jargon or acronyms. Other attributes that help the entomologist's credibility in court are good personal grooming and appearance, which show professionalism and respect for the court. Most importantly, forensic entomologists must be capable of integrating their personal skills and subject knowledge to show competence in the full spectrum of examination, analysis, interpretation, reporting, and testimonial support of physical evidence. This level of competence is achieved only with experience and that is not easy to acquire in an area such as forensic entomology, where casework and court appearances may be quite sporadic, either due to the presence of alternative, possibly more experienced forensic entomologists in a region, or just because the use of entomological evidence is as yet a rarity in a local jurisdiction.

30.7 STANDARDS

Although most forensic entomologists generally hold advanced degrees, there is no mandatory universal jurisdiction standard against which the merits of aspiring forensic entomologists are tested. Offsetting this, the ABFE has instituted a practice of inviting membership from forensic entomologists worldwide, providing a peer review certification process that includes a written and practical examination. Details of membership status, which varies according to the level of certification, that is, member or diplomat, can be found at www.forensicentomologist.org/.

In most other countries, recognition is generally achieved through the courts where, in addition to demonstrating educational qualifications, the supposed expert forensic entomologists must also be able to declare consistent involvement with research or teaching of forensic entomology. The usual way in which this is achieved is through peer-reviewed publications relevant to the subject in reputable journals along with entomological teaching or workshops. Evidence of involvement with such activities is used to show that the expert has hands-on, day-to-day involvement with forensic entomology rather than someone who is "book smart but without actual experience" (Hall 2009).

Worldwide, many researchers and practitioners recognize the advantages that would follow from establishment and enforcement of professional standards for the forensic entomology discipline. As discussed by Wells (2003), in other fields this is achieved by defining analytical procedures that, if followed, produce an accurate conclusion (quality control) plus a system of oversight to ensure those procedures are followed and that they are properly working (quality assurance). These processes may include the certification (internal process) or accreditation (external process) of individuals or an entire laboratory. As forensic entomologists are generally not employed in forensic science laboratories and are more or less freelancers who are involved in investigations on a consulting basis, rarely do they carry out their tests and research in a laboratory accredited under standards issued by the International Organization for Standardization, such as ISO/IEC 17025, which is the most common standard applied to government forensic science laboratories across the globe. Accreditation under ISO/IEC 17025 allows laboratories to carry out procedures in their own ways, but an independent auditor (assessor) may require the laboratory to justify using a particular method. In common with other ISO quality standards, ISO/IEC 17025 requires continual improvement and so accredited laboratories carry out their own regular internal audits to show not only the uptake of opportunities to make the test or calibration better but also awareness of scientific and technological advances in areas relevant to the laboratory.

For forensic entomologists, particularly those who belong to an association such as NAFEA or EAFE, it is probable they are up-to-date with respect to new ideas and that they spend large amounts of time conducting research or supervising that of graduate students, which is aimed at broadening the scope of the field. Such currency of knowledge is an essential component of helping the court to understand the reliability of tests carried out or opinions provided. Further, given the protocols available from NAFEA and EAFE, it is reasonable to assume conformance to currently accepted best practice, as judged by most members of the associations, in collection of evidence, accurate and complete documentation of the collection, and continuity of evidence process. Beyond that, an accreditation system is possibly still premature for forensic entomology because there are many areas for which there is no validated test of the efficacy and accuracy of procedures.

Determination of an acceptable range for PMI estimates and establishing levels of sensitivity of such estimates are obvious examples of the lack of consistency among forensic entomology practitioners and ones which cannot be resolved until the various methods and models used by practitioners are tested repeatedly. Undoubtedly, this shortcoming would be aided by access to a common database recording cases where PMI has known values but experience suggests that PMI may never be truly "reliable" given variation in a myriad of parameters for example, growth rates for a single species in different locations or understanding the many microclimate issues that can limit generalizing results to different localities. To come near to providing the necessary validation of techniques and models, cases would need to have occurred in a wide variety of environmental conditions and each record would require inclusion of extensive details of the prevailing micro- and macroenvironmental conditions, insects present, techniques used, and assumptions made. This approach to validation would be an ambitious and long-term project requiring cooperation from multiple individuals around the world. The singular nature of forensic entomologists and the low base of support funding suggest that the ambition will not be quickly realized. In addition, too often, scientists do not publish what they do or observe, for example, when collecting molecular data from sequencing or undertaking localized decomposition studies. These are missed opportunities to develop the intellectual capital of forensic entomology.

30.8 TRAINING NON-ENTOMOLOGISTS

In recent years, skilled and experienced forensic entomologists around the world have been providing instructions or training to the many individuals who attend crime scene investigations. This training is often done in combination with other forensic specialists. For example, Evidence Response Team members from Federal Bureau of Investigation (FBI) field offices across the United States, in partnership with the University of Tennessee Medical Center at Knoxville, annually attend the "Human Remains Recovery School" at the UTK Anthropology Research Facility where they are provided with both lectures and hands-on practice training by international subject experts in disciplines such as forensic entomology, forensic anthropology, pathology, and odontology. Figure 30.2 illustrates similar training being undertaken by police officers in Australia.

Generally, there is a regular turnover of police or other field officers participating in crime scene investigation and this requires training be continuously available for those new to evidence collection. Without this training, there are many missed opportunities for using entomological expertise in accounting for the presence or absence of insects. There are also many instances where entomological material is collected by those without training who unknowingly use poor methodology, suboptimizing the extent of analyses and information that might otherwise have been extracted from such material. Poor collecting is slowly being rectified in places such as Europe, Canada, United States, and Australia with standardized insect-collecting kits recommended to crime scene investigation teams to aid in collecting insects from a corpse (see https://www.bioquip.com/search/DispProduct.asp?pid=4227 and Figure 30.3 for examples). Further, EAFE has made available a protocol document for best practice in forensic entomology, which includes an overview of equipment used for collection of entomological evidence and a detailed description of the methods applied (Amendt et al. 2007). Despite efforts to have standard operating procedures available for practitioners, many deficiencies still occur in the methodology used for collecting material. Such deficiencies become particularly important when an experienced forensic entomologist is not directly involved from the outset of an investigation of unattended death. In addition to the potential for lost opportunities in

Figure 30.2 Training Western Australian Police, Perth, Australia (Courtesy of I.R. Dadour).

shedding light on a crime, errors and omissions in crime scene collection of entomological evidence by an inexperienced and untrained person have been known to cause later embarrassment in court for entomologists questioned on matters such as the lack of collecting data, poor preservation technique of the insect material, or whether the oldest insect larvae were collected.

As it is impossible to attend every crime scene, entomologists by necessity must provide training for non-entomologists to work on their behalf in collecting insect evidence and related data. By way of example, the University of Western Australia's Forensic Entomology Accreditation Course is offered regularly over 2 days for Police Crime Scene Practitioners in Australia and 5 days for a similar course in South Africa. The workshop allows participants to learn about the collection and preservation of insect material from a crime scene. On completion of the workshop, each police officer is certified by the University of Western Australia as a Practicing Forensic Investigator (Entomology). Forensic entomologists around the world provide similar experiences to police and other field investigation officers, ensuring participants have "hands-on" experience of the rudiments of forensic entomology sufficient for them to be aware of the discipline, crime scene procedures relevant to insects, and the latest research on time since death estimations, insect succession, and temperature-dependent development.

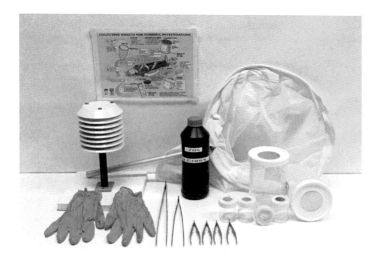

Figure 30.3 An insect collecting kit, as used in Australia (Courtesy of I.R. Dadour).

30.9 RESEARCH

Entomologists fit into a team of forensic scientists who collaborate to investigate unattended deaths typically, but also civil issues involving urban or commercial product infestations of insects. Marks et al. (2009) describe the case for pathologists, entomologists, and anthropologists being considered as part of a continuum or team, working together for improved information on estimating time since death. In their view, forensic pathology is about the process of autolysis, entomology with the early decay stages, and anthropology with the later stages. As such, the pathologist first relies on the entomologist and then the anthropologist to assist with time since death assessments. Regretfully, such interactions between subject specialists working on cases are still uncommon, often due to the scarcity of funds to support the forensic science interpretation of physical evidence associated with a case.

The patchy distribution of forensic entomologists, rapid pace of scientific and technological advances in various areas, and changing nature of public sector research funding and accountability in most countries have all been serious challenges to achieving any significant innovation and uniformity in forensic entomology. Lack of funding especially places forensic entomologists at risk of making inferences from single-event observations, which is poor scientific practice (Tibbett 2013). Although forensic entomology often falls outside the research funding remit for major granting bodies, there are signs of change. Tomberlin et al. (2011), for example, highlight the transformational research of forensic entomologists working in the ecological realms of decomposition ecology, generating the type of publications necessary for academic researchers to gain further competitive research grants and obtain promotion in academic and research institutions.

From DNA analysis to radiography, few of the technologies that have featured in forensic entomology articles in the past decades have been developed specifically for the field. However, entomological researchers, in common with other scientists and perhaps aided rather than disadvantaged by careers that commenced in related disciplines, maintain awareness of broader scientific developments, evaluating and adopting those technologies that offer the promise of resolving intractable problems in forensic entomology, such as quick and reliable identification of early-stage larvae of closely related species (Harvey et al. 2003). In many cases, new technologies are expensive to acquire and maintain, with many requiring specially trained personnel to operate them, as is the case with, for example, computed tomography (CT) scans, used for locating larval fly masses on corpses (Johnson et al. 2012). Thus, acquisition of such specialist equipment lies outside the reach of most forensic entomologists and access is generally negotiated through collaborative arrangements with other scientists, thereby leveraging scarce funds in exchange for joint publications. Although a new technique based on sophisticated technological analysis could improve case analysis, it may remain out of reach if the cost exceeds the budget of the researcher or their institution. Likewise, in an applied field such as forensic entomology, development of expensive technical analyses need to be balanced against the resources of the wider criminal justice system and provide demonstrable value for money to the taxpayers (House of Commons Science and Technology Committee 2013).

Fortunately, there is a wealth of research, which is still needed from all regions to generate background data necessary for assessing the evidential value of the entomological findings for the courts. As mentioned earlier in this chapter, climate change has a major impact on insects and hence, the need for continuing ecological and behavioral research is ongoing and much can be achieved with relatively modest levels of funding. Although areas for current and future research have been covered in other chapters, a summary of ongoing data needed from experiments and observations to provide a quality baseline from all regions is listed in Table 30.2.

Table 30.2　Common Research Topics for Forensic Entomology

Research Area	Issues	Potential for Research and Collaboration
Identification	Distinguishing between closely related necrophagous species at all stages of their life cycle	Molecular biology techniques for identification of a broad range of species, not just necrophagous
		Replication of DNA-based techniques
		Standardized molecular biology methodology and choice of markers
		Chemotaxonomy
		Classical taxonomic keys
		Trapping
Succession ecology of necrophagous species on carrion and corpses	Lack of site, season and specific baseline data, even with more or less identical insect fauna	Study of factors such as illumination, carcass/corpse size, time of the year, and type of habitat
		Ability to replicate results
		Statistical significance of variations
		Computer resampling methods
		Indoor and outdoor effects
		Various types of cover
Aquatic environments	Lack of species baseline data	Taxonomic keys for life stages of arthropods with forensic potential
	Site-specific life cycle data	Rates of development of aquatic forensic indicators
	Succession ecology	
		Seasonality effects on arthropod fauna
		Effect of carcass size and species on arthropod fauna
		Marine vs. freshwater differences
		Floating vs. submerged corpses
		Succession in water following death on land and vice versa
Movement of a corpse	Distribution	Carrion fly species found in urban and rural biomes
	Seasonality	Trapping in different biomes throughout the year
	Residues	Detection of toxins, drugs, and host DNA

Table 30.2 Common Research Topics for Forensic Entomology (*Continued*)

Research Area	Issues	Potential for Research and Collaboration
Estimating time elapsed since death	Knowledge inferred from too few environmental test sites (e.g., UTK "Body Farm")	Human decomposition research facilities in multiple habitats for cross-environmental research on PMI
	Maggot mass thermogenesis and behavior	Micro CT tomography
	Species-specific rates of growth and thresholds	Standardized laboratory techniques for recording rates of growth
	Effect of other arthropods and microbes	Statistics
	Presence of narcotics in corpses	Maggot mass effects
	Estimated age of immature insects within an instar	Effects of microbiology, predation, and parasitism
		Size, species, and clothing of model animals
		Effect of narcotics on development of insects of each region under various temperature regimes
		Gene expression with respect to immature insects triangulated with other methods such as morphological development assessment and steroidogenesis
Determining ambient temperatures	Estimating temperatures at the site of a corpse	Statistics
		Studies of microclimates and post-discovery temperatures for best estimates of conditions prevailing at a site pre-corpse discovery
		Use of satellite imagery and ground-truthing data to reconstruct weather conditions at a crime scene
Absence of insects	Corpses found in situations that exclude insects	Study of corpse/carcass fauna under conditions of burial, freezing, and wrapped/packaged
Quality assurance	Confidence that the integrity of specimens has been maintained, no matter which country or jurisdiction is involved	Standard protocols for all collection, preservation, examination, analysis, and reporting
	Comparison of analyses conducted by independent entomologists	When were the specimens preserved?
		What conditions prevailed for insects from the time of discovery of the corpse until delivery to the entomologist?
Animal welfare	Cases of neglect of companion animals	Identification and rates of growth of species in cases of nonhuman myiasis
	Illegal killing of protected wildlife	Succession ecology with respect to a range of native species in a region at different seasons
Wildlife forensics	Errors in generalizing about insect succession and development based on data from a narrow range of model animals	Use of a broader range of animal models to identify insects of forensic interest and determine rates of growth of immature insects
		Succession ecology with respect to a range of both native and domestic animal species in a region at different seasons

30.10 CONCLUSIONS

International collaboration today has an important role to play in forensic entomological research and training. However, given that forensic entomologists are largely a network of self-motivated, dedicated individuals, such collaboration is not easy to achieve. This is particularly true for those who operate in more isolated parts of the world, who may be limited by a lack of financial resources and the physical, technological, and human structures necessary for research. In this sense, national and international forensic entomology associations play an important role by facilitating contacts and the exchange of knowledge, and also by stimulating joint projects.

Overall, the research landscape that has developed in forensic entomology is varied and in some ways fragmented, a natural consequence of individual entomologists following personal research interests opportunistically as funds become available. Improvements in the degree of linkages and communication would drive innovation most effectively. As an example, closer links with the justice system and law enforcement agencies would ensure that research is relevant to the needs of the crime scene investigator and court. Such links would possibly provide advantages to the teaching and training programs for undergraduate and postgraduate students via internships, guest lectures, and access to equipment. Links with other forensic scientists, particularly pathologists and anthropologists, have shown improved analysis and interpretation of physical evidence from investigations, whereas collaboration with other fields of science, such as molecular biology, has taken insect identification from the often flawed examination of external features to the cellular level where DNA markers provide more reliable methods of separating immature insect species. Collaborations with other scientific, information technology, and engineering disciplines are emerging and opening up fascinating cross-disciplinary fields of interest to experts not currently involved in forensic entomology research.

Forensic entomology, like insects themselves, is truly globally transcending national borders. Individuals, with little more than a shared passion for the discipline in common, have developed and maintained the forensic entomology associations, established standard procedures, shared ideas through journal articles, trained new generations of forensic entomologists, and provided evidence and opinions to the judiciary that has assisted many civil and criminal investigations. Through global collaboration in education, research, casework experience, new technology, and innovation, forensic entomology has advanced and is now gathering pace toward being a mainstream forensic science capable of offering full-time career opportunities in many countries.

REFERENCES

Amendt, J., C. Campobasso, E. Gaudry, C. Reiter, H.N. Leblanc, and M.J.R. Hall. 2007. Best practice in forensic entomology—Standards and guidelines. *International Journal of Legal Medicine* 121: 90–104.

Anderson, G.S. 1999. Wildlife forensic entomology: Determining time of death in two illegally killed black bear cubs. *Journal of Forensic Sciences* 856: 44.

Australia New Zealand Policing Advisory Agency National Institute of Forensic Science. 2012. Criteria for the Specialist Advisory Groups. National Institute of Forensic Science.

Benecke, M. 2003. Neglect of the Elderly: Cases and Considerations. 1st EAFE meeting, April 2–4, 2003, Frankfurt, Germany. European Association for Forensic Entomology, 29–30.

Beyer, J.C., W.F. Enos, and M. Stajic. 1980. Drug identification through analysis of maggots. *Journal Forensic Science* 25: 411–412.

Byrd, J.H. and J.L. Castner (eds.) 2009. *Forensic Entomology. The Utility of Arthropods in Legal Investigations, Second Edition*. Boca Raton, FL: CRC Press.

Catts, E.P. and M.L. Goff. 1992. Forensic entomology in criminal investigations. *Annual Review of Entomology* 37: 253–272.

Crosby, T.K., J.C. Watt, A.C. Kistemaker, and P.E. Nelson. 1986. Entomological identification of the origin of imported *Cannabis. Forensic Science Society* 26: 35.

Dadour, I.R. and M.L Harvey. 2008. The use of insects and associated arthropods in legal cases: A historical and practical perspective. In: Oxenham, M. (ed.) *Forensic Approaches to Death, Disaster and Abuse.* Bowen Hills, Australia: Australian Academic Press.

Dunn, R.R. and M.C. Fitzpatrick. 2012. Every species is an insect (or nearly so): On insects, climate change, extinction, and the biological unknown. In: Hannah, L. (ed.) *Saving a Million Species: Extinction Risk from Climate Change.* Washington, DC: Island Press.

Godin, B. and M.P. Ippersiel. 1996. Scientific collaboration at the regional level: The case of a small country. *Scientometrics* 36: 59–68.

Goff, M.L., W.A. Brown, K.A. Hewadikaram, and A.I. Omari. 1991. Effect of heroin in decomposing tissues on the development rate of *Boettcheisca peregrina* (Diptera: Sarcophagidae) and implications to the estimations of postmortem intervals using arthropod development patterns. *Journal of Forensic Science* 36: 537–542.

Gunatilake, K. and M.L. Goff. 1989. Detection of organophosphate poisoning in a putrefying body by analyzing arthropod larvae. *Journal Forensic Science* 34: 714–716.

Haglund, W.D., D.T. Reay, and D.R. Swindler. 1989. Canid scavenging/disarticulation sequence of human remains in the pacific northwest. *Journal Forensic Science* 34: 587–606.

Hall, R.D. 2009. The forensic entomologist as expert witness. In: Byrd, J.H. and J.L. Castner (eds.) *Forensic Entomology: The Utility of Arthropods in Legal Investigations.* 2nd ed. Boca Raton, FL: CRC Press.

Harvey, M.L., I.R. Dadour, and S. Gaudieri. 2003. Mitochondrial DNA cytochrome oxidase I gene: Potential for distinction between immature stages of some forensically important fly species (Diptera) in western Australia. *Forensic Science International* 131: 134–139.

Hill, J.K., H.M. Griffiths, and C.D. Thomas. 2011. Climate change and evolutionary adaptations at species' range margins. *Annual Review of Entomology* 56:143–159.

House of Commons Science and Technology Committee. 2013. Forensic Science: Second Report of Session 2013–2014. London, UK: House of Commons.

Johnson, A., M. Archer, L. Leigh-Shaw, M. Pais, C. O'Donnell, and J. Wallman. 2012. Examination of forensic entomology evidence using computed tomography scanning: Case studies and refinement of techniques for estimating maggot mass volumes in bodies. *International Journal of Legal Medicine* 126: 693–702.

Katz, J.S. and B.R. Martin. 1997. What is research collaboration? *Research Policy* 26: 1–18.

Leclercq, M. 1969. Entomological parasitology. The relations between entomology and the medical sciences. In: Leclercq, M. (ed.) *Entomology and Legal Medicine.* Oxford, UK: Pergamon Press.

Lord, W.D. and J.R. Stevenson. 1986. *American Registered Professional Entomologists.* Washington, DC: Chesapeake Chapter.

Magni, P., S. Guercini, A. Leighton, and I. Dadour. 2013. Forensic entomologists: An evaluation of their status. *Journal of Insect Science* 13: 78.

Marks, M.K., J.C. Love, and I.R. Dadour. 2009. Taphonomy and time: Estimating the postmortem interval. In: Steadman, D. (ed.) *Hard Evidence Case Studies in Forensic Anthropology.* 2nd ed. Upper Saddle River, NJ: Prentice Hall.

Merck, M.D. 2007. *Veterinary Forensics: Animal Cruelty Investigations.* Oxford, UK: Blackwell Publishing.

Prichard, J.D., P.D. Kossoris, R.A. Leibovitch, L.D. Robinson, and W.F. Lovell. 1986. Implications of trombiculid mite bites: Report of a case and submission of evidence in a murder trial. *Journal of Forensic Science,* 31: 301–306.

Rodriguez, W.C. and W.M. Bass. 1983. Insect activity and its relationship to decay rates of human cadavers in east Tennessee. *Journal of Forensic Science* 28: 423–432.

Smith, K.G.V. 1986. *A Manual of Forensic Entomology.* London, UK: British Museum (Natural History).

Sumodan, P.K. 2002. Insect detectives: Forensic entomology. *Resonance* 7(8): 51–58.

Tibbett, M. 2013. Forensics: Step up funding to halt forensic folly. *Nature* 501: 33.

Tomberlin, J.K., R. Mohr, M.E. Benbow, A.M. Tarone, and S. VanLaerhoven. 2011. A roadmap for bridging basic and applied research in forensic entomology. *Annual Review of Entomology* 56: 401–421.

Tomberlin, J.K. and M.R Sanford. 2012. Forensic entomology and wildlife. In: Huffman, J.E. and J.R. Wallace (eds.) *Wildlife Forensics.* Chichester, UK: John Wiley.

Turchetto, M. and S. Vanin. 2010. Climate change and forensic entomology. In: Amendt, J., M.L. Goff, C.P. Campobasso, and M. Grassberger (eds.) *Current Concepts in Forensic Entomology.* Dordrecht, Netherlands: Springer.

Walsh, J.P. and N.G. Maloney. 2002. Computer network use, collaboration structures, and productivity. In: Hinds, P. and S. Kiesler (eds.) *Distributed Work*. Cambridge, MA: MIT Press.

Ward, D.F., J.R. Beggs, M.N. Clout, R.J. Harris, and S. O'Connor. 2006. The diversity and origin of exotic ants arriving in New Zealand via human-mediated dispersal. *Diversity and Distributions* 12: 601–609.

Wells, J.D. 2003. Quality control and quality assurance issues for forensic entomology. 1st EAFE meeting, Frankfurt, European Association of Forensic Entomologists.

Ynalvez, M.A. and W.M. Shrum. 2009. International graduate science training and scientific collaboration. *International Sociology* 24: 870–901.

CHAPTER **31**

Current Global Trends and Frontiers

Jeffery K. Tomberlin and M. Eric Benbow

CONTENTS

31.1 INTRODUCTION

A simple observation of the behavior of a fly in thirteenth-century China served as the seed from which a completely new discipline germinated and is today a globally established and recognized field of entomology and forensic science—forensic entomology. Basic research is continuing to be the foundation for practice in forensic entomology, and this book reflects the global representation and new dimensions that are defining the discipline as we move forward in the 21st century. This book is the product of both established and emerging scientists and practitioners of forensic entomology that represents 66 authors from 20 nations. The chapters represent the history and current status of forensic entomology worldwide and provide a synthesis of global progress in research and practice: the international dimensions of forensic entomology. The book also introduces some of the more recent and innovative research endeavors that are the result of a strong basic science foundation merging with state-of-the-art techniques and procedures: the frontiers of forensic entomology. In this chapter, the book is concluded with a summary of the latest publication trends in forensic entomology. Such an endeavor provides a coarse representation of the impact, current changes, and developments of forensic entomology through the analysis of peer-reviewed journal publications.

31.2 FORENSIC ENTOMOLOGY LITERATURE SURVEY (1973–2012)

To examine current trends in forensic entomology research, a large literature search was conducted using two scientific search engines: Web of Science® (WOS) and Commonwealth Agricultural Bureaux (CAB). For both search engines, the key words "forensic entomology" were used with default search settings, including all words, all categories, and all years for that search engine. For WOS the time frame was 1965–2012, and for CAB it was from 1973 to 2012 (1974 was the first year that the term forensic entomology was used). All citations from each search were imported into Endnote X5®, where all duplicate citations were evaluated using a search of author and year. This allowed for the identification and removal of theses, reports, books, and book chapters so as to focus on peer-reviewed scientific publications. The theses, reports, books, and book chapters have tremendous value; however, our objective for this chapter was to limit our search to peer-reviewed journal articles. Further, this search was limited to the search engine depth of coverage and undoubtedly missed literature published prior to these time periods. However, the goal of this search was to evaluate publication trends using commonly available scientific search engines that provide a robust evaluation of the literature over the last 40 years. This approach was intended to identify current patterns of forensic-entomology-related publications among years, countries, and journals.

The search revealed 920 forensic entomology publications from 48 countries that spanned 1973–2012 (Figure 31.1). Of this total, 95% were published from 1983 to 2012 and 78% in the past 11 years. A notable peak was in 2001, which seemed to be the beginning of a major upswing in overall forensic entomology publishing that has continued to this date. This may be the result of a combination of factors, namely, increased interest and resources for studies related to forensic entomology, including those based on basic carrion entomology and ecology; online publication practices by journals that have facilitated the identification and citation of papers from regional or more discipline-specific journals; increased accessibility of all papers; and a rise in the number of open-access journals.

Out of a total of 209 journals with publications related to forensic entomology, the top three journals with the most number of publications were the *Forensic Science International*, *Journal of Forensic Sciences*, and *Journal of Medical Entomology*, which published 37% of the total papers that we found in this search (Figure 31.2). The wide range of journal outlets for forensic entomological research suggests that the basic biological and ecological processes related to understanding variability in entomological evidence are founded in a broad range of subdisciplines including medical entomology, parasitology, insect physiology, and waste management, among others. These results also demonstrate the broad global interest and application of entomology in forensics, with many journals representing geographically specific foci for entomological and ecological research (e.g., *Revista Brasileira de Entomologia*). Clearly, there is broad interest in forensic entomological research that spans nearly every continent; however, the number of publications represented on a country-by-country basis is quite skewed, suggesting limitations related to sociopolitically dependent resources for scientific research in general (Figure 31.3).

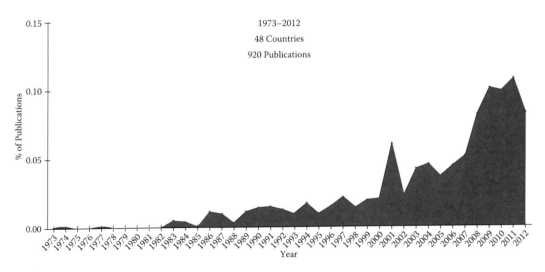

Figure 31.1 **(See color insert.)** Publications from 1973 through 2012 that reference "forensic entomology."

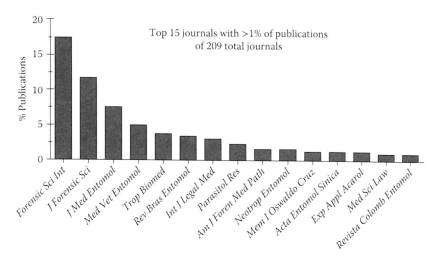

Figure 31.2 **(See color insert.)** Journals that published manuscripts from 1973 through 2012 referencing "forensic entomology."

The literature review also documented that although 48 countries were represented by forensic entomology publications from 1973 to 2012, 73% of them were from only 10 countries (Figure 31.3). The overall percentages ranged from 3.3% for Canada to 21.8% for the United States, with Brazil representing 11.8% of all publications. The percentage of publications from all other countries ranged from 0.1% to 2.5% (Figure 31.3). The reasons for this are unclear but may generally represent limitations to research opportunities based on available funding agencies with missions that include supporting studies related to forensic entomology (e.g., carrion ecology).

From 1983 to 2012, the proportion of publications from these 10 countries varied substantially on an annual basis (Figure 31.4). If the countries with the highest overall percentages (6.1%–21.4%) of publications (Figure 31.4a) are considered alone, before 1996 there were no publications from China and very few from Brazil and Australia, with the most coming from the United States and the United Kingdom except during 1983–1985. For the next five countries (3.3%–5.0%), there was also substantial annual variation, with most publications occurring after 2000 (Figure 31.4b). Before 2000, Canada and France had low, but consistent, numbers of publications. After 2000, there was increased representation from Thailand, Germany, and Malaysia (Figure 31.4b). These trends suggest consistent publication records from several countries over the last 30–40 years but substantial increases in research interests or opportunities for several countries in the last decade.

Researchers from these countries are publishing their articles in top forensic journals, such as the *International Journal of Legal Medicine* (Gagliano-Candela and Aventaggiato 2001), *Journal of Forensic Sciences* (Bharti and Singh 2003), and *Forensic Science International* (Turchetto et al. 2001), which are available to the forensic sciences community as a whole. In addition, papers related to forensic entomology are also now being published in journals that have not had a traditional forensic focus, such as *Trends in Ecology and Evolution* (Tomberlin et al. 2011a), *Animal Behaviour* (Tomberlin et al. 2012b), *Journal of Tropical Medicine and Parasitology* (Heo et al. 2008b), and *Tropical Biomedicine* (Heo et al. 2008a). These expansions of forensic entomology publications from around the world and into journals with more general scientific breadth suggest that the state of forensic entomology research may be changing and arguably expanding. This may be exemplified with recent changes within the United States, where the National Science Foundation has been mandated to become more actively involved in the forensic sciences. If this trend continues globally, over the next decade the field of forensic entomology will undoubtedly continue to diversify and bridge with other disciplines creating new frontiers for investigating how insects and other arthropods can be used in legal matters worldwide.

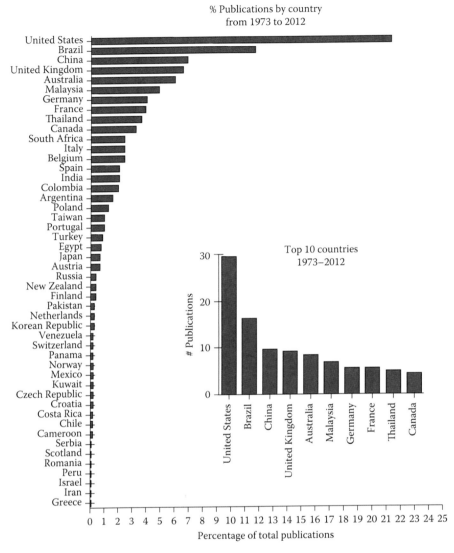

Figure 31.3 (See color insert.) Number of publications from each nation referencing "forensic entomology."

Forensic entomology continues to mature into a global science. However, much remains to be accomplished. Efforts to increase global communication between entomologists from various regions of the world could lead to refined standards, applications, and novel questions of global relevance. Furthermore, such communication could lead toward global efforts to continue to validate the applications of entomology in criminal investigations as well as determine sources of error and variability in the use of arthropods in legal matters. Currently, few entomologists are employed within crime laboratories. Continued networking among forensic entomologists could lead to increased opportunities for research (e.g., funding agencies taking note of the impressive research being conducted) and job opportunities within state, federal, and regional institutions. In addition, as demonstrated within this book, there are increased opportunities to bridge across scientific disciplines allowing for diversification of multidisciplinary research and the recruitment of students and young scientists from these other areas.

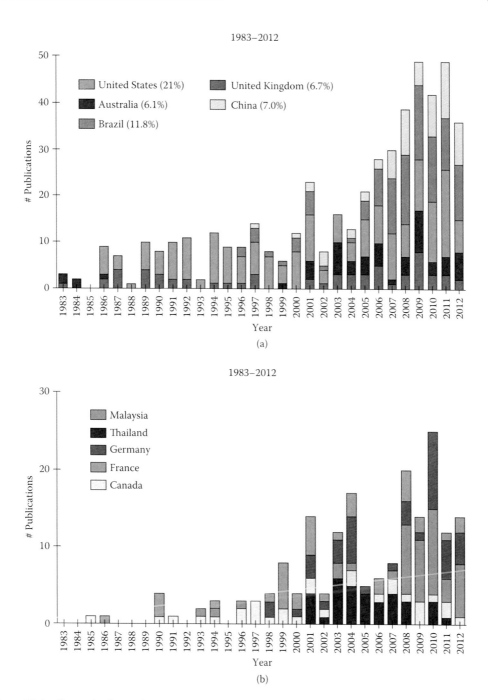

Figure 31.4 **(See color insert.)** Articles pertaining to forensic entomology related topics that were published from 1983 to 2012: (a) the top five nations and (b) the following five nations.

This book demonstrates that forensic entomologists should celebrate the accomplishments of the field as it has advanced significantly since Sung T'zu and Pierre Mégnin. The future is very bright for forensic entomology; however, much remains to be done to continue the international dimensions and new frontiers of this rapidly emerging science.

ACKNOWLEDGMENT

We would like to thank J. Pechal for his assistance in carefully editing this chapter.

REFERENCES

Bharti, M. and D. Singh. 2003. Insect faunal succession on decaying rabbit carcasses in Punjab, India. *Journal of Forensic Sciences* 48: 1–11.

Gagliano-Candela, R. and L. Aventaggiato. 2001. The detection of toxic substances in entomological specimens. *International Journal of Legal Medicine* 114: 197–203.

Heo, C.C., A.M. Mohamad, F.M. Ahmad, J. Jeffery, H. Kurahashi, and B. Omar. 2008b. Study of insect succession and rate of decomposition on a partially burned pig carcass in an oil palm plantation in Malaysia. *Tropical Biomedicine* 25: 202–208.

Heo, C., A.M. Mohamad, J. John, and O. Baharudin. 2008a. Insect succession on a decomposing piglet carcass placed in a man-made freshwater pond in Malaysia. *Tropical biomedicine* 25: 23–29.

Tomberlin, J.K., M.E. Benbow, A.M. Tarone, and R.M. Mohr. 2011a. Basic research in evolution and ecology enhances forensics. *Trends in Ecology & Evolution* 26: 53–55.

Tomberlin, J.K., T.L. Crippen, A.M. Tarone, B. Singh, K. Adams, Y.H. Rezenom, M.E. Benbow et al. 2012b. Interkingdom response of flies to bacteria mediated by fly physiology and bacterial quorum sensing. *Animal Behaviour* 84: 1449–1456.

Turchetto, M., S. Lafisca, and G. Costantini. 2001. Postmortem interval (PMI) determined by study sarcophagous biocenoses: Three cases from the province of Venice (Italy). *Forensic Science International* 120: 28–31.

Index

Printed and bound by CPI Group (UK) Ltd, Croydon, CR0 4YY

24/10/2024

01778309-0007